HISTORY
IN ASPHALT

HISTORY IN ASPHALT

THE ORIGIN OF
BRONX STREET AND PLACE NAMES

THE BRONX, NEW YORK CITY

by

JOHN McNAMARA

Third Revised Edition

THE BRONX COUNTY HISTORICAL SOCIETY
The Bronx, New York
1991

Library of Congress Cataloging in Publication Data

McNamara, John, 1913 -
History in Asphalt.

Reprint. Originally published: Harrison, N.Y.: Harbor Hill books, 1978.

Bibliography: p.

1.   Bronx (New York, N.Y.)-Streets. 2. New York (N.Y.)-Streets.
3.   Street names - New York (N.Y.)-History. 4. Names, Geographical -
New York (N.Y.)-History. 5. Bronx (New York, N.Y.) 6. New York (N.Y.)-
History. 1. Title.

[FI28.68.B8M3 1984] 974.7'275'00321 84-7078
ISBN 0-941980-154
ISBN 0-941980-16-2 (pbk.)

Manufactured in the United States of America

THE BRONX COUNTY HISTORICAL SOCIETY
3309 Bainbridge Avenue, The Bronx, New York 10467

Editors
Dr. Gary D. Hermalyn
Prof. Lloyd Ultan

CONTENTS

FOREWORD

History in Asphalt: The Origin of Bronx Street and Place Names was one
of the first, and one of the most successful, books produced by The Bronx
County Historical Society. The first and second editions in 1978 and 1984
sold out in only a short number of years.

Since that time, the enormous public demand for books on The Bronx and
New York City provided additional proof that John McNamara's master
work must be updated. For this third edition, the Society's editors kept the
general format of current names and McNamara's old guide, and updated
the listings.

The third edition was made possible through the generous support of Larry
Barazzotto and the author.

Dr. Gary Hermalyn
Executive Director
The Bronx County Historical Society

1991

INTRODUCTION

The Bronx has over 840 miles of streets, and during a lifetime spent in the borough. I have traversed them all – East to West, North to South – on foot, by bicycle and, in some cases, by canoe, in search of stories behind local place names. I found they provided fascination in the mystery and diversity of their origins. When official records were lacking, I had conversations with old-timers, while voluminous correspondence with descendants of early settlers, now living all over the United States, filled in more gaps. Even talks I had with children ascertained if local nicknames had survived from previous generations. From street names linguists can determine ethnic migrations and foreign language influence, while folklorists can amass a wealth of local legend in the choice of naming streets. Most streets received their name by solemn edict or ordinance, but quite a few owe their naming to a whim. The impact of events can be found in clusters of avenues whose names have a common denominator – so that the average browser is continually bemused by the history to be found in the street names of his neighborhood.

This book would never have seen the printer's press had it not been for the behind-the-scenes zeal, acumen and infinite patience of Professor Lloyd Ultan and Dr. Gary Hermalyn, and also Mrs. Kathleen Pacher, Ms. Mary Ilario, Mrs. Laura Tosi, Ms. Kay Gleason, Ms. Rose Politi, Dr. Elizabeth Beirne, Mrs. Mary Crapo, Ms. Kathleen McAuley and Mrs. Mildred Nestor – all of The Bronx County Historical Society. To two fellow explorers in the wilds of Onomastics – Arthur Berliner and Frank Wuttge – who helped me in a seemingly endless quest, I dedicate this book.

John McNamara

1991

9

CURRENT BRONX STREET AND PLACE NAMES

CURRENT BRONX STREET
AND PLACE NAMES

ABBOT STREET. This short street is in a tiny enclave, almost entirely surrounded by Mount Vernon. Formerly part of the C. J. Penfield Estate, it doubtlessly honors a Civil War Colonel H. L. Abbot who resided there.

ABRAMS BALL FIELD. See: Melville Abrams Ball Field.

ABRAMS PLAYGROUND. See: Ben Abrams Playground.

ACORN PLACE. This Place is a short thoroughfare in Silver Beach, arbitrarily given the first letter of the alphabet. It had been part of the Stephenson farm of the 1790s which was sold by daughter Glorianna to a Revolutionary War veteran, Abijah Hammond. It was listed as the lands of William Whitehead in the 1850s, and was later owned by two men, Havemeyer and Robert, in 1869. Two families, Peters and Sorgenfrei, operated the former estate as a Summer bungalow colony in the 1920s, calling it Silver Beach Gardens, and it is they who named its streets in alphabetical order, using the names of flowers and trees that were in profusion on the estate.

ADAMS PLACE. Reverend Carson Adams was pastor of the West Farms Presbyterian Church in 1876. On 1888 maps, this section was known as Adamsville, with the Place being called Adams Avenue. It was opened and extended from East 182nd Street to Crescent Avenue in 1897.

ADAMS STREET. Most streets in Van Nest carried the names of Presidents. President John Adams (1797-1801) was the first occupant of the White House.

ADEE AVENUE. The Adees, a wealthy merchant family, originally settled in Rhode Island before the Revolution, then migrated to Albany to form the firm of Timpson and Adee. Grandsons established themselves in New York City and bought considerable property in what was then Westchester County. John T. Adee owned land between White Plains Road and Boston Road in Civil War times. Before annexation to New York City, Adee Avenue was called King Street from Bronx Park to Cruger Avenue. From Gun Hill Road to the New England Thruway, Adee Avenue was once on the property of Robert Givan who bought the land in 1795. It remained in the family for over a century, and when streets and avenues were laid out, present-day Adee Avenue was called LeRoy Avenue after an Elizabeth Palmer LeRoy who was a great-granddaughter of Robert Givan.

ADEE DRIVE. The unofficial name for the entrance road into the Park of Edgewater, honoring a former landowner who built the mansion there in the mid-1850s on his estate, "Edgewater." George Townsend Adee, born in 1804, purchased the country estate from Edward LeRoy in 1851. In 1844 he had married Ellen Louise Henry who became the mother of four sons, all lawyers. Mr. Adele died November 20, 1884, and was buried in the family vault in St. Peter's graveyard off Westchester Square.

ADLER PLACE. In July 1967, when Co-Op City was laid out, American author and archeologist Cyrus Adler was honored (Local Law). It was originally on the Pell Grant of the 17th century on which Philip Pinckney, one of the settlers in 1662, agreed to "sett down at Hutchinsons." Pinckney's Meadows became the Freedomland of 1959-1965.

ADMIRAL COURT. Shorehaven is a 1100-plus condominium complex at the end of Clason Point, and Admiral Court is one of its private streets. It overlays the site of three tennis courts and forty-eight handball courts that belonged to the Shorehaven Beach Club (1947 to 1986). In the early 1900s, the Clason Point Amusement Park was situated there, with a mammoth carousel operating near present-day Admiral Court.

ADMIRAL LANE. Shorehaven Beach Club's salt water swimming pool, largest in New York City, once occupied this area. When the Clason Point Amusement Park was there in earlier years, a roller-coaster ran alongside today's Admiral Lane.

AGAR PLACE. The Hungarian word Agar means "Greyhound," and is sometimes used as a family name. John Giraud Agar was a lawyer and a Judge Advocate during the Spanish-American War. He was a member of the Westchester Polo Club, the grounds and playing fields of which were once the Ferris lands of Revolutionary times. The Place was laid out in 1920 when the Polo Club disbanded. Cherry Tree Point is at the foot of Agar Place, facing Eastchester Bay.

AILEEN B. RYAN RECREATIONAL COMPLEX. Pelham Bay Park's Playground Number 1, Baseball Fields Number 4 and Number 5, and the ten tennis courts adjoining them received this name in May 1990. Mrs. Ryan, New York City Councilwomen for 17 years, a former teacher, and charter member of The Bronx County Historical Society, died in 1987.

ALAN PLACE. This is a private street off Pennyfield Avenue running to Hammond's Cove. It was part of the pre-Revolutionary farm of the Stephenson family that became the extreme southern boundary of Willett Leacroft's land in 1794. Abijah Hammond was the owner in 1810. In the 1920s, a Summertime bungalow colony called Schuyler Hill was bounded by this street, and it was sold by J. P. Day to private home owners. Other private streets are laid out alphabetically: Bevy, Casler, Dare, Eger and Fearn. Mr. Graf, an elderly employee of J. P. Day, informed this writer that on several occasions Mr. Day let his office staff put in a hat, names to be used as street names on Throggs Neck and Staten Island. His only criterion was that the names be short and not conflict with existing City street names. Alan was a draftsman's grandson. Bevy was one man's wife's maiden name. Casler was another maiden name, but here Mr. Graf recalled it had been Kassler or Kessler and Mr. Day "Americanized" it. Dare and Fearn he could not recall. One draftsman was German, and might have been responsible for Eger, which is a town on the Czech-German border, which has since been changed to Cheb by the Czechs.

ALBANY CRESCENT. During Colonial and Revolutionary times, this was an important junction of the Albany Post Road and the Boston Post Road, and was the scene of many battles from 1776 onward. During the 19th century, Kingsbridge old-timers referred to it as Boston Hill. The present name was acquired in 1913. In 1922, the name Barry Street was proposed for Albany Crescent, but was not approved.

ALCOTT PLACE. Louisa May Alcott, American author, was honored by a Local Law in July 1967 when Co-Op City was built. Alcott Place is almost directly over the course of Rattlesnake Creek. See: Adler Place.

ALDEN PARK STREET. This is actually the waterfront itself. Around 1935 Alden Park was acquired and named by the Shaw family who owned adjoining Edgewater Park. The name has no particular significance, according to a few members of the family.

ALDEN PLACE. This marks the limits of the old Village of Upper Morrisania, which later became Tremont. The eastern end of Alden Place once touched the Mill Brook. James Alden was one of the original property owners. He belonged to a group of men living in lower New York (Bowery, Prince and Mulberry Streets) who pooled their money to purchase a tract from Gouverneur Morris II in the 1850s.

ALDER BROOK. This brook originated in two separate spring-fed ponds. Dodge Pond, which was filled in 1958, was located at the southeastern corner of West 247th Street and Arlington Avenue in the 1880s. Indian artifacts were found on its banks in the 19th century. This spring-fed pond was one of the two sources of Dogwood Brook, now known as Alder Brook. Goodridge Pond, filled in 1961, was a small body of water that was formerly located by Netherland and Riverdale Avenues near the junction with West 249th/250th Street and Goodridge Avenue. This was the northerly source of Dogwood Brook, an earlier name, on 18th and 19th century maps, of a stream that still courses through the Dodge estate and its annex, "Dodgewood Estate, Inc... Alderbrook, Inc." which still runs westward through private estates in a series of waterfalls, parallels West 248th Street and empties into the Hudson River. See: Alderbrook Road.

ALDERBROOK ROAD. Listed in the Bronx Post Zone Guide, this is a private road of an estate, "Alderbrook," established in 1932. The area embraces West 247th to West 249th Streets west of Riverdale Avenue.

ALDRICH STREET. Thomas Bailey Aldrich, American author, is remembered in this Co-Op City street. See: Adler Place.

ALDUS STREET. At times it was called Aldine Street. It is across the former estate "Brightside" belonging to Colonel Robert M. Hoe whose company invented the rotary press. When the Colonel's acres were cut into city streets in 1897, three of them were given names connected with the printing trade: Hoe, Guttenberg and Aldus. Guttenberg Street has since become East 165th Street, but Aldus Street remains, honoring Aldo Manuzio, a Venetian editor and printer (1449-1515), better known by his Latin name of Aldus Manutius.

ALEXANDER AVENUE. This is laid out on Jonas Bronck's original grant which eventually became the Manorlands of the Morris family. It was later part of the Long Neck Farm (313 acres) worked by the Scottish overseer of the Morris family named Alexander Bathgate. However the street is not named for him, but for a family named Alexander who owned property in the area in Civil War times. There are deeds relating to Robert and Ellen Alexander (1867) and later, to an Edwin I. Alexander, a realtor of the South Bronx. After its formal opening in 1891, it was a fashionable residential street of prominent families. It was nicknamed "Doctors' Row" and "Irish Fifth Avenue."

ALEXANDER HAMILTON BRIDGE. In 1963, the Harlem River was once more spanned by a bridge, this one commemorating the name of the early American statesman who had many ties to the area on the Manhattan side. The Bronx approach to this bridge goes through the former "Boscobel" estate of William B. Ogden. See: Villa Boscobel.

ALFRED LORETO PARK. See: Loreto (Alfred) Park.

ALLEN PLACE. Many streets and avenues in this area carry names of former Mayors of New York, and Stephen Allen served from 1821 to 1824. He was among over one-hundred passengers who died in the fiery wreck of the Hudson River steamer *Henry Clay* off Riverdale, on July 28, 1852. Anne Hutchinson's farm is believed to have existed in this vicinity. See: "Locust Grove" (2).

ALLEN SHANDLER RECREATION AREA. An area formerly known as Holly Park, it was officially named after a local boy in September 1967. It is adjacent to the Van Cortlandt Park golf course and is bounded by Jerome Avenue, 233rd Street and the Major Deegan Expressway. On August 9, 1966, Allen Shandler died at the age of 15 of a brain tumor. The boy had become known as an exceptional one when he intervened to seek mercy for two teenagers who burned his grandmother to death in 1960. Knowing he had not long to live, he continued to lead a useful life. He and his father staged a drive to obtain $300,000 from the City to establish the Holly Park recreational area, which, later on, was renamed in his memory.

ALLERTON AVENUE. Known as Bleeker Avenue, it now honors Daniel Allerton (1818-1877), a settler whose wife was a Hustace (Heustis) of equally old stock. His ancestor, Isaac Allerton, came in the *Mayflower*, but settled in Nieuw Amsterdam, where the firm of Allerton and Loockerman was listed in 1653. Thomas Haddon's Saw-Mill Lane (1720s) is covered by Allerton Avenue from Lurting to Tenbroeck Avenues. In 1914, Allerton Avenue was cut through what was the Hamersley Estate, which was mapped in 1856. The Allerton family plot is located on Central Avenue in Woodlawn Cemetery.

ALLERTON CIRCLE. At East 203rd Street inside the Botanical Garden, this is an extension of Allerton Avenue. See: Allerton Avenue.

ALPINE AVENUE. It is located in about the center of Woodlawn Cemetery, and the tomb of Mayor Fiorello La Guardia can be found in its vicinity. Its name might stem from its high elevation.

ALUMNI WALK. This promenade is on the Bronx Community College grounds, formerly New York University. Previously, it was the Schwab estate, and earlier the Archer lands and the site of Revolutionary War Fort Number 8. The first members of the alumni graduated from the School of Arts and Sciences in 1833 (in Manhattan), and the Bronx campus of New York University was opened in 1894.

AMENDOLA PLAZA. On March 18th 1983, Mayor Koch dedicated the intersection of Westchester and Burr Avenues to honor Anthony Amendola, and noted his civic spirit and many charities. Governor Nelson Rockefeller and President Ronald Reagan had been guests at his well-known restaurant, the Chateau Pelham.

AMETHYST STREET. The upper end of this Van Nest street, from present-day Rhinelander Avenue to White Plains Road was originally called Oakley Street on a 1905 map. There are two versions why this short street is so named, neither of which are official. The surveyor was given the opportunity to name the street and his name was Rudolph Rosa (the German name for "pink") and he chose a color. The second story is that a few not very valuable stones of amethyst were dug up by workmen when the City took title to the land in 1916.

AMPERE AVENUE. Andre Ampere, a French physicist (1775-1836), was prominent in the field of electricity. One story comes down to this day that he was to address an important meeting in Paris, and en route he jotted down equations on the wall of the hansom cab in which he was riding. After he had paid the driver, he realized he had no notes and had to chase the cab through the streets to regain them. The background of this avenue is identical with that of Ohm Avenue. See: Ohm Avenue.

AMUNDSON AVENUE. John Arnold Amundson was a real estate operator, prominent in many East Bronx realty developments around 1908-1920, particularly in Edenwald.

ANDERSON AVENUE. Two separate deeds of the Woodycrest property mention the Anderson family. In 1794 James Anderson purchased 60 acres from an Englishman, Medcef Edin. At a later date, 1820, Rachel Eden [spelled this way] and Aaron Buff conveyed additional land to James Anderson. An 1854 map shows an Abel T. Anderson farm extending across Cromwell Creek (now Jerome Avenue) and eastward up the slope to the Concourse, then called Mount Eden. The James Andersons lived in a stone villa on the east side of Anderson Avenue near West 164th Street. The old house was built of native stone, called gneiss, that was quarried within 200 feet of the site. The property was later called "Woody Crest" due to its high ground and thick woods.

ANDREWS AVENUE NORTH. This street runs athwart the former William Loring Andrews estate which extended, according to Civil War maps, from the Harlem River to Jerome Avenue, and from approximately West 181st to West 183rd Streets. Andrews was a wealthy, retired leather merchant. The avenue received its name in February, 1886, from the Board of Aldermen. In May of 1889, its lower stretch was renamed Montgomery Avenue.

ANDREWS AVENUE SOUTH. See: Andrews Avenue North.

ANGELO CAMPANARO PARK. Located at Eastchester and Gun Hill Roads, this park was dedicated on June 15, 1985. Mr. Campanaro had served in the Army Corps of Engineers in Alaska. When he returned to civilian life, he was President of the Chester Civic Improvement Association, and Chairman of Community Board 12. He was active in many other civic affairs throughout his life.

ANGELS' SPRING. Earlier names of the stream were Forrest Brook and Randolph Brook. The spring on the grounds of Mount St. Vincent was found on October 2nd, 1867, after fervent prayer prompted by a critical need of a water supply. It rises near West 261st Street and flows westward and downhill to the Hudson River. According to the Sisters of Charity, the name commemorates the Feast of the Guardian Angels.

ANITA PLACE. This Place was officially named on May 3, 1927 by the Board of Aldermen for a former resident of Baychester, Anita Demo.

ANTHONY AVENUE. First called Prospect Avenue for the fine view, or prospect, of the Mill Brook valley (Webster Avenue), in 1889 it was called Avenue C. A sizeable pond was one time situated at the intersection of Anthony and Burnside Avenues. Charles L. Anthony owned many tracts of land in the 1870s, principally above Kingsbridge Road and stretching from Jerome Avenue to Webster Avenue. The earliest record of this name occurs in 1668, when Abard Anthony was listed as a Dutch merchant.

ANTHONY J. GRIFFIN PLACE. See: Griffin Place.

AQUEDUCT AVENUE EAST. Obviously named for the Croton Aqueduct it parallels. Nearby University Avenue originally carried the name of Aqueduct Avenue until 1913.

ARCHBISHOP HUGHES CIRCLE. This is a circular drive leading to the Administration Building of Fordham University. It was named after the founder of St. John's College (1839), which later became Fordham University.

ARCHER ROAD. This short street in the western sector of Parkchester harks back to the Mapes/Archer property of the late 1800s. See: Archer Street.

ARCHER STREET. John Archer was one of the early settlers of the Town of Westchester, who later purchased the Manor of Fordham. The area crossed by this street was once the Mapes farm, which was subdivided into a realty venture called "Park Versailles." It is said the auctioneer, John S. Mapes, named the avenue after his mother, whose maiden name was Archer. There is an Archer/Mapes/Hunt family plot on White Oak Avenue, Woodlawn Cemetery.

ARDEN SQUARE. At Boston Road, between Tillotson and Givan Avenues, this was named after the Arden Estate which was originally part of the Givan property. Alison Givan married Thomas S. Arden, but was widowed at 23. She lived on the Givan lands to her 88th year, and is buried in St. Peter's graveyard near Westchester Square.

ARLINGTON AVENUE. This avenue was opened in 1922 and honors Arlington C. Hall who was descended from an old Westchester family. He was a builder in Riverdale who ran head-on against the early attempts (1930) to keep apartment houses out of Riverdale by zoning laws. The lower end of Arlington Avenue figured in American history. This is the crest of Spuyten Duyvil hill, known on Revolutionary War maps as Spike & Devil Hill. Some 200 feet south of West 230th Street and 230 feet west of Arlington Avenue was once Fort No. 2, built by Colonel Abraham Swartwout's American militia. It was later occupied by Hessians and demolished by them in 1778.

ARMAND PLACE. This dead-end street, some 40 yards long, with but two houses on it, runs south from Perot Street, laid across the former Perot farm, situated on Tetard's Hill, where Colonel Armand was prominent in action against the German mercenaries of the Revolutionary War. He managed some other notable exploits in The Bronx, particularly the capture of the Tory Major Bearmore at Oak Point. His broken English report to General Heath on this is interesting and amusing. His full name was Charles Armand, Marquis de la Rouerie. He was barely 20 at the opening of the American Revolution.

ARNOW AVENUE. Arnaux was a French Huguenot family that was conspicuous in the early annals of Westchester County history. Some members owned property in what is now the Pelham Bay area, Westchester Square and along Boston Road. Laid out as Union Street around 1900, it bisected the Adee Estate from White Plains Road to Boston Road. From there to Eastchester Road to the New England Thruway, it ran across the Givan property. Sometime before WWI, Union Street became the Americanized Arnow Avenue.

ARNOW PLACE. It was called Arnow Avenue until 1931. Beers' *Atlas* of 1868 shows a Mrs. Arnaux living at this spot. In November 1945, a motion was put before the City Council to change the name to Peskar Place, but it was defeated.

ARTHUR AVENUE. Originally called Broad Street in Tremont, and Central Avenue in Belmont. Catherine Lorillard Wolfe and her uncle, Jacob Lorillard, owned this tract in Belmont in the 1880s. She was an admirer of President Chester A. Arthur and donated to New York City, a statue of him which stands in Madison Square Park. There is a local story that an Arthur Hoffman was a City surveyor, and had this avenue and nearby Hoffman Street named after him, but there is no record of any such surveyor.

ARTHUR H. MURPHY SQUARE. See: Murphy (Arthur) Square.

ASCH LOOP. Sholem Asch, American Yiddish novelist and author was honored in July 1967, when Co-Op City was built. See: Adler Place.

ASTOR AVENUE. The last owner of record was William Astor, whose estate was surveyed and taken by the City around 1909. A brook ran through the property, marked now by a low spot at Matthews Avenue. This William Astor was distantly related to John Jacob Astor, who was born, 1768, in Waldorf, near Heidelberg.

AUGARTEN BALLFIELD. See: Sid Augarten Ballfield.

AUGER SQUARE. See: Lou Auger Square.

AUSTIN PLACE. This was once included in the extreme end of the Morrisania Manorlands. In 1852, Levinus Austin purchased property from Gouverneur Morris II, and the tract was bounded by Bungay Creek (now East 149th Street).

AVENUE ST. JOHN. This street marks the extreme edge of the Manorlands of the Morris family. In 1852, George and Catherine St. John purchased property from Gouverneur Morris II and lived there for many years. The principal street of their property was mapped in 1905 as St. John Avenue, with the alternate usage of Avenue St. John, which eventually won out. Another Avenue St. John in the village of Belmont led to St. John's College, which later became Fordham University. This Belmont avenue was then given the name Hughes Avenue to avoid confusion.

B STREET. This is a short dead-end street in what was the settlement of Schuylerville (Bruckner Boulevard and East Tremont Avenue), a predominately Irish village that evolved in the 1850s supplying maids, coachmen, gardeners and laborers to the surrounding estates and Fort Schuyler. See: Hobart Avenue. The Schliessman family lived at the end of this lane for four generations, and many records and artifacts of The Bronx County Historical Society were stored in their barn.

BABE RUTH FIELD. A baseball field north of Yankee Stadium, which had formerly been an athletic field and quarter-mile track where many Olympic runners competed.

BABE RUTH PLAZA. At East 161st Street, this plaza was named for the famous "Bambino," George Herman Ruth, the New York Yankees home run king. His name is synonymous with nearby Yankee Stadium, "the house that Ruth built." It was once part of the Gerard Morris estate, "The Cedars."

BAILEY AVENUE. From West 225th Street to West 230th Street, this Avenue was once the scene of Revolutionary skirmishes. From West 232nd Street northward, it was the Albany Post Road. Nathaniel Platt Bailey owned a large estate east of the avenue and south of Kingsbridge Road in the 1870s. Around 1922, during a real estate venture, Bailey Avenue at West 231st Street was proposed to be renamed Gallagher Avenue, but the move failed. See: Albany Post Road.

BAILEY PLACE. N. P. Bailey's mansion was once located on this site. He was a Vestryman of St. James Protestant Episcopal Church, and at one time (1884) was president of the aristocratic St. Nicholas Society of the City of New York.

BAINBRIDGE AVENUE. William Bainbridge commanded the American frigate *Constitution* (Old Ironsides) in the War of 1812. The avenue is laid out across the lands of the Varian family, Briggs farm and Jerome holdings. It was once known as 2nd Avenue in the settlement of Bedford Park.

BAIRD COURT. This plaza in the Bronx Zoological Gardens is one of the finest examples of landscaping and architecture in the East. The sea lions' pool is in the center of the Court. Named in 1900, it honors Professor Spencer Fullerton Baird, famed zoologist, founder of the U.S. National Museum and once Secretary of the Smithsonian Museum.

BAISLEY AVENUE. This was originally known as Elliott Avenue in the village of Schuylerville. It later became known as Baisley Street, after a family who owned land in the vicinity, but, in March 1943, the name was changed to Baisley Avenue. A James Baisley (died in 1843) is buried in St. Peter's P. E. churchyard, and 1904 deeds mention a woman named Schuylerine Baisley.

BAKER AVENUE. This was originally Jackson Street in Van Nest to honor one of our Presidents, but was changed to Baker Avenue in May 1915 when it was made a public street. Although there have been Bakers in the area for centuries, beginning with "1720 - William Baker of ye Burrow Town of Westchester," the street was named after Seward Baker, prominent lawyer, who lived on nearby Walker Avenue, now East Tremont Avenue. Mr. Baker was born in 1853, practiced law in Poughkeepsie and moved to Westchester Township in 1887. He was a City Magistrate, Vestryman of St. Peter's P. E. Church, head of the Bronx Borough Bar Association, and director of the North Side Board of Trade. His name is found on many West Farms indentures, wills and legal cases. The area was once part of the Hunt farm, which was subdivided into small estates prior to the Civil War. The Pierce Estate was one, and Seward and Helena C. Baker purchased part of it.

BALCOM AVENUE. This avenue runs through the former Seton Estate from East Tremont Avenue to Waterbury Avenue. The section that is now inside St. Raymond's Cemetery was part of the main road that led from Westchester Square to Ferry Point in the early 1800s. Balcoms have lived on the land for generations, a John Balcom being listed in an 1851 deed, Irving S. Balcom being a director of the Seton Homestead Land Company of the early 1900s, and as late as 1930, Marie Balcom being listed as a property owner. Balcom Avenue from Lawton Avenue to the East River was a traditional division line for centuries, having been the western boundary of the Livingston, Havemeyer and Huntington lands and the eastern bounds of Francis Morris' property and that of his grandson Alfred Hennen Morris. The short stretch from Bruckner Boulevard to the Cross-Bronx Expressway was once mapped as St. Joseph's Avenue due to its proximity to St. Joseph's School for the Deaf.

BALDWIN STREET. This is the topmost street in The Bronx in a tiny enclave surrounded by Mount Vernon. South Railroad Avenue was its name according to Mount Vernon maps of 1923, as it ran into Railroad Avenue in that city. Baldwins were and are numerous in the area. An 1883 deed refers to a Hall F. Baldwin; a William Baldwin was a surveyor in the 1890s; and Thomas Baldwin was Foreman of the volunteer Empire Engine Company of 1896 at East 240th Street. In 1867, William and Eliza Ann Baldwin purchased this property from Charles Darke of Yonkers, and an Ann Baldwin bought land in 1902 from William Penfield.

BALSAM PLACE. This is a short Place in Silver Beach Gardens, where all the streets are named for flowers and trees. See: Acorn Place.

BANTAM PLACE. This Place was known as Van Cortlandt Street in 1907, and mapped as Grove Street in 1911. It was laid out through the original holdings of Robert Givan (1795) which remained in the family until recent times. One theory is that Bantam is a synonym for "short," which would describe the Place off Eastchester Road. Another supposition is that it honors the Banta family, old settlers of Eastchester.

BANYER PLACE. It was part of the Black Rock farm of Clason Point, owned by the Ludlow family. Banyer Ludlow was last owner of record. His mansion boasted of an extensive art gallery. One oil painting was of the Deputy Secretary of the Province of New York, Goldsborow Banyar [spelled this way], who was related to the Ludlows.

BARKER AVENUE. The Barkers were old-time Bronx settlers with branches in every part of our present Borough. The original White Plains Road which figured in Colonial history crossed Barker Avenue from the southwest corner to the northeast corner of Waring Avenue.

BARKLEY AVENUE. At its western end was once the swamp surrounding Baxter Creek (now Huntington Avenue). Uphill to East Tremont Avenue the land was first owned by Thomas Baxter in the 1680s, by the Ferris family in the 1700s, and by Stillwells and Crosbys in the 1800s. After 1912 until 1919, Barkley Avenue was the eastern boundary of Hoffmann's Park, a famous picnic resort. Northward, it was part of Coster's farm (1857-1926) from East Tremont Avenue to Throggs Neck Expressway, when it was opened in 1929. James Barkley purchased land on Throggs Neck from Mary Coddington in January 1826, and most likely is the man whose name is remembered. The First Lutheran Church of Throggs Neck was started in a barn on Barkley Avenue in 1921.

BARNES AVENUE. This was an ancient Siwanoy Indian path (from Gun Hill Road to Bussing Avenue), which later was utilized by the early colonists. In pre-Revolutionary days, it was called the Kingsbridge Road. Later, it was the main thoroughfare known as White Plains Road. When the town of Wakefield was laid out (1850) it became 4th Street. In the 1870s, the Burke Estate below Gun Hill Road was plotted, and that stretch of present-day Barnes Avenue was called Cedar Street. In Van Nest, the inhabitants had a liking for Presidential names, and Barnes Avenue from Morris Park Avenue to the railroad lines was called Madison Street, a name that remained in use until around 1899. A 1692 Westchester deed refers to a Captain William Barnes, "High Sheriff of ye Burrough Town of Westchester," and a 1720 sale was consummated by Thomas Barnes, so the name is an old one in our past. However, I believe the avenue refers specifically to Samuel and Susan Barnes who owned a farm alongside the Kingsbridge Road (Barnes Avenue) in the 1840s.

BARNETT PLACE. Scarcely 40 feet long, it is located off Barnes Avenue, below Rhinelander Avenue. At one time, Barnett Place was five blocks long, running from White Plains Road to Matthews Avenue. Innholder William Barnett filed a will on March 7, 1741, which was witnessed by Jeremiah Fowler.

BARNHILL SQUARE. According to the Municipal Reference Library, it was named on July 15, 1941. Located at Kingsbridge Road and Reservoir Avenue, it is directly opposite the Fordham Manor (Dutch) Reformed Church in which Reverend Oliver Paul Barnhill served from 1924 to 1934. While conducting services, he collapsed and died of a heart attack.

BARRETT AVENUE. Avenues in this section of Clason Point carry the names of Judges. This street honors Supreme Court Judge George Carter Barrett, and not Barrett's Creek that once flowed in the area. Barrett Avenue's curve follows the present course of yet another stream called Pugsley Creek.

BARRETT PLACE. This Place, partly on Locust Point, was named for County Judge James M. Barrett of the 1920s, contemporary with Justices Giegerich, Glennon, Hatting, Mullan and Tierney. It adjoins the historic Throggs Neck Road (now Pennyfield Avenue), and was once part of Willett Leacroft's land, 1793, and then purchased by Abijah Hammond around 1810. A William Whitehead was the landowner in the 1850s. In the 1920s, Barrett Place was tentatively laid out across a small hill occupied by Summertime bungalows called Schuyler Hill.

BARRETTO STREET. Part of this avenue, from Fox Street to Intervale Avenue was changed to Fox Street in 1899 by Municipal Ordinance. Francis J. Barretto was a merchant and an Assemblyman of Westchester County. His wife was Julia Coster (see nearby Coster Street) and his former estate, known as "Blythe," was crossed by both Barretto and Coster Streets. He was a Vestryman of St. Peter's P. E. Church in Westchester 1830-36, and a Warden from 1836 to 1871.

BARRY PLAZA. At Clay and Anthony Avenues and East 174th Street, this tiny island of .09 acres and three trees acquired its name in December 1940. Though the Parks Department had no information, Arthur Berliner, a student of Onomastics, states the plaza honors William J. Barry, who died in World War I.

BARRY STREET. Formerly part of Leggett Avenue, it was briefly mapped as Wenman Street in the 1890s. It is in an area where naval men are commemorated. The street honors Commodore John Barry (1745-1803).

BARTHOLDI STREET. The Statue of Liberty was unveiled in New York harbor on August 5, 1884, and one of the principals was the sculptor, Frederic Auguste Bartholdi. (His mother posed for the head of the statue.) In 1886, this event was still fresh in the minds of all New Yorkers, and when a street was surveyed in Williamsbridge, it was named for the sculptor; this also pleased the French colony that was considerable in the Gun Hill Road area. It was not until 1909 that the street was finally cut through.

BARTOW AVENUE. Reverend John Bartow came to America in 1702, was the first Rector of St. Peter's P. E. Church in Westchester village, and served until his death in 1726. Originally the family was of French Huguenot descent, and spelled their name "Bertaut." Reverend Bartow purchased land along the Eastchester Road extending to the Hutchinson River. He was interred in the family burial plot near the present crossing of Pierce and Tomlinson Avenues, but the exact location of his grave has been lost. A son, Theopholis Bartow, married Bathsheba Pell and inherited land now Pelham Bay Park. It was this Bartow who gave his name to the creek, mansion, settlement and estate on the Shore Road.

BARTOW-BAYCHESTER MALL. Outside Co-Op City, this Mall was opened in the early 1970s and its name is borrowed from two nearby avenues.

BARTOW CIRCLE. Theopolis Bartow was the nineteenth century landowner of all the surrounding countryside. See: Bartow Avenue.

BASSETT AVENUE. This was the northern extent of the Bayard farm of the 1820s which passed into the hands of John Hunter III. Robert Bassett is mentioned in a 1695 deed wherein he sold six acres of meadowland to the Hunt family, the property now in Pelham Bay Park, that portion later to be owned by John Hunter III.

BASSFORD AVENUE. The name is a hark-back to a large estate of pre-Civil War times in Tremont. In the 1840s an Abram Bassford was listed as the property owner, and a Colonel Stephen Bassford was prominent in early Tremont. The Abram Bassford family plot is located in Maple Plot off White Oak Avenue, Woodlawn Cemetery.

BATH GATE. This is not a street, but the entry gate to Fordham University at the junction of Bathgate Avenue and East 191st Street. Since it marks the end of Bathgate Avenue, the name is probably a play on words. See: Botany Gate, Dixie Gate, Veterans' Gate.

BATHGATE AVENUE. This Avenue is a reminder of the extensive Bathgate farm from which Crotona Park and portions of Tremont were carved. The first Bathgate was an overseer of the Morris Manorlands. The farm was conveyed to his three sons in 1847 by Gouverneur Morris II. At one time, this once-fashionable tree-lined street was known as Cross Street and also as Madison Avenue. Another early name was Elizabeth

Avenue. Both William and James Bathgate were on the membership rolls of the Westchester Agricultural Society in 1820.

BATTERY HILL. The historic hill on which the British built Fort No. 8. Later, it was part of the Archer property, then passed through the ownership of Andrews, Mali and Schwab before being acquired by New York University in 1893. Dr. Henry M. MacCracken stood on Battery Hill and said (according to the college publication *Forum*), "This is the place!" and construction began.

BAY SHORE AVENUE. Once part of the shoreline of Eastchester Bay, filled in around 1955 to support this avenue. It is on the Colonial holdings of the Underhill family, which passed to the Ferris, Waterbury, Edgar, Furman and Spencer families of later centuries.

BAY STREET. Formerly Adams Street, this street runs across the widest part of City Island. The name doubtlessly refers to Eastchester Bay.

BAY VIEW AVENUE. The self-explanatory name was approved by the Board of Estimate on October 20, 1922. It is on part of the Lorillard Spencer Farm, mapped in the late 1800s, and the bay it refers to is Eastchester Bay.

BAYCHESTER AVENUE. This was originally called South 18th Avenue in South Mount Vernon, and Comfort Avenue in the Edenwald section. It was named after the real estate venture, "Baychester," of the 1890s.

BEACH AVENUE. It is not named because it ends at a beach on the East River, but for Dr. Wooster Beach, who was a landowner on Clason Point in Civil War times. The Beach Estate measured 35 acres and remained in the family to around 1922, when a descendant named Cox sold it. When the *Great Eastern* steamship moored in the East River in the 1860s, its captain used to stay overnight at the Beach home. On the north, this street originates at East Tremont Avenue (once Walker Avenue) and was part of the lands of Lewis Guerlain in 1804. This became the Mapes farm that was auctioned off around 1910, which became Park Versailles. At East 172nd Street, the original Indian path leading to Clason Point crossed Beach Avenue, and this path later became Cow Neck Road that led past the McGraw farm. In the 1800s, Beach Avenue was crossed by Pugsley Creek at Story Avenue. In 1887, Lucas and Catherine Van Boskerck had a farm at Beach and Lafayette Avenues, which is now covered by James Monroe Houses.

BEACH STREET. As there is no record of a Beach family in deeds or old records of City Island, it can be assumed the name refers to a small beach at the foot of the street.

BEACON CIRCLE. This private turnabout covers what had been the picnic area of the Shorehaven Beach Club. In earlier times, when the Clason Point Amusement Park was situated there, Gilligan's dance Pavilion was built out over the waters of the East River. See: Shorehaven.

BEACON LANE. A children's pool was located here when the property belonged to the Shorehaven Beach Club. Gilligan's Hotel stood here in the early 1900s and in colonial times it incorporated parts of the original buildings built by the pioneer settlers, Cornell and Willett.

BEATTY PLAZA. It was acquired as a park by condemnation in November 1890, at East 169th Street and Franklin Avenue. It received its present name in December 1940, honoring Arthur G. Beatty, an A.E.F. sergeant of World War I.

BEAUMONT AVENUE. This was mapped in 1889 as Jackson Street from Grote Street to Crescent Avenue. The origin of this name is unknown, but Frank Wuttge theorized it honored Francis Beaumont, poet and author of a 19th century book, *The Knight of Malta.*

BEAVER POND. A 1/2 acre pond in almost the exact center of Bronx Zoological Park. Its name is self-explanatory.

BECK STREET. Originally Beck Street began at Jackson Avenue, but that stretch has since been changed to East 151st Street. The Beck family owned property in the vicinity of Cauldwell and Westchester Avenues in the 1850s to 1870s. In 1871, Anna Beck sold the property to Charles Staedler, President of the Malt and Hops Company. At one time, East 156th Street from Third Avenue to Cauldwell Avenue was known as Beck Street.

BEDFORD PARK BOULEVARD. In 1868, it was mapped as Corsa Avenue when it ran from the present-day Concourse to the northern end of Fordham University, which had been Corsa property. After annexation to New York City, it was numbered 200th Street, but acquired its present name in 1906. As it once formed part of the extensive Leonard Jerome estates, it is conjectured an associate of his, Edward Thomas Bedford, is so honored. Mr. Bedford was a director of Standard Oil and president of the Bank of the State of New York.

BEECH AVENUE. In the northeastern section of Woodlawn Cemetery, it runs past the graves of R. H. Macy, and also a former Clason Point landowner, Charles Leland.

BEECH PLACE. A short street in Silver Beach, given the second letter of the alphabet. See: Acorn Place.

BEECH TERRACE. Forming the uplands of the original Bronck grant of 1642, it might well be the "Bronck's Hill" mentioned in a later deed. It was a spot familiar to the Westchester Guides of the Revolutionary War as it afforded a fine view of the Morris mansion and the British forces that were stationed across the Bronx Kill on Randall's Island. When Gouverneur Morris II sold parts of his manorlands beginning in 1850, this became a small settlement known as Wilton, nicknamed "Actorville"because of the theatrical people who lived there. It was noted for the neat homes, shady gardens and tennis courts. Beech trees were common to the area, and this would account for the name.

BEECH TREE LANE. This residential street is not included in the Bronx Postal Zone directory, as 7/8ths of the Lane is in Westchester County. Only a 20-yard stretch is in The Bronx, at Pelham Bay Park.

BEEKMAN AVENUE. Beekman is an ancient Dutch name in New York history, and a W. Beeckman [spelled this way] attended to the legal transactions when the widow Bronck disposed of her holdings. In 1715, a Gerardus Beekman was listed as one of "the Chirugeons practicing in the City" and the family was prominent in later civic affairs. Because of its proximity to St. Mary's Park, the avenue most likely honors Henry R. Beekman, commissioner of the Parks Department in 1886, about the time the park was acquired. Its background is identical to that of Beech Terrace, having been part of "Actorville."

BELDEN POINT. Geodetic charts going, back to 1887, give this name to the southernmost tip of City Island. William Belden purchased the land from George Washington Horton and his son, Stephen Decatur Horton. He then operated a picnic grove, amusement park and beach resort catering to the steamboat trade.

BELDEN STREET. William Belden's home was on this short lane. According to an 1853 map of City Island, it was called Windmill Street.

BELL AVENUE. E. Y. Bell was listed as a resident of South Yonkers in 1875, and his family owned property along the city line. Henry Bell and his son, Lowerre Bell, were also well known in the Wakefield vicinity. The street was opened in 1914.

BELLAMY LOOP. Edward Bellamy was a Socialist writer, whose name was given to this Co-Op City street in July 1967. Laid out across the eastern end of Pinckney's Meadow of the 1600s, the Loop parallels the former bed of Rattlesnake Creek (Long Pond Creek). This waterway was filled in during 1958 to accommodate Freedomland, which existed until 1965.

BELMONT AVENUE. Although some Bronxites believe this street was named to honor the nineteenth century financier, August Belmont, it really commemorates the Jacob Lorillard estate, "Belmont." After 1850, the area was known as the Village of Belmont, and this avenue was first called Cambreleng Street after a townsman of that name. The lower end at Crotona Park was called Ryer Place after landowners there. The street was authorized to be laid out in 1896 as Belmont Avenue.

BELMONT STREET. Jane Street was its former name. This was originally a road leading to "Belmont," the estate of the Lorillards.

BEN ABRAMS PLAYGROUND. This community leader was honored when Mayor Edward Koch signed a bill naming a playground within Comras Mall. The date was December 1985, and the occasion marked the passing of a well-known civic and religious activist in the Pelham Parkway area for over 50 years. Ben Abrams' son, Robert, had been Borough President of The Bronx and, at the signing ceremony, was the Attorney General of New York State.

BENCHLEY PLACE. Robert Benchley was an American humorist. See: Bellamy Loop.

BENEDICT AVENUE. The name appears in records involving the settlement of Centerville, located around Castle Hill and Westchester Avenues in the 19th century. As late as 1904, James H. Benedict is mentioned in a deed involving riparian rights.

BENSON STREET. This was once known as Madison Avenue in the village of Westchester. It was named for John Benson, born in Yorkshire, England, in 1805. In existence is an 1838 deed to John Benson, miller, to "Hope Mills," and an 1849 map places his home on Westchester Turnpike, near present-day Havemeyer Avenue. He died in 1858 and is buried in St. Peter's P. E. churchyard.

BERGEN AVENUE. This street is located on the upper end of Jonas Bronck's tract which passed to the Morris family. In the 1870s, this was a meadow, part of Karl's Germania Park, a noted picnic resort. When cut through, it was named Retreat Avenue, but then renamed Bergen Avenue in honor of Michael Bergen, Trustee of the village of Morrisania, Town Clerk, later Commissioner of Surveys and Gradings, and Chief of the Morrisania "Lady Washington" volunteer Hose Company. He died in 1869. His son, John H. Bergen, was a real estate developer whose office was at East 147th Street and Willis Avenue.

BERGEN (William C.) PARK. Acquired by condemnation in July 1897, it came under the jurisdiction of the Department of Parks in 1931. William Bergen was a longtime resident whose home was at East 181st Street and Anthony Avenue, nearly opposite the tiny park.

BERNARD S. DEUTSCH PLAZA. See: Deutsch Plaza.

BERTHA KELLER CORNER. On May 16, 1990, the southwest corner of East 196th Street and the Grand Concourse was designated Bertha Keller, who devoted a half century of volunteer work as a neighborhood activist and advocate of tenants' rights.

BETHUNE PARK. A vest-pocket park at Tinton Avenue and Home Street that was dedicated in May 1969, by Morrisania's Active Pioneers, a senior citizens' group. Mary McCloud Bethune was active in Black civic affairs.

BETTS AVENUE. On February 13, 1667, Governor Nicolls granted the first patent to the settlers of Westchester. One of the original five patentees was William Betts. He left acreage near Castle Hill to Samuel Barrett, who married his daughter Hannah, and 20 acres in addition to his grandson.

BEVY PLACE. See: Alan Place.

BICENTENNIAL VETERANS MEMORIAL PARK. It was laid out in 1960 over Weir Creek, a mixed salt/fresh water course flowing into Eastchester Bay on a line with Schley Avenue. In centuries past, Indians stretched woven nets (weirs) across the creek to catch fish. It was initially called Weir Creek Park, but in May, 1976, Mayor Beame signed a bill changing the name to Bicentennial Veterans Memorial Park to signal our nation's 200th birthday, and to honor veterans of all the nation's wars.

BIG TOM. This rock is in Eastchester Bay, approximately a quarter mile south of Belden Point, and might be considered to be in the Sound. Seldom visible, it nevertheless is referred to in old-time journals. It is thought to refer to Thomas Pell, first Lord of the Manor of Pelham and owner of City Island and surrounding lands (1654).

BILLINGSLEY TERRACE. It is laid across the western slope of the original Archer tract that was owned by the founders of Fordham Manor. The estate was called "Archer Manor" and passed to the Morris family. It figured in the Revolutionary War as the edge of Fort No. 8. It was the Gould estate of the late 1880s. Logan Billingsley, realtor, was president of 30 corporations, including the East Fordham Syndicate. In 1927, he spearheaded the drive to extend the Concourse south from East 161st Street to East 138th Street. He defeated a plan to have the Concourse cut across Franz Sigel Park to connect with Seventh Avenue in Manhattan.

BINGHAM ROAD. Since December 1989, Independence Avenue, from West 252nd Street to Spaulding Lane, carries the name of the late Jonathan Bingham. The Congressman left a legacy of distinctive service to the people of The Bronx and the nation.

BIRCH AVENUE. This driveway is in the north-central section of Woodlawn Cemetery.

BIRCHALL AVENUE. The Birchall family came to America from England in 1818. They settled in Bronxdale, along with the Bolton family that had arrived in the same sailing vessel. Ann Birchall married Thomas Bolton. The avenue is laid out across the Neal (Neill) estate of 1868.

BISHOP PERNICONE SQUARE. Belmont Street signs flanking the Church of Our Lady of Mount Carmel, honor Bishop Joseph M. Pernicone, pastor of the church from 1944 to 1966. He later served as vicar for the New York Archdiocese in Dutchess County. He died in 1985, aged 81, after a long illness.

BISSEL AVENUE. Students of the American Revolution might contend this short street hard by Bussing Avenue (the original road to Boston) honors Israel Bissel who galloped down that very road spreading the news of Concord and Lexington. Actually, an 1838 map shows "Squire Bissel's lands" as 45 acres on the city line, of which tract Wilson Bissel is listed as owner in 1894. The family included George E. Bissel, who was Paymaster in the Civil War and a noted sculptor. He created statues of General Horatio Gates, of President Chester Arthur and also of Abraham DePeyster in Bowling Green Park. The family is of French Huguenot extraction, and maintains a family plot in Woodlawn Cemetery. In 1868, the area was known as Washingtonville, and the street was on the Searing Estate. It was only in 1923 that the street was made a continuation of East 241st Street.

BIVONA STREET. It is laid out across the vanished Holler's Pond where Rattlesnake Creek emptied into it. It was named in 1966, honoring Richard "Dick" Bivona, well-known Bronx realtor and civic leader of Eastchester, who died in August 1964. Borough President Pericone suggested the name at the time the Boston-Secor Housing Project was being planned.

BLACKROCK AVENUE. This was 8th Street in Unionport until 1917. It was named for the Black Rock farm on Clason Point, which was the property of the Ludlow family. Some of these black rock outcroppings remain at Soundview Avenue, Morrison Avenue and Seward Avenue, but the specific black rocks mentioned in the old records are directly on the Bronx River at Lafayette Avenue.

BLACKSTONE AVENUE. This Riverdale street was formerly mapped as Park Avenue. Evidently, the present name is descriptive, as there are exposed strata of hard Fordham gneiss, a banded or speckled rock with black crystals.

BLACKSTONE PLACE. This is a private road servicing three homes in Riverdale. A homemade sign was noted at West 246th Street near Independence Avenue. This short lane is laid out in the Dodge estate known as Dodgewood. See: Blackstone Avenue.

BLAIR AVENUE. This Throggs Neck avenue honors Civil War Major-General Frank P. Blair. It is laid out across the original Pennyfield that already had its name in Colonial records. This was the nineteenth century property of Ogden Hammond, John T. Wright, Herman Newbold and finally Francis Wissmann.

BLAIR COURT. The plaza facing the seals' pool in the Bronx Zoo (officially the New York Zoological Garden) carries the name of a former director.

BLAUZES, The. These are four sharp, mussel-covered reefs slightly northwest of Hart Island in Long Island Sound and were the scene of many shipwrecks throughout the centuries. The origin of the name is in doubt. One theory is that it derives from "Blazer," an Old English word for a marker or guide, and that the mariners used these reefs as guides into City Island. The blue color of the islets might have caused the early Dutch mariners to dub them *de Blauwtjes* (the little blue ones).

BLONDELL AVENUE. This Avenue was originally the back lots of residential homes that fronted on Westchester Square. An 1890 deed mentions a Helen L. Blondel [spelled this way] who purchased plots of land in the neighborhood. In 1914, the City vested title to the land, and the street that was cut through was named Franklin Avenue. To avoid confusion with the Franklin Avenue of Morrisania, it was then renamed Blondell Avenue.

BOGART AVENUE. It was laid out in 1916 from Sackett Avenue to Van Nest Avenue, and was extended to Rhinelander Avenue in 1920. The land it crosses was the pre-Revolutionary land of the Underhill family. Maps of 1778 show it as the Hunt and Fowler farms, and in 1860, as part of the Sackett property. Bogardus was a prominent name in the early annals of Nieuw Amsterdam, and it was later anglicized to Bogart. A good possibility is that the street was named for John Bogart, Chief Engineer in the Department of Parks (1872-77) and later Construction Engineer of Washington Bridge.

BOLLER AVENUE. This Eastchester street once went by the misleading name of Sea View Avenue. Alfred Pancoast Boller was a Civil Engineer and bridge architect who designed five Harlem River bridges.

BOLTON AVENUE. Originally called Beech Avenue, this Clason Point avenue honors Robert C. Bolton, who was a Vestryman of St. Peter's P. E. Church. He is not to be confused with the Boltons of "The Bleach" on the Bronx River.

BOLTON STREET. James Bolton, born in England in 1780, came to American territory in 1818 on the same sailing vessel that brought the Birchall family. He established the bleach mills in the village of Bronxdale, commonly called "The Bleach." His mill site and dam form part of Lake Agassiz in Bronx Park. Originally, it was Siwanoy Indian domain that passed into Dutch possession, became English and, finally, American. In 1868, it was part of the Neill estate.

BONNER PLACE. This land was originally General Staats Morris' tract of 1750 and used as a racing course. The Dater Brothers leased it in 1860 and built grandstands, called it Fleetwood Race Track. Robert Bonner, founder and proprietor of the *New York Ledger* and noted turfman, had a home at the foot of Bonner Lane (now College Avenue at East 164th Street).

BOONE AVENUE. The former name was Boone Street, but it received its present title on March 14, 1904. No trace of this name has been found in records from 1670 to 1904.

However, in the West Farms deeds of 1886, a John H. Bones was listed as a property owner near the Bronx River at present-day East 176th Street. A small cemetery was noted on 1888 maps, and a conjecture is that Bones Street would not be a suitable name, hence a slight change in spelling.

BOSCOBEL PLACE. A reminder of William B. Ogden's estate (1840s). See: Villa Boscobel.

BOSTON CLOSE. This is a semi-circular street inside the Boston-Secor Housing project off Boston Road. The word "close" denotes an entry or passage leading to a court and the houses within.

BOSTON ROAD. In 1672, Governor Lovelace issued "A Proclamacion for a Post to goe monthly from this city to Boston and back againe." The post rider took a well-defined path that had been an Indian trail running from lower New York to Kingsbridge, across the northern Bronx and up into present-day Westchester County. This original Boston Road is now Gun Hill Road, Barnes and Bussing Avenues in that order. The present-day Boston Road was laid out after the Revolutionary War because of the efforts of (General) Lewis Morris. [He is referred to as "the Signer" because he is often confused with his father, also Lewis Morris.] He had been an officer in the Revolutionary army and signed the Declaration of Independence. He sought to re-route the Post Road so that it ran through his Manorlands in the lower Bronx. This entailed building a bridge across the Harlem River from Harlem, and in 1794, John B. Coles did construct such a span and dam and continued with a road from Morrisania to Eastchester. It was called Coles' Boston Road to distinguish it from the older road that ran across the upper Bronx. Coles' Boston Road began at Third Avenue from East 135th Street to East 164th Street, then pioneered modern day Boston Road to the City line.

BOTANICAL SQUARE. First called Depot Square, it was renamed by the Board of Aldermen on December 27, 1926 due to its proximity to the Bronx Botanical Gardens.

BOTANICAL SQUARE NORTH. See: Botanical Square.

BOTANICAL SQUARE SOUTH. See: Botanical Square.

BOTANY GATE. This is the entrance to Fordham University at Bronx Parkway, east of Webster Avenue. It is a possibility that the name could be a reference to the nearby Botanical Garden. See: Bath Gate, Dixie Gate, Veterans' Gate.

BOUCK AVENUE. Streets in this section carry names of New York State Governors. William C. Bouck served in 1843-44. He was of German ancestry and frequently mentioned he had been born in a log cabin. At DeWitt Place and Bouck Avenue, a spring on the former Kidd Estate was the origin of Stoney Brook that flowed southeasterly, taking the name of Abbott's Brook, into Westchester Creek.

BOWNE STREET. Along with the Hortons, Fordhams and Scofields, the Bowne family was counted among the pioneer families of City Island. They owned considerable property, particularly on the northern end. William Bowne and Elnathan Hawkins share a family plot in Woodlawn Cemetery at Spruce Avenue.

BOYD AVENUE. From East 233rd Street to Bussing Avenue this avenue was once called Coster Street. A Mary Boyd purchased property from George Penfield in 1890 in the vicinity of Bussing Avenue.

BOYNTON AVENUE. This is on the original land grant of the Cornell family. Barrett's Creek once wound in from the Bronx River and cut across this avenue between Bruckner Boulevard and Watson Avenue. Its original name was Genner Avenue, named after John S. Genner, one of the early Westchester colonists who swore allegiance in 1656 to the New Netherlands. In 1868, the avenue was part of William Watson's estate, "Wilmount," that was subdivided in 1911 by the A-Re-Co, American Real Estate Company. The street was renamed in February 1911 by the Board of Aldermen to honor Edward S. Boynton, vice-president of the A-Re-Co and a founder of the Bronx Beautiful Society, an organization that raised funds to improve the parks.

BRADFORD AVENUE. There is a possibility that this avenue is named for William Bradford, first Governor of Plymouth colony, as other avenues in the area are in the same vein: Puritan, Pilgrim and Mayflower. However a William Bradford is mentioned in the Town Deeds of Westchester in 1768 and 1771, and a Nathaniel G. Bradford owned some 63 acres of land in the immediate vicinity prior to the Civil War.

BRADLEY STREET. David Bradley is listed in 1879 and 1880 in minor real estate transactions in Westchester County. The land, surrounded by Mount Vernon on three sides and the Bronx River on the fourth, was part of the C. J. Penfield Estate in 1890. When this street was cut through, it was first known as Mechanic Street.

BRADLEY TERRACE. A flight of stairs leading from Palisades Avenue to the Spuyten Duyvil station of the New York Central railroad, this appears to be just an extension of Palisades Avenue. It acquired its name in 1927, but neither local residents, nor employees of the Bronx County Engineering office, knew who Bradley was.

BRADY AVENUE. This was formerly the southern portion of the Astor property. Still earlier, it was included in the village of Bronxdale. The Dunn family of Fordham claim it honors their grandfather, Justice John J. Brady, who owned land in the area traversed by the avenue. This writer feels the avenue, next to Lydig Avenue, is named for Lydig's son-in-law, William V. Brady, major of New York in 1847-48.

BRANDT PLACE. No amount of research and inquiry has brought forth any information on this Place, except that it was named in 1900. Numerous Brandt families were contacted without success.

BRIDGE STREET. Since Colonial days, ferries from Rodman's Neck had connected City Island with the mainland, but in 1873 a toll bridge was built across the strait. David Carll was its owner. The present bridge, costing $250,000, replaced the old wooden toll bridge in 1898-1901.

BRIGGS AVENUE. This is a family prominent in Bronx civic life for almost two centuries. Daniel C. Briggs was Sheriff in 1801, and a landowner was Josiah Briggs (born 1852), whose homestead was located near Jerome Park. The avenue runs across Peter Briggs' farm, and on this same tract was a pond, giving rise to the name of Pond Place. Southward was the Walter Briggs land, mapped in 1888.

BRINSMADE AVENUE. This avenue runs through the former Charlton Ferris farm, the swamps now covered by the Throggs Neck Houses and the former Morris Estate, south of Lawton Avenue. An 1868 map shows the estate of N. F. Brinsmade to consist of 15 acres in the area around Lafayette and East Tremont Avenues sloping westward to the swamps of Baxter Creek, now blotted out by Ferry Point Park and the new St. Raymond's Cemetery. As of 1930, Mrs. Henry Brinsmade was listed in the City Record as a property owner.

BRISTOW STREET. This short street is a reminder of a resident of Morrisania in the 1870s, George Bristow, who was a musician and composer.

BRITTON STREET. This street marks the northeastern margin of Pierre Lorillard's land, according to an 1868 map. Originally mapped as Sheridan Street, the name was changed in 1917 to honor Dr. Nathan Britton, a noted botanist, and his wife, who was Honorary Curator of the Bronx Botanical Garden. He wrote *Illustrated Flora of the Northern States and Canada* in three volumes.

BROADWAY. From the Harlem River north to West 230th Street, this thoroughfare runs through Manhattan territory, as the ancient boundary of Spuyten Duyvil Creek was filled in 1894-95 at West 230th Street. Around 1808, the present road was filled in along the side of Paparinemo Marsh (West 230th to West 236th Street) by the Highland Turnpike Company, which charged tolls. It was part of Kingsbridge Road, a prominent highway in Revolutionary times. The thoroughfare derives its name from the Dutch, *Breede Weg*.

BRONX and PELHAM PARKWAY. See: Pelham Parkway.

BRONX BOULEVARD. From Gun Hill Road north to East 219th Street, it was called 2nd Avenue in the village of Olinville. From East 219th Street to Nereid Avenue, it was Bronx Terrace, and from there to East 241st Street, it was Marion Street. On May 2nd, 1916, by a Board of Aldermen Ordinance, the entire stretch was renamed Bronx Boulevard.

BRONX KILL. This waterway, originally spelled Bronck's Kill (Dutch for "creek"), figured in the first settler's grant. Jonas Bronck, a Dutch citizen of Scandinavian origin, together with his family and servants, settled on the southern shoreline of what is today The Bronx. Although he purchased some 500 acres from the Dutch West India Company, he also paid off the local Indians with farming tools, household utensils, clothing and a barrel of cider. This latter transaction occurred in December of 1639. Jonas Bronck died in April of 1643, probably of natural causes. Bronx Kill is mentioned frequently in connection with the Morris family whose holdings it bounded. Various Revolutionary War annals refer to it as well. Once a meandering waterway from the Harlem River to the East River, it has since become constricted between concrete banks and straightened along its one-mile length. It is navigable to small craft, but has a strong and dangerous tidal current.

BRONX LAKE. A twenty five-acre one-mile-long lake in the Zoological Park extending from Boston Road south to the Lower Falls, which were also known as Delancey's Falls and Lydig's Falls in the 18th and 19th centuries respectively. It is part of the Bronx River. In the center of the lake, the patent and manor lines of Fordham, West Farms and Westchester formed a corner.

BRONX PARK. Almost two miles long, it contains 720 acres. In Revolutionary times it was the property of the DeLancey family, some of whom were Tories while the others fought on the American side. The lands passed into the possession of the Lydig family until in 1888 it was acquired as a public park. The park contains a world-famous zoo, a botanical museum and conservatories. See: Lydig Avenue.

BRONX PARK AVENUE. Thomas Pell claimed all the land from Long Island Sound westward to the Bronx River, so that the land across which this avenue runs was the extreme end of his grant which later became part of Westchester County. In pre-Revolutionary times the Delancey family owned the property, which later passed to the Lydigs. In 1850, the approximate eastern boundary of James Devoe's tract followed the line of present-day Bronx Park Avenue. When the street was surveyed in 1887, it was given the obvious name as it ran directly to the newly-opened park.

BRONX PARK EAST. This thoroughfare overlays a far older road leading to the now vanished village of Bronxdale. See: Bolton Street.

BRONX PARK SOUTH. This was once an important Indian trail from Fordham, through West Farms to Hunts Point. Later called Kingsbridge Road in the 1800s, it was then numbered 182nd Street until 1917. It was part of the Delancey property and then part of the lands of Philip Lydig. A stagecoach inn was maintained by the Hunt family at the corner of this street and Boston Road. It was called the *Planters' Inn*.

BRONX PARKWAY. This is a thoroughfare marking the northern limit of Fordham University extending from Webster Avenue to Southern Boulevard.

BRONX RIVER AVENUE. This is a road in the Soundview sector that parallels the Bronx River from Story Avenue to Bruckner Boulevard.

BRONX RIVER PARKWAY. This important artery came into being in 1925 when the Bronx River Parkway Commission finished years of condemnation proceedings on properties on both banks of the Bronx River. Their work extended from The Bronx to White Plains and resulted in a Parkway noted for its beauty. William W. Niles was vice president of the Bronx River Parkway Commission. It was through his intercession that the Old Soldier in the Bronx River was left undisturbed after all private properties had been taken over for park purposes.

BRONX SCIENCE BOULEVARD. In May 1988, to celebrate the 50th anniversary of the Bronx High School of Science, commemorative street signs were installed on Goulden and Paul Avenues at West 205th Street.

BRONX STREET. This Street is laid out across the Delancey lands and was the site of a blockhouse that saw much action during the Revolutionary War. When West Farms Square became an important river port and mercantile center, Alexander Smith's carpet works was established on Bronx Street in 1844. Halcyon Skinner, a mechanic, invented a new type of power loom and was granted a patent. During the Civil War his factory manufactured blankets and uniforms. His mills were destroyed by fire, and he moved the business to Yonkers.

BRONXDALE AVENUE. In Indian times, this was a trail skirting the Bear Swamp and leading to Castle Hill Point. It carried the name of Bear Swamp Road in the 1800s, although an 1853 map showed it had an alternate name of Snuffmill Lane, as it led to Lorillard's snuff mill (now preserved in Bronx Botanical Garden). It received its present name from the fact it ended in the village of Bronxdale.

BRONX-WHITESTONE BRIDGE. This bridge was opened in 1939 over the East River to Whitestone, Queens, providing easy access to the New York World Fair. The Bronx approach to the bridge is laid out over the former diZerega lands of the 19th century.

BRONXWOOD AVENUE. It is named after Bronxwood Park, a real estate development which was laid out between Olinville and Wakefield in 1893. It was 5th Street from Gun Hill Road to Bussing Avenue.

BROOK AVENUE. It was officially opened in 1876, and overlays in part the Mill Brook of Morrisania Manor. Below East 149th Street, Brook Avenue was a prominent residential stretch of the 1890s. Above East 149th Street, small homes predominated.

BROUN PLACE. Heywood Campbell Broun, American journalist, was honored in July 1967 by having a Place in Co-Op City named in his honor. See: Bellamy Loop.

BROWN MALL. See: Thomas Brown Mall.

BROWN PLACE. This was part of the original Bronck farmlands, which later became the Morris Manorlands. Four men named Brown were connected with this area. Gouverneur Morris II conveyed property to Lewis B. Brown in 1858, and in 1866 another tract was purchased by Clarence S. Brown in the vicinity of St. Ann's Avenue and East 134th Street. By coincidence, a Major Alexander H. Brown lived at this spot, but he had moved there in 1891 after the Place and nearby Alexander Avenue had been named. Local residents thought the Place honored a Judge William F. Brown of the 23rd Ward. The second Brown is the most likely candidate for the honors.

BROWN'S LANE. Listed in the Bronx Postal Zone Guide, this City Island alley has no official street sign. Brown's Hotel formerly stood on the corner.

BRUCKNER BOULEVARD. From the lower Bronx to the Bronx River this important artery was once called Eastern Boulevard. Above the river, where it ran across the Ludlow property, it carried the name of Ludlow Avenue until it entered the village of Unionport. The street was not continuous because of intervening swamps. In Unionport, before the 1895 annexation, it was merely 6th Street, although some maps denote it as Commerce Street where it touched Westchester Creek. From the Unionport bridge, it was called Willow Lane, overlaying an ancient Indian path to East Tremont Avenue. Willow Lane ran past Schuylerville and ended in the estates now incorporated into Pelham Bay Park. After annexation, the designation Eastern Boulevard was extended to include all the other regional names. On July 29, 1942, the name was changed to honor Henry Bruckner, Bronx Borough President from 1918 to 1933. The family vault is in the northwestern corner of Woodlawn Cemetery.

BRUCKNER BOULEVARD BRIDGE. See: Eastern Boulevard Bridge.

BRUCKNER EXPRESSWAY. This is the latest designation of the heavily trafficked elevated and depressed highway of the East Bronx. See: Bruckner Boulevard.

BRUNER AVENUE. P. Bruner owned an estate on the Eastchester Road below Allerton Avenue, in the 1860s, called "Rosedale." It had been part of the Givan Estate (1794) and Reverend Bartow's land of 1725. From Allerton Avenue north to Boston Road, the avenue follows the course of Black Dog Creek, which was the boundary between the ancient townships of Eastchester and Westchester. As part of the Arden Estate (1898), it was called Ash Street. From Edenwald Houses to Bussing Avenue, Bruner Avenue was once known as Oakes Avenue. Above Bussing Avenue, it was called Willis Place.

BRUSH AVENUE. This street overlays the original swampy riverfront of the Grove Farm or "Grove 'Siah." The name appears in early Westchester records. John C. Brush witnessed a Deed in 1789. Ichabod Brush married Euphemia Wilkins of Castle Hill Point (opposite Brush Avenue), and in 1831, David Brush and Charles Ferris were contestants in a lawsuit. Brush Avenue cuts across the ancient lands of the Ferris families, flanking the Westchester Creek.

BRUST SQUARE. At West 242nd Street, Manhattan College Parkway and Fieldston Road, this small plot was marked "public park" in 1911. It was acquired in 1932 by the City, and named in honor of a World War I soldier in March 1940 (Local Law 19). Corporal Charles Brust was not a local resident, but had lived on Hull Avenue. He was killed in action on August 30, 1918.

BRYAN PARK. It is located at the junction of East Fordham Road and Kingsbridge Road, and was once the extreme corner of "The Elms" farm of the Keary family. Acquired by the city in November 1913, it was made a City park in 1933 and named to honor a Fordham resident and World War I veteran, John Fraser Bryan.

BRYANT AVENUE. Its proximity to streets dedicated to American poets Longfellow, Whittier and Drake would seem to be reason enough to honor William Cullen Bryant. At various times it was known as Walker Street, Hunter Avenue and Oostdorp Street. In the Hunts Point sector, this avenue is on land first granted to Richardson and Jesup in 1663. In the West Farms area, it runs across the former Minford and Walker estates. Still, it should be noted that some Revolutionary War buffs believe the street honors a Captain Bryant who served under Washington in this part of The Bronx.

BUCHANAN PLACE. In 1865, Andrew and Mary Buchanan purchased property near Prospect Hill, which, on maps of that year, was located at the Grand Concourse and East 182nd Street. There is no connection with President James Buchanan.

BUCK STREET. The Pugsley family owned considerable property in this area since Colonial days. (Parkchester is partially located on former Pugsley farmlands). In October 1876, Charles P. Buck purchased this tract of land from Hiram Pugsley. The street was opened in 1914.

BUCKLEY STREET. It was named in the 1920s after a longtime City Island resident and physician. It was formerly known as Palmer Street when a small real estate development was planned and mapped at this point and given the name Palmer Park. Benjamin Palmer was the owner of entire City Island before the American Revolution.

BUHRE AVENUE. Conrad Buhre was a farmer with large holdings outside the settlement of Middletown prior to the Civil War. He is buried in St. Peter's P. E. Churchyard. His son, Henry, served in the Civil War and fought in many battles. In the late 1890s, Buhre Avenue was known as Libby Street from the Hutchinson River Parkway to Westchester Avenue and from there to Bruckner Boulevard, it was Madison Street.

BULLARD AVENUE. This was once known as 1st Street in Wakefield. Some surveying was done in 1874 by George F. Boulard, and in 1889, he was Resident Engineer of the New York Central. He raised the grade of nearby Woodlawn station. A more likely candidate would be John E. Bullard, whose property was in Mount Vernon at the top of this avenue. His name occurs frequently in 1889 and later.

BURKE AVENUE. This is a reminder of an extensive estate belonging to a J. M. Burke in the 1860s. Its lower boundary was present-day Burke Avenue from Bronxwood to Throop Avenues, and its upper limit was Gun Hill Road. On later maps, dated 1884, it was called the Winifred Masterson Burke Foundation. When laid out, the main road running from the Bronx River to Boston Road was called Morris Street, but later was renamed Burke Avenue.

BURNETT PLACE. Opened in 1904, it may well be the only reminder of a small landowner of East Morrisania named John L. Burnett. He and his wife, Elizabeth, owned property in 1869 off Prospect Avenue.

BURNSIDE AVENUE. See: East Burnside Avenue and West Burnside Avenue.

BURR AVENUE. Aaron Buff, attorney, purchased land in that vicinity in 1790. In 1868, a C. Burr was listed as landowner near Middletown Road. It was first called Parkview Avenue as it overlooked Pelham Bay Park.

BUSH STREET. The area was called Prospect Hill. An 1854 deed describing this hill listed George Bush as the proprietor. Jeremiah Bush, in 1856, bought 8 lots situate, lying and being at Fordham, Westchester County. Surveyed in 1867 by S. E. Holmes.

BUSINESS PLACE. Park Drive at Pelham Bay Park to City line.

BUSSING AVENUE. Arent Harmanse Bussing came to Nieuw Haarlem from Westphalia in 1658 and was one of the soldiers who went to fight the Indians at Esopus. With Dyckman, he paddled a canoe upriver, and both men were Corporals upon their safe return. In the 1800s there was still a Bussing farm on the Manhattan side of the Harlem River, and about that time, a branch of the family settled in what is now Wakefield. Their mansion was built at East 233rd Street between the Old White Plains Road and Kingsbridge Road (now Bussing Avenue). This last-named avenue was the original Indian footpath leading from the tribal headquarters at Poningo (Rye, N.Y.) down to Gun Hill Road and beyond. Bussing Avenue then became part of the first Boston Post Road and was used by Paul Revere, Generals Washington and Rochambeau, Aaron Burr and others. Later it was used by the stagecoaches to Boston.

BUSSING PLACE. Running one block from DeReimer Avenue to Wilder Avenue, this curving Place represents a short stretch of Colonial road, part of the Kingsbridge Road that figured in the Revolutionary War. See: Bussing Avenue.

BUTLER PLACE. Isaac Butler was an English carpenter who settled in West Farms in 1849, but later moved to Westchester Village. He became a casket maker and mortician in an establishment opposite St. Peter's P. E. Church, and served as sexton for 56 years. He was also treasurer and vestryman for many years. Mr Butler was the first member of Wyoming Masonic Lodge to receive a 50-year jewel. Butler Place crosses historic Seabury farm, touching what was once Seabury Brook (Indian Brook), a scene of the first Quaker meeting.

BUTTERNUT AVENUE. This is a roadway located in the northwestern section of Woodlawn Cemetery, and is named after the butternut tree.

BUTTRICK AVENUE. Major John Buttrick is credited with firing the first shot at the battle of Concord Bridge, 1775. From the East River to Ferry Point Park, this avenue runs across what was once the Livingston Grant of pre-Revolutionary days. Later it became the Bayard Clark property (1840s), which was purchased by Francis Morris in the early 1850s. His grandson, Alfred Hennen Morris, retained the property until the 1920s. See: Hosmer, Emerson, Davis and Robinson Avenues.

BYRON AVENUE. It was formerly Byron Street and owes its name to the Byron Realty Corporation (1914) that was headed by a local resident. The few papers surviving studiously omit his first name.

CAESAR PLACE. This Place and other "classic" streets in the Castle Hill neighborhood owe their names to Solon L. Frank, a real estate developer active around 1904. Named after the Roman, Julius Caesar, the Place is laid out across the former lands of the Wilkins and Screvin family.

CAHILL PLACE. This short Eastchester Place was formerly known as Elwood Place, and acquired its present name in 1973. Monsignor Cahill was Pastor of Sts. Philip and James R. C. Church. He was born in New York in 1898 and ordained in 1923. He was a professor of Greek in Cathedral College, from 1924-1949. He taught philosophy and ethics in Fordham University, and was assigned to the Eastchester parish in 1953.

CALANDRA FIELD. This Little League field is laid out on the New York State Hospital grounds, east of Eastchester Road and north of Waters Avenue, the former swamplands of Westchester Creek. It was named for Senator John Calandra who was instrumental (1964-67) in setting up the field.

CALHOUN AVENUE. The avenue is named for John C. Calhoun who served as Vice President under John Quincy Adams (1825-1829). At its southern end, Calhoun Avenue was an Indian site on the East River, since numerous artifacts were found at Schurz Avenue. This location was also the site of Throckmorton's colony in 1643. Later it was in turn the estate of Philip Livingston, Mitchell, Post, Ash, Havemeyer and Huntington. From Lawton Avenue to the Cross-Bronx Expressway it was the John A. Morris estate from the 1860s to the early 1890s. At Philip Avenue it runs across the DeEscoriaza farm (1865). At Lafayette Avenue, the avenue cuts across the former Brinsmade and Overing estates and from there to Bruckner Boulevard, the 1868 Stillwell property and the (1900s) Hoffmann's Park. At this point, the avenue was called Augusta Place, Augusta Flanagan being one of two daughters of a small property owner in the 1890s. The First Lutheran Church of Throggs Neck was started in a barn west on Barkley Avenue in 1921. See: Revere Avenue.

CAMBRELENG AVENUE. Stephen Cambreleng owned the land in the years 1850 to 1873 in the area known as Adamsville. He was a brother-in-law to Reverend William Powell who inherited the Bayard farm (East 188th Street to Fordham Road). Mr. Cambreleng was also a manager of the Home for Incurables in 1866. At various times the street was called Monroe Street, Fulton Street and Pyne Street.

CAMBRIDGE AVENUE. This Avenue is part of the northern end of the John Ewen estate of Civil War times. It was named in honor of the English university town, as is adjoining Oxford Avenue.

CAMERON PLACE. It was mapped as Elizabeth Place in 1878, and Elizabeth Street in 1889 when this section of West Tremont was largely rural. Adam Cameron was an 1850 landowner near Prospect Hill.

CAMP STREET. Hugh Nesbitt Camp (1827-1895) was a financier and investor in upper Bronx real estate. He was president of the School Board, Elder of St. James P. E. Church, a trustee of Woodlawn Cemetery and a resident of Fordham.

CAMPANARO PARK. See: Angelo Campanaro Park.

CAMPBELL DRIVE. This was originally part of the upper farmlands of the Ferris family, who were there at the time of the American Revolution. A Duncan Pearsall Campbell was a Vestryman of St. Peter's P. E. Church and the son-in-law of Samuel Bayard whose property (1821 map) was directly north of this Drive in what is now Pelham Bay Park. An 1817 deed refers to a Bess Rock as the boundary marker between their lands on Eastchester Bay. Mrs. Campbell was Bess Bayard before her marriage and this boulder may have been named for her, or for her mother who was also named Elizabeth. Part of this Drive was in the Van Antwerp estate of the 1860s, and this later was a section of the Westchester Country Club property which included a polo field and golf course. See: Woof Field.

CANAL PLACE. It is a reminder of the Mott Haven Canal that once extended from the Harlem River to East 136th Street. It was opened in 1903 when the canal itself was partially filled in.

CANAL STREET WEST. This street parallels and is west of Canal Place from the Major Deegan Expressway to East 138th Street. See: Canal Place.

CANNA AVENUE. Located in Woodlawn Cemetery, this private road runs across the southern end of the cemetery. A canna is a flower.

CANNON PLACE. It was so named in 1913 because of its proximity to the site of Fort Independence from which Revolutionary cannons and cannonballs were excavated. At one time it was known as Montgomery Place as the Montgomery farm was located there before the Revolution. Montgomery threw in his lot with the American colonists, and was killed at Quebec.

CAPUCHIN WAY. Wallace Avenue from Bartholdi Street up to Gun Hill Road was renamed Capuchin Way in June 1977, to honor the Capuchin Fathers who serve the nearby Roman Catholic Church of the Immaculate Conception.

CARDINAL COOKE PROMENADE. The sidewalk along the south side of Bruckner Boulevard from Edison to Logan Avenues was so designated in June 1984 at the request of the members of Community Board 10. Council member Michael DeMarco stated, "This was the scene of Terence Cardinal Cooke's fond boyhood memories. He was a pupil of St. Benedict's school, and an altar boy in St. Benedict's Church where we are now standing."

CARDINAL SPELLMAN PLACE. A section of Needham Avenue was renamed to honor Francis Cardinal Spellman whose name was already on the Catholic High School nearby. The ceremony took place on May 18th 1990.

CARDINALS' WALK. Circling Fonthill Castle on the grounds of Mount St. Vincent is the Cardinals' Walk, so named because Cardinals Farley, Hayes and McCloskey often came up from the city to spend a few days at Fonthill Castle, and enjoy the quietude and beauty of the Hudson River.

CARLISLE PLACE. This was once part of the James Burke estate in Williamsbridge. Christopher Carlisle became a landowner of this tract in 1891, having purchased the property from William B. Duncan. Duncan Street is nearby.

CARPENTER AVENUE. This was originally laid out as 2nd Street from East 219th Street to East 226th Street in the village of Wakefield. From East 236th Street to the city line, it was Catherine Street. The Carpenter name is frequently encountered in old records, but the one remembered by this avenue is Stephen V. R. Carpenter who purchased lots from Mary Duryea in 1868 in this vicinity.

CARROLL PLACE. Colonel William D. Carroll of the Civil War was a resident of Claremont Heights. He was honored by an Ordinance of the Board of Aldermen in April 27, 1937, by the renaming of East 164th Street.

CARROLL STREET. Once known as Prospect Street at about the widest part of City Island, this street is about the halfway point from either end. In Civil War times, a hotel, The Bayview House, stood at the eastern foot of Prospect Street. The Carll family has been on City Island for many generations, and possibly this is a misspelling of their name.

CARTER AVENUE. There is a good possibility it honors a former landowner named William H. Carter who was described as "a resident of Westchester" in 1846. In those years, the area was part of that County. This street was also known as William Street from Claremont Park to the former Echo Park (now Julius J. Richman Park).

CARVER LOOP. Laid out across the seventeenth century land grant of Thomas Pell, its swampy composition kept it uninhabited until 1968. It was not a part of Freedomland, a short-lived amusement center. Up to 1965, a small tidal creek ran under what is the western arm of this Loop and was utilized by canoeists to "shoot the rapids." In July 1967, this Loop was named for George Washington Carver, a noted botanist.

CASALS PLACE. The famous Spanish cellist was honored in July 1967 when Co-Op City was opened. The former swamplands were not a part of Freedomland.

CASANOVA STREET. The name is that of a Cuban importer who owned property on Hunts Point facing the East River. He called his mansion (the former Whitlock mansion) "Costello de Casanova,"and it was rumored the house was used as an armory for Cuban rebels who smuggled ammunition and weaponry out of New York. He died in Havana in May 1890, leaving a will in Spanish that was probated in New York City. It was signed Ynocencio Casanova y Fagundo.

CASLER PLACE. See: Alan Place.

CASS GALLAGHER NATURE TRAIL. A longtime resident of Kingsbridge, Florence Cass Gallagher was a devoted environmentalist, a photographer of Van Cortlandt Park in all seasons, and a lecturer on bird and plant life. After her sudden death in April 1983, the Nature Trail was formally opened on October 17,1984.

CASTLE HILL AVENUE. Lafayette Avenue was its original name from East Tremont Avenue to Westchester Avenue. Proceeding toward the East River, it carried the name of Sheep Pasture Road in the 1700s, and in 1850, Parsonage Road. When Unionport was laid out in 1854, it received yet another signpost, Avenue C. It was the original Indian path leading out to the point of land where European voyagers reported seeing a palisaded fort *(kasteel)* at approximately Lacombe and Castle Hill Avenues. The nineteenth century road led to the estate, "Castle Hill," owned by Gouverneur Morris Wilkins, who passed it to his daughter and son-in-law John Screvin.

CASTLE HILL PARK. This tiny park at the foot of Castle Hill Avenue carries the name of a former estate, "Castle Hill"belonging to Gouverneur Morris Wilkins. See: Hart Street.

CATALPA AVENUE. This is a driveway located in the northeastern section of Woodlawn Cemetery. A catalpa is a tree. Herman Melville, author of numerous sea stories including *Moby Dick*, was buried there in 1891.

CATALPA PLACE. This is a short street in Silver Beach Gardens bearing the name of a tree. See: Acorn Place.

CAULDWELL AVENUE. It was once called Woodlawn Street and later mapped as Park Street as it led to St. Mary's Park. It was named in 1916 for Senator William Cauldwell, who was head of the Board of Education and who worked for annexation of The Bronx by New York City. This came about on January 1, 1874. His daughter Emily married Thomas Rogers, whose mansion stood on the site of Morris High School. Senator Cauldwell's father, Andrew, was the first resident (the Fall of 1848) of the village of Morrisania, having erected a house on Washington Avenue near East 170th Street. This avenue runs through what were the villages of Woodstock (sometimes called Deckerville) and Forest Grove. Pudding Rock, a glacial boulder, once was located at East 167th Street and served as an Indian lookout, and later, a Huguenot resting place. At Westchester Avenue, it was once included in Henry Purdy's 1847 purchase from Gouverneur Morris II. Next, it was part of the Ursuline Academy grounds until 1888, when the avenue was blasted through the solid rock.

CAVANAGH PARK. This is not a park in the accepted sense of the word, but a triangular area on the Eastern side of Givan Square. It is bounded by Eastchester and Gun Hill Roads and Adee Avenue and was named in October 1974 to honor Emile J. Cavanagh on his 79th birthday. Mr. Cavanagh organized the Eastchester Home Civic Association in 1925, and, in 1941, was awarded a plaque designating him as "Protector of the North Bronx."

CAYUGA AVENUE. This Riverdale avenue carries two different reasons for its naming, neither of which can be found in official records. One story is that the early residents of the heights found a resemblance between Van Cortlandt Lake and Lake Cayuga. Another bit of folklore is that the surveyor was a graduate of Cornell University on Lake Cayuga.

CEDAR AVENUE (1). This area is the ancient Manorlands of the Archer family to whom was granted the original Fordham Patent. It figured in Revolutionary battles, and was later part of the Morris and Schwab estates. In the 19th century, the avenue ran through the estates of Popham, Cammann and Andrews. It was called Commerce Avenue in the 1890s, but a 1910 map lists it as Riverview Terrace. Cedar Avenue became its name in 1938.

CEDAR AVENUE (2). Located in the northeastern section of Woodlawn Cemetery, it runs past the diZerega family plot. A tall shaft marks the last resting place of the old-time shipping magnate, and grouped around it are tombstones of his sons, daughters, business associates and even a slave or two. It is named after the cedar trees so prolific in the cemetery.

CEDAR LANE. This is all that remains of the former roadway that led to the Gerard Morris estate "The Cedars." Later, it was modified to an entrance to Cedars Park, which later was renamed Franz Sigel Park.

CEDAR PLACE. See: Acorn Place.

CENTER STREET. This is the central street in Edgewater Park.

CENTRAL AVENUE. This is one of the principal roadways of Woodlawn Cemetery extending from the western side to the northeastern corner. The mausoleum of F. W. Woolworth can be seen from this main artery, as can the Allerton and Scribner family plots, and that of John B. Haskin. The Cammann family plot numbers 28 graves and represents a large family that once owned most of West Fordham.

CENTRAL STREET. This street runs down the center of Hart Island from north to south, and is bisected by five numbered streets that run east to west.

CENTRE STREET. Despite its name, this street does not mark the center of City Island. (Fordham Street is the halfway mark between Belden Point and Carey's Point). There is a strong possibility this street marks the former centerline dividing the lands of Schofield on the north, from Bowne on the south, according to maps of 1868 and 1872.

CHAFFEE AVENUE. This avenue was laid out on what used to be Wright's Island, and now called Locust Point. The western end touches an ancient road that led to Frog Point (Pennyfield Avenue) and was once the property in 1794 of Willett Leacroft, and later of the Hammond family. The avenues in the neighborhood are devoted to military men, so the only candidate would be General Adna Chaffee of the Civil War, who died in 1914.

CHAPEL HILL AVENUE. This is a roadway in the northeastern section of Woodlawn Cemetery. There is a chapel and receiving vaults on the hillside. The Arctic explorer, George Washington DeLong, is buried there, as well as Collis Potter Huntington, the railroad financier.

CHARLOTTE STREET. This street, laid out across the former Fox estate, honors Charlotte Leggett who married William Fox in 1808. Her sister, Rebecca, married George S. Fox. Both families were of pioneer Quaker stock. Charlotte, her husband and children, are buried in Section 9 of Woodlawn Cemetery.

CHARLTON PLAYGROUND. This small park is bounded by Boston Road, East 164th Street, Cauldwell Avenue and Teasdale Place honoring a Black sergeant, Cornelius Charlton, who was killed in action in Korea, 1952. The playground was named on May 12, 1952.

CHATTERTON AVENUE. Once known as 7th Street in the village of Unionport. The street honors early settlers of Westchester, although the name was ofttimes spelled Chadderton in records of 1734 and 1759.

CHATTERTON FIELD. At the intersection of Bruckner Boulevard and Chatterton Avenue, this lot was made over into a Little League ballpark around 1954-1955. It was named simply for its location.

CHESBROUGH AVENUE. Ellsworth Chesebrough [spelled this way] was a small property owner alongside Williamsbridge Road in the 1850s. He purchased land that had originally been Cornell property, later Arnow, and eventually that of Isaac Valentine. The first name of this avenue-from Williamsbridge Road to Blondell Avenue-was Mary Street.

CHESTER STREET. Originally this small street was called Chester Avenue, the main thoroughfare of a real estate venture called Seneca Park. Some residents claim this name is a reminder when the area was part of Westchester County prior to 1900. It most likely honors Lt. Colby M. Chester of the Spanish-American War, who was associated with nearby landowner Frederic Penfield on financial and legal missions to the Near East. He was decorated by the Sultan of Turkey with the Order of Osmanie. See: Osman Place.

CHESTNUT AVENUE. This roadway in the northwestern part of Woodlawn Cemetery is named after the chestnut trees that are still plentiful there.

CHESTNUT HILL AVENUE. This winding roadway in the northwestern section of Woodlawn Cemetery is doubly descriptive because it refers to both the hill and the trees upon it.

CHESTNUT STREET. Once called North Chestnut Drive, it is part of a real estate development, "Bronxwood Park," of which John T. Adee was one of the financial backers. The streets received names such as Oak, Chestnut and Hickory.

CHIMNEY SWEEPS. Originally included in Thomas Pell's purchase from the Indians, these small islands off City Island were owned by the Delancey and Hunter families in turn. In 1896, they were owned by Henry D. Carey, and in 1914, his widow sold them to a Charles Swan. In 1915, Franz Marquardt bought them. These reefs have been the scene of marine wrecks since Revolutionary times. The City formally acquired them in 1939. Together, the two islets total less than a half acre. When, why and how they acquired the name has never been cleared up, and there have been many improbable stories as to the origin of the name. A random one is that chimney sweepers traditionally carry two stones tied together to scour out the chimneys, and as seen on a marine map, the two islets resemble the stones. Another one touches on the humped appearance of the two islands with their dirty grey look, which makes them "drab as chimney sweeps."

CHISHOLM STREET. An 1875 map of the village of Woodstock shows the property of Walter Chisholm running along present-day Prospect Avenue. He was listed as a registered voter (1877) of the 23rd Ward. Mr. Chisholm, formerly a teacher in Sir Walter Scott's family, was brought to this country by Mr. Faile to run a private school for the children of the wealthy families on Hunts Point.

CHOCTAW PLACE. This Place, along with adjacent Pawnee Place and Seminole Avenue, has no historical connection with the land, which was part of the Denton Pearsall estate called "Woodmansten."

CHURCH SQUARE. It refers to the Catholic church of the Holy Family on Castle Hill Avenue off Watson Avenue.

CICCARONE PLAYGROUND. The property was purchased by the Police Department of New York in 1934 to provide a memorial for policemen killed in World War I. It was formally opened in the same year, and rehabilitated in 1952. Belmont has had an Italian-American population since 1900, and Vincent Ciccarone was a local resident. He was a private in Company B, 305th Infantry, 77th Division and was wounded in the Argonne forest in 1918, and later died as a result of his injuries.

CICERO AVENUE. Named for the famous ancient Roman orator and writer, Cicero. See: Caesar Place.

CINCINNATUS AVENUE. This street bears the name of the Roman farmer, Cincinnatus, who was chosen Dictator to lead the army against an invader, and who returned to his plow once he completed his task. See: Caesar Place.

CINTRON FIELD. On September 9, 1984, the Throggs Neck Rapid Soccer Club honored its organizer, Wil Cintron, by giving his name to the playing field at Bruckner Boulevard and Balcom Avenue.

CITY ISLAND. The first European to sight the island was Adriaen Block who claimed it for the Netherlands, so when Thomas Pell settled in the area he was served with a writ of ejection ordered by Governor Peter Stuyvesant (1655). Pell had received a grant from his English sovereign to all that territory, and had also purchased from five Siwanoy sachems all that land "and the islands in the Sound." He refused to leave, and was finally allowed to remain after taking an oath of allegiance to the Dutch. It became colonized quite early by fishermen. The island was called Mulberry, Minnifers, Minneford's and Minnewits which, some say, were the names of the Indian inhabitants. Samuel Rodman became owner in the early 1700s, followed by the Delancey family, and in 1761, Benjamin Palmer. In his memoirs, General Heath uses the name "New City Island," so the name must have been in use during the Revolutionary War. After the war, Palmer parcelled the island into 4,500 lots and sold them. In an attempt to make a new city of the port, the name New City Island was adopted around 1800. In the 19th century, when it became an important maritime center, resort spot and oyster farm, the name was shortened to City Island.

CITY ISLAND AVENUE. This is the main artery of City Island over which the stagecoaches ran, later to be replaced by horsecars, trolleys and finally buses. For many years it was simply Main Street.

CITY ISLAND BRIDGE. In 1873, a toll bridge was opened to the public, and made a free bridge in 1895, when City Island was annexed to New York City. Six years later, the wooden bridge was replaced by the present steel structure over the Narrows from Rodman's Neck to City Island.

CLAFLIN AVENUE. Horace B. Claflin was a wealthy Quaker (1811-1885) who established a dry goods business in 1843 and became the foremost wholesaler in the city. He only lived on his country estate in Fordham in the Summertime, being a Brooklyn resident the rest of the year. His son, John, an extensive traveler in the Amazon basin, sold out in 1903. This street was acquired by the city in 1922.

CLAREMONT PARK. "Claremont" was a large estate owned by the Zborowski family. Thus, when the village of Claremont was established in 1852, it took the name from the estate, which is now Claremont Park. At one time, the property of Elliott and Anna Zborowski de Montsaulain, this estate, covering thirty-eight acres of landscaped greenery, was mapped in May 1888, extending from Morris Avenue to Brook Avenue, from East 170th Street to East 174th Street. Their mansion, overlooking Webster Avenue, was built in 1859 and was famous for its raised sculpture in white marble. Beautifully landscaped lawns descended in terraces to the Mill Brook (before Webster Avenue was cut through). Once the estate became Claremont Park, the mansion was used by the Parks Department, and it was razed in 1938. Legend has it that, because of a "curse" no male member of the clan ever died peacefully in bed: Martin Zborowski died in a wheelchair; Elliott Zborowski (who married Anna Bathgate) was killed by a train; Francis Zborowski drowned in the Bronx River; Max Zborowski was thrown from his horse and died of his injuries; and Elliott Zborowski, Jr., fell from his motorcar and was killed. Elliott Zborowski married Anna Bathgate, who lived on the adjoining estate, now Crotona Park. See: Claremont Parkway, Crotona Park.

CLAREMONT PARKWAY. In 1850, the thoroughfare was called Bathgate Place. In 1890, it was changed to Wendover Place to honor Congressman Peter H. Wendover who regulated the size of American flags, and who lived on the street, which was a carriage road between the adjoining estates of the Zborowskis (now Claremont Park), and the Bathgates (now Crotona Park). Elliot and Anna Zborowski's estate was called "Claremont." lending its name to the little village, the park and finally, in 1913 the parkway. See: Bathgate Farm (1), Claremont Park.

CLARENCE AVENUE. This avenue was surveyed and laid out through what was the former Turnbull estate and the lands of Bruce Brown. Around 1906, it was part of the Century Golf Course. The avenue was opened in 1927 from Shore Drive to Layton

Avenue and in 1932 was extended to Fairmount Avenue. There is some speculation on the origin of this name. It might honor Clarence Blair Mitchell, who, prior to 1910, owned parcels of land on Throggs Neck. As there was already a Blair Avenue and a Mitchell Place, this might account for Clarence Avenue.

CLARKE PLACE. See: East Clarke Place, West Clarke Place.

CLASON POINT. This was originally known as Cornell's Neck after the proprietor of 1688. Its Indian name was Snakapins, which has two divergent translations, Place of Groundnuts and Land by Two Waters. Isaac Clason was a wealthy merchant whose lands were later subdivided into smaller estates. He purchased the eastern side of Cornell's Neck in 1793-94 and lived there for many years. A son, Augustus Washington Clason, had a nearby home which eventually was sold in 1855 to Joseph J. Husson along with 15 acres of land. Isaac Clason's mansion, which incorporated a part of the earlier Willett and Cornell home, became an inn in the 19th century. A subsequent landowner, Clinton Stephens, preserved it as an historic landmark, but it was razed in the 1930s to make way for a beach club.

CLASON POINT LANE NORTH. It is a street in Clason Point Gardens shaped like a crescent running from Story to Noble Avenues.

CLASON POINT LANE SOUTH. It is a street in Clason Point Gardens running from Story to Lafayette Avenues shaped like a scallop.

CLASON POINT PARK. This small park at the foot of Soundview Avenue was the terminus of Clason Point Road, a road no longer in existence.

CLAY AVENUE. The lower end of Clay Avenue from East 164th Street to East 167th Streets was part of General Staats Morris' racetrack of the 1700s. At the top of the hill, East 168th Street, was the mansion of William H. Morris in the 1870s. It was known as Elliott Street where it ran alongside Claremont Park, which had been the Zborowski estate "Claremont." It was known as Crestline S treet from East 169th Street to East 171st Street, and as Anthony Street to East 174th Street. It is thought to honor Senator Henry Clay, but one old-time story is that the Black Swamp, west of the Zborowski estate, held up construction so much that when the excavators finally found clay instead of mud, they gratefully named the avenue that ran across it.

CLEMENTE (Roberto) PLAZA. The intersection of East 149th Street, Third, Melrose and Willis Avenues was given the name of Roberto Clemente, a Puerto Rican ballplayer who lost his life on a mercy mission in 1972. His plane, carrying supplies to survivors of a Central American earthquake, crashed into the Caribbean, leaving no trace. The bill to call the intersection Clemente Plaza was approved by Mayor Lindsay in 1973. For almost a century before that, it had been called "The Hub." See: Roberto Clemente State Park.

CLIFFORD PLACE. It is shown as "Clifford's Place" on an 1883 map. Mr. G. Clifford was a landowner through whose estate the Grand Concourse was laid. Once, around 1900, it was mistakenly labeled Clifton Place. West of Jerome Avenue, Clifford Place is a flight of steps.

CLINTON AVENUE. When the village of Morrisania was settled, this avenue was one of its finest. At one time, it was aptly named Rustic Avenue. It was renamed to honor DeWitt Clinton, mayor of New York in 1803 and 1811, and governor of the State when the Erie Canal was opened in 1825.

CLINTON PLACE. Clinton farm once occupied the area of West 182nd Street from Aqueduct Avenue to approximately Jerome Avenue. A story dating back to the 1850s tells of the farmer who sold 'moonshine' whiskey to the Irish laborers working on the Croton Aqueduct. He kept a hay wagon, loaded with whiskey kegs, in a nearby meadow that was beyond the control of the Aqueduct police.

CLOSE AVENUE. John Cloes was one of the early Westchester colonists who, under Lt. Thomas Wheeler, pledged allegiance to the New Netherlands in 1656. Other colonists were William Fenfell (see: Evergreen Avenue), John Genner (see: Boynton Avenue), William Ward, and Sherrood Damis (see: Colgate Avenue). The name was changed by Cloes' descendants, one of them being Odell Close, a delegate to Albany (1867- 1873).

CODDINGTON AVENUE. Once known as 3rd Street from Edison to Crosby Avenues, it was officially opened in 1926. It is the southern boundary of "the Quaker Tract" deeded to the Friends in 1808 by Josiah Quinby. The avenue is named after the numerous Coddingtons of Westchester Township, of which Jonathan was an early postmaster. Silas is mentioned in an 1854 deed, and a Horatio and a Robert Coddington figured in other land transactions. The hilly avenue was once part of the sloping property of the Harrington School for Boys that flourished in the 1800s to early 1900s. In Revolutionary times, this was the side of Bridge Hill up which the British and Hessian troops dragged cannons in October 1776.

COFFEY FIELD. This is an athletic field at the northern end of Fordham University, named after a famous graduate and athletic manager, Jack Coffey.

COLDEN AVENUE. It is laid out across the former Morris Park racetrack and, above Pelham Parkway, the Astor and Mace properties. It honors Cadwallander Colden, a mayor of New York from 1815 to 1818, whose grandfather wrote *History of the Indian Nations* in three volumes. His daughter inherited his love of botany, corresponded with the Swedish naturalist Linnaeus, and catalogued many North American plants.

COLES LANE. Jacob Cole was Postmaster of Fordham in the 1860s. He was listed as residing at the corner of Marion Avenue and Fordham Road. William and Andrew J. Cole were registered voters in 1877, their address being Coles Lane.

COLGATE AVENUE. This was once part of the extensive William Watson estate that was surveyed and subdivided around 1900. At first, it was called Damis Avenue, an early settler being named Sherrood Damis who had to swear allegiance to the Dutch authorities in 1656 in order to retain his land. On February 7, 1911, the street was officially named Colgate Avenue, after James Boorman Colgate, founder of Colgate University. He was a banking associate of Ferdinand Thieriot and Spencer Trask, and both men have their names affixed to nearby avenues.

COLLEGE AVENUE. The lower end was part of the Indian camping grounds called by the Dutch and English "Ranachqua," its Indian name. It was part of Jonas Bronck's lands which became the southern end of the Morris Manorlands. It became part of the tract sold to Jordan L. Mott from which he formed Mott Haven around 1853. Rider and Conkling bought 600 lots from Jordan Mott, and this street was first known as Rider Street. The only college in that region was the Convent of the Sacred Heart, mapped in Civil War times, at East 133rd Street. Matthew O'Brien and Owen Cullen were the stonemasons. They specialized in bluestone, a superior grade of slate. When the college was demolished in the 1890s, the curbing and retaining walls were of such fine workmanship that they were allowed to remain. The upper end of College Avenue, from East 164th Street northward, was opened in 1893 on lands that had been used as the Fleetwood Racetrack.

COLLEGE ROAD. Running northeasterly from Waldo Avenue and West 246th Street to the Post Road at West 252nd Street, this street is within walking distance of Manhattan College, the Barnard School, Horace Mann School and the Riverdale School for Boys.

COLLIS PLACE. It was named after Collis P. Huntington, owner of the land around 1887. His mansion is now Msgr. Preston High School. Collis Place was once the lands of Philip Livingston in Revolutionary times. In 1822, it was part of the property of Dr. Wright Post, who leased it to his brother, William. William was a prominent butcher and used the Throggs Neck meadows to raise cattle. In the 1860s, the tract was owned by Peter Van Schaick who donated a free library and reading room to the Westchester townspeople, who refused it. See: Huntington Avenue.

COLONIAL AVENUE. It runs directly through the old settlement known as Stinardtown (1868 map) and what was the J. Appleby estate of 1905. This name might well be a companion piece to other avenues in the neighborhood, such as Mayflower and Pilgrim.

COLUCCI (Florence) PLAYGROUND. Located at Wilkinson Avenue and Hutchinson River Parkway, this plot was named for Florence Colucci, a civic worker of that neighborhood. Groundbreaking ceremonies took place on April 12, 1969. The land once belonged to the Paul farm, which dated back to Revolutionary times, General George Washington one time visiting "the widow Paul."

COLUMBUS SQUARE. This was once part of the Hoffman farm, mapped in 1888 as "Cedar Grove Farm." Located at Bathgate Avenue and Lorillard Place in the heart of Belmont, the Italo-American community was gratified when this Square was named in April 1922.

COMMERCE AVENUE. This avenue runs along the southern boundary line of the Benjamin Ferris property, off Westchester Square. The lane originally ran to Pasture Hill. The Ferris burial plot, mentioned as early as 1702, is still situated alongside this avenue, having been restored in 1971-74.

COMMONWEALTH AVENUE. In the Clason Point section, part of this avenue overlays the original Cornell Patent, which later became (1793) the holdings of Isaac Clason. Dairy Creek once ran along this avenue to what is now Seward Avenue. An Indian trail from Clason Point cut across this avenue at approximately 930 Commonwealth Avenue. West of Westchester Avenue, Commonwealth Avenue was once part of the Mapes farm that was subdivided into Park Versailles. This writer's theory is that the naming of this avenue is connected with nearby avenues, Stratford and Virginia.

Stratford is a town in the Commonwealth of Virginia, and might allude to the surveyor's birthplace, or to that of one of the landowners.

COMPTON AVENUE. Once the landing place for Siwanoy Indians whose village lay at Lacombe and Leland Avenues on Clason Point. It was part of the original Cornell grant that passed to Isaac Clason. An 1893 map shows A. G. Compton owned 14 acres, and the curve of the street reflects that of Pugsley Creek.

COMRAS (Oscar) MALL. See: Oscar Comras Mall.

CONCORD AVENUE. It was part of the eastern bounds of the vast Morrisania Manorlands that was sold to the Dater brothers, John Crane, and others. It was surveyed in 1853-57, and subdivided into building lots around 1893. In 1900 its upper stretch, above East 149th Street, was opened. The name honors the town of Concord, Massachusetts, where the famous Revolutionary skirmish took place, for nearby East 147th Street was once Lexington Avenue.

CONCOURSE, The. See: Grand Boulevard and Concourse.

CONCOURSE VILLAGE EAST. This street is formerly parts of Park and Morris Avenues touching the eastern edge of the Concourse Village Cooperative housing development south of East 161st Street and north of East 153rd Street. It is named after the Concourse Village development, which, in turn, gets its name from the Grand Boulevard and Concourse, which is west of it. The name was given by the City Council after the developers petitioned for the change.

CONCOURSE VILLAGE WEST. This street is formerly Sheridan Avenue touching the western edge of the Concourse Village Cooperative housing development south of East 161st Street and north of East 153rd Street. See: Concourse Village East.

CONNELL PLACE. In the 17th century, this was part of the Underhill family lands which was, during Revolutionary times, owned in turn by the Baxter and Ferris families. In the mid-1800s, it was on the estate of Pierre Lorillard, passing to his daughter and grandson, Lorillard Spencer, in 1868. J. Van Antwerp was the next owner, and in the following decade it was the edge of the Westchester Polo Club grounds, known as Country Club. The Place itself was surveyed in 1922, and the map filed by Philip E. Connell.

CONNER STREET. It was once mapped as Eastchester Lane. J. Conner lived at the head of Westchester Creek at Baychester Avenue, 1798-1861. The settlement was known as Connersville and so mapped in 1871, and a horse-trolley line ran out to an amusement park, "Eagle Grove," on the Hutchinson River. The Conner family has a plot in St. Peter's P.E. Churchyard.

CONTINENTAL AVENUE. This avenue is laid out across the northern end of a Revolutionary farm known as "Stoney Lonesome." Stoney Brook flowed west of this avenue, where the State Hospital is located. The land was later owned by Aaron Burr and then Simon Paul. The name evidently refers to the Continental road that led across Stoney Brook, but it is not certain if the present-day avenue overlays the ancient road. In October 1776, a party of English and Hessian soldiers attempted to ford upper Westchester Creek, but were repulsed by Americans that were stationed on the higher ground of Eastchester Road. As there is now no trace of the road, both Continental and Colonial Avenues can only be approximate routes.

COOKE (Cardinal) PROMENADE. See: Cardinal Cooke Promenade.

CO-OP CITY BOULEVARD. The most important artery of the large project known as Co-Op City is built on what was once Pinckney's Meadows of 1662. It was officially opened in 1968. See: Cooper Place.

COOPER PLACE. Located on a land grant from the King of England to Thomas Pell, the swamplands were part of Philip Pinckney's lands on which he settled in 1662. It covers what once was a tidal creek - still navigable in 1965 by kayak or canoe. James Fenimore Cooper was honored in July 1967 by Local Law, when his name was affixed to this Co-Op City Place.

COOPER STREET. William Cooper was a nineteenth century squire whose land ran from Ponton Street to Westchester Creek in the vicinity of Blondell Avenue. His grandfather, George D. Cooper, was Village Collector and at one time (1816) was vestryman of St. Peter's P.E. Church.

COPE LAKE. This small lake is inside the Bronx Zoological Park in the northwest corner, near Fordham Road. It is about 2 acres in area and very shallow. Professor Cope, a paleontologist, is honored.

CORLEAR AVENUE. In the 19th century it was called Water Street from West 230th Street to West 232nd Street and overlays the course of Tibbett's Brook, which once flowed southward to Spuyten Duyvil Creek which also once flowed along West 230th Street at the foot of Corlear Avenue, which is why an earlier name was Water Street. Later, it was known as Ackermann Street after an old-time Kingsbridge family. Anthony Van Corlaer, or Corlear, was the official messenger sent by Peter Stuyvesant to get reinforcements from the mainland when Nieuw Amsterdam was attacked by the British fleet. He was drowned in the Spuyten Duyvil Creek or, according to a legend recounted by Washington Irving, was "pulled under by a giant mossbunker."

CORNELL AVENUE. Thomas Cornell, a native of Essex in England, was the earliest landowner of Clason Point, and in consequence, its name was Cornell's Neck (1646). An important Indian trail terminated at Cornell and White Plains Avenues.

CORNELL PLACE. A Cornell farm in Middletown according to an 1851 map, and this is a relic of it.

CORPORAL IRWIN FISCHER PLACE. See: Fischer (Corporal Irwin) Place.

CORPORAL WALTER J. FUFIDIO SQUARE. See: Fufidio (Corporal Walter J.) Square.

CORSA AVENUE. This street, running through what had been the nineteenth century Kidd estate, is named for the pre-Revolutionary family that supplied Westchester Guides to Washington and his staff. Another Corsa had an estate on which Fordham University is located. Sometimes spelled "Curser" the name is traced back to a Dutch ancestor named Hendrik Cortinsenze.

CORTLANDVILLE WALK. This is an unpaved path in Van Cortlandt Park. The name was probably made up by the Parks Department, with the obvious allusion to the park itself.

COSTER STREET. This is on part of the original (1663) purchase by Richardson and Jessup of the Twelve Farms Tract. The area became the property of Thomas Hunt, who married Elizabeth Jessup (or Jesup). An Indian trail crossed this street at Lafayette Avenue which evolved into the original Hunts Point Road, the supply route during the Revolutionary War for ships that docked at Hunts Point. In the 19th century, the street was on the estate, "Blythe," owned by Francis J. Barretto, and partly across Paul Spofford's land, "Elmwood." The street was named for Mrs. Barretto, the former Julia Anna Maria Coster, whose ancestors migrated from Haarlem, Holland, as prosperous flax merchants. Her brothers, Daniel J. and John S., and a nephew, Henry A. Coster, owned a large tract on Throggs Neck. See: Logan Avenue.

COTTAGE PLACE. This Place is at the extreme northern limit of the Manorlands of Morrisania. Alexander Bathgate, foreman to the Morris family, purchased a considerable tract which is now Crotona Park, and it remained in his family until 1883. The Bathgate mansion was in the center of the park-and this Place was named for a groundskeeper's cottage that was demolished when Crotona Park South was cut through.

COUNTRY CLUB ROAD. This is a winding reminder of what was once the fashionable Westchester Country Club, patronized by yachtsmen and polo horsemen. It encompassed 26 acres. It was part of the Underhill lands in the 17th century, Ferris lands of the Revolutionary times, and Van Antwerp estate of the 19th century. Originally called North Road, it was the division between the Country Club and the Lorillard Spencer estate.

COURTLANDT AVENUE. This name appears as early as 1850 when Andrew Findlay surveyed North Melrose. Courtlandt is apparently an anglicized version of Major Stephenus Van Courtlandt who signed his name S:V:Cortlandt, at times adding the "u." He witnessed documents dated 1686 and 1687 signed by Governor Dongan whose name was originally affixed to East 163rd Street. When the villages of Melrose and Melrose South came into being in 1850, this avenue on the crest of the hill was lined with small homes. It gained the nickname of "Dutch Broadway" due to its German beer gardens, *Turnvereins* and Teutonic population. East 151st Street had a gymnasium and social hall that was frequented by General Franz Sigel. The Haffen house was at East 152nd Street, the Arion Singing Society had its headquarters on East 154th Street, and the First Lutheran Church was organized at East 156th Street.

CRAMES SQUARE. It was named in March 1923 to honor Pvt. Charles Crames, who died in World War I. The Square was once crossed by an Indian trail that led to a village out on Oak Point Avenue.

CRANFORD AVENUE. It was once called Bronx Place for its four-block stretch to the City line. The avenue takes its name from the Kenneth Cranford estate, mapped in 1905 and 1913 as a tract of 25 acres.

CRAVEN STREET. The street in East Morrisania was named to honor Admiral John L. Worden, but later the name was changed to Craven Street. A few streets immediately flanking Craven Street honor Naval men such as Truxton and du Pont, so this street recalls Admiral Tunis A. MacDonough Craven.

CRAWFORD AVENUE. A 1760 deed lists Robert Crawford, resident of Eastchester, and the name recurs in 1853, 1894 and 1913, when the Crawford Realty Company held a tract of land along Baychester Avenue.

CRESCENT AVENUE. This curving avenue roughly resembles a crescent and was so noted on early maps of Belmont. It is laid out on the lower end of the (1810) Bayard farm which became the (1840) Powell farmlands, and the (1880) Hoffman farm.

CRESTON AVENUE. At its northern end, once it was part of the Haskin estate of Fordham. A supposition is that this avenue, on the crest of an elevation shared by the Concourse, received its name from that fact. In the 1800s, the hills were called Mount Sharon (Fordham Road) and Prospect Hill (East 180th Street).

CRIMI ROAD. A stretch of the Shore Road was renamed to honor Charles Crimi, a local resident, active in many civic, educational and religious affairs throughout his lifetime. Mayor Koch signed the bill into law on June 15, 1982.

CRIMMINS AVENUE. This was a wealthy and well-known family that lived in East Morrisania in the 19th century. Thomas Emmet Crimmins was president of the Westchester Electric Railroad and also a director of the Merchants' and Traders' Bank (1885). John D. Crimmins was responsible for building most of New York's early transit system, including elevated and subway train lines, when these were private enterprises. He financed the construction of Corpus Christi Monastery on Hunt's Point. A noted researcher on the American-Irish, he published two books.

CROES AVENUE. It was a former crossroad at Westchester Turnpike leading out to Dominick Lynch's lands, the Ludlow family's "Black Rock Farm" and Isaac Clason's land. The upper end of this avenue runs through what had been Mapes' farmland which later (1890s) became the real estate development, "Park Versailles." A few streets in this neighborhood were named for civil engineers of which C.L. Croes was one.

CROES PLACE. This is a small Place in Clason Point Gardens. See: Croes Avenue.

CROMWELL AVENUE. James Cromwell, descendant of early settlers in America, established a mill on Cromwell's Creek in 1760. Their farmlands extended from the winding creek (Jerome Avenue) eastward to present-day Walton Avenue and from East 157th Street to West 168th Street. Cromwell's Creek flowed into the Harlem River where the Yankee Stadium is situated, following Jerome Avenue and River Avenue in part. This creek, partly spring-fed and partly tidal, formed the boundary between the Morrisania Manorlands and Daniel Turneur's eighty-acre tract in the vicinity of Highbridge.

CROSBY AVENUE. This was formerly Schuyler Street in the old settlement of Schuylerville. Horace Crosby was a surveyor of City Island and Throggs Neck in the 1850s, and once lived on the estate south of St. Raymond's Cemetery. His son, Alpheus Dixi Crosby, was born in the mansion (demolished in 1970) and was a professional Army officer. He died in retirement in 1956.

CROSBY (Howard) WALK. This footpath on the campus of Bronx Community College honors the fifth Chancellor of New York University who served in a time of financial difficulties (1870-1881). He was a graduate of the University in 1844 and later a professor of Greek there.

CROSS STREET. The narrowest waist of City Island, enabling mariners to pick up their craft and cross over, thus eliminating a long haul around Belden Point or Carey's Point, gives the street this name. A Michael Smith of Eastchester told this writer of using the portage in the 1870s.

CROSS-BRONX EXPRESSWAY. This largely depressed highway was cut entirely across The Bronx in the 1950s at tremendous cost as long stretches had to be blasted through solid rock and hundreds of apartment houses had to be demolished in its path.

CROTONA AVENUE. This thoroughfare from Crotona Park South to Bronx Park passes through Tremont and forms the eastern boundary of Belmont. Its width at the northern end once made it a favorite promenade for townspeople. See: Crotona Park.

CROTONA PARK. Laid out in 1883 on the upper end of Morrisania Township, the property was obtained from the Bathgate family. It was the intention of the Park Commission to name it Bathgate Park, but after a dispute with the family, the Commission's engineer manufactured the name "Crotona" from Croton, a Greek colony in southern Italy, noted for its athletes who excelled in the Olympic Games.

CROTONA PARK EAST. This avenue faithfully follows the eastern curve of Crotona Park. See: Suburban Place.

CROTONA PARK NORTH. As its name denotes, it forms the northern side of Crotona Park. See: Crotona Park.

CROTONA PARK SOUTH. Marking the northern edge of the former Manor of Morrisania, this street was once called Julia Street. Julia Huertel was the owner of a tract from Jefferson Place to this street in 1889.

CROTONA PARKWAY. A strip of land bought by the City in 1888 to connect Crotona Park with Bronx Park. It was formally opened in 1910, having been an unpaved passage with the local name of Penfold Street. This overlays the ancient "Hassock Meadow" mentioned in the deed of 1664 when Edward Jessup and John Richardson purchased from the Indians what is now West Farms.

CROTONA PLACE. This was once part of the Bathgate farm which became Crotona Park, and is believed to have had its origin as a lane leading to an ice pond. In the 19th century, it formed the western side of Zeltner's picnic grounds.

CROWNINSHIELD STREET. On the Fort Schuyler peninsula, the New York State Maritime College has a short road across its campus. Commander Crowninshield is remembered for his term as Superintendent from 1887 to 1890 .

CRUGER AVENUE. John Cruger, Sr., and John Cruger, Jr., were both mayors of New York. The father served five consecutive terms, from 1739 to 1744, and had emigrated from England at an early age. In 1698 he was engaged as supercargo* of a slave ship called the *Prophet Daniel*, which was captured by pirates and sold to an African chieftain. He, in turn, sold it to other white men who sailed it to South Africa, taking Cruger along as passenger. Eventually, the young man reached New York after two years and wrote an account of his adventures in *A Voyage to Madagascar in the ship "Prophet Daniel."* Cruger became a sober businessman, although his connection with the slave trade in those days was no reproach. The avenue is laid out through what was the Hitchcock estate of 1870 around Boston Road, the Mace estate of 1868, and the Astor estate of the same date. At Bronxdale, it was called Brown Avenue and in Williamsbridge, it was called Timpson Avenue. *A term dating from 1697 -- the officer of a merchant ship in charge (supervisor) of the commercial concerns of the voyage; from the Spanish *sobre-* over (from Latin *super-*) + *cargo* = (cargo).

CUBAN LEDGE. Boatmen know this sandy isle with rocks at its southern tip that only appears at low tides. It is between Rodman's Neck and Country Club in Eastchester Bay. The exact origin of the name is not known but there are several posited. Version (1) A lumber sloop from Eastchester plied between that Hutchinson River port and the Little Corn Islands off Honduras, transporting lumber. One trip, however, the cargo was rum, and the drunken crew ran the ship aground on this reef. The ship was the *Cuban Lady* and Michael Smith told the tale. Version (2) The Hutchinson River was being dredged and widened in 1898, and a crew aboard a stone barge was passing the Pelham bridge with a cargo of rocks. The bridge tender shouted the news that the battleship *Maine* had just been blown up in Havana harbor. The sailors on the barge dumped their rocks, and headed for port to enlist. The rocks remained and formed a reef that is now called Cuban Ledge. Version (3) Throggs Neck was subdivided into city streets around the time of the Spanish-American War (1898) and the avenues there acquired names of military and naval heroes. The reef received its name at that time, as some of the war's action took place in Cuba-and the surveyors had utilized the island as a reference point. Version (4) - Looking at a map, the ledge bears a strong resemblance to the outline of Cuba.

CUNNION-HOURIGAN-McCARTHY FIELD. North of the Bronx Botanical Garden on Allerton Avenue, between Mosholu Parkway and Bronx Park East, this playing field was named in December 1967. Michael Cunnion was downed in a helicopter at Quang Tri, Viet Nam, on July 15, 1966. Private First Class William Hourigan was killed in action on December 11, 1966 in DaNang. Peter McCarthy was killed on April 8, 1966 in Viet Nam. As boys, all three had played baseball at this field.

CYPRESS AVENUE. From Bronx Kill up to St. Mary's Park, it was formerly called Trinity Avenue. Gouverneur Morris' estate below East 138th Street was enriched with many rare ornamental trees, and two or three specimens of his American cypress were said to be the finest in America. When the street was laid out, the lower end of this avenue was lined with cypress trees.

CYRUS PLACE. Former Borough President Cyrus Chace Miller was a Fordham resident, and an unusual combination of a practical politician, a dedicated historian, and an expert archaeologist. He was the first official Historian of The Bronx and managed to save some landmarks from demolition. He helped straighten out streets and insisted Fordham Road be a wide thoroughfare instead of the narrow stretch which had been proposed. He died in 1956, at the age of 89.

D'AURIA-MURPHY SQUARE. This Square, at East 183rd Street, Crescent Avenue, and Adams Place, was named on May 16, 1931. An American Legion Post was organized in 1918 and was to be named John Purroy Mitchel Post in honor of the Mayor, but that dignitary did not appear at the installation, so the veterans voted to name their Post after two former Belmont residents, John D'Auria and Henry J. Murphy. John D'Auria was born in Italy and came to this country at age 15. He became a stonecutter and lived at 2503 Hughes Avenue. He was a machine gunner in the 105th Infantry, and was killed at the age of 29, on October 12, 1918. Private. Murphy, 27, was killed in action on September 30, of the same year. He had lived at 2421 Cambreleng Avenue and was a Post Office clerk.

D'ONOFRIO SQUARE. This Square at East 213th Street and White Plains Road was acquired in November 1900 as a public place, and came under the jurisdiction of the Department of Parks in 1926. Sgt. Salvatore D'Onofrio, a local resident, died in World War II on June 24, 1944.

DAFFODIL HILL. A hillside in the Bronx Botanical Garden that is used for Summertime band concerts. The name is used by the staff to designate this hill in the southwest sector, and it occurs in all their publications. In season, the slope is a sea of daffodils.

DAISY PLACE. This is a short thoroughfare in Silver Beach. See: Acorn Place.

DALY AVENUE. Judge Charles P. Daly has had his name affixed to this avenue, as his father-in-law, Philip Lydig, was the last owner of what is now Bronx Park. For a time, the avenue was called Catherine Avenue, presumably to honor another landowner, Catherine Lorillard Cammann. Another name was Elm Street.

DANIEL STREET. Daniel Buhre was the last owner of record in 1904. Around 1900, his farm in Middletown was subdivided to form a real estate development, "Tremont Terrace."

DARE PLACE. See: Alan Place.

DARK STREET. Charles Darke [spelled this way] owned property in different parts of The Bronx from the 1850s to the 1870s.

DARROW PLACE. Clarence Seward Darrow, American lawyer, was born in Ohio in 1857 and died in 1938. His name was given to a Co-Op City street in July 1967. See: Adler Place.

DASH PLACE. This steep, crooked Riverdale street overlays nineteenth century Dash's Lane. Bowie Dash was a wealthy coffee merchant.

DAVIDSON AVENUE. It honors Colonel Mathias Oliver Davidson, Chief Engineer of Streets from 1867 to 1872 and a Fordham landowner. On an 1868 map, his estate was bounded by Kingsbridge Road, Davidson Avenue, West 190th Street and the Veterans' Hospital grounds. This avenue runs along the slope of what was the Townsend Poole property-through an ice pond (now Goble Place) and the Andrews estate, eventually to end up in the Valentine farm, north of Fordham Road. Some of the Davidson property was sold to Adam Cameron, which accounts for Cameron Place.

DAVIS AVENUE. It is laid out over what was once swamplands of Baxter's Creek (Morris Cove) and the original Livingston Grant that passed through the ownership of Bayard Clark to the Morris family (1850-1922). Along with Buttrick, Hosmer and other Concord Minutemen, Captain Isaac Davis figured in the battle of Concord bridge and was killed there.

DAWSON STREET. Henry B. Dawson, a noted historian of the 19th century, was a lifelong resident of Morrisania. East 155th Street, when it was surveyed and cut through, was originally named Dawson Street in his honor.

DE KALB AVENUE. This Avenue was once part of the Valentine farm which became Woodlawn Cemetery. As an Aide to Lafayette, this German general took part in the Revolutionary War. He was killed at Camden in 1780. (German historians call him Baron Kalb, as the "de" was considered a French affectation.)

DE KRUIF PLACE. Paul De Kruif, born in 1890, was an American author and bacteriologist honored in July 1967 by Co-Op City planners. The Place is situated a short distance north of what had been Reid's mill, a tidal mill that stood from 1739 to 1900 when it was blown down in a storm. From 1959 to 1965, this Place was the edge of Freedomland parking lot. See: Reed's Mill Lane.

DE REIMER AVENUE. The name means "the saddler." Isaac DeReimer was Mayor of New York in 1700-01. He was a descendant of an old Dutch family whose name appeared on church rolls in 1653. A Margaretta DeReimer was the wife of another Mayor of New York, Cornelius Steenwick. This avenue formerly carried the name of Monaghan Avenue and St. Mary's Avenue from Bartow Avenue to Mace Avenue. From Bartow Avenue to the city line it was South 17th Avenue.

DE ROSA-O'BOYLE SQUARE. Situated at East Tremont Avenue over the Cross-Bronx Expressway, it marks the division line of the nineteenth-century estates of Catherine Lorillard Wolfe and Frederick Wendell Jackson. The Square honors two young men of Throggs Neck who lost their lives in World War II, Peter DeRosa and Andrew O'Boyle.

DE WITT PLACE. This is a remnant of Judge George DeWitt's estate.

DEAN AVENUE. It was officially opened in 1927. A Christopher Dean was mentioned in 1687, and the family included town clerks, farmers and a merchant or two. Nicholas Dean was an early settler, whose daughter Phoebe married Joseph Pell in 1736. The avenue passes through Dr. Turnbull's land (1868-1900), a noted pleasure resort and picnic area at Layton Avenue. Proceeding northward, the land was once part of the land owned by James and Charity Ferris (1776), which passed through several ownerships: Van Antwerps, Laytons, Ellis. Last owner of record at that spot was an Elmer Dean Coulter in 1917, who had bought the land from the Lohbauer Realty Corporation.

DEBS PLACE. This Place in Co-Op City, honors Eugene V. Debs, American Socialist. See: Adler Place.

DECATUR AVENUE. This avenue was part of Jan Arcier's (John Archer's) mid-1600 Manor of Fordham, which he lost by foreclosure to the wealthy Steenwick family. The land passed to the ownership of the Dutch Reformed Church, and, around Revolutionary War times, was sold to the Valentines, Archers and Varians. The lower end of Decatur Avenue was on the Hugh N. Camp estate, while the upper end was part of the Briggs farm. When the street was laid out, it was first called Prospect Avenue because of the fine view (prospect) it afforded of St. John's College, which is now Fordham University. Later, in 1894, it was named after Stephen Decatur, the Naval hero, and around that time was also known as Decatur Street. Above Mosholu Parkway, it was laid out as Norwood Avenue, but, in 1897, Decatur Avenue included this avenue. See: Norwood.

DEEGAN EXPRESSWAY (Major William F.) This is an important West Bronx six-lane expressway laid out alongside the Harlem River on partially filled-in land, and overlaying an earlier street called River Road. Laid out in 1944, it was extended from East 144th Street to the city limits in 1956. William F. Deegan was born in The Bronx and established himself in the building trade. At the age of 25, when World War I broke out, he was commissioned a captain, assigned to General Goethal's staff and put in charge of constructing Army bases in and around New York. He was made a Major by the time World War I was over. He became State Commander of the American Legion in 1921, a personal friend of Mayor James Walker and, in 1928, was appointed Commissioner of Tenement Housing. Following an appendectomy, he died at the age of 39 on April 4, 1932 and had a most impressive funeral. (Twenty-five airplanes circled overhead.) Five years later, the Expressway was named in his honor.

DEFOE PLACE. Daniel Defoe, English author and famous for his *Robinson Crusoe*, is remembered by this Co-Op City Place. Long Pond Creek once flowed under this Place, and was navigable to small craft until 1963. See: Adler Place.

DELAFIELD AVENUE. Major Joseph Delafield was the 1829 landowner of Fieldston Manor. It was sold for city lots in 1873 and one of the streets was first christened Field Street. Later on, it took the more grandiose title of Von Humboldt Avenue. However, in 1916, the Major was remembered.

DELAFIELD LANE. A private road paralleling West 246th Street from Arlington Avenue to a private development known as "Dodgewood Estates." See: Delafield Avenue.

DELANCEY FALLS. A waterfall of the Bronx River 50 yards north of East 180th Street. It is named after the Colonial family that ran a mill on the right bank of the stream.

DELANCEY PLACE. This French Huguenot family migrated from Caen in the 17th century and were already on church rolls of Nieuw Amsterdam in 1703. The family was prominent in affairs prior and during the Revolution in West Farms and Hunts Point. At one time they owned Hart Island. John Delancey owned 12 acres in this area and sold them to the Hunt family in 1771. During the 19th century it was part of the Neill farm, which descended to a grandson, John DeLancey Neill, who called it "Round Meadow." This is evidently the owner who is honored in Delancey Place.

DELANOY AVENUE. Peter Delanoy, whose name was sometimes spelled DeLanoy, was born in Holland. He was a successful merchant, and was the first Mayor of New York elected directly by the people in 1689-90. The avenue overlays the east branch of Stoney Brook [spelled this way], which afforded the Indians and early colonists a portage to Black Dog Brook (which led to the Hutchinson River.) In 1727 it was part of Reverend Bartow's holdings, known as "Scabby Indian," which was willed to his widow. In the mid-1800s, this was the extreme western edge of John Hunter's land.

DELAVALL AVENUE. Sometimes spelled De La Valle, Thomas Delavalle was a wealthy merchant and active in military affairs. He served three terms as Mayor of New York, 1666, 1671 and 1678. The avenue was part of the original land grant of Thomas Pell, 1654, which later became the colony of Eastchester. It is thought to be the site of Ane Hutchinson's farm.

DELL PLAZA. See: James Dell Plaza.

DEMARCO SPORTS COMPLEX. See: Michael DeMarco Sports Complex.

DEMEYER STREET. Along with other streets in this area, this thoroughfare is named for a former Mayor of New York. According to old maps, this was a swampy area through which it was possible to pass from the head of Westchester Creek eastward to an arm of the Hutchinson River called Givan's Brook or Black Dog Brook. This was the limit of John Bartow's land (willed in 1727 to his wife) and passing through various ownerships-Furman, Stillwell, Bayard and Hunter. Nicholas DeMeyer was a baker by trade, and served as Mayor in 1676. He was related to the Schuylers.

DEPOT PLACE. The Highbridge station of the New York Central Railroad is at the foot of this street. Around 1875 to the early 1900s, it was a popular stop for excursionists who visited the amusement parks in the vicinity. It overlays ancient Crab Island, mentioned in the 1676 Patent dividing the Manorlands of Morrisania from Daniel Turneur's Land.

DEPOT SQUARE EAST. This is the former name of Botanical Square East.

DEUTSCH PLAZA. This Plaza on University Avenue was named in 1937 as a tribute to Bernard S. Deutsch, President of the Board of Aldermen, and President of the American-Jewish Congress.

DEVANNEY SQUARE. Located on the west side of Burnside Avenue at the Concourse, the small square honors Private. Patrick J. Devanney, who was killed in World War I. Originally from the Hub section, the Devanney family moved to West Tremont prior to his enlistment.

DEVOE AVENUE. Old records show this family intermarried with another early family named Mapes. From the 1780s onward the name was variously spelled DeVaux, DeVoo and DeVoe. James R. and Ann Eliza Devoe were listed as inhabitants of the village of West Farms in 1850, and Smith DeVoe is mentioned in 1899 transactions. Paralleling the Bronx River, this street once formed the extreme western edge of the Siwanoy tribal lands. In later centuries, it was the last lane in Westchester Township, the river being the boundary line.

DEVOE PARK. It lies on former Devoe property and an earlier Valentine farm. Valentine's Brook is under the north border, now Father Zeiser Place. The five-acre park was purchased by the City in 1907 and opened in 1910. See: Devoe Terrace.

DEVOE TERRACE. The Devoe family were descendants of Daniel Turnier who acquired the lower section of the Fordham Patent in 1676. In 1868, shortly before streets were laid out in this area, at least four Devoe families were landowners, including the one who owned what is now Devoe Park.

DEWEY AVENUE. West of East Tremont Avenue, this was a lane that led to a farmhouse located on the John A. Morris estate. East of East Tremont Avenue, it was on part of the Doric Farm (sold in 1827 to John D. Wolfe), and an orchard was located at Logan Avenue. In 1776, a British brigade was stationed at Dewey and Revere Avenues to guard the Throggs Neck Road (now East Tremont Avenue). It was named for Commander George T. Dewey, an Admiral in the Spanish-American War.

DEYO STREET. The name is French, and is also spelled Deyoe and Dejou. The Deyo family came to America about 1674 and were among the French Huguenot families that settled New Paltz, N.Y. This street, which opened in 1926, is named for Solomon LeFevre Deyo (1850-1922) who came to New York from New Paltz and became chief engineer for the first subway line. He built the first tunnel under the Hudson River and was Vice President of the American Society of Civil Engineers. See: Fteley Avenue.

DICKINSON AVENUE. Although this avenue was on the blueprints for over 30 years, its unofficial name was Harald Avenue. This had a definite link with the adjoining avenues named Norman and Saxon: Harald, (also Harold) the Saxon king, fought the Normans. On May 2nd, 1916, by Ordinance of the Board of Aldermen, the avenue was named Dickinson Avenue to honor John Dickinson, through whose estate the street was finally opened. See: Hillman Avenue, Saxon Avenue.

DIGNEY AVENUE. Formerly called Hobart Street, this is a reminder of John M. Digney of White Plains, who purchased much North Bronx real estate when this Edenwald area was still part of Westchester County. He was a Notary whose name is found on many deeds of the 1890s.

DILL PLACE. It is on the southern boundary of a nineteenth century farm mapped as the Escoriaza Farm. It later marked the boundary of a tract known as the Rhinelander Plot on Throggs Neck. James Brooks Dill was a partner of the real estate firm that developed Tremont Heights *(circa* 1900) under the presidency of Nathan Lamport. See: Lamport Place.

DITMARS STREET. The western half of this City Island street was once called Oyster Street, while the eastern half was Prospect Street "because of the fine prospect it afforded eastward, overlooking Rat Island, Green Flats and Hart Island" according to an old guide book. Among old-timers the story is current that the street received its name from the Ditmar Explosive Company that used the dock to load or unload their dynamite cargo. The arsenal was located on the Hutchinson River. For years, there were street signs "Ditmar" and "Ditmars" until the latter spelling prevailed. In the late 1870s, William Byles operated a small boat yard at the foot of Ditmar Street on Eastchester Bay.

DIXIE GATE. This is the eastern entrance to Fordham University leading onto Southern Boulevard. Jesuit teachers, the first instructors of St. John's College (now Fordham University) came from Kentucky. Might this be the reason?

DOCK STREET. The unofficial name of a short road in Edgewater Park leading to the dock. This dock overlays a smaller one built by the Adee family of Civil War days.

DODGEWOOD ROAD. This is a private road in a real estate development known as "Dodgewood Estates." This tract, west of Riverdale Avenue at West 246th Street, was carved out of the extensive Dodge estate.

DOGWOOD DRIVE. This is a privately maintained street inside Parkchester, an extensive housing development that was built 1938-1942 on lands once occupied by the Catholic Protectory. Dogwood trees were common on the land and no doubt inspired the naming of this drive.

DOLEN PARK. Originally, this was a block of wooden houses in the village of Westchester which were razed around 1900 to form part of an open square. During World War I, a monument to the servicemen of the neighborhood was unveiled in the square, and Owen F. Dolen, longtime resident and history teacher of P.S. 12, spoke at the ceremonies. He then sat down and died of a heart attack. The park was named in his honor in April 1926.

DONIZZETI PLACE. Records of 1662 show this was the property of Philip Pinckney who "sett down at Hutchinsons," referring to an earlier settler, Anne Hutchinson. For almost 250 years, a mill lane ran approximately under this Place leading to a mill built in 1739. "Gaetano Donizzeti was an Italian composer whose works included *La Fille du Regiment* and *Lucia di Lammermoor*. He was born in Bergarno in 1797, and died in 1848. His name was given to this Place in Co-Op City in July 1967. See: De Kruif Place, Reed's Mill Lane.

DOREY PARK. Due to its semicircular form, this small park at Hall Place and East 165th Street was known locally as Horseshoe Park. It was acquired in March 1897, and placed under the jurisdiction of the Department of Parks. No one seems to know what Albert Dorey did to have his name given to the park.

DORIS STREET. This was formerly part of the E. Haight estate, mapped in 1868 as "Hopedale." An 1894 map shows a Dore Lyon as the owner of a small estate on the same site, which he called "Lyon Court," and there is a possibility that Doris is a misspelling of his first name. Another map listed him as Darius Lyon. A vague reference was made of a James Doris, but no record has been found of this individual.

DOROTHEA PLACE. This is a sloping Place only 90 feet long, and was laid out in 1887 on what was once the Hugh N. Camp estate, merely to give access to two homes at the top of a bluff, off Fordham Road. It is local legend that the Place commemorates his daughter, but none of his three daughters were named Dorothea. However, his wife's maiden name was Elizabeth Dorothea McKesson.

DORSEY STREET. It is named after homeowners of a century ago who lived near Zerega Avenue. Their tombstones can be seen in St. Peter's P.E. churchyard. When the street was opened in 1914, it was called Carroll Lane.

DOUBLE ROAD. This is a paved road skirting the outside of the penitentiary and (plant) nursery on Riker's Island, a part of The Bronx. This perimeter road is double the width of the other thoroughfares.

DOUGLAS AVENUE. The 1893 land books lists a James Douglas, but at the time this Riverdale thoroughfare was opened in May 1916, the owner was a W.E. Douglas. This wealthy family made its mark in mining, government posts and in New York's social life. Their estate is now the lower half of Riverdale Park.

DRAINAGE STREET. This oddly-named street is in reality a short public path from Longfellow to Boone Avenue near East 172nd Street. In the 1860s, it was on the lands of Thomas E. Minford. Records show it was vested by the City in 1905 and was used as a natural drain as it leads downhill to the Bronx River.

DRAKE PARK. It opened in 1910 to encompass an ancient burial ground that belonged to the Hunt family. Joseph Rodman Drake was a physician and poet who spent his happy youth on Hunt's Point and wrote odes in its praise. He died in 1820, age 25, mourned by all, and is buried in the cemetery. His friend Fitz-Greene Halleck wrote his epitaph. Incidentally, "Drake" derives from the Elizabethan English word for Dragon.

DRAKE PARK SOUTH. See: Drake Park.

DRAKE STREET. It is laid out through the former estates of Barretto, Spofford, Dickey, Faile and Hoe, which were subdivisions of the earlier Leggett and Hunt lands of Revolutionary times. The street is named for the poet (see Drake Park) and this led to the naming of adjoining avenues for Halleck, Longfellow and Whittier.

DRAPERS WALK, The. This is a footpath across the campus of former New York University (now Bronx Community College), the former Schwab estate and the still earlier Archer lands and the Revolutionary Fort Number 8. It honors Professor John W. Draper and his two professor sons, John C. and Henry. The father was a Professor of Natural History and a pioneer (1839) in photography, while his sons were equally illustrious in Chemistry and Physiology.

DREISER LOOP. Theodore Dreiser, American journalist and author, was honored by the founders of Co-Op City in 1967. A navigable branch of Rattlesnake Creek once ran under the southern side of the Loop. The land was part of the Freedomland parking lot (1959-1965).

DUDLEY AVENUE. This avenue ascends the sharp hill which figured in the Revolutionary battle of October 1776, when the British mounted their cannons atop "Bridge Hill" to shell Westchester village. In the 19th century, the slope was known as "Crow Hill." In 1855, a wooden church, the First Presbyterian church, was built and it acquired the nickname of "the Soup Church" in the Civil War because returning Union soldiers were given soup as they passed the church on their way home. It burned down in 1875 and was replaced by a brick church on the same site. The avenue was opened in 1927 and took the name of Percy S. Dudley, who bought the land in 1906 from Thomas Ballard Harrington.

DUNCAN STREET. William Butler Duncan, Vice President of the New York City Chamber of Commerce, was also a realtor in the Williamsbridge area. An 1891 deed, shows he sold land to a Christopher Carlisle (see Carlisle Place). Duncan Street was opened in 1925 across what had been part of the J.M. Burke estate.

DUNCOMB AVENUE. David S. Duncomb was listed as a property owner in Olinville in 1852. He and his brother, Alfred H., were wealthy brush manufacturers.

DUNE COURT. This is a private street in the Shorehaven complex on Clason Point, alongside White Plains Road. The land was once the edge of a meadow called Aviation Field, said to have been used by amateur balloonists and aviators in the early 1900s.

DUPONT STREET. This short street was first laid out in 1887 in an area claimed by the Morris family for almost two centuries. Due to a disputed boundary with the Leggett and Hunt families, it was known as "the Debatable Lands." Samuel Francis Du Pont, (1803-1865) a hero of both the Mexican and Civil Wars, employed a variant spelling of the family name, which was then condensed for the street name and sign. He was a Captain who played an important role in devising the decisive naval strategy of the war, the Southern blockade. After winning a critical battle he was made a rear admiral. He was a nephew of the famous pioneer Frenchman Eleuthére Irénée du Pont de Nemours, who established a manufacturing plant in 1802, near Wilmington, Delaware, to produce black powder. E. I. du Pont, who had been a student of Antoine Lavoisier, the father of modern chemistry, brought to America new ideas for manufacturing consistently reliable gun and blasting powder. Unlike products available up to this time, his ignited when and how it was supposed to. The citizens of the new republic appreciated it, and Thomas Jefferson even thanked him for his quality powder, which was used to clear the land at Monticello, his home in Charlottesville, Virginia.

DURYEA AVENUE. The Duryea family is an old one in the north Bronx. Duryea's Zouaves of the Civil War were recruited in Fordham, and the men distinguished themselves in many battles. The name is equally prominent on City Island. An earlier name of this street was Wright Avenue.

DWIGHT PLACE. Back in the 1600s, this land belonged to the Underhill family, and was called "Kennelworth" after their family seat in England. It passed through a few proprietorships until it became the estate of Lorillard Spencer in the 19th century. When the tract was subdivided around World War I, all the streets were given names with electrical connotations, such as Ohm, Watt, Ampere, etc. This Place was named for Stanley Dwight, President of the C & C Electric Company, on October 20, 1922. Oddly enough, it was passed by the Board of Estimate.

DYRE AVENUE. The avenues in this area carry names of New York Mayors. William Dyre was Mayor of New York in 1680-81. An Englishman by birth, he migrated to New England, then to New York where he served the customary one year term as Mayor. After that, he moved to the island of Jamaica and there died in 1685. This man's name is occasionally spelled Dyer. National landmark status has been given to The little Red Schoolhouse at 4010 Dyre Avenue. This gem of architecture was built in 1877.

DYRE AVENUE LINE, The. The Dyre Avenue Line was originally built by the New York, New Haven & Hartford Railroad and was operated by the New York, Westchester and Boston branch that opened in May 1912. The line closed down in December 1937 in bankruptcy. In 1941, the Board of Transportation took over the rolling stock, the stations, signal towers and the four-track right-of-way at a cost of $1,783,577.03.

EAGLE AVENUE. This was once a particularly rocky plateau of the Morris Manorlands that was sold in 1847 to William Carr and Henry Purdy. Sold to the Ursuline Sisters in 1854, this avenue was part of their grounds. The upper end of the avenue (later known as West Ray Street and Westray Street) was favored by brewers as the hillside was ideal for vaults. At East 156th Street (Carr Hill), Ebling's Brewery was in operation from the 1880s to the early 1940s. Off this corner were charmingly ornamented, and richly furnished, sturdy brick homes of the German brew masters. At East 160th Street (Hupfel's Hill), was the Hupfel Brewery, established in 1865. At one time, during Prohibition, this brewery was converted into a thriving mushroom farm. The origin of the name is debatable. Although it was mapped in 1853, it was many years later that the street was actually cut through. The Minutes of the Parks Department pinpoint the date as June 13, 1890. This rules out the story that Civil War veterans who maintained a cemetery nearby (Rae Place) honored our national symbol, the eagle. Commodore Henry Eagle, buried in Woodlawn Cemetery, may have had some connection with that area in the 1850s, but that link has never been found.

EAMES PLACE. This Place, off Kingsbridge Road, was laid out on what had once been the Claflin estate of the 1870s. The area was subdivided in 192021 and lots were sold by Joseph P. Day. John C. Eames was Vice President of the H.B. Claflin Company, and a Trustee of the Hudson-Fulton Celebration.

EARHART LANE. Amelia Earhart, once a world famous aviatrix, is remembered by this winding thoroughfare in Co-Op City, named in 1967. Until 1965, the land was occupied by some frame houses, a few barges and boat liveries. It was once part of the extensive Givan property, purchased in 1802 from the Bartow family. Earliest mention of this land was found in 1662 when Thomas Pell invited colonists to "sett down at Hutchinsons."

EARLEY STREET. Martin J. and Margaret Earley purchased land on City Island from Cyrus Pell, in 1893. The latter traced his lineage back to Thomas Pell, First Lord of the Manor of Colonial times.

EAST 132nd STREET. The site of Jonas Bronck's "Emmaus" settlement was in the vicinity of Lincoln Avenue. The spot was later occupied by the manor house of Lewis Morris which was demolished in 1891. East of the Mill Brook (now Brook Avenue) was the mansion of Gouverneur Morris that was razed in 1905. Even before the Annexation of 1874, there was talk of joining this Bronx territory to New York City. In the mid-1860s, realtors named Campbell, Willis and Brown purchased 100 acres of land from the Bronx Kill to present-day East 138th Street and, in the laying out and mapping of these plots, the cross streets were numbered in continuation of the streets in Manhattan. This numbering system was adhered to in progressive steps northward from 1874 to the early 1900s.

EAST 133rd STREET. This was originally an Indian settlement that became part of Jonas Bronck's farm. In the 1890s, there were two hotels at East 133rd Street and Third Avenue: John Hubert's "N.Y., N.H.&H. Hotel" and "The Palmer House."

EAST 134th STREET. This was known as Mott Street in the village of Mott Haven. Jordan L. Mott purchased a tract of land from Gouverneur Morris in 1850 from the Bronx Kill to East 148th Street, west of Third Avenue, which he called Mott Haven.

At Third Avenue and East 134th Street was a famous sea food restaurant at the turn of the century known as The Oyster Bay.

EAST 135th STREET. This was formerly called Orange Street from the Mott Haven Canal to Rider Avenue. Legend has it that it commemorated the Dutch House of Orange. In Old Morrisania, it was called Edsall Street, recalling Samuel Edsall, owner of the land from 1668 to 1670. He was fluent in the Indian language and served as interpreter for the Dutch authorities. He sold his land to the Morris brothers.

EAST 136th STREET. This was formerly known as Cherry Street in the village of Mott Haven. Gouverneur Morris' cherry orchard was at St. Ann's Avenue. East of Third Avenue (Old Morrisania), it was called Smeeman Street as Harman Smeeman bought the land from C. Hendrick in 1662 and sold it to Samuel Edsall. Jordan L. Mott's mansion once stood on the southeast corner of Third Avenue and East 136th Street. In Port Morris, in the 1870s, the area was developed for a port of entry and the streets were numbered; this was 6th (Sixth) Street. See: East 137th Street to East 141st Street, Mott Haven Canal, Rider Avenue.

EAST 137th STREET. This was Hendricks Street in Old Morrisania. C. Hendrick(s) bought the land from J. Van Stoll in 1662 and sold it to Harman Smeeman the same year. The Mill Brook crossed under the northwestern building of Mill Brook Houses, 520 East 137th Street. In Port Morris, this street was called 5th Street as laid out in the year 1877.

EAST 138th STREET. It was laid out as Van Stoll Street in the first platting of Old Morrisania. Jacob Van Stoll (sometimes spelled Stohl) bought the land from Arent Van Corlear in 1651, and sold it a few years later to C. Hendrick. The Mill Brook crossed at the southwestern corner diagonally to 501 East 138th Street. It was 4th Street in Port Morris.

EAST 139th STREET. This was first known as Van Corlear Street, harking back to one of the original owners of the land: Arent Van Corlaer [spelled this way], who bought the land in 1644 from Jonas Bronck's widow and sold it to Jacob Van Stoll in 1651. The Mill Brook crossed from 500 East 139th Street diagonally to the northwestern corner. This was 3rd Street in Port Morris (1877 map).

EAST 140th STREET. It was Cheever Place from the Harlem River to what is approximately the Grand Concourse. John H. Cheever was listed as a landowner in 1859 and the name was retained on street signs until 1911. East 140th Street was called Bronx Street in Old Morrisania and named after the first settler of 1641. The Mill Brook crossed from 508 East 140th Street diagonally to 507. It was 2nd Street in Port Morris.

EAST 141st STREET. It was Lowell Street until 1901, and 1st Street in Port Morris. The Mill Brook crossed from 490 East 140th Street diagonally to 481.

EAST 142nd STREET. Though mapped as Concord Street, it is not clear if there were concord grapevines there, or if it honored Concord village of Revolutionary fame. The *New York Eagle* newspaper (1884-1896) was located on this street between Rider and Morris Avenues. The Mill Brook crossed East 142nd Street from 482 to 475.

EAST 143rd STREET. On old maps, this street was named Garden Street. The Mill Brook crossed East 143rd Street from 470 to 471 East 143rd Street.

EAST 144th STREET. It was Main Street in Old Morrisania. Martin Norz ran a thriving market at the corner of Third Avenue in the 1860s, with 32 delivery carts. The Mill Brook crossed from 468 to 465 East 144th Street. At Brook Avenue stood the "North Side Hotel," headquarters of Konstantinopolischerdudelsackpfeifers, a German drinking club. East 144th Street, from St. Mary's Park to Southern Boulevard (East Morrisania) was called Joseph's Street, for a time, then Grove Street.

EAST 145th STREET. Villa Place was its original name from College Avenue to Third Avenue, named after the Van Doren villa at the corner of Third Avenue. The Mill Brook crossed this street at an angle from 452-458 to 457. In East Morrisania, it was called Crane Street from St. Mary's Park to Southern Boulevard (after John J. Crane, merchant) and later, in 1889, it became Elm Street.

EAST 145th STREET BRIDGE. From The Bronx viewpoint, this name is a misnomer as it touches The Bronx at East 149th Street. However, it comes from East 145th Street in Manhattan, reflecting the relative importance of the two boroughs in 1905, when the bridge was opened over the Harlem River.

EAST 146th STREET. It was once Grove Street in Mott Haven, from the Harlem River to about the Concourse, then called Mott Avenue. Another name was Cottage Street. The Mill Brook crossed from 434 East 146th Street diagonally to the eastern end of 433 East 146th Street.

EAST 147th STREET. From Bergen Avenue towards Brook Avenue, it was once the south side of Karl's Germania Park-later Loeffler's Park. The Mill Brook crossed from the center of 424 East 147th Street and curved to 421. Orchard Pond lay at that street in St. Mary's Park and was drained in 1892 by Park Commissioner Louis F. Haffen. The section from St. Mary's Park to Southern Boulevard once constituted the outlying area of Jonas Bronck's holdings, which became part of the Morris family Manorlands. In the 1870s, it was mapped as Lexington Avenue. It formed part of the (1888) Dater estate, and when it was surveyed and cut through in 1894, the street was called Dater Avenue.

EAST 148th STREET. This street marks the northern limits of Bronck's farm, and also became, in 1850, the northern limit of Mott Haven. Its first name was Mott Street. Frog Hollow, a notorious neighborhood of the 1890s, was located around East 148th Street and Morris Avenue. East of Third Avenue, it was Henry Street, probably for Henry L. Morris, who owned one-third of Old Morrisania. The street was extended eastward from Willis Avenue in the 1890s, destroying a picnic grove and amusement park variously known as Karl's Germania Park, the Twenty-Third Ward Park and Loeffler's Park. The Mill Brook crossed from 432 to 433, and the building line of 431 East 148th Street shows the course of this ancient stream.

EAST 149th STREET. At the Concourse (Mott Avenue), the elevation was occupied by wealthy families, and called "The Ridge." In the village of Melrose South (1864), Kirchof's Brewery was located at Third Avenue. The Mill Brook crossed west of Bergen Avenue. In the 1850s, this street was the lower boundary of Benjamin Benson's purchase, "Bensonian" and was called Benson Street. During the Revolutionary War,

St. Ann's Avenue and East 149th Street was the site of Tory quarters known as the Refugee Huts. The stretch from Third Avenue to Southern Boulevard was Westchester Railroad Street and was used by stagecoaches bound for Hunt's Point. Later, horsecars went over this route to take holiday crowds to Oak Point. The Hopkins East Morrisania Hose Company firehouse stood at the corner of Jackson Avenue. From Southern Boulevard to the East River it was called Bungay Street, as it followed the course, more or less, of Bungay Creek.

EAST 150th STREET. In the village of Mott Haven, on the slope west of the Concourse, this was the New York Athletic Ball Grounds of the 1880s. In Melrose, it was called Denman Street, and on it, the R.C. Church of the Immaculate Conception was built in 1853. Hampton B. Denman had purchased lots from Lewis Morris III in that era of expansion. Over in East Morrisania, East 150th Street was first called Fox Street, after William W. Fox, a wealthy Quaker. The property adjoining Southern Boulevard was purchased by the Dater brothers from the Morris family, and when this street was laid out it was named Ungas Street. It was changed to East 150th Street in 1895.

EAST 151st STREET. This street was opened in the village of Melrose South around 1851-52 and originally was named Gouverneur Street in honor of Gouverneur Morris, who sold the land. A German Redemptorist order of priests organized a Catholic school in 1873, and several German nuns of Christian Charity arrived to form the teaching staff In East Morrisania (Jackson to Tinton Avenues), it was known as Beck Street after an early landowner, and it later became Pontiac Street.

EAST 152nd STREET. It was cut through in 1852-53 from Morris Avenue to Third Avenue and was called Elton Street, after a Robert Elton, former landowner. The Haffen family house stood on the northwest corner of Courtlandt Avenue and the J & M Haffen Brewery (1856-1917) was on the southwest corner of Melrose Avenue. From approximately 1910 to 1938 it carried the nickname of "The D.S.C. Hill" as the Department of Street Cleaning maintained wagon lots, stables, harness repair shop and a blacksmith shop between Courtlandt and Melrose Avenues. From Third Avenue to Brook Avenue, it was called Rose Street until 1905. Further east, from Westchester to Prospect Avenues, it was called Willow Street, after the Dater estate was surveyed and subdivided. Early maps list it as Kelly Street, as the last owner was Capt. Samuel Kelly.

EAST 153rd STREET. In the village of Melrose South, it was first called Schuyler Street. Beyond Third Avenue to the Mill Brook (now Brook Avenue), it was Grove Street.

EAST 154th STREET. It was originally named Springfield Street.

EAST 155th STREET. From Park to Third Avenues, this thoroughfare went by the name of Mary Street, possibly for the wife of General Lewis Morris. In East Morrisania, it was called Dawson Street, from Wales to Tinton Avenues, in honor of the historian Henry B. Dawson. The street was numbered in 1894 to conform to the grid pattern.

EAST 156th STREET. This was the original boundary between Melrose and Melrose South, and was called Melrose Street. The Winona Hotel was once a prominent inn on Courtlandt Avenue, and the Public Baths were at Elton Avenue. The hotel was razed in the 1940s, and the baths were demolished in 1953. From Third to Cauldwell Avenues, past Aurora Park, an earlier ballpark and picnic grove, it was known as Beck

Street (after Anna Beck, a landowner in 1871). This was called Carr's Hill. Ebling's Casino was a famous landmark on St. Ann's Avenue. It stood on the site of Aurora Park. At Hunt's Point, East 156th Street was known as Craven Street until 1900, honoring an Admiral T.A. Craven.

EAST 157th STREET. It was Prospect Street in the village of Melrose. The dictionary defines a prospect as a view, and a good one was to be had from Courtlandt Avenue overlooking the valley westward, or from Elton Avenue looking eastward down to the Mill Brook.

EAST 158th STREET. It was once Milton Street in the village of Melrose. The blind poet Milton might be remembered, as literary taste sometimes dictated the choice of names. Melrose (East 156th Street) and Waverley Streets (East 159th Street) harked back to Sir Walter Scott. For many years, the Melrose Turnverein was situated off Elton Avenue. From Third Avenue to Westchester Avenue, it was once called Cedar Place.

EAST 159th STREET. It was Waverley Street from Park to Third Avenues. Waverley is a section in the Scottish town of Melrose and it is quite likely the Bronx street got its name from Surveyor Andrew Findlay, he of Scottish descent. From Brook Avenue up to Eagle Avenue, it carried the name of John Street, although the popular name for the steep incline was St. John's Hill. An 1888 map has this short street called Cedar Street. East 159th Street from Forest to Union Avenues was Denman Place in an 1879 Atlas.

EAST 160th STREET. Findlay Street was its original name, honoring Andrew Findlay who surveyed the tract in 1848. At Third Avenue, it was known as Grove Hill, the estate of the DeGraaf family. Further on, in the village of Woodstock (Union Avenue), a Woodstock Park picnic grounds was located there in 1891. George Dettner was the proprietor.

EAST 161st STREET. This was Cedar Street from the Harlem River to the Concourse, harking back to the 1840 property of Gerard Morris, "The Cedars." In the village of Melrose, it was known as William Street, William H. Morris being the former landowner. On Melrose Avenue, the original Bruckner brewery stood, but in later years the sons specialized in soda water. From Third Avenue to Prospect Avenue, it was Grove Hill, later to be named Cliff Street. At Forest Avenue, it was the estate of Peter P. Decker that gave the area name of Deckerville (map of 1872). In 1891, in the village of Woodstock, the famous Tallapoosa Clubhouse stood at Tinton Avenue and East 161st Street. See: Tallapoosa Club.

EAST 162nd STREET. From approximately the Concourse to Sherman Avenue, it was once Grove Street. From there to Park Avenue, it was called Halsey Street. Louis Haffen, first Borough President, lived at 304 East 162nd Street from 1896 to 1935. Prior to the numbering system, it was Union Street from Park Avenue to Washington Avenue, and Union Lane on to Third Avenue.

EAST 163rd STREET. When Fleetwood was laid out from the former Morris estates, this street was named Helen Street from the Concourse to Park Avenue. Henry Haffen, Commissioner of Highways, once lived at 306 East 163rd Street around 1900. In the incorporated village of Morrisania, it was 1st Street. From Westchester Avenue to Southern Boulevard, it was called Dongan Street until 1908. Governor Dongan was an official in Colonial times.

EAST 164th STREET. This was first called Ella Street from the Concourse to Park Avenue after the wife of William H. Morris, landowner of the tract. In the vicinity of Third Avenue it was 2nd Street (also 2nd Place) in the village of Morrisania.

EAST 165th STREET. This was known as 3rd Street in Morrisania, running from Brook Avenue to Boston Road. It was called Wall Street from that Road to Prospect Avenue, where it ran through Eltona, mapped in 1860. From Westchester Avenue to Whitlock Avenue (the former Hoe estate), it was first called Guttenberg Street from 1897 to 1902, whereupon the name was changed to Bancroft Street, 1902-1911 and is believed in honor of Hubert Bancroft, a historian.

EAST 166th STREET. It was 4th Street in the village of Morrisania from the Mill Brook (Brook Avenue) to Third Avenue. Washington Hall, 497 East 166th Street, was a famous meeting place and social center. At Franklin Avenue, it was Spring Place, opened in 1849, as the spring was on the Allendorph estate that is now occupied by the Armory. In Woodstock (from Jackson to Tinton Avenues), it was George Street and might hark back to a George Dettner, who owned a picnic grove and tavern there.

EAST 167th STREET. In West Morrisania (Jerome Avenue to the Concourse), it was first mapped as Overlook Avenue as the slope overlooked Cromwell's Creek where it wound through the wooded nineteenth century estate of James Morris. In Fleetwood (Concourse to Morris Avenue), it was James Street after the former owner. The Mill Brook crossed along the southeast and northeast corners of Brook Avenue and East 167th Street forming the boundary of Morrisania. There it was 5th Street, the main shopping district of the 1880s, and Sylvan Park was popular with the German element. From Southern Boulevard to Westchester Avenue, this street was once called Lyon Street as it crossed the former Lyon estate off Home Street.

EAST 168th STREET. It was Charles Place from Gerard Avenue up to the Concourse after Charles Arcularius, a nineteenth century hotel keeper. From the Concourse to Webster Avenue, it was once part of the William Morris estate, which ended at 405 East 168th Street (the Mill Brook). From there, eastward, the small settlement was called New Village, but then was renamed Morrisania. This was 6th Street and continued as such to Boston Road. From there to Prospect Avenue, it was Glen Street.

EAST 169th STREET. This was Arcularius Place, from Jerome Avenue up to the Concourse, named after Charles P. Arcularius who ran the Jerome Park Hotel. It was 7th Street in the village of Morrisania, where, in the 1890s, Feigel's Park was situated at Brook Avenue. In 1856-1859, a stagecoach line to West Farms ran along this street from Third Avenue to McKinley Square whence it continued along Boston Road.

EAST 170th STREET. Called 8th Street in the village of Morrisania. The Mill Brook crossed in the center of Brook Avenue, running due north at this point. At Third Avenue, Urbach's Hotel was advertised in 1887, 1891.

EAST 171st STREET. From the Concourse eastward to the limit of Claremont Park, this was once the estate of Elliott Zborowski, who was related to the Bathgates. At Morris Avenue was the infamous Black Swamp where cattle had been lost since the time of the Indians. For years it defied all efforts by contractors to fill it up. Zborowski's estate, called Claremont, extended to Brook Avenue. From there, Morrisania commenced where East 171st Street was 9th Street.

EAST 172nd STREET. From the Concourse to Claremont Park, this was once the Black Swamp, known from the earliest days as a dangerous morass. See: Clay Avenue. It was called Martindale Place in 1873, and from Webster Avenue eastward, it was 10th Street in the village of Morrisania.

EAST 173rd STREET. This was 11th Street in Morrisania.

EAST 174th STREET. This was once known as Spring Street from the Concourse to Webster Avenue to become 12th Street in Morrisania. Where it touches the Bronx River, the original name was Beacon Avenue.

EAST 174th STREET BRIDGE. Spanning the Bronx River at that numbered street, the eastern end of this bridge is in the former Watson estate, while the western end is on the long-vanished grounds of August Woodruff's estate.

EAST 175th STREET. Before the numbering system took effect, this was Oxford Street from Jerome Avenue to the Concourse. From there to Webster Avenue, it was Gray Street, and the most important building was J.W. Katzenberger's Webster Hotel (1880) on that avenue. It was Fitch Street from Webster to Third Avenues, after James T. Fitch, local resident and tax assessor. It was Fairmount Street from Marmion Avenue to Southern Boulevard as it led to the Fairmount estate of Colonel Dunham, an Englishman.

EAST 176th STREET. This was once known as Orchard Street from the Concourse to Anthony Avenue and Mole Street from Webster to Third Avenues. It was called Prospect Street where it overlooked the Bronx River, but later the name was changed to Woodruff Street to honor August Woodruff, landowner.

EAST 177th STREET. It was known as Locust Avenue from Third to Arthur Avenues and referred to the locust trees on the Ryer farm. The Ryer brothers gave the name of Monterey to their holdings, as they had fought in the battle of Monterey, Mexico.

EAST 178th STREET. This was known as Grove Street from Creston Avenue to the Concourse. From that point to Third Avenue, this street ran through the village of Upper Morrisania, which was established in 1851. Later the name was changed to Tremont. From Webster to Third Avenues, this was part of the original Quarry Road that sometimes was mapped at Marble Street. Between Webster and Park Avenues was Ittner's Park, a picnic resort of the 1870s. John Ittner's hotel was located at Tremont and Park Avenues. Another picnic spot, Crotty's Woods, was located in the 1890s at Third Avenue. Oakley Grove, an estate belonging to Miles Oakley, second Mayor of Westchester (1730), was once in this approximate spot.

EAST 179th STREET. This is an historic old lane that evolved from an Indian path leading from Kingsbridge to Hunt's Point. Berry Street was the first name of East 179th Street from Jerome Avenue to Anthony Avenue. From Third Avenue to Southern Boulevard, it was mapped at times as Lebanon Street as well as Monroe Street, and Elm Street (1890) in the village of East Tremont. Glover Street was used in the village of West Farms (1889 map). Central Street is the former name from Webster Avenue to Brook Avenue. It was laid out over a brook that formerly flowed into the Mill Brook from a pond that was situated at Burnside and Anthony Avenues. Later, the street was extended to Third Avenue, finally to Southern Boulevard, retaining its name of Central Street until 1891. Centre Street is a name by which East 179th Street was known from Southern Boulevard to Boston Road, actually being an extension of Central Street. An 1892 map shows that at Boston Road, the northeast corner was the site of Colonel Delancey's blockhouse, later Mapes' Temperance Hotel, which became the Peabody Home for Aged Women.

EAST 180th STREET. It was Talmadge Street from Webster to Third Avenues and Samuel Street from Third Avenue to Southern Boulevard. This street formed the boundary of Samuel Ryer's farm with that of the L.S. Samuel property, according to Bromley's map of 1891.

EAST 181st STREET. No less than seven names disappeared when this street was numbered. It was Irene Place from University to Jerome Avenues, 5th Street from Jerome Avenue up to the Concourse, Irving Street from Park Avenue to Third Avenue and John Street from there to Southern Boulevard. From the boulevard to Vyse Avenue, it was once designated as Ponus Street. It was Clover Street running to Boston Road and finally Ann Street at Bronx Park.

EAST 182nd STREET. From Jerome Avenue up to the Concourse, it was once called 4th Street. During the 1850s, this was the Adam Cameron estate, on Prospect Hill. It was known as Fletcher Street, from Park to Washington Avenues, and as Grove Street as it continued to Third Avenue. It was Elm Avenue from Crotona Avenue to Southern Boulevard. It is part of the ancient Kingsbridge Road, alongside Bronx Park. The street was widened, paved and officially opened on November 29, 1901, when it was numbered. The length alongside Bronx Park was renamed Bronx Park South in 1917. A Colonial tavern, the Planters Inn, once stood at East 182nd Street and Boston Road and was a rest-stop for stagecoach passengers. It was maintained by Levi Hunt and, later, by a Robert Hunt, who may have been his grandson.

EAST 183rd STREET. It was 3rd Street from Jerome Avenue up to the Concourse, across the (1853) land of Adam Cameron. It was known as Columbine Avenue from Quarry Road over to Southern Boulevard.

EAST 184th STREET. This street marks the southern line of the Colonial holdings of John Archer (Arcier), landowner of Fordham Manor. At Morris Avenue (1906) stood the crumbling mansion of Maria Corsa, a descendant of a pioneer family, who died in abject misery, even though money was under carpets and hidden in furniture. The street overlays an Indian trail from Third to Belmont Avenues, which later was called Kingsbridge Road (1820). It was called Adam Street when surveyed in 1855, and Belmont Place in 1886.

EAST 185th STREET. This was once part of the Bassford farm of the 1860s. When streets were cut through, this received the name of Wetmore Street from Park to Third Avenues.

EAST 186th STREET. It was formerly Bassford Street from Park to Third Avenues.

EAST 187th STREET. On an 1884 map, this was Parole Place from the Concourse to Tiebout Avenue. It was also known as Sanford Street, down to Third Avenue. From there to approximately Hughes Avenue (1868), it was called Jacob Street, after Jacob Lorillard, the landowner there. For a short period, 1900-1905, from Cambreleng Avenue to Southern Boulevard, it was Clay Street.

EAST 188th STREET. It was the slope of Mount Sharon (1853 map) at Grand Concourse and East 188th Street. From the Concourse down to Webster Avenue, this street was once Walsh Street, and was Lafayette Place from Park to Third Avenues. It was Bayard Street from Third Avenue to Beaumont Avenue as it had been part of the Bayard farm (1800).

EAST 189th STREET. It was known as Welch Street from Webster to Park Avenues until 1911. It also went by the name of Powell Place from Third Avenue to Washington Avenue. This was once part of the (1800) Bayard farm that was inherited by the Reverend William Powell (1830-1850). When it was incorporated into the village of Belmont, and when streets were cut through, this was Webster Avenue from Beaumont Avenue to Southern Boulevard. All this countryside was annexed to New York City in January 1874.

EAST 190th STREET. Pipe Street was the earliest name of East 190th Street from Jerome Avenue to Grand Avenue, referring to the aqueduct pipe that ran eastward from the Croton Aqueduct. When St. James Protestant Episcopal church was built in 1884, the entire stretch (West 190th Street and East 190th Street) was renamed St. James Street. This was the glebe of St. James' P.E. Church, off what is today Jerome Avenue. In 1894-95, the street was cut through, and was numbered shortly thereafter. See West 190th Street.

EAST 191st STREET. It was originally mapped as College Street, from Bathgate to Hughes Avenues, alongside former St. John's College (now Fordham University). It overlays an ancient Indian footpath.

EAST 192nd STREET. From St. James Park to Kingsbridge Road, it was once called Primrose Street, according to an 1897 land book.

EAST 193rd STREET. It was once known as Brookline Street from Marion Avenue down to Webster Avenue as it followed the line of a brook that flowed from Briggs farm to join the Mill Brook at Webster Avenue. The source of the brook was a spring on the Valentine estate, to the west.

EAST 194th STREET. This was formerly called Cole Street, as Jacob Cole was the 1860 landowner, and the Postmaster of Fordham. It had been Valentine property for almost 100 years. East 194th Street off Pelham Parkway was cut through the Paul farm of earlier times. See St. Paul Avenue.

EAST 195th STREET. From Marion down to Webster Avenues, this street was once known as Tappen Street, part of the extensive nineteenth century Berrian holdings. East 195th Street off Pelham Parkway was cut through the Paul farm of earlier times. See St. Paul Avenue.

EAST 196th STREET. Once called Ridge Street and later, Sherwood Street, a former name of East 196th Street from the Concourse to Marion Avenue. The landowners there in the 19th century were Daniel and Margarette Sherwood, two of the few civilians that are buried in the West Farms Soldiers' Cemetery. When the Jerome Park Racetrack was in operation, Donnybrook Street was the name of this thoroughfare from Jerome Avenue up to the Concourse. It was also mapped as Wellesley Street. Donnybrook Fair was a famous Irish horse show of the 1700s, but became notorious for its brawls. The Jerome Park Racetrack was likened to the Irish horse show in its heyday. See: Donnybrook Street, Ridge Street, Wellesley Street.

EAST 197th STREET. This was part of John Archer's Manor of Fordham. In 1683, it was lost by foreclosure to Cornelius Steenwick, whose widow endowed it to the Dutch Reformed Church. It was bought and developed by Isaac Varian in the late 1700s. When first plotted from Bainbridge to Decatur Avenues, the street was called Rosa Street after the surveyor, Rudolph Rosa. Briefly it was called William Street. From Decatur Avenue down to Webster Avenue, the street was called Isaac Street.

EAST 198th STREET. It once crossed the wide estate of Leonard W. Jerome. The original name of the thoroughfare was Travers Street after a former landowner named Maria L. Travers who lived there in 1887.

EAST 199th STREET. This was once known as Garfield Street from the Concourse to Bainbridge Avenue.

EAST 201st STREET. This was part of Jan Arcier (John Archer's) Manor of Fordham that became the property of the Dutch Reformed Church. This portion was bought by Isaac Varian and his grandson subdivided and sold plots in 1885-1889. This street was called Gambrill Street from the Concourse down to Webster Avenue. It also was mapped as Suburban Street in 1894. The name was changed to East 201st Street in May 1916.

EAST 202nd STREET. It was formerly called Summit Street, from the Concourse down to Webster Avenue. From there to the railroad (one block), it was called Lower Place.

EAST 203rd STREET. It was once known as Rockfield Street, from the Concourse to Mosholu Parkway. From Webster Avenue to the railroad (one block), it was mapped as Signal Place. At first glance, this would seem to indicate a railroad signal had been situated at this point, but Civil War authority Tom Mullins believed it commemorated the *U.S.S. Signal*, a Union gunboat very much in the news at the time East 203rd Street was being surveyed. See: East 209th Street.

EAST 204th STREET. Before this street was numbered in 1911, it was Potter Place from Jerome Avenue to the Concourse where the Potter brothers had lived in the 1870s. It was also Scott Street from the Concourse to Webster Avenue in 1888. It was also part of Woodlawn Road.

EAST 205th STREET. It was once known as Ernescliffe Place from Bainbridge Avenue to Webster Avenue.

EAST 206th STREET. Grenada Place was its earlier name from the Concourse to Mosholu Parkway when it was on the lands of George P. Opdyke.

EAST 207th STREET. According to *Robinson's Atlas* of 1888, this street had the more picturesque name of Eclipse Street from Reservoir Oval to Webster Avenue.

EAST 208th STREET. At Jerome Avenue, there was once a small lane called Gilroy Place which was eliminated by the building of the subway. Mayor Thomas F. Gilroy served in 1893-1894, and is buried in Woodlawn Cemetery. East 208th Street was then substituted when the construction was over.

EAST 209th STREET. This section was laid out during Civil War days, and the *U.S.S. Ozark*, a Federal gunboat in the Mississippi Squadron, was very much in the news. This street originally went by the name of Ozark Street, from Perry Avenue to Webster Avenue. See: East 203rd Street.

EAST 210th STREET. This street runs across the Colonial farmlands of the Valentine family, which passed into the hands of the Varians in 1791. Some Revolutionary skirmishes took place there.

EAST 211th STREET. On an 1884 map, it was known as Ridge Street along the southern boundary of Woodlawn Cemetery. Over in Williamsbridge, East 211th Street was called Ruskin Street.

EAST 212th STREET. Once known as Logan Street in the settlement of Olinville No. 1 which ran from the Bronx River to White Plains Avenue.

EAST 213th STREET. This was Flower Street in Olinville No. 1, honoring Governor Roswell Pettibone Flower. Beyond White Plains Avenue to Barnes Avenue, it carried the name of Arthur Street and, later on, Randall Avenue.

EAST 214th STREET. It was Avenue A in the village of Olinville No. 1, prior to 1896. From White Plains Avenue eastward to about Bronxwood Avenue, it was Shiel Street, named for a lawyer, Peter A. Shiel, active in Williamsbridge and Wakefield affairs.

EAST 215th STREET. This was known once as 1st Avenue in the village of Wakefield until this area east of the Bronx River was annexed to New York City in 1895, and the streets renumbered to conform to the general city plan. This street was the boundary between Wakefield and Olinville No. 1.

EAST 216th STREET. It was 2nd Avenue in the village of Wakefield. See: East 215th Street.

EAST 217th STREET. It was 3rd Avenue in the village of Wakefield. See: East 215th Street.

EAST 218th STREET. This was 4th Avenue in the village of Wakefield. Until the middle 1920s this was a region of open fields and few houses. The original European population was German and Irish, with a small percentage of Polish. Later, Italians moved in. See: East 215th Street.

EAST 219th STREET. It was 5th Avenue in the village of Wakefield. See: East 215th Street.

EAST 220th STREET. This was 6th Avenue in the village of Wakefield. There were a few small farms at the turn of the century on this slope. One of the larger tracts belonged to the Catterson family who ran a tombstone and mausoleum business at nearby Woodlawn Cemetery. See: East 215th Street.

EAST 221st STREET. This was 7th Avenue in Wakefield. See: East 215th Street.

EAST 222nd STREET. This was 8th Avenue in Wakefield from the Bronx River to Bronxwood Avenue. See: East 215th Street.

EAST 223rd STREET. This was 9th Avenue in Wakefield. See: East 215th Street.

EAST 224th STREET. This was 10th Avenue in Wakefield. See: East 215th Street.

EAST 225th STREET. This was 11th Avenue in Wakefield. See: East 215th Street. See: 226th Drive at end of listing.

EAST 226th STREET. This was 12th Avenue in Wakefield. See: East 215th Street.

EAST 227th STREET. This was 13th Avenue in Wakefield. See: East 215th Street.

EAST 228th STREET. This was 14th Avenue in Wakefield. See: East 215th Street.

EAST 229th STREET. This was 15th Avenue in Wakefield. See: East 215th Street. See: 229th Drives North/South at end of listing.

EAST 230th STREET. This was 16th Avenue in Wakefield. See: East 215th Street.

EAST 231st STREET. In 1853, the former Governor of Touraine, France, lived with his family in a small house west of White Plains Road. He was the Marquis de Chateauneuf. Perhaps the countrified aspect of 17th Avenue in Wakefield reminded him of St. Ouen, a suburb of Paris for, at that time, both places had small houses and tree-lined streets. See: East 243 Street.

EAST 232nd STREET. This was 18th Avenue in Wakefield. See: East 215th Street.

EAST 233rd STREET. It was known as Eastchester Street from Van Cortlandt Park to Webster Avenue as it led to the village of Eastchester. The name was changed in 1894. It was 19th Avenue in Wakefield, where it was the southern boundary of the old settlement of Jacksonville which later was absorbed by Wakefield. In Eastchester, East 233rd Street was called Fisher's Landing Road where it ran to the Hutchinson River, and the ancient records so read: "November ye 11th 1714, Eastchester - One Highway laid out where William Fisher liveth." There was no better remembered tavern on the road to Boston than Aunt Hannah's. Hannah Fisher was famous among the turnpike travelers and the Hutchinson River boatmen for her size and strength, good nature and (according to eminent historians) her beard.

EAST 234th STREET. From Woodlawn Cemetery to Webster Avenue, it was called Clifford Street until the name was changed in 1894. In Wakefield, it was 20th Avenue. In the 1880s to the early 1900s this area was known as Jacksonville.

EAST 235th STREET. In Woodlawn, it was Williard Street until 1894. The street honored E.K. Williard, who helped lay out the streets in 1873. It was the scene of Colonel Simcoe's Raiders' encounter with the Stockbridge Indians in 1778. See: Peters Place.

East of the Bronx River, in Wakefield, it was once 21st Avenue after the small settlement of Jacksonville had been absorbed by Wakefield.

EAST 236th STREET. In the village of Woodlawn, it was once known as Opdyke Street, running from Van Cortlandt Park to Webster Avenue. It was named after Mayor George Opdyke, whose term in office was 1862-1864. In 1894, the street was numbered. Fast 236th Street from the Bronx River to White Plains Avenue was formerly 22nd Avenue in Wakefield, after it had absorbed the smaller settlement of Jacksonville.

EAST 237th STREET. Given the number 1st Avenue when the Village of Woodlawn Heights was laid out in 1873, it next received the name of Oakley Street (1894) from Van Cortlandt Park to Vireo Avenue. It was the 1778 scene of a clash between British cavalry and Indian allies of the patriots. East of the Bronx River, it was Elizabeth Avenue, after Elizabeth DeMilt, who owned property in this section, Jacksonville.

EAST 238th STREET. The street was laid out in the Village of Woodlawn Heights in 1873 and called Kemble Street, which lasted until 1894. This was once the scene of a battle between Indians and English cavalrymen during the Revolutionary War. East of the Bronx River, this was a boundary line between Jacksonville and Washingtonville. Both settlements eventually were absorbed by Wakefield.

EAST 239th STREET. This was Knox Street in the village of Woodlawn Heights, until 1894. East of the Bronx River to White Plains Road, it was Kossuth Avenue (1868 map of Washingtonville). A 1900 map shows a Grotis Park facing the Bronx River. It was a popular retreat that had been established sometime around 1874, with punts and rowboats for hire.

EAST 240th STREET. In Woodlawn Heights, it was known as Holly Street and, later, 4th Avenue. In the Washingtonville sector, it was called Westchester Avenue from the Bronx River to White Plains Avenue. The Empire Engine Volunteer Company was located at the junction of this street and Richardson Avenue prior to 1896.

EAST 241st STREET. This was once called 5th Avenue, then Hyatt Street, in Woodlawn Heights. The latter name was a reminder of the Hyatt farm from which the Heights was plotted in 1873. East of the Bronx River to White Plains Avenue (Washingtonville), it was called Becker Avenue. It was part of the W.H. Searing estate on to the city line.

EAST 242nd STREET. It was called DeMilt Avenue from the Bronx River to White Plains Avenue, a reminder of Elizabeth and Sarah DeMilt's property there. They were related to the Penfield family. This was the margin of Grotis Park. See: East 239th Street.

EAST 243rd STREET. This was formerly the Penfield estate, east of the Bronx River. The original name was Huguenot Street and might have been so named by the Penfields who traveled extensively in France. It might have been this cosmopolitan family that named nearby St. Ouen Street. See: East 231st Street.

EAST AVENUE. This is a private street in Parkchester, on farmlands once owned by the Catholic Protectory. It traverses the east portion of the huge complex.

EAST BAY AVENUE. Laid out across what had been a shallow bay in Indian times, it was filled in during the 19th century when a commercial enterprise, known as the East Bay

Corporation, was to transform Hunt's Point into a shipping center. The intersection of East Bay Avenue and Tiffany Street marks the location of a vanished island, Long Rock (Duck Island), that lay in the mouth of Leggett's Creek. See: Viele Avenue.

EAST BORDER AVENUE. This is a private road alongside Webster Avenue inside Woodlawn Cemetery which borders the eastern side of the cemetery. The Pearsall family vault is on this road.

EAST BURNSIDE AVENUE. This street originally ran only from the Concourse east to the Mill Brook (Webster Avenue) and was called Transverse Road on a 1900 map. It was later named in honor of General Ambrose E. Burnside of the Civil War, who popularized the usage of, and the nickname of, "sideburns."

EAST CLARKE PLACE. Maps & Profiles of the 23rd and 24th Wards (1895) lists a Clarke family residing in this lower section of Fordham. In 1874, Benjamin Clarke was listed as a landowner. It was briefly known as Gerard Avenue.

EAST FORDHAM ROAD. Considering how ancient is the name of Fordham in Bronx history, this road's name is comparatively new. Originally an Indian path, it became the Kingsbridge Road of Colonial days (from Marion Avenue to Third Avenue) whereas the length from Jerome Avenue to Marion Avenue was once Highbridge Road. Around 1850 a railroad station was built near St. John's College (now Fordham University) and was given the name of Fordham, as it was on part of the Fordham Grant. The station attracted businesses, and a village of Fordham grew up around it. East Fordham Road and West Fordham Road were widened to become the principal thoroughfare. See: West Fordham Road.

EAST GUN HILL ROAD. See: Gun Hill Road.

EAST KINGSBRIDGE ROAD. See: Kingsbridge Road.

EAST MOSHOLU PARKWAY NORTH. See: Mosholu Parkway North.

EAST MOSHOLU PARKWAY SOUTH. See: Mosholu Parkway South.

EAST MOUNT EDEN AVENUE. See: Mount Eden Avenue.

EAST NONATIONS. These are two adjoining reefs in Long Island Sound, the easternmost bits of Bronx territory. They are in the vicinity of Hart Island. The name is believed to derive from an old story that neither Holland nor England was willing to fight over such worthless bits of land, hence they belonged to no nation. Some historians believe the original spelling was "Notations," such as a footnote, or abbreviation.

EAST SHORE DRIVE. This is a driveway along the eastern shoreline of Hart Island. The easternmost building on this former prison isle was a synagogue, but Warden Ed Dros pointed out it was closed due to lack of Jewish prisoners.

EAST TREMONT AVENUE. Upper Morrisania was laid out in the 1850s and Hiram Tarbox, its first postmaster, suggested the name of Tremont as there were three mounts inside the limits: Mount Hope, Mount Eden and Fairmount. From the Concourse to Webster Avenue, East Tremont Avenue was first called Waverley Street. It was Morris Street in Tremont, and West Farms Road at the Bronx River with the local nickname of "Plant Road." From the Bronx River to Castle Hill Avenue it was long known as Walker Avenue, and Bear Swamp Road from there to Williamsbridge Road. From Westchester Square out to Lawton Avenue, it was the Throggs Neck Road or the Fort Schuyler Road. At Lawton Avenue, once called Ashes Corner, East Tremont Avenue stopped at the estate wall of what was once Philip Livingston's estate, later the smaller holdings of Mitchell, Ashe, Van Schaick and Huntington. Around 1920, the avenue was extended to the East River, where there had been an Indian site and the location of Throckmorton's colony, It was also the landing place of General Howe's troops in 1776. A Cedar of Lebanon, planted in 1791 by Livingston, was in danger of being cut down when East Tremont Avenue was extended, but Cyrus Miller, borough president, is credited with having the angle altered to save the tree. It was toppled in a storm in 1945, and taken away in 1959 by Theodore and Ronald Schliessman of The Bronx County Historical Society.

EAST TWIN ISLAND. A favorite hunting and fishing spot of the Siwanoy Indians, judging from the arrow points and spears found there, this cigar-shaped isle off Hunter Island passed through a series of ownerships from Thomas Pell to James D. Fish, a financier of the 1880s, who built a handsome red sandstone mansion on a rocky ledge. It was acquired by the Parks Department in 1888, and the mansion was demolished around 1930. East Twin Island was joined to the mainland in 1947-48 by the extension of Orchard Beach.

EASTBURN AVENUE. An earlier name was 1st Avenue, from the Concourse to Claremont Parkway. The southern end of this avenue marks the lower boundary of Fordham Manor. The land passed to the Steenwicks, to the Reverend Selyn (Selwyn) who endowed it to the Dutch Reformed Church. After the Revolution, it was in the possession of various farmers, among them, an Elisha Monroe.In the 1850s it formed the upper end of the Zborowski estate. The name may refer to its location-"East" and "Burn," a brook.

EASTCHESTER BRIDGE. The name is geographical, as the bridge is in the ancient settlement of Eastchester and spans the Hutchinson River, which was also known as Eastchester Creek. The bridge, carrying Boston Road over the stream, was completed in 1965, replacing a smaller bridge that had been there since 1922.

EASTCHESTER PLACE. Opened in 1957, this Place, alongside the New England Thruway at the Hutchinson River, is a reminder when the river was known as Eastchester Creek. See: Hutchinson River (Current).

EASTCHESTER ROAD. This ancient road is mentioned in the 1667 Nicolls grant of the Ten Farms as "the Westchester Path." This was the earliest connecting road between Westchester and Eastchester. After the Revolutionary War, the term "Westchester Path" fell into disuse and Eastchester Road took its place. Before the days of the Oostdorp settlers in what is today Westchester Square, it was a trail used by the Siwanoy Indians. Just below Pelham Parkway, the Americans had an outpost to prevent the British from crossing the shallow Westchester Creek in October 1776. A wooden bridge spanned the creek in the 19th century at about Morris Park Avenue.

EASTERN BOULEVARD BRIDGE. Located at the point where Bruckner Boulevard crosses the Bronx River, this bridge still bears the former name of Bruckner Boulevard.

ECHO PLACE. Formerly Ludlow Street from the Concourse to Anthony Avenue, and Buckhout Street from Anthony Avenue to Julius Richman Park, (formerly Echo Park) one name was a reminder of a former owner of the property, as James Buckhout was the owner of a wooded glen on Webster Avenue, famous for its echoes. There were Buckhouts mentioned as far back as 1670, and the generations included musicians, realtors, merchants and surveyors. The City acquired the property in 1888, and named it Echo Park, a title it kept until 1973. The street names were changed to Echo Place in 1903. The Place has retained its name – an echo of the former park and glen.

EDENWALD AVENUE. It was formerly mapped as Jefferson Avenue. It is named after the Edenwald estate bounded by Light and Conner Streets, north of Boston Road, and owned by John H. Eden. The estate appears on maps from 1900 to 1913.

EDGE STREET. An unofficial name of a roadway dividing "A" and "B" sections of Edgewater Park on Throggs Neck, which appears on Park maps, but not on City maps.

EDGEHILL AVENUE. This is a very fitting name for an avenue that is on the edge of Tippett's Hill, overlooking the valley to Marble Hill. It is formerly the site of Revolutionary Fort Number 3.

EDGEMERE STREET. About 20 yards of this street lies in Bronx County, just north of Pelham Bay Park. A map, filed in 1884, showed the northern city line forming the top of Pelham Bay Park; but in 1892 a bill was passed authorizing the line to be slightly shifted to the north, which created a half-block gap in which terminated some roads belonging to Pelham Manor. The name is descriptive, as there were and are swampy pools (meres).

EDGEWATER PARK. See: Park of Edgewater.

EDGEWATER ROAD. Laid out from 1893 to 1894 along the edge of the Bronx River.

EDISON AVENUE. The avenue is obviously named after Thomas Alva Edison, the electrical genius, in connection with Ohm, Ampere and Watt Avenues. Laid out in the village of Schuylerville from Baisley Avenue to LaSalle Street, it was first called Seton Street, as the Setons were property owners of note (1868 map). Later, the avenue was extended southward through the Coster farm, Lorillard Wolfe estate and Frederic Jackson's lands, ie., from Bruckner Boulevard down to Dewey Avenue.

EDSALL AVENUE. Thomas Henry Edsall wrote *History of the Town of Kingsbridge* in 1887. His estate overlooked Spuyten Duyvil Creek, according to an 1888 *Robinson Atlas*. His ancestors once owned part of the lower Bronx, and sold their holdings to the Morris family in 1681. The family was originally Dutch, and Samuel Edsall knew the Indian language and functioned as interpreter in Nieuw Amsterdam.

EDSON AVENUE. Streets in this region carry names of New York Mayors, and Franklin Edson served his term in 1883-1884. It was once called Sycamore Street, and also South 19th Avenue, when it was part of Mount Vernon. Before 1900, it was planned to name the avenue Bracken Avenue after Henry Bracken, a prominent member of Tammany in the Annexed Districts, and some land maps carried that name.

EDWARD L. GRANT HIGHWAY. See: Grant (Edward L.) Highway.

EDWARD P. LYNCH TRIANGLE. See: Lynch (Edward P.) Triangle.

EDWARDS AVENUE. Clarence R. Edwards was one of the directors of the Seton Homestead Land Company, which had bought up the Alfred Seton estate, between present-day East Tremont Avenue and Westchester Creek. See: Ellison Avenue.

EDWARDS PARADE. This is a large field occupying the center of Fordham University, and named after Major-General Clarence Edwards, an early commandant of the R.O.T.C.

EFFINGHAM AVENUE. This is most likely named for a family that intermarried with families on nearby Clason Point. This avenue is laid out on the site of the Indian "castle" that gave the Point its name. In 1685, it was called Cromwell's Neck. Later, three generations of the Wilkins family owned the land, which passed eventually to a son-in-law, John Screvin. Around 1900, the region was known as Screvin's Point. Effingham Schieffelin was a landowner on Clason Point in Civil War times. His estate, called "Ashburne," was originally the property of Dominick Lynch, who owned one side of the peninsula, while Isaac Clason owned the other. The Effinghams married into the Drake and Lawrence families, too, for the name keeps cropping up in those genealogies as a first name.

EGER PLACE. This is a private street running off Pennyfield Avenue to Hammond's Cove. See: Alan Place.

EIGHTH STREET. This is a narrow path in the Park of Edgewater, eighth in order from the waterfront. See: Park of Edgewater.

EINSTEIN LOOP. Albert Einstein, German-born physicist, was honored in July 1967 when his name was remembered by the planners of Co-Op City. The eastern side of the Loop overlays what was Givan's Creek (navigable until 1967 by kayak and canoe), while the opposite side covers a smaller tidal stream mapped as Barrow Creek, and filled in, in 1963. The land was part of the 14-acre Rose Island conveyed by John and Theodosius Bartow to Robert Givan in July 1802. In 1905, it was the property of Jemima Watson.

ELDER AVENUE (1). It was first known as Watson's Lane, so called because it once led to the Watson mansion, near James Monroe High School. John P. Elder was a director of the A-Re-Co (American Realty Corporation) which subdivided the Watson estate around 1894. Mr. Elder lived on Longwood Avenue.

ELDER AVENUE (2). This is a roadway in the northwestern section of Woodlawn Cemetery. Charles Evans Hughes is buried near this lane. The elder is a shrub of the honeysuckle family, a name in keeping with the Woodlawn Cemetery practice of using floral and arboreal names.

ELEVENTH STREET. This is a street in the Park of Edgewater, eleventh in order from the waterfront. See: Park of Edgewater.

ELGAR PLACE. This Co-Op City Place, named for an English composer, Sir Edward Elgar (1857-1934), is situated over what had been Givan's Creek that flowed into the Hutchinson River. See: Einstein Loop.

ELLIOT PLACE. Originally, this street was only open east of the Concourse in 1893-94, when it was known as Stebbins Place. The Stebbins family were property owners in the neighborhood at the time. Later, the street was extended westward to Jerome Avenue. A former owner of "Claremont," an extensive estate of the 1800s, was Elliott [spelled this way] Zborowski, and a jocular account of the day was that it was far easier to spell his first name than his last.

ELLIS AVENUE. This avenue is laid out through what was once the 17th century grant to the Underhill family. In 1794, it was the Cow Neck farm of Talman Pugsley, which passed to the Cobb and Larkin families of the 1850s. It was laid out as 13th Street in the village of Unionport, and was changed to its present name around 1902, when Unionport's numbering system conflicted with New York City's grid-pattern. Dr. James E. Ellis practiced medicine for many years in old Westchester, and was a Vestryman of St. Peter's P.E. Church, (1888-1889).

ELLIS PLACE. This Place is on the former grounds of the Westchester Polo Club, and an 1881 deed mentions a sale of land by Peter Lorillard to John and Julia Ellis. Their mansion faced Rawlins Avenue at Lohengrin Place, and their tract was listed in a Bromley map of 1897.

ELLISON AVENUE. Cut through the former Seton Estate (1868 map) that was subdivided under the name of Seton Homestead Land Company. Charles Ellison was a director of this company. See: Edwards Avenue.

ELLSWORTH AVENUE. The avenue was laid out across the extensive swamps alongside Weir Creek, and runs across the former Turnbull estate, Bruce Brown estate and a section of the Century Golf Course, on Throggs Neck. It is named for Colonel Elmer Ellsworth, friend of President Lincoln and heroic leader of the Fire Zouave regiment, composed of New York volunteer firemen. He was shot dead by a citizen of Alexandria, Virginia, as he was removing a Confederate flag from the house.

ELM DRIVE. This is a privately maintained street inside Parkchester. Elms were prominent when this land belonged to the Catholic Protectory.

ELM PLACE (1). The Place was laid out in 1916, and named after an old homestead, "The Elms." The extreme northern tip of this farm was bisected to straighten out Fordham Road, leaving a small island at its junction with Kingsbridge Road. The owners of "The Elms" in the 1890s were named Keary.

ELM PLACE (2). This is a short thoroughfare in Silver Beach, arbitrarily given the fifth letter of the alphabet. See: Acorn Place.

ELSMERE PLACE. It was laid out across an estate, "Fairmount," once belonging to a Colonel Dunham, but passing to Robert Cochran (1854) and to Thomas Minford of the 1870s onward. It was vested in 1903. It was the local belief Elsmere Place owed its name to a Polar expedition when Admiral Robert E. Peary reached Ellesmere Island in 1909, but the Place had already been named in 1905. More likely the name is a reminder of Ellesmere, a town in Cheshire, England, the birthplace of one of the former owners, all of whom were English. See: Peary Gate.

ELTON AVENUE. Originally Washington Avenue in the village of Melrose. Robert H. Elton was a prominent landowner in the early 1850s and East 152nd Street was first called Elton Street. He purchased property from Gouverneur Morris II in 1848, adjoining the new settlement of Morrisania, and called it Eltona. Its approximate bounds were Third to Forest Avenues, East 163rd to East 165th Streets. Elton Avenue retained its quiet, residential aspect for almost a century. The inhabitants were mainly German, and there were three German churches along its length, one of them with a Swiss minority. A Public Bath was once situated at East 156th Street and was a marvel of its day, resembling a Roman temple and featuring all sorts of luxuries.

ELY AVENUE. It is named for Smith Ely, Mayor of New York 1877-78. At its southern end, Ely Avenue overlays the former Givan's Creek and mill that were landmarks of the 19th century. Directly after the Revolutionary War, Robert Givan purchased the land from the Bartow family, the tract extending to Boston Road, east of Eastchester Road. The name of the estate was "Ednam," and it remained in the family for four generations, later to be sold as a real estate venture called "Pelham Bayview Park" (1895-98). Ely Avenue in part overlays the ancient boundary watercourse called Black Dog Brook that separated Eastchester from Westchester. Above Boston Road, the avenue was once called Doon Avenue. Above Nereid Avenue, its older name was Bayard Street, honoring Nicholas Bayard, a Mayor of New York in 1685. Smith Ely died in 1911.

EMERSON AVENUE. This avenue did not come into existence until the 1940s when Morris Cove was filled in to form Ferry Point Park. A 1718 deed alludes to Ox Meadow Cove, and in 1811 there is Ox Pasture Cove. Where Emerson Avenue meets Harding Avenue the promontory was once called Morris Point. Four or five avenues here honor Minutemen of the Revolutionary War. Reverend William Emerson, grandfather of Ralph Waldo Emerson, was pastor of the church in Concord. He was chaplain of the Provincial troops, and left Concord in 1776 to join the army at Ticonderoga. See: Buttrick, Hosmer, Davis and Robinson Avenues.

ERBEN AVENUE. On the Sound side of the New York State Maritime College at Fort Schuyler is a straight road dedicated to Captain Erben, Superintendent of the college from 1879 to 1882.

ERDMAN PLACE. The American novelist and poetess Loula Grace Erdman was honored in July 1967 by having her name given to this Co-Op City street. A few yards south of Erdman Place was once Givan's Creek. See: Einstein Loop.

ERICSON PLACE. Ericson Place cuts across the slope of Bridge Hill (Crow Hill), the scene of a Revolutionary artillery battle. It was opened in 1912, and recalls John M. Ericsson [spelled this way] who owned lots on nearby Middletown Road from 1911 to 1914.

ERSKINE PLACE. This Place overlays the western shoreline of Givan's Creek where it flowed into Hutchinson River. It was the seventeenth century holdings of Thomas Pell, Lord of the Manor. Later landowners were the Bartows and the Givans. John Erskine, the American educator and author, was remembered by the planners of Co-Op City in July 1967.

ESPLANADE. See: Westchester Esplanade.

EVELYN PLACE. Due to its proximity to Fort Number 8 of Revolutionary War fame, situated on the grounds of the Bronx Community College, this Place might memorialize Captain William Glandville Evelyn of the King's Own regiment who died in the battle of Pell's Point in 1776. Naming streets for gallant fighters on the opposite side has a historic precedent in this area, as Colonel Emmerich [the correct spelling] of the Hessians once had Emmerick [spelled this way] Place named in his honor. It has since become incorporated into Sedgwick Avenue on the edge of Devoe Park. Evelyn might well have been the name of the wife of either Thomas Partridge or Robert Craighead who had the land surveyed in 1869. Evelyn Place was opened in 1903. See: Emmerick Place.

EVERGREEN AVENUE. This street was evidently named for the type of trees that abounded on the William Watson estate, "Wilmount," of the 1870s. At first, it was called Fenfell Avenue from the Bronx River to Bruckner Boulevard. Fenfell was one of the Westchester colonists who pledged allegiance to the New Netherlands in 1656, along with Lt. Wheeler, William Ward and John Cloes (Close), whose names were given to adjoining avenues.

EWEN PARK. Resident of Spuyten Duyvil hill, Civil War general John Ewen's property became this park and received his name. It is a steep incline from Johnson Avenue down to Irwin Avenue. See: Ewen Street.

EWEN STREET. Daniel Ewen was a city surveyor, *circa* 1825, but it was his son, John Ewen, civil engineer and Civil War general, who is remembered. See: Ewen Park.

EXTERIOR STREET. Most of its length, flanking the Harlem River, was originally the river itself, or bogs. From West 193rd Street to West 230th Street, Exterior Street follows the original course of the river, before the Ship Canal was cut through in 1895. The street is well named, being on the western rim of the borough. West 225th Street marks the site of the Free Bridge erected in 1759 to afford toll-free passage from Manhattan to The Bronx mainland. See: West 225th Street or Palmer Avenue. Old-timers recall Kyle's Park, north of High Bridge. Exterior Street crosses the site of the wooden pier leading to the famous old picnic grounds. Steamers from Mott Haven and from lower Manhattan docked at this pier to disgorge their thousands of excursionists, from the 1880s to about 1905.

FAILE STREET. This Hunts Point street runs through the former estate "Woodside" that belonged to Edward G. Faile (1824-1870). His home was atop the summit of Garrison and Lafayette Avenues and he was in the habit of commuting daily to lower Manhattan by coach. He owned thoroughbred horses which were imported from Argentina at $1000 apiece in transportation alone. Squire Faile always had a fresh team of horses waiting for him in Harlem, which was the halfway mark in his daily trip. An 1890 map shows this street carrying the name of Hillsdale Avenue. The Faile family plot is prominent in Woodlawn Cemetery at Poplar Avenue.

FAIRFAX AVENUE. The first skirmish of the Civil War took place at Fairfax Courthouse, Virginia, in 1861. Most avenues in this area carry Civil War names. After Siwanoy Indian occupation, this sector was the lower end of the Underwood holdings of 1685 called "Kenilworth." During the Revolutionary War, it was the property of James and Charity Ferris. Nineteenth-century ownership went first to William C.

Layton, and then a Dr. Ellis. The avenue was formerly one property line of the Lohbauer Park picnic grounds, a noted beach resort of the 1900s.

FAIRFIELD AVENUE. Many of the English settlers of what is today our Bronx came from Fairfield, Connecticut, which was also known by its Indian name of Onkway. John Archer, proprietor of Fordham Manor, hailed from Fairfield. The Manor took in this part of Spuyten Duyvil.

FAIRMOUNT AVENUE. This was the Revolutionary holdings of James and Charity Ferris, which in the mid-1880s was part of the estate of the Van Antwerp family. In the 1890s, a Dr. Ellis was owner. Lohbauer's Casino was located on this avenue (1899-1920 approximately) a little west of Wilcox Avenue. Around 1920, Fairmount Avenue was cut through from Bruckner Boulevard to the Eastchester Bay shoreline. A local historian, William Tarbox, believed the name stemmed from the estate's designation by the Van Antwerps, but it was never mapped as such.

FAIRMOUNT PLACE. "Fairmount" was an estate in the possession of a Colonel Dunham, then Robert Cochran (1854) and, later, Thomas Minford. A small settlement near Southern Boulevard eventually took the same name. On an 1874 map of West Farms, this had been a village lane known as Waverley Place, but when it was cut through in 1902, paved and curbed, it received its present name to perpetuate the name of the estate.

FAIRVIEW AVENUE. Located in about the center of Woodlawn Cemetery, from which a fair view can be had.

FARADAY AVENUE. This Riverdale avenue was officially named on May 2nd, 1916, for scientist Michael Faraday.

FARKAS (George) SQUARE. Located at the intersection of Fordham Road and the Grand Concourse, this Square honors the founder of the Alexander's department store chain. The original store opened on Third Avenue and East 150th Street in 1928. In 1994, the firm closed all its retail outlets to concentrate on its more profitable real estate business.

FARRAGUT STREET. This street was cut through what was once "Foxhurst," the estate of George S. Fox. It was also called Fox Woods. It was named in December 1918, in honor of David Glasgow Farragut, first Admiral of the United States Navy, 1801-1870. The Admiral is buried in Woodlawn Cemetery.

FATHER ZEISER PLACE. See: Zeiser (Father) Place.

FEARN PLACE. This is the northern slope of what was called Schuyler Hill in the late 1890s and early 1900s at the end of Throggs Neck. Its historical background is identical with that of Alan Place.

FEATHERBED LANE. On an 1868 map, it is Feather Bed Lane, and the lower end was formerly Belmont Street until 1889. There are three well-known versions of the origin of this name. During the Revolution, residents padded the road with their feather beds to muffle the passage of the patriots. Another story is that the spongy mud gave riders the effect of a feather bed. Still another tale is that the farmers found the road so rough, they would use feather beds on their wagon-seats to cushion themselves. There is a fourth supposition advanced by a native of Highbridgeville that Featherbed Lane was a sly allusion to ladies of easy virtue who lived there. In short, it was the local Red Light district during the 1840s when work on the nearby Croton Aqueduct was going on. Unsuspecting real estate developers of a later time liked its quaint name and retained it.

FEISS BOULEVARD. See: Rose Feiss Boulevard.

FENTON AVENUE. Reuben Fenton was Governor of New York from 1865 to 1869. From 1795 to around 1840, the avenue was part of the Givan farm. A Captain J. Watson owned a tract west of Eastchester Road (1860s) and his home was at the junction of Fenton and Allerton Avenues. Later, the section below Allerton Avenue was subdivided and became the Sullivan, Landon and Stern estates of the later 1890s. The section there was a wooded one, and a local nickname for it, around WWI, was "Tanglewood."

FERN AVENUE. This is a roadway in the northwestern section of Woodlawn Cemetery. Mayor William Wickham, who has a Bronx street named for him, is buried near Fern Avenue. The name is purely descriptive of the plant life found there.

FERN PLACE. This is a short, private thoroughfare in Silver Beach that was given the sixth letter of the alphabet. See: Acorn Place.

FERRIS AVENUE. This service road or the Hutchinson River Parkway runs across farmlands that were tilled by the Ferris families for generations. The Parkway obliterated the original Ferris Lane in 1938.

FERRIS PLACE. Once called Dock Street, it led to the Westchester Creek wharf. It also was called by the name of Bowne Street from the Quaker shopkeeper who had a large store on this street. The original John Ferris was one of the five patentees of Westchester in 1667. James Ferris ran a nearby mill in the 1820s, and Benjamin Ferris owned what is now Westchester Square in 1839. He conveyed it to the School Board. Ferris Place runs through his property.

FERRY POINT PARK. This is a reminder of Ferry Point, an early name derived from an ancient ferry system that touched this peninsula en route from Westchester Village to Whitestone. The ferry ran from Westchester Square (Village), stopped at Unionport Dock, Ferris Dock and Whitestone Landing on the opposite shore of the East River. Ferry Point was part of the Throckmorton Grant of 1642, which passed to Spicer and Brockett, the Hunt family, the Ferris families, which was similar in sound to Ferry, and Ferris Point really was a former name of Ferry Point, since farmlands on this peninsula had been occupied by the Ferris families from the 1750s to the 1900s before being sold to the Prime, diZerega and Lorillard families. The last owner was the Roman Catholic House of the Good Shepherd that sold out to the City in 1937.

FIELD PLACE (1). At different times it carried the names of 2nd Street and Kirk Street. It most likely honors Benjamin H. Field of Fordham, President of the House of the Incurables. Two other suppositions are possible. In the 1850s, this was a meadow or field between Prospect Hill (East 182nd Street) and Mount Sharon (Fordham Road) and merely received a descriptive name. Another story is that at the time-1856-when streets were being laid out, Cyrus Field had just completed the laying of the first Atlantic cable, and the Place commemorates his name.

FIELD PLACE (2). Directly under the Throgs Neck Bridge, on the Fort Schuyler peninsula, this short Place is a reminder of lieutenant Commander Field, Superintendent of the New York State College from 1894 to 1897.

FIELDING STREET. This short street off Eastchester Road had been called Lowell Street until March 1950, when it received its present name at the behest of Congressman Paul Fino. One reason for the change was to avoid confusion with another Lowell Street near Whitlock Avenue. Henry Fielding was an English novelist, but the Congressman once remarked the street honored Anna Fielding, a poetess and lifelong resident of The Bronx. Originally, the street was part of the Pell Purchase, which in the early 18th century belonged to the Bartow family. After 1795, this area was part of the Givan farm. In the mid-1880s, it was the dividing line between the Watson and Bruner estates. P. Bruner built a mansion there and called his estate "Rosedale." Israel Watson died, and both holdings were then bought by the Valentine family around the 1880s, and formed a 27-acre estate extending from Eastchester Road downhill and eastward to Givan's Creek. The former Valentine house is located at Fielding Street and Woodhull Avenue and still has remnants of its past impressiveness.

FIELDSTON ROAD. It was laid out through the former Delafield property in 1917. Major Joseph Delafield bought 250 acres in 1829 on what had been Upper Cortlandt, and named his estate "Fieldston," which was the name of his family seat in England.

FIELDSTON TERRACE. Upon being acquired by the City in 1915, the former Delafield property, "Fieldston," was cut through by streets and places, and this Place is a reminder.

FIFTH AVENUE. This is one of the widest streets in the Park of Edgewater, fifth in order from the waterfront. It was once used for athletic contests on the Labor Day weekends. See: Park of Edgewater.

FIFTH STREET. This street runs from east to west across Hart Island, and is the last of five numbered streets.

FILBERT AVENUE. This is a road located in the northwestern section of Woodlawn Cemetery. A hazel is a small native tree, and another name for a hazelnut is a filbert. Franklin Simon's grave can be seen from this road.

FILLMORE STREET. Van Nest residents favored Presidents when naming their streets, and Millard Fillmore served as Chief Executive from 1850-1853. The street itself runs across what was once Hunt property.

FINDLAY AVENUE. Andrew Findlay, surveyor of West Farms, Morrisania, Throggs Neck and many other sections of what was then Westchester County during the 1840s and 1850s, is remembered. He was born in the village of Westchester in 1811, and died in 1892, being buried in Woodlawn Cemetery. His parents came from Wigton, Gallowayshire, Scotland. The surveyor owned property west of Claremont Park on what is today the Grand Concourse. He is credited with originating the name "Melrose." See: Marcy Place.

FINK AVENUE. John Finch, or Fink, was one of the three settlers who signed an oath of allegiance to the Dutch authorities: "This first January A° 1657: in east towne in the N. Netherlands Wee hose hands are vnder writen do promes to oune the gouernor of the Manatas as our gouernor. . . ." Richard Ponton and Edward Waters were the other two men, and adjoining avenues are named for them. "East towne" was the English version of Oostdorp, which became Westchester Square. An Alexander Fink, butcher (and possibly a descendent of the town father), purchased land near Eastchester Road in 1813, and is again mentioned in an 1821 deed. A John B. Fink is noted in 1825 records of land sales in and around Westchester village.

FIRST STREET. This street runs from east to west across Hart Island and is the first of five numbered streets.

FISCHER (Corporal) PLACE. Known as Highbridge Street, listed in 1871, running across what had been the Anderson farm of the 1800s, for many years (so the story goes), the Place got another name from a milkman who used to station himself, his wagon and horses in a little lane. He was there so many years, it was named after him and became known as "Hennessy's Place." This tale is offset by a Joseph Hennessy, who was Commissioner of Parks in 1918. Around 1948, a local law changed Hennessy Place to James R. Murphy Place for an able, but minor, politician who had founded the Bronx Grand Jurors' Association. The name lasted a few days, at least, until the local inhabitants strenuously objected. Hennessey Place was restored until May, 1949, when it was renamed Corporal Irwin Fischer Place, for a soldier who probably died in WWII.

FISH AVENUE. Hamilton Fish, 1808-1893, gave his name to this avenue solely because all avenues in this neighborhood are named after New York Governors. He served from 1849 to 1850, and was also a Senator before becoming Secretary of State. Stony (Stoney) Brook once ran southeast along this avenue and gave rise to a latter-day story that the avenue was named after a Fish Creek.

FLEET COURT. A private street in the Shorehaven condominium which was the extensive picnic area of the Shorehaven Beach Club (1947-1986). When this same location was inside the long-ago Clason Point Amusement Park, the Midway was nearby.

FLETCHER PLACE. This small Place in the former village of Tremont harks back to an early resident who lived in the 1880s on what is now East 182nd Street (then called Fletcher Street) off Washington Avenue.

FLINT AVENUE. Surveyed and named in 1935, this short avenue occupies the extreme northeastern corner of The Bronx. It is on the original Pell Grant, and possibly part of Ane Hutchinson's farm. An 1886 deed lists Charles H. Ropes as owner of this tract, and in 1900 it was part of the Peter Prove estate. The street was named for Edward Flint Campbell, Pelham Manor Village Engineer for more than 25 years, who laid out the streets as access roads to the Pelham incinerator.

FLOOD SQUARE. It was named in March 1940, at a time when the American Legion was active in having squares named in honor of World War I casualties. The neighborhood just south of Fordham Road once had a strong Irish element, which would give credence that a local lad was being honored. No Christian name has been forthcoming.

FLORENCE COLUCCI PLAYGROUND. See: Colucci (Florence) Playground.

FLYNN (John and Michael) PLAYGROUND. At East 158th Street and Third Avenue, this small park was once the site of the oldest school in Melrose. It was named on May 24, 1949, and honors two Army flyers who were killed in World War II.

FOLIN STREET. Caius Folin purchased land near Webster Avenue from the Weeks family in 1854. His home was on nearby Park Avenue. There is a mention of a LaFayette Folin in the 1880s in the Business Directory.

FORD STREET. A short steep street that runs down into Webster Avenue, it was opened in 1907, many years after the Mill Brook was covered over. Therefore, it is not named for a ford over the stream, nor is it an abbreviation for Fordham-as it lies in Tremont. Frederic G. Ford was a small landowner of Upper Morrisania (now Tremont) in the 1880s.

FORDHAM PLACE (1). This Place was once the banks of the Mill Brook below Fordham University at East 189th Street.

FORDHAM PLACE (2). At about the middle of City Island, this Place was formerly called Fordham Street. The Fordhams were among the earliest settlers on the island and have both a street and a place named after them. See: Fordham Street.

FORDHAM PLAZA. This is the current popular designation of that portion of Fordham Square south of East Fordham Road. Although the name is not yet an official one, and is not carried by street signs, the buses which have their terminus in Fordham Square use this name to signify their destination. See: Fordham Square [Current].

FORDHAM ROAD. See: East Fordham Road, West Fordham Road.

FORDHAM SQUARE. This was once part of the Corsa farm when Fordham Road was called Kingsbridge Road (early 1800s to 1860s). A manor house was built in 1838 and this 200-acre estate later passed through the ownership of the Watts and then Brevoort families, and then became the Mowatt estate called "Rose Hill." The New York Central Railroad cut through the estate, and the square was called Rose Hill Place. The site was sold to the Catholic Diocese and became St. John's College, at which time the manor house served as the administrative building, and the circular garden behind St. John's Hall, next to the chapel, covered the unmarked cemetery of the Corsa family. The college was formally opened in June 1841, by the Right Reverend John Hughes. At one time, Emmet Street was absorbed by the University north of Fordham Road, and is now sodded over as a lawn directly opposite Washington Avenue. A good possibility this short street was all that was left at the time, to remind us of Emmet Powell, who was one of three men who purchased the land in 1851, before the college expanded. In 1905, the corporate name of the college was changed to Fordham University, and the name Rose Hill Place was changed too, probably to honor the newly rechristened college.

FORDHAM STREET. It was formerly Fordham Avenue, the true center of City Island. Its western end was once the boundary of the Vail estate of the 1860s and the Barker property. Orrin Fordham came from Connecticut in 1790 and started oysterbeds in the shallows off City Island, and pioneered an industry. Among the Fordhams was the first schoolmistress, the foremost historian and a few who were well-versed in Indian archeology. A Fordham was the last town clerk of Pelham, of which City Island was a part, before its annexation by New York City. In 1875, a shipyard at the foot of

Fordham Street was run by one "Hubie" and afterwards by John Hawkins. It specialized in major repairs to the Sound schooner trade.

FOREST AVENUE (1). This is all that is left to remind us of a little settlement called Forest Grove. Originally, this land was called the Shingle Plain of Gouverneur Morris' farmlands. It was laid out in the village of Morrisania (1856) on the top of Grove Hill. This section was also called Woodstock (1868 map). A Peter Decker lived on East 161st Street and Forest Avenue in 1872 and on some maps the area was called Deckerville. See: Woodstock.

FOREST AVENUE (2). This is a private street located in the northeastern section of Woodlawn Cemetery, hence not really a duplicate name. It evidently harks back to the nineteenth-century woods in that area when it was farmland. The Haffen family plot is alongside this road as is that of the Frederick Wendell Jacksons' (the family of a historian and relatives of the Havemeyer family). Also found there are plots of the Sackett, Loomis, Jessup and diZerega families - all nineteenth-century landowners.

FOREST ROAD. About 20 yards of this street lies in The Bronx, where it extends down from Pelham Manor. The name is descriptive. In 1884, the city line coincided with the northern limit of Pelham Bay Park, but in 1892 a bill was passed authorizing the New York City line to be shifted a little north. This revision created a half-block gap between the boundary of Pelham Bay Park and the present city line. The lower end of Forest Road was thus annexed to The Bronx.

FORSTER PLACE. The Honorable George H. Forster was listed in 1888 as owner of an estate at the junction of the Old Post Road and Broadway (Robinson's *Atlas*).

FORT INDEPENDENCE PARK. On August 7, 1987, the residents of Kingsbridge celebrated the dedication of the 3.2 acre park on the shoreline of Jerome Park Reservoir. The name harks back to the Revolutionary War fort that stood on Giles Place. See Fort Independence.

FORT INDEPENDENCE STREET. This is a short street honoring the famed Revolutionary fort once located on nearby Giles Place. See: Fort Independence.

FOUNDERS' ROAD. This road on the Bronx Community College campus now exists in small segments, after its initial laying-out in the 1890s. The land it crosses was once the Schwab estate, and before that, Archer lands and also part of the area covered by Fort No. 8 of Revolutionary War fame. The road honors the founders of New York University - former occupant of the college grounds - and the list included Morgan Lewis, Governor of New York 1804-1807, Isaac S. Hone, brother of Mayor Philip Hone, Valentine Mott M.D., probably the most prominent physician in New York at the time, and other notables.

FOURTH STREET (1). Running from east to west across Hart Island, it is the fourth in five numbered streets.

FOURTH STREET (2). This is a footpath in the Park of Edgewater, and is the fourth row from the waterfront. See: Park of Edgewater.

FOWLER AVENUE. Fowlers have figured in early Bronx history since pre-Revolutionary days. Jeremiah Fowler was a Sheriff under the State Constitution of 1777, and the family was noted as small landowners in the Van Nest section on a British Headquarters map of 1780. In 1805, a Richard Fowler owned property, a portion of which is now Parkchester. This avenue is on the bed of Downing's Brook.

FOX STREET. The Fox family was of pioneer Quaker stock, and well-to-do landowners. William W. Fox married a Charlotte Leggett, and their daughter married H.D. Tiffany. The extensive Fox estate was surveyed by Louis A. Risse (a Frenchman who was later to become Chief Topographical Engineer of Greater New York) in the mid-1880s. The Fox mansion was demolished in 1909. Mr. Fox and his family are buried in Woodlawn Cemetery, Section 9.

FOX TERRACE. This private street is listed in the Bronx Postal Zone Guide, but not found on maps. It parallels the Dyre Avenue subway line at Eastchester Road between Hammersley and Burke Avenues. The property was once owned by a Mr. Fox (first name not given) prior to 1920. It has 13 houses, located along both sides of its 200-foot length.

FRANK SIMEONE SQUARE. See: Simeone Square.

FRANKLIN AVENUE. The avenue honors Benjamin Franklin. This was once the most fashionable avenue in the village of Morrisania. The finest mansions lined this thoroughfare, according to old accounts, including that of Henry B. Dawson, historian, Lucy Randall Comfort, a writer of children's books, and Dr. John Condon who figured in the famous Lindbergh baby kidnapping trial.

FRANZ SIGEL PARK. This 17.5 acres of cedar woods was named in 1902 after the German-born Civil War General who lived on Mott Avenue (now Grand Concourse) and East 147th Street. The tract was laid out in 1880 as a public park, "Cedars Park," out of the former estate of Gerard W. and Mary Morris whose property ran prior to the Civil War from approximately Park Avenue to the Harlem River. This estate was called "Cedar Grove." Franz Sigel was born in 1824 and attended a military academy in Germany. He took part in the Rhineland revolution of 1848 and had to flee to neutral Switzerland. He made his way to London in 1851 where he played piano for a living. In New York, he taught mathematics, Italian and fencing. He was also the editor of a German-language newspaper, *Baltimore Wecker*, and an instructor in gymnastics. He was appointed a General in the Civil War. He retired on pension, and died in August 1902. The funeral cortege stretched over a mile, containing units of gymnast clubs, Civil War veterans, civic leaders, military bands and thousands of German-Americans. He is buried in Woodlawn Cemetery.

FREEMAN STREET. Dr. Norman K. Freeman lived in a mansion facing present-day Southern Boulevard near Freeman Street. His estate is mapped on early records of Westchester County (now Bronx County-West Farms sector). Dr. Freeman taught at St. John's College (now Fordham University) from 1845 to 1850. One of his ancestors was a Revolutionary War soldier who escaped from the notorious prison ships.

FRENCH CHARLIE. Located at approximately East 203rd Street inside the Botanical Garden of Bronx Park, this name harks back to a French restaurant and picnic grove that was popular in the 1890s. "French Charley" [spelled this way] Mangin was the proprietor in those days.

FRISBY AVENUE. Once it was 2nd Street in the village of Westchester. On another map, 1852, the area behind Westchester Square was marked "Frisby's Land." The street was opened in 1912, and supposedly honors the publisher of *The Westchester Times* in the 1860s who spelled his name DuBois B. Frisbee. See Overing Street.

FRISCH FIELD. Frankie Frisch was a baseball player of national fame in the 1920s, and this field in Gully Park is a reminder of a local boy, for his nickname was "the Fordham Flash." See: Gully Park.

FTELEY AVENUE. It is mistakenly ascribed to a Fort Ely, or Ft. Ely, an imaginary Revolutionary fort on Clason Point. Streets in this neighborhood are named after Civil Engineers, and Alphonse Fteley was President of the American Society of Civil Engineers, and also Chief Engineer of the Aqueduct Commission from 1888 to 1900. He invented a device for geological surveys called Fteley meter. He was born in France, but died in New York in 1903, and is buried in Woodlawn Cemetery. The upper end of Fteley Avenue was on Mapes farmland, which later became the real estate development called "Park Versailles." See: Deyo Street.

FUFIDIO SQUARE. Originally, this was called Garrison Square after a Hunts Point family who lived in the vicinity. It was renamed in April 1953 to honor a resident of nearby Casanova Street who died on Iwo Jima in World War II. Corporal Walter Fufidio was born September 24, 1924, and died, March 23, 1945.

FULLER STREET. Edward S. Haight was an owner of considerable property that extended from the Westchester Turnpike to Bear Swamp Road (now Bronxdale Avenue). Hiram L. Fuller purchased a small tract from Edward Haight in June, 1875.

FULTON AVENUE. Fulton Avenue was a pleasant tree-lined street of early Morrisania, that is, prior to the Civil War. Parades were held along its length, and the various civic organizations were located there, i.e., schools, churches and headquarters of the volunteer firemen. The Lady Washington Fire House (where the present-day FDNY house is located) owned a famous pumper called "the White Ghost" which is now housed in the museum of the New York Historical Society. The avenue honors Robert Fulton, artist, engineer and inventor. See: White Ghost, Lady Washington Hose Company No. 1.

FULTON PARK. These 0.94 acres, are located at the west side of Fulton Avenue at East 169th Street. See: Fulton Avenue.

FURMAN AVENUE. It was first known as Garden Street when laid out, and evidently referred to Furman's garden. George Furman sold his small estate, off White Plains Road, to the Yonkers Saving Bank in 1902.

GALE PLACE. Henry D. Gale purchased 190 acres in 1835 in what was called South Yonkers. In 1853, Loring Gale was listed as a small landholder in the same area. The Place was opened in 1937.

GALE WALK. This footpath on the campus of the Bronx Community College (formerly New York University) honors Leonardo D. Gale, Professor of Geology and Chemistry, whose experiments helped in the invention of the telegraph. The walk extends across the former Schwab property, which had been the Revolutionary battlements of Fort No. 8, and the ancestral property of the Archer family, pioneers of Fordham.

GALLAGHER NATURE TRAIL. See: Cass Gallagher Nature Trail.

GARDEN PLACE. This short Place at East 240th Street and White Plains Road was part of the DeMilt lands of the 19th century. It must be assumed there was a garden there before the street was cut through.

GARDEN STREET. This was originally an Indian trail. Although the street was on maps of West Farms Township as early as 1874, it was officially opened in 1903. Mr. Garden's home was situated on the western end of the street. Hugh Richards Garden was a lawyer from South Carolina who moved to lower Westchester County (now The Bronx) just prior to the Civil War. He was appointed Chairman to the Special Reception Committee by Mayor Fernando Wood, who was a Southern sympathizer.

GARFIELD STREET. Garfield Street in Van Nest was once crossed by Downing's Brook at its intersection with Morris Park Avenue. It was opened in December 1918 in honor of President James Garfield who was born in an Ohio log cabin in 1831. He was a Civil War Major-General, and was assassinated in 1881 while serving as President.

GARRETT PLACE. This was once the limit of John Eden's property called "Edenwald." Later, it became the Halsey estate from 1868 to 1897, when Louis B. Halsey sold the tract to Thomas and Annie Garrett.

GARRISON AVENUE. The Garrison family, headed by Commodore John H. Garrison purchased the land from the Fox Estate in the late 1820s. He later became Sheriff (1856) of West Farms, which included Hunts Point. When the avenue was laid out in the 1880s, it was known as Mohawk Avenue, but received its present name in 1896. There is no truth in the story the name memorializes a Civil War garrison.

GATES PLACE. It was laid out in 1903 across the nineteenth-century holdings of the Dickinson family. Where the Place meets Mosholu Parkway, there was once a spring, the origin of Schuil (School) Brook that flowed eastward to Webster Avenue. This formed the headwaters of the Mill Brook that flowed south down to the Bronx Kill. The Place honors General Horatio Gates, and other nearby avenues honor other Revolutionary War heroes.

GEHRIG (Lou) PLAZA. See: Lou Gehrig Plaza.

GENTILE SQUARE. David W. Gentile was killed in action in World War 1, and the intersection of East 142nd Street, College and Morris Avenues was given his name in 1927. He was a resident of that neighborhood.

GEORGE FARKAS SQUARE. See: Farkas (George) Square.

GEORGE MEADE PLAZA. See: Meade (George) Plaza.

GEORGE MEANY SQUARE. See: Meany (George) Square.

GEORGE MIELE PARK. This is a sitting area at Bruckner Boulevard and Buhre Avenue, named for a retired Parks Department labor foreman.

GEORGE STREET. The land was originally Siwanoy Indian territory, which passed to Dutch possession to be called "Vredelandt." In 1860, it was on the property of Dr. Brainard T. Harrington, whose holdings extended to Cornell Place, Coddington Avenue and East Tremont Avenue. George and William Jorgenson bought plots in the 1920s and built homes on them. See: William Place.

GERANIUM PLACE. A short thoroughfare in Silver Beach given the seventh letter of the alphabet. Geraniums were and are commonplace on Throggs Neck. See: Acorn Place.

GERARD AVENUE. Gerard Walton Morris, lineal descendant of the early owners of Morrisania Manorlands had an estate, "Cedar Grove," west of the present-day Grand Concourse (1869 map) to the Harlem River. Cedar Park was eventually laid out by the city surveyors, to be renamed Franz Sigel Park in 1902.

GERARD STREET. This street still exists on some maps, but has been swallowed up by street developments until only a bare 4 yards off East 149th Street remain. It is thought to be a reminder of yet another Gerard Morris.

GERBER PLACE. This short street cuts across the seventeenth-century lands of Thomas Baxter, whose house was a few streets northward between Sampson Avenue and the Throggs Neck Expressway-according to Mr. Ferriera whose specialty was Throggs Neck. Records of 1840 show this Place on the Ludlam farm, and in 1880 it formed part of the Frederic W. Jackson estate. The eastern end of Gerber Place was an apple orchard until the 1920s. John G. Gerber owned property in Unionport (1888 deed) and Williamsbridge (1889 deed) and in this area in 1913.

GERTLAND PLACE. About 20 yards of this Place lies in The Bronx above Pelham Bay Park. See: Edgemere Street. No Gertlands appear in Bronx records, so it must be assumed the Place is a reminder of some official in Pelham Manor, possibly the surveyor.

GIEGERICH PLACE. This area was charted as an island during the Revolutionary War and on subsequent maps of 1874. In the 19th century, it was known as Wright's Island. Streets were surveyed on Locust Point in 1914, but only laid out in 1941. These streets were given the names of Justices Glennon, Hatting, Mullan, Tierney and Giegerich. Leonard Anthony Giegerich was born in Bavaria, Germany in 1855 and was brought to New York City as a small boy. During the latter part of his life, he lived in the Riverdale section of The Bronx (West 252nd Street) and he officiated at the formal opening of Henry Hudson Park.

GIFFORD AVENUE. This was the swampy end of the Charlton Ferris farm in the days before the Civil War. Judge Silas D. Gifford of the Supreme Court (under the State Constitution of 1846) is thought to be honored.

GILBERT PLACE. W.W. Gilbert was a landowner in the 1870s, and his daughter Lucy married Colonel Richard Hoe. His estate, "Sunnyslope," was bounded by the Hunt's Point Road, the Bronx River, Lafayette and Seneca Avenues. Gilbert Place was opened in 1912.

GILDERSLEEVE AVENUE. It is on historic Cornell's Neck, which later was called Willett's Point at the time it was mapped by Major John Andre during the Revolution. After 1800, it was named Clason Point. A Siwanoy Indian trail cut across Gildersleeve Avenue at Beach Avenue. Henry Alger Gildersleeve was a captain in the Civil War and took part in the battle of Gettysburg, and also in Sherman's March to the Sea. He was a Judge in the Supreme Court in 1897, and is buried now in Woodlawn Cemetery.

GILES PLACE. Though named Giles Street on an 1886 map, the Place itself is part of the original driveway leading to the Giles mansion that was built upon the site of Fort Independence of Revolutionary War fame. In the 1860s, William Ogden Giles, a vestryman of St. James. P.E. Church in Fordham purchased the land and built a mansion on the foundations of this fort. From this vantage point he enjoyed a sweeping view of the upper Bronx, and his mansion was a showplace and landmark for many years. The historic grounds held cannonballs, rum bottles, military buttons and rusted bayonets which turned up in the ensuing decades. As recently as 1957-1958, when the Giles mansion was demolished to make way for modern housing, several Bronxites unearthed additional relics of the Revolutionary War. See: Fort Independence.

GILLESPIE AVENUE. Old records indicate this short street was on farmlands belonging to the Society of Friends (1840) and later to a farmer named Koch (1870). A spring close to present-day Zulette Avenue and Gillespie Avenue was the source of Middle Brook that wound southward to become Weir Creek at Kearney and Layton Avenues. When the area was being surveyed for city streets (1910-1917) Richard H. Gillespie was the Chief Engineer.

GILLESPIE SQUARE. Located at East 166th Street and Webster Avenue, this square was named in 1923 in honor of a local resident. Eugene F. Gillespie, born in 1898, was killed in action on September 30, 1918 when his ship, the USN Transport *Ticonderoga* was torpedoed in 'mid-ocean. He was one of several sailors who were machine-gunned by German submariners as they tried to escape by lifeboats. His Requiem Mass was celebrated in the Immaculate Conception R.C. Church on Melrose Avenue and East 150th Street.

GIVAN AVENUE. It was once called Kingston Avenue. Robert Givan and wife Agnes (Thompson) came to America in 1795 from Kelso, Scotland, to establish a mill in what was then Westchester County. Their seven children were all born in Scotland and in later life married into the Palmer, LeRoy and Morgan families. As there was no Presbyterian church as long as Squire Givan lived in Westchester, the family drove to the Harlem River in pleasant weather, crossed in an open boat and attended the Scottish Church in Harlem. Later, they attended the Dutch Reformed Church in West Farms. After Mr. Givan's death in 1830, the younger generation attended St. Paul's in Eastchester and St. Peter's P.E. Church in Westchester. The land was chosen for its elevation and navigable creek, and remained in the family almost 100 years. In laying out the Givan tract, preparatory to a sale, the family names were given: Palmer Avenue, Morgan Avenue and Givan Square. The estate had been called "Ednam" after Mr. Givan's home county in Scotland, and it lay east of Eastchester Road from Boston Road to approximately Mace Avenue.

GIVAN SQUARE. It was named for Robert Givan whose estate lay directly east. "Givan's Drive," a circular roadway, is shown on 1868-1872 maps starting at this Square. It led to a lavish mansion that overlooked the Hutchinson River and its branch, called Givan's Creek (now under Co-Op City). His mill was reached by the Old Saw Mill Lane from Williamsbridge Road.

GLADSTONE SQUARE. At Westchester and Hoe Avenues, West Farms Road, the west side of this square was once an Indian trail. The property was acquired in 1896 as a public place and given the name Fox Square. In 1937, the name was changed to honor World War I soldier and New York State Assemblyman, Benjamin Gladstone, 1897-1935.

GLEASON AVENUE. This was formerly 12th Street in Unionport. Joseph J. Gleason owned the tract of land east of Westchester Avenue from Beach to approximately Olmstead Avenues (1880-1900) which had been the Cobb farm as mapped in Bromley's *Atlas* of 1897. The genial story has it that Gleason was a gambler who wisely invested his winnings in real estate.

GLEBE AVENUE. Reverend John Bartow reached Westchester Village in 1702 on his way to Rye, but was persuaded by Colonel Heathcote to remain as pastor of St. Peter's Protestant Episcopal Church at Westchester Square. Reverend Bartow was given a grant of 20 acres of glebe (church land) and 3 acres for his personal use. In 1848, to raise money for a Vestry, the church auctioned off a part of the Glebe farm. Three churchmen, Wilkins, Newbold and Van Cortlandt, purchased the lots and turned them back to the church. The three acres reconveyed have since been known as the Glebe Avenue Lot, touching Manning Street and Starling Avenue. Where Glebe Avenue angles sharply at Glover Street once flowed Indian Brook/Seabury Creek on its way to join Westchester Creek. In 1868, the avenue was the boundary line of E. Height's estate, "Hopedale." The avenue was opened in 1912.

GLENNON PLACE. The streets in this area were named around 1920, honoring Supreme Court Justices, one of whom was Edward J. Glennon. Locust Point was originally an island with earlier names of Horse Point, the Point and Locust Island. Three generations of the Wright family lived there from the 1830s onward, when it was known as Wright's Island. Captain George Wright's home was situated near Glennon Place and East 177th Street. See: Giegerich Place.

GLENZ ESTATES. This is a small tract at White Plains Road and Gildersleeve Avenue containing eight bungalows and two alleys. It is noted on several Postal Zone directories in the 1960s, and was the property of George and Kathe Glenz in 1912.

GLOVER STREET. Originally part of the church property of St. Peter's Protestant Episcopal Church, in 1868, it was part of E. Height's property called "Hopedale." Historic old Indian Brook crossed this street at its junction with Glebe Avenue. Glover Street was opened in 1912, honoring Charles Glover, schoolmaster of the town of Westchester, 1713-1719. "Paid by the Propagation Society of London, he was recommended as a person sober and diligent, well affected to the Church of England, and competently skilled in reading, writing, arithmetic, psalmody and the Latin tongue."

GLOVER'S ROCK. A large granite boulder that evidently served as an Indian lookout, as it stood alongside a Siwanoy Indian trail in what is now Pelham Bay Park. The colonists called the subsequent lane that ran past the rock City Island Road. As near as can be determined, Colonel John Glover and his Marblehead regiment first established contact with the advancing British and Hessian troops on October 18, 1776, near this rock. In 1901, a tablet was affixed to the rock by the Daughters of the American Revolution commemorating the battle, but it was stolen by vandals in the late 1930s. One of the first official acts of The Bronx County Historical Society was to re-dedicate the rock with a slightly larger bronze tablet on November 11, 1960. See: Split Rock Road.

GOBLE PLACE. George S. Goble and Austin Chandler were proprietors of an ice pond and ice house, listed in 1868. Cromwell's Creek (now Jerome Avenue) was led into a pond on Goble's land at this point. An 1871-72 Directory lists George S. Goebel [spelled this way] living on nearby Anderson Avenue.

GODWIN TERRACE. It was first surveyed as Tackamack Place, but when it was opened in 1922, it was named after a former proprietor, Joseph H. Godwin who owned a large tract of Kingsbridge, as well as Godwin's Island in the Spuyten Duyvil Creek. Godwin's mansion was situated at West 230th Street, but was demolished in 1920. The site had been occupied by Macomb's mansion (1800s), Cox's Tavern (1700s), and Verveelen's inn (1600s).

GOLDEN ROD AVENUE. This is a roadway located in the northwestern section of Woodlawn Cemetery. This plant is still prominent in the cemetery.

GOOD PLACE. This was originally part of the Pell land grant of 1654, which, in turn, was deeded to the ten families who "sett down in Eastchester." An 1866 deed lists Charles H. Ropes as owner of this tract, and a 1900 map shows it as part of the Peter Prove estate. The street was named for Thomas B. Good, real estate broker, who arranged the sale of an incinerator tract to the Village of Pelham Manor. The Place was opened in 1935.

GOODRIDGE AVENUE. This high lying land was part of "Upper Cortlandt," or the Cortlandt Ridge of Revolutionary days. The avenue was formerly called Alamo Avenue, but the name was changed in 1916 to honor a former resident and Arctic explorer. Dr. Frederick Grosvenor Goodridge purchased the property in 1873 from the Delafield family. He accompanied the Peary (Admiral Robert E. Peary) expedition in 1897. See: Peary Gate.

GOOSE ISLAND. The first reference to this small island was in 1679 in the records of Eastchester and evidently it was given this name by the colonists who noted the flocks of geese there. Canoeists and fishermen on the Hutchinson River used this islet since the earliest times. In the 1920s, to protect boaters and swimmers, the Goose Island Division, a standard frame building belonging to the Volunteer Life Saving Service of New York, was manned by four officers and eight surfmen. Surprisingly, the City only acquired title to the island in 1935.

GOULDEN AVENUE. Congressman Joseph A. Goulden lived on Creston Avenue. He was a Committeeman on the North Side Board of Trade in 1913, a trustee of the Soldiers' and Sailors' Memorial Monument, head of the Board of Taxes and Assessments, etc. The avenue was named in honor of his civic zeal and interest in public education. Fittingly enough, the avenue runs through an educational complex.

GOUVERNEUR AVENUE. This street is laid out over the original Van Cortlandt holdings, and members of the Gouverneur family had married with members of the Van Cortlandt family. See: Gale Place.

GOUVERNEUR MORRIS SQUARE. At Bruckner Boulevard and East 138th Street, this Square memorializes one of the earliest landowners of the South Bronx.

GOUVERNEUR PLACE. This is a short street running from Park Avenue to Washington Avenue across lands once owned by the Morris family. Sarah Gouverneur was the mother of Gouverneur Morris. In the late 1800s, the population of this Place was almost entirely German-born, and many men were employed in the nearby breweries.

GRACE AVENUE. This avenue was once the boundary line of the Bathgate estate (1900 map). From Boston Road southward, Grace Avenue runs across part of the Palmer estate of Civil War days. The Palmers were descendants of Robert Givan, who bought the extensive tract in 1794 from the Bartow family. It remained in the Givan family's possession for about a century, and in or around 1887, Elizabeth Palmer LeRoy, a granddaughter, sold the property. It was to be a real estate development called Pelham Bayview Park, and all the streets carried names of the descendants. Grace Avenue was first called Jones Avenue after a fourth generation descendant. William Russell Grace was an Irish immigrant boy who went to South America and prospered in Peru and Chile. He was Mayor of New York in the 1880s, and founded the shipping line that bore his name. He is remembered best in politics for his battle against Tammany.

GRAFF AVENUE. From the East River to Ferry Point Park, this was once the Livingston Grant of the pre-Revolutionary days. In the 1850s, it was part of the Francis Morris estate, which eventually passed to his grandson. Alfred H. Morris sold the property in the late 1920s. The northern end of Graff Avenue was not in existence until the 1940s, when landfill was used to support Throggs Neck Houses and form Ferry Point Park. The Morris mansion, in Greek Revival style, owed its design to a noted architect of the mid-1800s by the name of Frederick Graff. Graff Avenue points directly to the site of the mansion on Schurz Avenue.

GRAHAM SQUARE. It forms part of the Long Neck farm of 1847 belonging to Colonel Lewis Morris. Later it became the business and social center of Old Morrisania. Located at Third Avenue and East 138th Street, it was named in 1919 to honor a John B. Graham, who died in World War I. His is one of the approximately 100 names on a monument at that spot. The Square was named December 23rd, 1919. By happenstance, Judge James Graham was Attorney-General of the Province in 1740, and his daughter married into the Morris family, on whose manorlands the Square is located.

GRAHAM STREET. A short street, barely 50 feet long, leading north from Morris Park Avenue, west of Bronxdale Avenue. Van Nest old-timers vaguely recall a family of that name.

GRAND AVENUE. It takes its name from the Grand Boulevard and Concourse, which it parallels. It once formed the eastern boundary of J.D. Poole's estate, "Rose Hill" (1868), in the vicinity of West 174th Street. It was known as Edenwood Avenue from Fordham Road to Kingsbridge Road and was 6th Avenue in Bedford Park.

GRAND BOULEVARD AND CONCOURSE. The idea of this boulevard originated in
1890 with Louis Aloys Risse, an engineer under Louis J. Heintz, Commissioner of
Street Improvements. Ground was broken in 1902 by Borough President Louis Haffen,
and it was officially completed in 1909. It ran from East 161st Street up to Mosholu
Parkway, but, later on, a lower section was added down to East 138th Street (which had
been called Mott Avenue). Eventually the entire stretch became the Grand Boulevard
and Concourse in 1927. In 1926, the Gold Star Mothers put on a determined drive to
have the boulevard called Memorial Parkway. It did not succeed. A year later, the
Claremont Heights Civic Association tried to have the Concourse renamed Woodrow
Wilson Boulevard. For many years, trees on this thoroughfare carried metal nameplates
of Bronxites killed in World War I, in effect, a memorial parkway. When the
Independent subway was constructed under the Concourse, the trees were removed to
Pelham Bay Park.

GRAND CONCOURSE. See: Grand Boulevard and Concourse.

GRAND VIEW PLACE. This short street is atop a broad plateau and, before the area was
built up, afforded a grand view of Cromwell's Creek (now Jerome Avenue) and the
wooded slopes of Highbridgeville. Eastward there was a slope to the Mill Brook valley
(Webster Avenue), so the Place was well named. When the City took title to the land
in 1896, the street was cut through and named Morris Avenue.

GRANT AVENUE. The land across which this avenue runs was once the tract of General
Staats Morris in the 1700s. The meadow was used in those days as a race course for
the landed gentry. In 1870, wealthy man named Dater leased it as the Fleetwood Race
Course, and it lasted until 1898. The 11/4 mile track crossed Grant Avenue at East
164th Street. The avenue honors Ulysses S. Grant, President, 1882-1885.

GRANT (Edward L.) HIGHWAY. In 1887, this street was laid out from Undercliff
Avenue and the site of Washington Bridge to Sedgwick Avenue and was called Boscob
el Avenue. In 1893, it was extended from the bridge to Jerome Avenue. By Local Law
25 in 1945, it was renamed Edward L. Grant Highway for a member of the New York
Giants baseball team who had died in World War I.

GRANT (Hugh J.) CIRCLE. It is situated on what had been the Leggett lands of the
1700s, and on the Pugsley tract of the early 1800s, which became the Cobb and Larkin
farms later on. It was named in 1911 for a former New York Mayor who died in 1910.
Hugh Grant had a poor reputation as Sheriff, and his elevation to the higher office did
nothing to change his ways. When he took office in 1889, he was only 31, the youngest
Mayor ever. He is remembered best for his opposition to acquiring Pelham Bay Park
and Van Cortlandt Park, saying they were too inaccessible to the masses.

GRAUER FIELD. Conrad "Cooney" Grauer's long years of volunteer work with the
Kingsbridge Little League was recognized in July 1989 when the ballfield at West
233rd Street and Bailey Avenue was formally named in his honor. He was 85 years old
at the time.

GREEN AVENUE. A short street on what was the Colonial farm of the Seabury family.
Later, it was part of the Cebrie Park realty venture. It is an extension of what once was
called Green Lane (now Zerega Avenue). No one is certain how it received its name.

It may be merely a descriptive name, referring to its former tree-lined aspect. It may hark back to a G.W. Green, who was a surveyor in that district during 1879 and 1880. Or, it might even honor Andrew Green, "Father of Greater New York," who conceived the idea of combining all the boroughs into one city-which became a reality in 1899. Green Avenue received its name in the 1890s when Andrew Green was most active.

GREEN FLATS. These are a low-lying cluster of rocks due east of City Island at Cross Street. They appear only at low tides. Harvey Hauptner, lifelong islander, could give no clues to the name except that "flats" is a nautical term for shallows.

GREENE PLACE. This was once part of the Coster farm on Throggs Neck. As both Revolutionary War heroes and Civil War generals are honored by streets in this area, Greene Place might be named for General Nathaniel Greene of the Revolutionary War, or his direct descendant, Brevet Major-General George S. Greene, who commanded the New York troops at Gettysburg.

GRENADA PLACE. This was originally Nelson Place in Edenwald, but on May 2nd, 1916, the name was changed by Act of the Board of Aldermen to avoid confusion with Nelson Avenue. No reason was given for naming it Grenada Place.

GREYSTONE AVENUE. This Riverdale avenue is named after the William E. Dodge "Greyston" estate from which it is carved. At one time, Samuel J. Tilden, Democratic Governor of New York State (1874), lived at "Greyston."

GRIFFIN PLACE. Anthony J. Griffin Place received its name in 1936, displacing the former name of Spencer Place. At one time, Mr. Griffin was Collector of Internal Revenue, and also was instrumental in obtaining the Mott Haven post office for the borough. He was a State Senator in 1914. The Place was part of the 872 acre Manorlands of Lewis Morris (1746) which remained in the hands of his descendants until 1950. Lewis Spencer Morris, who died in 1944, was the last owner.

GRINNELL PLACE. This once formed part of the western boundary of Richardson and Jessup's West Farms Grant of 1663. A swampy section, it contained several creeks which went by varying names: Wigwam Brook, Indian Brook, Bound Brook, Leggett's Creek and Bungay Brook. This caused much contention between Richardson and Jessup on one side and the Morris family on the other as they could not agree which creek was the dividing line between their properties. This was the famed "Debatable Lands." Leggett's Creek once ran diagonally across the western end of Grinnell Place. In the 19th century, it was part of the extensive lands of G.S. Fox, which he called "Springhurst," later becoming the property of E.T. Young. Moses Grinnell was a wealthy New York shipping magnate, important in politics of pre-Civil War days, and one-time landowner in what is today The Bronx.

GRISWOLD AVENUE. This was originally a salty meadow on the Lorillard Spencer estate which, when flooded in the Wintertime, was used for ice-skating. On old maps it is designated as Palmer's Cove, named after a pre-Revolutionary owner of that land. The name most likely honors Griswold Lorillard, an uncle of the landed proprietor. Griswold was a son of Pierre Lorillard, who owned a larger tract next to the Pelham bridge. His mother was a Griswold of a noted 19th century shipping firm. Her father was one of three men who financed John Ericsson to build the ironclad, Monitor, during the Civil War.

GROSVENOR AVENUE. Situated near the highest point in The Bronx - 284 feet, 5 inches above sea level-this area was once part of the extensive Van Cortlandt holdings, known as "Upper Cortlandt." After the Revolutionary War, it was owned by the Delafield family, who sold it in 1873 to Dr. Frederick Grosvenor Goodridge.

GROTE STREET. Frederick Grote was a musician, composer, unpaid official of the Mayor's reception committees and a resident of Belmont. His home, between Belmont and Cambreleng Avenues, faced East 182nd Street (former Kingsbridge Road) according to maps of the 1860s and 1870s. He was also a trustee of the Board of Education of West Farms.

GUERLAIN STREET. This street was once part of the extensive Underhill lands of the 1600s. Around Revolutionary times, a Lewis Guerlain purchased 174 acres from Nathaniel Underhill, and this tract included a part of Parkchester. He built an attractive chateau for his wife, Miriam, who died shortly after in the West Indies. He sold out to Richard Fowler in 1805. After the Civil War, this was known as the Leonard Mapes farm and was subdivided and auctioned off by John Mapes as a real estate venture called "Park Versailles."

GUION PLACE. This was once laid out as Guion Street in the development "Park Versailles," once the Mapes farm. The name is frequently encountered in tales of Colonial and Revolutionary events in the area. The Guions were originally French Huguenot settlers from New Rochelle.

GULLY PARK. This park is located at Webster Avenue and East 201st Street. The gully referred to is the bed of a former brook. See: Frisch Field; Mosholu Parkway South.

GUN HILL ROAD. In Indian days, the Bronx River crossing at this road was called Cowangongh (the boundary beyond). The Indian trail was widened to accommodate carts and drags of the Colonial times, and was called the Kingsbridge Road as it led westward to the settlement of that name. During the Revolutionary War, this road was an important artery, fought over by both sides. On January 25, 1777, a small party of Americans dragged a cannon to the top of the hill west of the Bronx River and fired on a British force. This hill is now inside Woodlawn Cemetery (Summit sector) and was called Gun Hill long afterwards. Around 1875, the road received its present name, but did not extend much beyond White Plains Road on the east. By segments, it was advanced to the Hutchinson River by 1938.

GUNTHER AVENUE. Adhering to the policy of naming streets in this area after Mayors, this avenue honors Charles G. Gunther (1864-1866), who is chiefly remembered for his determined effort to build the first steam railroad to Coney Island, despite much opposition from the Long Island farmers. The avenue itself is laid out through what was once the Givan homestead east of Eastchester Road from Boston Road to Mace Avenue and running alongside the Black Dog Brook which was the dividing line between the ancient townships of Eastchester and Westchester. At Boston Road, this was once called Walnut Street. Above East 233rd Street, Gunther Avenue was first mapped as Fox Street to its junction with Bussing Avenue. In 1905, it was on the Bathgate farm and beyond that, on the 11-acre property of James Russell.

HACKETT PARK. At Riverdale Avenue, West 253rd to West 254th Streets, it was dedicated on May 30, 1956 to Frank S. Hackett, founder of the Riverdale schools.

HAFFEN PARK. It was named in 1961 at the suggestion of this writer, to honor Louis Haffen who was born in Melrose in 1854. He attended Immaculate Conception R.C. School, St. John's College (now Fordham University), Niagara College and Columbia University. He was a City surveyor, later Superintendent of Parks, first Borough President and served from 1898 to 1910. He helped plan the Grand Concourse with Louis A. Risse and Louis Heintz, and was instrumental in securing considerable parkland for the Bronx. He died on Christmas Day, 1935.

HAIGHT AVENUE. In the 1800s this land was conveyed from Eliza Macomb to John A. Morris, which was made into a racing track. Between Morris Park Avenue and the railroad was once the track that, when abandoned in 1904, became a field used for early airplanes and balloons. Stephen Haight was a landowner of three separate tracts - this one in Van Nest, one in Middletown and another below Westchester Square.

HALL OF FAME TERRACE. It is laid out on the ancient manorlands of the Archer family, who were the landowners of Fordham in Colonial times. It was surveyed in 1886 and mapped as University Avenue, as it was next to New York University, but, in 1913, it took its present name from the nearby Hall of Fame.

HALL PLACE. Judge Ernest Hall was born in London, but, as an American citizen, fought at Gettysburg in the Civil War. His home was located at 1089 Boston Road at the southwestern corner of East 166th Street.

HALLECK STREET. Proceeding from the East River, it runs through the former estates of Francis Barrette's, "Blythe," Paul Spofford's, "Elmwood," and terminated at the northern border of C.D. Dickey's, "Greenbank" (the present Lafayette Avenue). Fitz-Greene Halleck was a close friend of Joseph Rodman Drake, and the avenues honoring them run side-by-side. Halleck often visited Drake's grave and once remarked that he wished to be buried beside his friend. Halleck died in 1867 and was buried in Guilford, Connecticut. Around 1903, plans were entertained to exhume Drake's body and transport it to Guilford and so re-unite the old friends. However, protests from Bronx literary societies and the North Side Board of Trade prevailed.

HALPERIN AVENUE. It was opened and named on March 17, 1936, cutting across what had been Arnow property at the junction of Eastchester and Williamsbridge Roads. No one knows who Halperin is or was.

HALPIN (Monsignor John) PLACE. In July 1983 the pastor of St. Frances de Chantal R.C. Church was honored by having Silver Beach Place renamed. The occasion marked his 60th year in the priesthood. The western slope yielded Indian artifacts when George Younkheere, historian, conducted "digs" there in the 1950s. The eastern slope once was occupied by a Summer tent colony called Harmony Lane, in the 1920s.

HALSEY STREET. This small street is a reminder of a family that was well-known in Westchester village for generations. An 1842 deed mentions a Stephen Halsey, and two others dating back to 1873 and 1874 credit Benjamin Halsey with some land in Unionport. The street itself is laid across the pre-Revolutionary holdings of the Seabury family. Seabury Creek once flowed past the western end of Halsey Street. A 1905 map shows this area as "Cebrie Park." The City vested title on February 14, 1941.

HAMMERSLEY AVENUE. J. Hooker Hamersly [spelled this way] owned many acres of woodland according to an 1868 atlas. Bounds were Williamsbridge Road on the west to Fish Avenue on the east - Boston Road on the north to Saw Mill Lane (Allerton Avenue) on the south. This was the Revolutionary lands of the Honeywell Watson family.

HAMMOND'S COVE. In 1696, it was known as Scuttle Duck Harbor. It gained its present name when Abijah Hammond purchased much property on Throggs Neck around 1805, including present-day Fort Schuyler, Silver Beach and Pennyfield. Abijah Hammond owned a large part of Greenwich Village, was a pallbearer for Alexander Hamilton and a Colonel in a Boston contingent during the Revolutionary War. On January 22nd 1791, he married Catherine L. Ogden, and built an impressive mansion for her on Throggs Neck. Two years after her death in 1816, Abijah Hammond married a Margaret Aspinwall, aged 44, and the daughter of a well-known shipping magnate. Colonel Hammond died on December 30th 1832 at his country seat. His son, Ogden Hammond, purchased additional Pennyfield estates in 1840. Localisms for this cove were Hammond Creek and Hammond's Flats (shoals).

HAMPDEN PLACE. This Place is laid out through the former Cammann estate. The City gained title to the street in 1892. G.W. Hampden was listed as a partner in the real estate developments of the East Fordham Syndicate. Its president was Logan Billingsley, whose name is on a nearby terrace.

HANUS STREET. Around the tip of Fort Schuyler runs this road that was named for Commander Hanus, Superintendent of the New York State Maritime College from 1902 to 1908.

HARDING AVENUE. The stretch from Hollywood Avenue east to Pennyfield Road constitutes the original Throggs Neck Road, as mentioned in ancient deeds. Later it was called the Fort Road when Fort Schuyler was built in the 1830s. From Hollywood Avenue west to Balcom Avenue, Harding Avenue runs through the former estates of Philip Livingston (1700s) and, in the 1800s, the lands of Captain Post, Mitchell, Ash, Havemeyer, Van Schaick and Collis P. Huntington. From Balcom Avenue to Ferry Point Park, it ran through the estate of Bayard Clark (1840s), Francis Morris and his grandson, Alfred Hennen Morris. His wife was Jessie Harding, whom he married in 1889. Her father owned *The Philadelphia Inquirer*. When the Morris estate was subdivided in 1922, Harding Avenue was extended through the tract, but this was only a coincidence. All the avenues on the Neck were given names of Spanish-American war heroes sometime around 1900, so that Harding Avenue is presumed to honor General Chester Harding, who later was appointed Governor of Panama.

HARDING PARK. Located near the Snakapins of Indian times, this was the seventeenth-century property of Thomas Cornell. After 1790, it became the lands of Dominick Lynch and his successors. This tract was the Stevens property of the late 1800s. At the turn of the century, the Higgs family maintained a beach there, "Higgs' Beach," while an adjoining picnic area and pavilion was called "Killians' Grove." In the 1920s, both were incorporated into the bungalow colony which, was named in honor of President Warren G. Harding, who died in the summer of 1923. The Volunteer Fire Company No. 3 was inaugurated that same year. In lieu of city street names, Harding Park has eleven numbered rows of houses not listed in the Postal Guide.

HARLEM RIVER TERRACE. The name would suggest a view of the Harlem River from this high point of West 191st Street and Bailey Avenue.

HARLEM SHIP CANAL. See: United States Ship Canal.

HARPER AVENUE. This was laid out as Bland Avenue around 1904, and where it crosses into Mount Vernon it became George Street after a former landowner. The Bronx stretch, in the 1860s, was part of the B.S. Halsey property, later to be considered part of John H. Eden's estate, "Edenwald." In the 1890s, it was part of the Robert George estate. Harper Avenue owes its name to the plan of naming streets in this area after former Mayors, and James Harper was a Mayor of New York in 1844-45. With his brother, he founded the printing firm of Harper Brothers.

HARRINGTON AVENUE. In Revolutionary times, this was the side of Bridge Hill up which the British and Hessian troops dragged cannons in October, 1776. Dr. Brainard T. Harrington was head of a finishing school for young men from 1849 to 1901. His lands extended back from the Throggs Neck Road (East Tremont Avenue) to about Cornell Place, and the square wooden school itself was midway between Harrington Avenue and Coddington Avenue, some 75 yards up the slope. His son, Thomas, sold the property in 1906 to Percy Dudley.

HARRIS PARK. At Goulden and Paul Avenue, West 205th Street, this park acquired its name in December, 1940. Reverend John Harris, O.S.A. of St. Nicholas of Tolentine R.C. Church, was on an outing to Bear Mountain with a group of children. He suffered a heart attack while swimming.

HARRISON AVENUE. A William Harrison was a small property owner - listed in 1888 - in the vicinity of present-day Sedgwick Avenue and West 184th Street.

HARROD AVENUE. Once part of the Mapes farm of Civil War days, this avenue overlays the former Hunt Ditch, a name that harks back to the 18th century Hunt family that owned most of the land extending eastward to present-day Parkchester. Harrod Avenue was laid out in 1927. One theory is that it was named after a Joseph Horridge, small dealer in real estate of the 1880s. Another is that the stretch of land was harrowed to mark the division between the Mapes farm and "Wilmount," the estate of William Watson. Most likely is that the street honors Bartholomew Harrod, an early settler of Westchester who figured in a lawsuit against John Archer in 1659. "Barthalomeu hurrode," it was sometimes spelled in the ancient records.

HARROD PLACE. See: Harrod Avenue.

HART ISLAND. It was once called Little Minneford to distinguish it from Great Minneford (now City Island). The earliest owners were the Pells; then it was acquired in 1774 by Oliver Delancey and mapped as "Spectacle Island" because of its resemblance to a pair of spectacles. It was mapped in 1777 as Hart Island and is believed to refer to its use as a game preserve (Hart is an earlier word for deer). A 1775 map has it "Heart Island." It was subsequently owned by the Rodman, Hunter and Haight families. During the Civil War, the island was used by the Federal Government to house troops. Potters' Field was established around 1865 when some soldiers, victims of an epidemic, were interred there. This so-called "Soldiers' Plot" remained until 1941, but then the bodies were removed to Cypress Hills National Cemetery. The City acquired the island in 1869 from Charles C. Haight, architect and designer, and the same year the first civilian burial was that of an orphan, Louisa Van Slyke. Since then, there have been over half a million burials. Around 1872, the island was used for the detention of "vicious boys" and since then has housed German war prisoners, minor criminals, vagrants and drug addicts.

HART STREET. On the extreme tip of Castle Hill Point, this might be the very spot where the hostile Siwanoys appeared when Adrian Block's *Onrust* sailed past. This was the Cromwell Grant of 1685 which passed into the hands of the Wilkins family (1770s) and to a son-in-law, John Screvin (1890s). As early as 1703, a Hart family was listed as inhabitants of New York. A Mary Elizabeth Hart lived on a nearby estate crossed by Seward Avenue. She married Lt. Governor Luther Bradish.

HASKIN STREET. John B. Haskin was an early representative in Congress from Westchester County (1778), which is now Bronx County. An 1801 deed credits a Benjamin Haskin with owning some land in this area, once the village of Schuylerville. Another John B. Haskin figured prominently in Bronx politics during the Civil War Draft Riots. He is buried in Woodlawn Cemetery next to Central Avenue. If pressed, the author would pick Benjamin as the man for whom it is named.

HASWELL STREET. It was once mapped as Meadow Lane, the meadows now being covered by the grounds of the State Mental Hospital. No known landowner of this name is found in records of Westchester village, although a Mrs. Haskell is listed in 1868. It might honor Gouverneur K. Haswell, a Civil War naval commander. Haswell Street leads off historic Eastchester Road, a former Indian trail.

HATTING PLACE. Peter A. Hatting was a Supreme Court Justice, contemporary with Justices Glennon, Giegerich and Tierney. This street touches the pre-Revolutionary road (Pennyfield Avenue) to the tip of Throggs Neck on the west, and the eastern end is on ancient Wright's Island which is now Locust Point. See: Giegerich Place.

HAVEMEYER AVENUE. Hermann Hoevemeyer helped organize a bakers' guild in Germany in 1644. Dietrich Wilhelm Hoevemeyer, born 1725, was a master baker and served in the Seven Years' War. Wilhelm, his son, came to America in 1799 and learned sugar refining. He died in 1851, aged 81. His grandsons became known as "the Sugar Kings." They simplified the name to Havemeyer. William Frederick Havemeyer was Mayor of New York three times. He built the mansion on Throggs Neck that became Collis P. Huntington's showplace. Theodore Augustus Havemeyer owned the Hammond mansion in Silver Beach. Henry O. Havemeyer lived in the former Brinsmade mansion off Lafayette Avenue. Samuel Lowerre's farm in Unionport was sold to the Odd Fellows who erected there a Home. Lowerre's Lane was renamed Avenue 9, but, around 1900, received the name of Havemeyer Avenue to honor the family who had been so generous in its contributions to the Home.

HAVILAND AVENUE. A pioneer family of Westchester, John Haviland, was listed as a vestryman of St. Peter's P.E. Church in 1702-03, and a Joseph Haviland was a trustee and freeholder of the Town of Westchester in 1703. Ebenezer Haviland was listed as a farmer of Cow Neck (vicinity of Grant Circle) in 1799. This was part of the Pugsley farm (1770-1854), which passed to Marcus Cobb and Francis Larkin in 1880. Pugsley Creek crossed Haviland Avenue at Virginia Avenue.

HAWKINS SQUARE. See: Reverend Edler G. Hawkins Square.

HAWKINS STREET. This was originally Orchard Street. John Hawkins was one of those who petitioned Pell in 1784 to settle on City Island. Elnathan Hawkins, sea captain of the Marmion, brought back the first flowering quince from the Orient in 1811. He is buried in Woodlawn Cemetery. The Hawkins Shipyard sheltered the America's cup defender Columbia, and Jim Hawkins was the last survivor of the City Island Volunteer Firemen.

HAWKSTONE STREET. Although opened in 1908, the name harks back to an earlier generation when a family of this name owned a small tract running westward and downhill to Cromwell's Creek (now Jerome Avenue). They were believed to have been in the foundry business.

HAWTHORN AVENUE. This road is located in about the center of Woodlawn Cemetery, and is named after the hawthorn tree.

HAWTHORNE DRIVE. A privately maintained street inside Parkchester. Despite its spelling, it is named after the hawthorn tree, which is found abundantly in The Bronx.

HAWTHORNE STREET. This short street off Eastchester Road touches, at its western end, an ancient Indian path. Part of the Pell Patent, it was sold by the third Lord Pell to Reverend John Bartow sometime before 1725, and the land had the peculiar name of "Scabby Indian." Robert Givan in 1795 became the landowner, and the tract extended northward to about Grace Avenue and Boston Road. This street was on the property of Israel Honeywell Watson in the 1850s, and he was a lifelong friend and neighbor of the Givan family. In Civil War times, it was P. Bruner's estate called "Rosedale," and around 1900, it was part of the Valentine 27-acre tract. According to records in the Borough President's office, Hawthorne Street was laid out on city maps in 1926, and the owner of the property requested it honor Nathaniel Hawthorne.

HAYES SQUARE. At East 167th Street, Bryant and Westchester Avenues, this plot was acquired on January 15, 1893 for park purposes. It is named in honor of a World War I soldier, in 1940. Local Law 15 does not give his first name.

HAYWARD (Agnes) PLAYGROUND. In October 1985 her Olinville community of friends and family memorialized her for her lifelong commitment.

HAZEL PLACE. This is a Silver Beach thoroughfare that was given the eighth letter of the alphabet. Hazel trees were found on the former estate. See: Acorn Place.

HEATH AVENUE. This hilly street was once called Darke Lane, or Darke Street, as a Charles Darke was a property owner there in the 1880s. Later, it was named in honor of Major General William Heath of the American Revolutionary forces.

HEATHCOTE AVENUE. This avenue runs across what is thought to be the site of Ane Hutchinson's farm. In 1662, it was part of the land granted to settlers by the Pell family. During Civil War times, this was inside an 11-acre estate of a Thomas Galway or Galwey. He built a substantial stone sea wall on the Hutchinson River at about the spot where the (new) Boston Road bridge crosses over. Caleb Heathcote was Mayor of New York from 1711 to 1714, and was prominent in Episcopal affairs, having appointed Reverend John Bartow as first minister to the village of Westchester in 1702. At different times, Heathcote was also a military commander, one of His Majesty's council for the Province, surveyor-general of customs for the Eastern District of North America, and a judge of the Court of Admiralty. When Heathcote was Mayor of New York, his elder brother was Lord Mayor of London.

HEATHER AVENUE. This is a quiet tree-lined road in the southeastern section of Woodlawn Cemetery. The name is purely descriptive, obviously referring to the heather once found on the farmlands.

HEGNEY PLACE. Originally this was a slope overlooking the Mill Brook, and was reputed to have been an Indian burial grounds. It was part of the Morris Manorlands from 1671 to 1848, when it was purchased by Benjamin Benson, who called the, area Bensonia. This Place was then a lane forming the west boundary of Bensonia Cemetery (later called the Morrisania Cemetery) and was first known as Balcom Avenue in 1858. Later, it gained the name of German Place. There was so much anti-German feeling during World War I that the name was abolished and renamed after Arthur Vincent Hegney, first Bronx soldier killed in action in France.

HELIOTROPE AVENUE. This road is located in the southeastern section of Woodlawn Cemetery, close to the site of Heath's Redoubt of Revolutionary War times. This fragrant flower is still in evidence in the cemetery.

HENDRICK (Tim) PLACE. See: Tim Hendrick Place.

HENNESSY PLACE. Ellen Hennessy was the last property owner of the land "overlooking the Harlem River" as an early map notes. It was some 20 years later (1905) that State Senator Joseph P. Hennessy was active, especially in the establishment of Fordham Hospital - so, though coincidental, it is obvious the Place was not named in his honor.

HENRY HUDSON BRIDGE. This bridge commemorates the visit of Henry Hudson in the *Half Moon* in 1609 to a point quite close to the present-day span over the Harlem River. The year of its opening is 1936.

HENRY HUDSON PARKWAY. This is the latest road in Riverdale, having been completed in 1938. It memorializes, of course, the English explorer who was in the employ of the Dutch when his ship, *Half Moon*, anchored off Spuyten Duyvil Creek. Close by is the Henry Hudson Memorial, a bronze statue erected on January 6, 1938 by the Parkway Authority. See: West 239th Street.

HENRY HUDSON PARKWAY EAST. See: Henry Hudson Parkway.

HENRY HUDSON PARKWAY WEST. See: Henry Hudson Parkway.

HENRY HUDSON RIVER. See: Hudson River [Current].

HENWOOD PLACE. It was originally called North Street and possibly was the northern end of some estate. Numerous surveys were made in 1863 by R. Henwood, principally in North New York, or Old Morrisania. An 1872 directory of Morrisania lists Richard Henwood. Henwood Place got its name in May, 1916.

HERING AVENUE. Hercules Herring (sometimes Hering) is mentioned in land grants dating back to 1814 and 1820. His lands, according to an 1809 map, were bounded by the Saw Mill Lane (Allerton Avenue) Stoney Brook (now overlaid by Sexton Place) and extended north to Gun Hill Road. Hering Avenue, below Pelham Parkway, runs through what was the Denton Pearsall estate, called "Woodmansten," of the 1850s. Around 1892, Messrs. Pinchot and Morrell laid out a 70-acre development called "Westchester Heights." H.H. Spindler was the surveyor. Hering Avenue was part of it, but was not intended as a thoroughfare - its course was depicted as a lawn.

HERKIMER PLACE. This street runs alongside the Revolutionary road to Mile Square, which was the scene of Simcoe's attack on the Stockbridge Indians. At approximately East 235th Street and Herkimer Place, the first encounter of the Indians and the British cavalrymen took place. Herkimer Place was opened in 1930 and named for the Revolutionary War general who won the battle of Oriskany (Fort Stanwix) in 1777. Some German textbooks call him Nicholaus Herchheimer.

HERMANY AVENUE. This street is laid out on the ancient Sheep Pasture of the village of Westchester, and later owned by Gouverneur Morris Wilkins. It was then incorporated into the town of Unionport. Anton Hermanny [spelled this way] was a landowner in the 1890s, and a William L. Hermanny was listed in 1913, shortly before the avenue was laid out.

HERON LANE. This private road in the Shorehaven complex on Clason Point covers what was formerly an extensive marina, operated by the Shorehaven Beach Club (1947 - 1986). Prior to that time, White's boat yard was located on this site. The swampy area once abounded in aquatic birds, among which were many herons.

HERSCHELL STREET. This was once called Washington Avenue when it was opened in 1914 in a real estate development called Cebrie Park. Herschell (no first name given) is supposed to have been one of the promoters of Cebrie Park. It was once part of the 18th- and nineteenth-century Seabury farm, which was partially bounded by Seabury (Cebrie) Creek, now covered by Zerega Avenue. In 1672, George S. Fox, founder of the Society of Friends, conducted his first meeting in America on the banks of this small stream - then called Indian Brook.

HEWITT PLACE. This was originally mapped as Whitman Street, but officially opened as Hewitt Place in 1902. It was once part of the original Leggett lands of Revolutionary times, which in the 19th century became country estates. This was located on the estate known as Longwood Park. There is a possibility this Place honors a former Mayor of New York, Abram S. Hewitt, who served in 1887-1888, although it is by no means a certainty. There are no other candidates for the honor, but it seems odd to have a Mayor's name so far from the other mayors.

HICKORY AVENUE (1). This was opened in 1903 as part of a real estate development called Bronxwood Park. Adjoining streets, named after trees, are Oak and Chestnut.

HICKORY AVENUE (2). This is a road that is located in the northwestern sector of Woodlawn Cemetery. Loring R. Gale (honored by two Bronx streets) is buried beside this driveway. The avenue got its name from the hickory trees.

HICKS STREET. This lies on the nineteenth-century Burke estate, the north portion of which was sold to Blodgett & Tilden, realtors, in 1889. Known as "Laconia Park," it was subdivided in the early 1900s and the streets were named after former Mayors of New York. Whitehead Hicks served as Mayor from 1766 to 1776.

HIGH BRIDGE. Projected in 1839, a functioning aqueduct in 1842, and a landmark since its official opening in 1848, the span at West 170th Street and University Avenue was originally known as "the Aqueduct Bridge." It was planned as a water level span, but the State Legislature ruled that the structure must be built so as not to interfere with river traffic. More than 180 feet above the water, it got the name High Bridge, and transferred its name to the area on The Bronx side, Highbridgeville.

HIGH ISLAND. This is an 8-acre island off Carey's Point of City Island which was included in Pell's territory of 1654. Its Indian name is unknown. In 1713, it was mapped as Lesser Minniford [spelled this way]. Early reference to it as High Island was when Captain John Wooley purchased it in 1762, after Benjamin Palmer had parcelled out all City Island and sold the lots to the inhabitants. Elisha King owned the island in 1829, and it remained in the King estate until 1872. During this period, rock from the island was used for the foundations of the Marshall mansion, which later became the famous Colonial Inn, on the mainland opposite City Island. Dwight Palmer owned the island in 1877. Later, a private organization, "The Midtown Club," owned the island until the early 1900s, then it was owned by the Miller family, who rented out bungalows. The Summertime population was 100 in 20 houses, while the Wintertime count was exactly two caretakers. Indian artifacts have been found on the shore. High Island was joined to City Island by a sandbar, but in 1928-29 a wooden footbridge was built. This sufficed until 1962, when the island was sold to a radio broadcasting company and the company widened the wooden span for automobiles. The Summer bungalows were then dismantled and burned. The island is comparatively high, being shaped like a gumdrop, but there is the possibility the original European name was Haai Eylandt, the Dutch for Shark Island. The island faces shallow Pelham Bay, once a favorite basking spot for sharks that prefer warm water. See: Colonial Inn, Marshall's Corner.

HIGHBRIDGE PARK. This narrow, steep park came into being sometime around 1888, and took its name from the High Bridge that overshadows it. See: High Bridge.

HIGHLAND AVENUE. This is a well named road in the eastern section of Woodlawn Cemetery, overlooking the Bronx River valley.

HILL AVENUE. The location of this street is on the side of a hill, but it is more likely that Governor David Bennett Hill is honored, as surrounding streets are named after Governors of New York State. Governor Hill (1885-1891) is known for signing legislation abolishing hanging in State prisons. Electrocution was substituted. This avenue was called South 14th Avenue when this area was part of South Mount Vernon.

HILLMAN AVENUE. Originally named Norman Avenue, it was laid out and partially opened in 1897 on the Dickinson estate below Van Cortlandt Park. Along with the adjoining avenues Saxon, and Harald - honoring Harald (or Harold) the Saxon king who fought the Normans, it was named Norman Avenue in 1900, and officially opened in 1937. Although it is on a hill, the current name was acquired in 1950, to honor the labor leader and organizer of union-sponsored homes, Sidney Hillman. See: Dickinson Ave.

HINES PARK. This small plot, Fulton to Franklin Avenues at East 167th Street, was ceded to the City for a park on November 8, 1864, and at first was called St. Augustine's Park. The Roman Catholic church was located, at that time, on Jefferson Place. It was renamed in 1929 for Colonel Frank Hines, 1868-1929, "A Sterling Citizen Soldier."

HOBART AVENUE. It was first called A Street in the village of Schuylerville, at the Baisley Avenue dead-end,1858. It was laid out across the former Koch farm and was surveyed through low lying swamplands. Also called Amsterdam Avenue. John Sloss Hobart (1738-1805) was a Revolutionary statesman whose name appears on Westchester deeds in 1786 in his capacity of Justice. On April 1, 1794, he purchased 80 acres on Throggs Neck for 1,000 pounds and lived there until his death. The Hobart farm was bought from Joshua Hunt and was located off present-day East Tremont Avenue running westward to Ferry Point Park.

HOE AVENUE. There were two branches of the Hoe family in the area of Southern Boulevard. Mr. P.A. Hoe owned an estate called "Sunnyslope" whereas Colonel Richard Morris Hoe lived on "Brightside" estate. The latter was the inventor of a printing press. The "Sunnyslope" mansion was demolished in 1909 to make way for apartment houses, but the "Brightside" mansion, in Gothic Revival style, survived on Faile Street at Lafayette Avenue for many years as a synagogue, Temple Beth Elohim, and now an African Methodist Episcopal church, Bright Temple. Members of the Hoe family are buried in vaults of historic St. Ann's Church of Old Morrisania. The name is the Saxon word for "Hill."

HOFFMAN STREET. William B. Hoffman was a vestryman of West Farms Episcopal church from 1840 to 1860. His "Cedar Grove" farm was mapped in 1888, south of present-day Fordham University. He was a member (1821) of the Westchester Agricultural Society. There is a local story, without basis, that an Arthur Hoffman was a city surveyor at the time Belmont was laid out, and this street and nearby Arthur Avenue were named in his honor.

HOG ISLAND. This northernmost bit of Bronx land was Siwanoy Indian territory that was sold to Thomas Pell and passed through various ownerships concurrent with Hunter Island directly south of it. It became New York Parks Department property in 1888, but so isolated was it that a beachcombers' colony flourished there into the 20th century. Mrs. Marion Laing remained on this rocky isle as a squatter, invoking Squatters' Right until her death in 1930. A tablet to her memory is on the island: "Marion H. Laing, died September 12, 1930. This Island was her paradise." This island and the smaller isle next to it are the "Pass Rocks" described as the boundaries in *The New Parks Beyond The Harlem* (1887) as lying above Hunter Island. The smaller island is 1/4 acre in area, and fishermen simply dub it "Little Hog Island." No origin to the name has been forthcoming. It is obviously not a place where hogs were penned as it is far too small.

HOLLAND AVENUE. Laid out through the development known as Van Nest, it was first called Lincoln Street. In Williamsbridge, from Adee Avenue to South Oak Drive, it was known variously as Post Avenue and Pine Street. In South Wakefield (Gun Hill Road to East 214th Street), it was mapped as Maple Avenue. Edward Holland was born in England, and was elected Mayor of New York in 1747. He died in office in 1757.

HOLLERS AVENUE. The Hollers were an old-time Eastchester family, in the ice business. Hollers' Pond was in existence until 1951. It was formed by damming Rattlesnake Brook, as it flowed from Seton Falls Park. Old Point Comfort Park, a popular picnic grounds at the turn of the century, is covered by Hollers Avenue. This amusement center later became Dickert's Park and Breinlinger's Park until around 1936, when it closed.

HOLLY LANE. A small Park Department road from Jerome Avenue to East 233rd Street into Van Cortlandt Park, leading to what used to be Holly Park. It is a floral name, of course, and referred to the bushes of holly there. See: Allen Shandler Recreation Area.

HOLLY PLACE. This street in Silver Beach was named sometime around 1924 when the former Havemeyer estate was converted to a Summertime bungalow colony. Holly once abounded there. See: Acorn Place.

HOLLYWOOD AVENUE. It was mapped as Grant Avenue until 1913, renamed Hollywood Avenue and opened in 1930. From Silver Beach to Harding Avenue, this street was an eighteenth-century rural road that led to a ferry slip on the East River, through the private lands of the Stephenson family (1790s), later Abijah Hammond (1810 to 1840s), then William Whitehead (Civil War times). The stretch from Harding to Lawton Avenues was part of the original Throggs Neck Road which was mentioned in the 1600s. This block-long stretch was also a consistent division line between estates from the Revolutionary times to the early 1900s. From Lawton to Miles Avenues once was the (1870) lands of Theodore Havemeyer. Beyond Miles Avenue it was on the Jackson farm, whose proprietor married a Havemeyer daughter. Above the Throgs Neck Bridge approach, Hollywood Avenue ran through the Wolfe estate (1840s), which later became the Bruce Brown estate. From Lafayette Avenue to Bruckner Boulevard, it was part of the orchards of the Coster farm. It can only be assumed the name refers to the holly shrubs, once common to the area.

HOLT PLACE. This Place bears the name of old-time settlers, for Henry A. Holt was a property owner in Fairmount (1840s), and there are Holt tombstones in the tiny cemetery of the First Presbyterian Church of Throggs Neck. This short Place on what was formerly Benjamin Berrian's land was named for Thomas and Ann Holt, who, in July 1872, purchased land "near the Fordham Depot" from the Berrian heirs.

HOME STREET. It was once called Lyons Street (1868) as it was on the James L. Lyons property. He was a Civil War veteran, and a charter member of the Bronx Old Timers. Home Street runs through Siwanoy Indian territory that eventually was the extreme eastern end of the Morrisania Manorlands - the boundary following Intervale Avenue. Stephen Wray in *The Village of West Farms* states the road led to the home of William H. Fox, hence its name.

HOMER AVENUE. This Castle Hill street was named by Solon L. Frank, a realtor who was active around 1900-1910. The avenue overlays the extreme edge of the grazing lands, or sheep pasture, belonging to the ancient town of Westchester. It was included in the sale of these commons to Isaac Wilkins. His son, Gouverneur Morris Wilkins, deeded the land to his son-in-law, John Screvin, around the turn of the century. Other streets in the area carry Greek or Latin names, thanks to Solon L. Frank. Homer was the Greek poet who wrote the *Iliad* and the *Odyssey*.

HONE AVENUE. In 1888, Eliza Macomb conveyed 158 acres to John A. Morris who then built the Morris Park Race Track. When the track went out of business, the land was subdivided and Hone Avenue cut through the field. All avenues in the neighborhood carry names of former Mayors of New York, and Philip Hone served in 1826-1827. His diary, which he kept for many years, is the best source of information of early New York days.

HONEYWELL AVENUE. It was once known as Orchard Avenue, and most likely was a reminder of the rural retreat once belonging to the wealthy Lorillard family. The Honeywells are an old family as their name occurs many times in pre-Revolutionary records. Sometimes the name was spelled "Hunnewell." A family plot is maintained alongside Poplar Avenue in Woodlawn Cemetery.

HORIZON COURT. This private street in the Shorehaven complex covers the site of Kane's Casino - a popular resort of the early 1900s that was famous for its political outings, clambakes, civic rallies, gymnastic exhibitions, dance marathons and brass band concerts.

HORIZON LANE. This Shorehaven street is not on part of the old-time Clason Point Amusement Park but was once the picnic grounds of Kane's Park. A houseboat colony was located along Pugsley Creek but landfill has obliterated any trace of it.

HORNADAY PLACE. This was first known as West Street, but was renamed in 1911 to honor William Temple Hornaday, who was director of the nearby New York Zoological Park (Bronx Zoo) for 30 years.

HORTON STREET. It was originally laid out as Washington Street. George Washington Horton settled near the end of City Island in 1818, and owned this property. The next owner was Stephen Decatur Horton, who sold out to William Belden. A grandson, Rochelle Horton, was a marine pilot, thus accounting for four adjoining streets being called Rochelle, Horton, Marine, Pilot.

HOSMER AVENUE. The Hosmers date back to the year 1016 in Kent, England. James Hosmer was the first of his family to settle in the New England colonies in 1635. His direct descendant, Abner Hosmer, was one of the first casualties of the Revolutionary War, being killed at the battle of Concord Bridge, 1775. See: Buttrick Avenue.

HOUGHTON AVENUE. When Unionport was laid out in the 1850s, this was 5th Street. A Francis Houghton of Groton, Connecticut, owned the tract in 1873, and this avenue remained a country lane until 1916, when it was first opened as a public road, and was suitably paved and curbed.

HOWARD CROSBY WALK. See: Crosby (Howard) Walk.

HOWE AVENUE. This may be the exact site of the fortified Indian village sighted by the early European explorers, Block and Dormer. In 1685, it was mapped as Cromwell's

Neck, and purchased by the Wilkins family prior to the Revolution. Their mansion stood near Lacombe Avenue, and the estate was called "Castle Hill." John Screvin, a son-in-law of Gouverneur Morris Wilkins, was the last owner. Howe Avenue was the boundary line between the Screvin estate and that of Catherine V.R. Turnbull. Charles J. Howe was listed in Castle Hill in 1905 as a landowner.

HOXIE STREET. This short street owes its name to the Hoxie Realty Corporation, listed in 1904. The name is not a common one, and only in the 1830s is it found when Joseph Hoxie witnessed many Westchester deeds in his capacity of Court Clerk.

HUBBELL STREET. It was named after Dr. Marvin D. Hubbell, who was active in the real estate development of this area below Westchester Square. See: Marvin Place.

HUDSON MANOR TERRACE. It runs from West 236th to West 239th Streets between Douglas and Independence Avenues, and harks back to a Hudson Park development, laid out in 1853 by a land company. "Manor" is a real estate term. At that time the terrace was part of a nineteenth-century farm of Samuel Thompson, and prior ownerships included Betts and Tippett (1668), Doughty (1666) and Adriaen van der Donck (1645-46).

HUDSON RIVER. As different cultures "discovered" or explored the Hudson River, they gave it their own name. SHATEMUC is an Algonquin Indian name for the Hudson River, which scholars accept in lieu of Cha-Ti-E-Mac, though Schoolcraft, an Indian authority, gives this latter name to the lower Hudson River, meaning "Stately Swan." The accepted spelling is Shatemuc. However, SKANEHTADE was yet another Algonquin name for it that was recorded by the early Dutch historians. It appears to have been used by the Weckguasgeeck Indians who inhabited The Bronx, west of the Bronx River. (The Siwanoys lived east of the Bronx River.) MAHICAN appears to be one other Algonquin name for the Hudson River, while MAHKANITTUK is yet another variant of the Algonquin name. RIVIERE VENDOME was an early name given to the Hudson River by Verrazano in 1524. RIO DE GUAMAS was an early name of the Hudson as charted by the sea explorer Gomez in 1525, who also mentioned it as RIO SAN ANTONIO in that same year. GREAT RIVER is an early name of the Hudson River given by Henry Hudson himself in 1609. GROOTE KILL was the Dutch name for the Hudson River, meaning the Big River, the term Kill meaning river. GREAT NORTH RIVER is so charted on DeLaet's map of the Hudson River in 1625. Similarly, GREAT KILL (sometimes, Ye Great Kill) was another early name. MAURITIUS RIVER was a Dutch name given to the River by cartographers in honor of Prince Maurice of Nassau. Thus it is not too odd that it was also called NASSAU RIVER by the Dutch, which is found chiefly on early maps, but seldom used by the colonists themselves, who chose to name it NOORDEN KILL, which means "The North River. Thus, NORTH RIVER is a former name of the entire Hudson River, which was a direct translation of the Dutch, Noorden Kill, in use in the 1600s.

HUDSON RIVER ROAD. It is a private road listed in the U.S. Postal Zone Guide, but not found on city maps. It services six homes, and runs parallel to and overlooks the Hudson River in the vicinity of West 254th Street and Independence Avenue.

HUGH J. GRANT CIRCLE. See: Grant (Hugh J.) Circle.

HUGHES AVENUE. It cuts across the nineteenth-century Lorillard estate called "Belmont." Below East 187th S treet, this street was known as Jefferson Avenue. In 1887, it was extended north of East 187th Street and called Avenue St. John because it led to St. John's College (now Fordham University). Reverend (later Bishop) John Hughes bought the property for the Catholic Diocese of New York, and the street was named in his honor in 1895, as well as to avoid confusion with another Avenue St. John in the lower Bronx.

HUGHES (Archbishop) CIRCLE. See: Archbishop Hughes Circle.

HUGUENOT AVENUE. This street honors the Huguenots who settled in the neighborhood in Dutch times. It overlays the former Codling Island in the Hutchinson River. The street was opened in 1935. The origin of the word is interesting. A Calvinistic religious movement, it was called "Eidgenossen," a Swiss-German word for Oath Brothers. "Eguenots" was a French corruption of the term, and eventually "Huguenots."

HULL AVENUE. This avenue runs through the nineteenth-century holdings of the Varian family who had it for three generations. Commodore Isaac Hull commanded the U.S. Frigate *Constitution* in the War of 1812, finishing off the *Guerriere* in a famous battle off the Grand Banks. He succeeded Commodore Bainbridge.

HUNT AVENUE. The Hunts were landowners in this region even before the Revolutionary War, and their farms are listed on maps of 1778. The Hunt burial ground was once located at Victor Street and Van Nest Avenue. Downing's Brook crossed Hunt Avenue at White Plains Road. At one time, this short thoroughfare was called Louise Street, then Cruger Avenue before carrying its present name. Marianna Hunt, who was related to the Archer and Mapes family, is buried in their family plot on White Oak Avenue in Woodlawn Cemetery.

HUNTER AVENUE (1). John Hunter III, grandson of the owner of Hunter's Island, elected to sell the island and live on the Bayard farm, an estate on Eastchester Bay which he called "Anneswood." An avenue extending to Givan's Creek was given his name. In 1969, it was blotted out by Co-Op City. Only the northern end of Hunter Avenue remains, touching the Boston Road. It was laid out across filled-in Hollers' Pond.

HUNTER AVENUE (2). It was formerly known as John Street. In 1870, John Hunter Ill with August Belmont sought to buy considerable property on City Island on which to erect a seaside pavilion, yacht club and a race track. The plan never materialized.

HUNTER CLOSE. See: Boston Close, Hunter Avenue.

HUNTER ISLAND. In early Colonial days the island was inhabited by the Siwanoy Indians who gave it the name "Laap-Ha-Wach-King" (the place of stringing beads), for its coves and beaches yielded huge amounts of shells used to make wampum. The Siwanoys sold their island to Thomas Pell of Fairfield, Connecticut, in 1654, and so it became Pell's Island. In 1669, the island passed to Sir John Pell, second Lord of the Manor, who wrote of "My Island called & known by the name of Pelican Island. . . ." Thomas Pell, third Lord of the Manor, was the next proprietor, and his son, Joshua, was the fourth and last Lord. When the American Revolution broke out, he joined the British army. Numerous skirmishes and some naval warfare took place in and around the island, and the lovely wooded game preserve was bought and sold a number of times until John Hunter and his wife Elizabeth, nee DesBrosses, purchased the 200-acre island for $40,000. Then he spent another $40,000 to landscape it. His mansion was fabulous and tales of his art gallery, wine cellar and lavish entertainment have survived to this day. Three generations of Hunters lived there, until it was purchased by Ambrose Kingsland, then by three or four other owners until the final purchaser was New York City. For many years Hunter Island remained isolated and undeveloped until the 1930s, when the island lost its identity by becoming attached by landfill operations to the mainland. In 1937, the Hunter mansion was razed to the ground, but the fieldstone carriage house survived until 1949.

HUNTER ISLAND MARINE ZOOLOGY AND GEOLOGY SANCTUARY. A local law, dated September 14, 1967, set aside tidelands off Hunter Island, Twin Island, Cat Briars Island or One Tree Island as a geological haven and preserve due to the efforts of Bronx historian, Dr. Theodore Kazimiroff. It is named for John Hunter, an early nineteenth-century merchant, politician, art collector and one-time owner of the island.

HUNTINGTON AVENUE. Collis Potter Huntington owned an estate along the East River comprising 33 acres, with additional acreage that is now inside Ferry Point Park. Born in poverty in 1821, he made a fortune in California and became a railroad magnate. He died in 1900 and was buried in Woodlawn Cemetery where his mausoleum is a showpiece. His widow paid for the elaborate funeral in gold coins. Collis Place is named for him, and nearby Milton Place was named for his son, Archer Milton Huntington. The Huntington mansion on Schurz Avenue became a Catholic high school.

HUNTS POINT AVENUE. This avenue overlays part of the ancient Hunt's Point Road. Thomas Hunt was one of the original settlers of 1670.

HUSSON AVENUE. This is formerly the land grant of Thomas Cornell, which was purchased, after the Revolutionary War, by Isaac Clason. In the early 1800s, it was part of the Stephens property. Joseph Husson owned 55 acres (1867-1893) on the right bank of Pugsley Creek. He was related to the Rudd family. He had seven children, and upon his death, the property was divided. A daughter, Susan, and her husband, Colonel A.G. Rudd, inherited the family home and part of the estate, which was called "Sans Souci." Executor of the will was Matthew Husson, whose name also appears as a member of the Reception Committee - August 6, 1921 - at the opening of the Municipal Ferry System between Clason Point and College Point.

HUTCHINSON RIVER. Called AC-QUE-HO-UNK by some of the Native Americans in the area. Translations differ. It could mean "High Bank" or "Red Cedar T ree." Aqueanouncke was the Siwanoy Indian name. Eastchester Creek was a 19th-century name for it.

HUTCHINSON RIVER PARKWAY. Nicknamed "The Hutch," this important road was constructed from 1936 to 1938, to coincide with the opening of the Bronx-Whitestone Bridge and the World's Fair of 1939. Where it passed through the village of Westchester it absorbed the Old Pelham Road and also Appleton Avenue. On Ferry Point, it overlaid ancient Ferris Lane. The Parkway gets its name from the Hutchinson River, along which it runs for a stretch in Pelham Bay Park. The river, in turn, harks back to the region's earliest settler, Anne Hutchinson.

HUTCHINSON RIVER PARKWAY EAST. This street Ranks the eastern side of the Hutchinson River in Co-Op City. See: Hutchinson River Parkway.

HUTCHINSON RIVER PARKWAY EXTENSION BRIDGE. This important highway has been open to motorists since 1941. It takes its name from the Hutchinson River over which it is arched.

HUTCHINSON RIVER PARKWAY WEST. This street flanks the western side of the Hutchinson River Parkway in Co-Op City. See: Hutchinson River Parkway.

HUTTON SQUARE. A tiny plot at East 182nd Street, Arthur Avenue and Quarry Road, it received its name by Local Law 64 in 1940. It had been acquired on June 1, 1897, and placed under the jurisdiction of the Department of Parks. None of the local inhabitants knew who Hutton was, nor was there any information available in Park Departmental records. However, it is a certainty the Square honors a Bronxite who died in World War I.

HUXLEY AVENUE. The Riverdale street was opened in 1927, but the resolution to name it after the English scientist had been passed on May 2nd, 1916 by the Board of Aldermen. Thomas H. Huxley, on a visit to America, had been the houseguest of the publisher William Appleton who lived at "Wave Hill" (West 248th Street). Some avenues in this area carry names of other scientists such as Liebig, Spencer and Tyndall. See: Wave Hill.

INDEPENDENCE AVENUE. Formerly called Yonkers Avenue and Palisades Avenue, this thoroughfare runs through the heart of Riverdale. Three streams pass under it. This avenue owes its name to an error. During the Revolutionary War, a British cartographer mistakenly located Fort Independence above Spuyten Duyvil. Nineteenth-century realtors opened a development, and called one of the streets Independence Avenue, and perpetuating the error.

INDIAN FIELD. In the northeastern part of Van Cortlandt Park a small meadow is dedicated to the seventeen Stockbridge Indians who, with their chief Nimham, were killed in the Revolutionary War by British cavalrymen on August 31, 1778. A bronze tablet is set on a cairn of stones, marking the spot.

INDIAN ISLAND. A local name for a small island in Bartow Creek, Pelham Bay Park, that is accessible at low tides from the south bank. Glacial pot-holes are found there, and Indian artifacts have been dug up in quantity.

INDIAN LAKE. A prominent lake in Crotona Park that was sometimes called Indian Pond. It was once located on the Bathgate farm, and later was called Zeltner's Pond as it was used by a family of that name for ice-storage. According to the late Frank Monohan, Assistant Borough Park Director, local boys gave the pond its name. It is spring-fed, and was the source of the Sacrahung or Bungay Brook (see Intervale Avenue) that flowed into the East River.

INDIAN POND. See: Indian Lake.

INDIAN ROAD. This is a private road into the former "Fieldston" estate owned by Major Joseph Delafield, who gave it the name. In the 19th century, social clubs, athletic teams and political organizations liked to adopt Indian names, and Squire Delafield indulged in the fancy.

INDIAN TRAIL. A scenic walk in Silver Beach Gardens, which was used by Indians in their day. It overlooks the East River from atop a high bluff, and was briefly occupied during the Revolutionary War by a detachment of British Rangers. The Americans also used Indian Trail as a vantage point from which to spy upon British vessels entering Long Island Sound. It was part of the Hammond property which passed to Havemeyer, the Sugar King, who erected a gazebo at one point. See: Acorn Place.

INDUSTRIAL STREET. Off Eastchester Road near its junction with Williamsbridge Road, this street services a few warehouses near the subway yards. The name is descriptive. Originally a semi-swamp, it overlays part of the eighteenth-century Cornell mill road. It is now flanked by the N.Y. State Mental Hospital grounds that were filled in from 1960 to 1964.

INTERVALE AVENUE. This street constitutes the boundary of Morrisania on the east, as it overlays Sacrahung or Bungay Brook that separated Morrisania from the West Farms Patent; A third name was Bound Brook (sometimes, Boundary Brook), and this led to a centuries-long dispute. The brook rose in Indian Lake, Crotona Park, and flowed into the East River, and its course - in the valley (inter-vale) -is closely followed by Intervale Avenue.

INWOOD AVENUE. Inwood was the name for West Morrisania in the 1840s and 1850s, and the avenue was already listed in the 1871 directory of Morrisania and Tremont. The origin of the name, Inwood, is obscure. The area in the vicinity of Jerome Avenue and East 167th Street was owned by Julia (Morris) Stebbins as shown on maps of 1858 and 1867.

IRVINE STREET. It was named in 1906, but on a 1912 land map it carried the name of Extra Street. This short Hunts Point street might honor Brigadier General James Irvine, who was with Washington on the Delaware, for a nearby avenue honors Lafayette, who also accompanied Washington. Frank Wuttge believed it had a connection with Irvine, Scotland, where Joseph Rodman Drake spent his honeymoon. In 1897, Joseph A. Irvine and his wife Prudence owned property on the far side of the Bronx River, and this might be the more prosaic explanation.

IRWIN AVENUE. William E. Irwin was one of the very few World War I servicemen to have an avenue named in his honor. He died in the Argonne Forest. The Irwin family lived in the neighborhood. This avenue overlays ancient Tibbett's Brook that once flowed into Spuyten Duyvil Creek.

IRWIN FISCHER (Corporal) PLACE. See: Fischer (Corporal) Place.

ISELIN AVENUE. It leads across the former Delafield property and is named after a wealthy landowner, C. Oliver Iselin, who owned Hunter Island. The family can be traced back to the year 1364 in Wuertemberg, Germany, as Yselin, and as Ysele when they migrated to Basle, Switzerland. Isaac Iselin, born in Basle in 1783, of this Protestant family, came to America as a youth and worked in the mercantile firm of LeRoy, Bayard & Co. Eventually he shipped out as supercargo (cargo supervisor) on the *Maryland* on a two-year voyage around the world. He kept a diary in English. He became a wealthy man, was wiped out in the Great Fire of 1835 and amassed another fortune. He retired to Basle, but fell into the swift Rhine currents and drowned in 1841. His son and grandsons remained in America, and were socially and financially prominent in the 19th century.

ITTNER PLACE. John Ittner's Hotel at Tremont and Park Avenues was a well-known stopping place in Civil War times. He was also the owner of a picnic grounds in this vicinity. The Place was officially opened in 1904, and was almost obliterated by the Cross-Bronx Expressway.

IVES STREET. This is a tiny street off Eastchester Road, once the early colonists' Westchester Path, and, even earlier, an important Indian trail. D. D. Ives is listed as a landowner in 1866, and was the manager of the Home for Incurables in the village of Belmont. He may have been related to Silliman Ives who was also a landowner at nearby East Tremont Avenue.

JACKSON AVENUE. No mention can be found that President Jackson was honored by the avenue. Rowland Robbins' farm was east of present-day St. Mary's Park in the 1850s, and when home owners moved in, *circa* Civil War, the one and only street in East Morrisania (the East Ward) was Robbins Street. This was also known as Robbins Avenue when it was extended to Woodstock (East 163rd Street area). In 1905, the entire length was renamed Jackson Avenue, which is a rather late date to think of Andrew Jackson. The property of what had been the Robbins' farm was owned by Washington Jackson and his wife, Rosetta, "of the East Ward" and it is a fair assumption they, and not the President, were remembered by the city planners.

JAMES BURKE PLAYGROUND. Dedicated in the early 1970s, this playground is located at Morrison and Watson Avenues. See: Watson Avenue.

JAMES DELL PLAZA. Members of the Spencer Estate Civic Association met at Middletown Road near Robertson Place to honor a past President who had served over 11 years. James Dell was also instrumental in closing down the Pelham Bay Park landfill, building Lehman High School and in the opening of the Pelham Bay library. The plaza was dedicated on August 29, 1985.

JARRETT PLACE. This is on a part of the 1800 Cornell farm that passed to Andrew Arnow in 1807. Matson S. Arnow sold a portion of the tract in 1877 to George F. Jarrett. During Civil War times, the settlement clustered around a railroad depot, "and had as many saloons as homes" according to old-timers.

JARVIS AVENUE. Nathaniel Jarvis is mentioned in Westchester deeds in 1843, and Judge Jarvis was a Justice of the Peace in the middle 1850s, but most likely candidate to have his name affixed to the street was his son Henry J. Jarvis, who was President of the Chester Taxpayers' Alliance before World War I. He was on the committee that attended the groundbreaking ceremonies at Hugh Grant Circle on May 27, 1916, and he was prominent in other civic affairs. When he died, a portion of Country Club Road was given his name. He is buried in St. Peter's P.E. churchyard.

JEFFERSON PLACE. This was formerly Jefferson Street in the village of Morrisania when it was opened in 1864. The original wooden Catholic church of St. Augustine was on this Place before 1900. The Place honors Thomas Jefferson, 1743-1826, the third President of the United States.

JENNINGS STREET. It was once called Charlotte Place, presumably after Charlotte Leggett, who married William Fox. One of the earliest settlers in this region was a John Jennings, mentioned many times in the records of the 1680s-sometimes rendered as Genones, Geninges, Jenings, etc.-and the family was intermarried with the Hunts and Leggetts and Foxes. Men of the Jennings family served in the American Revolution, War of 1812, the Mexican War and Civil War. Their old homestead, on the west side of Boston Road south of Jefferson Place existed into the 20th century. It became an inn, "The Old Stone Jug Tavern," and was demolished in 1909.

JEROME AVENUE. It was laid out as a plank road at a cost of $375,000 in 1874, just before annexation by New York City. It followed, in part, the bed of Cromwell's Creek, and appeared on 1888 maps as Central Avenue, for it led from Central Bridge (Macomb's Dam Bridge) up to Jerome Park Racetrack. It was paved about that time, and was a broad tree-lined boulevard. In 1866, Leonard W. Jerome helped organize the American Jockey Club at the racetrack. He was of Huguenot extraction, an important financier, stockholder in the *New York Times* and founder of the Academy of Music. His daughter, Jennie Jerome, became the mother of Winston Churchill. Legend has it that when Central Avenue became an important road, it was to be renamed in honor of an Alderman whose name is not given; but Kate Hall Jerome, wife of the gentleman horseman, was enraged when she learned of it. At her own expense, she had bronze street signs cast, hired workmen to put them up on poles, and personally supervised the work. The Board of Aldermen discreetly dropped the matter, and Jerome Avenue it remains to this day. Once the subway line was erected overhead, the avenue lost its rural look and became a commercial artery. Jerome Avenue serves as the dividing line between East Bronx and West Bronx, although the Bronx River would seem more logical.

JEROME PARK RESERVOIR. It is located on the site of the famous Jerome Park Racetrack that was constructed in 1865 and existed until 1894. The track was founded by Leonard W. Jerome, whose mausoleum is in Greenwood Cemetery, Brooklyn. Prior to the Civil War, the land was part of the Bathgate farm. See: Jerome Avenue.

JESUP AVENUE. This street cuts through the former "Rockycliff" estate of Mrs. Marcher as mapped in the 1860s. It is named for a landowner named Morris Ketchum Jesup, banker, philanthropist and President of the American Museum of Natural History in the 1890s. He was also on the Hudson-Fulton Celebration Committee of 1906.

JESUP PLACE. This Place is laid out on the former estate of the Poole family of pre-Civil War days. It was sold to Rebecca Marcher, who called the land "Rockycliff." Morris K. Jesup was listed in Robinson's 1888 *Atlas* as owning property there. See Jesup Avenue.

JOHN KIERAN NATURE TRAIL. In December 1987, the Parks Department opened a Nature Trail in Van Cortlandt Park, from the abandoned Putnam Division railroad north to Tibbett's Brook. John Kieran was a noted sportswriter in the New York Times for many years, but was equally well-known as a naturalist and bird-watcher

JOHN McCORMICK LANE. This 43-year-old Police sergeant was killed in the line of duty during a raid on narcotic dealers. He had lived on East 194th Street between Mayflower and Hobart Avenues, so the street sign carries this alternate name on that stretch.

JOHN McNAMARA SQUARE. What had once been a meadow on Squire Morris's estate on Throggs Neck eventually became a triangular plot at Calhoun and Randall Avenues. Dedicated on August 29th 1985, the Square honors a local man who has devoted a lifetime amassing a history of The Bronx and passing this knowledge on to following generations. This took the form of lectures, walking tours and thousands of articles, including this book. He is also a founder and Charter Member of The Bronx County Historical Society.

JOHNSON AVENUE. It is named in honor of the Johnson family, who ran an iron foundry on half of a 13-acre peninsula in the Harlem River, although purists might argue it memorializes the founder, Isaac Gale Johnson. He and his four sons manufactured shot and shell for the Union armies during the Civil War. The five Johnson mansions were built on the Spuyten Duyvil promontory. In 1890, the Johnson family, now expanded to five sons, bought the rest of the peninsula for rolling mills. In 1938, the buildings were razed.

JOHNSON SQUARE. Located at East Fordham Road, Crotona Avenue and Southern Boulevard, it honors Charles Johnson, a casualty of World War I less than a month before the Armistice. He had been a fireman of Engine 58 (West 115th Street) and died at the age of 24.

JOHNSTON WALK. This is a footpath on the campus of former New York University, now Bronx Community College. It was the grounds of the 19th-century Schwab family, earlier the Archer lands, during the Revolutionary War a part of Fort Number 8, and, before that, the Fordham Grant to John Archer (or Arcier). John Taylor Johnston, founder of the Law Library in 1863, received three degrees: A.B., 1839; A.M. in 1842 and LLD in 1890. He served as President of the Council of New York University from 1874 to 1886.

JOYCE KILMER PARK. See: Kilmer (Joyce) Park.

KAPPOCK STREET. Opened in 1895 on Spuyten Duyvil hill, it was first called Warren Avenue. It is believed that Thomas H. Edsall, the eminent historian of that area, was the man responsible for having the street renamed to preserve its Indian past. "Shorakkappock" was the Algonkian name of a settlement on that height, and meant "Sitting Down Place."

KATONAH AVENUE. Until March 1896, this was 2nd Street in the village of Woodlawn Heights. Richard Lederer, place name authority of Westchester County, cites an Indian Sachem, Ketatonah, meaning 'Great Mountain'. The surmise is that Heights residents thought the name was appropriate to the elevation. The area figured in a Revolutionary War battle between British cavalry under Simcoe, and the American patriots assisted by the Indian, Nimham, and his Stockbridge tribesmen.

KAZIMIROFF (Dr. Theodore) BOULEVARD. On May 9, 1981, the northern end of Southern Boulevard, from Fordham Road to Allerton Avenue received this name to honor an official Bronx Historian who helped found The Bronx County Historical Society. He was a consultant to the Bronx Zoo, Museum of the American Indian and the Museum of Natural History. He was also a driving force in preserving the ecology of Pelham Bay Park. See: Thomas Pell Wildlife Sanctuary.

KAZIMIROFF NATURE TRAIL. Dr. Theodore Kazimiroff was further honored on June 19, 1987 when a trail bearing his name was dedicated on Hunter Island. He died in 1980 at age 66.

KEANE SQUARE. Marine Lawrence Keane was a neighborhood youth who was killed in action on Guadalcanal on November 1, 1942. He earned the Order of the Purple Heart, and (posthumously) the Navy Cross. The intersection of Buhre Avenue and Westchester Avenue is in the old settlement of Middletown.

KEARNEY AVENUE. In the Country Club sector, the avenue was originally mapped as West Road. It was laid out across the former swamps of Weir Creek's southern arm when they were filled in the 1950s. South of Kearney and Lawton Avenues was once an Indian burial ground that was investigated by archaeologists from the Haye Foundation in the 1920s. The avenue is named after General Philip Kearney. He fought in the French army in the Algerian campaigns of 1839 and 1840 before coming to America, where, as a Second Lieutenant in the U.S. Dragoons, he fought in the Mexican War and in the Indian wars of California. He returned to the French army and took part in its war on Austria. Then he returned to the States and distinguished himself in the Civil War, serving under General McClellan. He is buried in Trinity churchyard.

KELLER CORNER. See: Bertha Keller Corner.

KELLY FIELDS. In May 1973, two ball fields were dedicated to the memory of Frank Kelly, a Riverdale civic leader. Mr. Kelly died from a heart attack at the age of 33, prompting members of various organizations to seek an appropriate memorial to him. The fields, at West 259th Street and Broadway, were dedicated by Parks Administrator Clurman and Borough President Abrams.

KELLY STREET. An 1818 deed mentions Captain Samuel Kelly's farm adjoining "Longwood Park." There are frequent references to this man in deeds of the Leggett family, as well as the Fox and Tiffany landowners.

KELLY STREET PARK. In February 1986 this brand new 7.5-acre park, off Dawson Street, was opened to the public, thanks to the efforts of local citizens. The name was a final compromise, as groups from South America, Central America and the West Indies each wanted to honor their own national hero.

KELTCH MEMORIAL PARK. At Jerome Avenue and Macomb's Road, this plot was acquired on May 31, 1899 by the Departments of Parks. In 1943, it was named for Ensign Rubin Keltch, who was killed in action that year. He held the Navy Cross, Purple Heart and a Presidential Citation.

KENNELWORTH PLACE. The last private owner of record was Lorillard Spencer. Proprietors of the land throughout the centuries included the Ferris, Palmer and Underhill families. Nathaniel Underhill (1685) was a son of the Indian fighter, Captain John Underhill, whose family ties went back to Kenilworth, Warwickshire, England. Captain

John called his lands on Oyster Bay "Kenilworth," and his son did likewise in his holdings on Eastchester Bay. Over the centuries, the spelling was altered.

KEPLER AVENUE. At one time in the 1890s, this street was mapped as Quail Avenue in Woodlawn Heights, then was given the number 3rd Street. It acquired its present name in March, 1896, and is thought to honor Johan Kepler, an astronomer, but for no apparent reason. The area figured in the Revolutionary War when, in 1778, Chief Nimham and his Stockbridge Indians fought the British cavalry under Simcoe. The cavalry circled around present-day Kepler and Katonah Avenues to encircle the Indians at Van Cortlandt Park East. See: Indian Field.

KIERAN NATURE TRAIL. See: John Kieran Nature Trail.

KILMER (Joyce) PARK. The tract of land extending north from the Bronx County Courthouse along the "Grand Concourse" was once part of the Gerard W. Morris estate. In the 1880s, Butternut Lane ran diagonally across its northern end from Walton Avenue. The Grand Boulevard and Concourse officially began at the stretch of lawn at East 161st Street, a park named the Concourse Plaza, sometime after 1902. However, because a statue of Louis J. Heintz was erected there, Heintz Park was a popular name with Bronxites, but never officially adopted. Louis J. Heintz became the first Commissioner of Street Improvements, when he was voted in by both Democrats and Republicans after he succeeded in providing separate powers for the "Annexed Districts." Born in Manhattan in 1861, he moved to The Bronx where he was secretary and treasurer of the Eichler Brewery. Mr. Eichler was his uncle. Louis Heintz married brewer Ebling's daughter, and was president of the Brewers' Board of Trade. He helped found the Schnorer Club and was its second president (1885-1891). He first proposed a Grand Concourse in 1890, which is one reason his statue stands at the southern end of the park. His remarkable career was cut short in his 30th year when he died from an unsuccessful appendectomy. He is buried in Woodlawn Cemetery. With a similarity in name, the park was also known, though to a lesser degree, as Heine Park for the Lorelei fountain and a bas-relief of Heinrich Heine had been put there in 1899, commemorating the German poet Heinrich Heine and his poem "Lorelei." In 1926, the the 4.3-acre park was named in honor of Joyce Kilmer, a World War I soldier killed in France, and known for his poem "Trees." See: Ebling listings, Grand Boulevard and Concourse.

KILROE STREET. This was formerly Elizabeth Street on City Island and was named after Elizabeth King, whose husband owned High Island. Postal authorities had tried several times to have the name changed, as it conflicted with an Elizabeth Street in Manhattan. It received its present name in June, 1950, to honor Reverend James M. Kilroe, who had been pastor of the nearby church of St. Mary, Star of the Sea. He had also inaugurated the Catholic parochial school in 1931. Father Kilroe was born on June 19, 1878 and died on February 28, 1945.

KIMBERLEY PLACE. This short street was once Verveelen's holdings in Colonial days near the original King's Bridge. It passed into the hands of Abraham Lent, then Alexander Macomb and, finally, Joseph Godwin. See: Godwin Terrace. There is an off chance the surveying of this Place occurred at a time in the 1890s when Rear Admiral L. A. Kimberley was very much in the news. A local story is that the digging took so long and was so deep it was likened to the famous Kimberley mine in South Africa. It was officially opened in 1922.

KINDERMANN PLACE. On December 22, 1962, a short stretch of unnamed street in Unionport was given this name, reviving the memory of an old-time Bronx family, long active in civic endeavors. It is between Blackrock and Haviland Avenues, 360 feet east of Havemeyer Avenue. In the immediate vicinity of the Odd Fellows' Home, it was once the farmlands of the Lowerre family in the 1870s and 1880s. An earlier Kindermann Place was located off Webster Avenue, and was obliterated by a housing project.

KING AVENUE. Elisha King owned property on the mainland opposite City Island, and there built his mansion in 1829. He also owned High Island, plus a small plot of land on the City Island shore to give him landing privileges. King Avenue leads to that landing lot. See: Kilroe Street.

KING (Martin Luther) MEMORIAL PARK. See: Martin Luther King Memorial Park.

KING'S COLLEGE PLACE. This is a short street running from Gun Hill Road to Woodlawn Cemetery, and was formerly mapped as Ochiltree Street. In Revolutionary times this was part of the Valentine farm, and sometime before 1900 it was the site of Columbia College's athletic field. King's College is the original name of Columbia College. See: Tryon Avenue.

KINGSBRIDGE AVENUE. This street represents the spine of the ancient island of Paparinemo, which eventually was attached to The Bronx mainland and called "Island Farm" and Kingsbridge. It is appropriately named for the historic King's Bridge, which figured so prominently in Revolutionary War days, and which is now buried beneath its southern end at West 230th Street. At one time, this avenue was called Church Street.

KINGSBRIDGE ROAD. From West 225th Street at Bailey Avenue eastward to its junction with present-day Fordham Road, this overlays the original Indian path. This trail was widened to the Revolutionary road that figured in many encounters in the years 1776 to 1782. It took its name from the fact it led to the King's Bridge. See: Kingsbridge Avenue.

KINGSBRIDGE TERRACE. Once known as Natalie Avenue, it may owe allegiance to a former property owner, Natalie Lorillard Morris. The name of this street was changed in 1903 after the neighborhood in which it is located. Harry B. Thayer owned most of the property, and is chiefly remembered as being the first man to place a transatlantic telephone call from New York to London in 1923.

KINGSLAND AVENUE. So far as is known, Mr. Kingsland did not own land here, but his name given to this avenue because all streets in the area are named for former Mayors. Ambrose Kingsland was Mayor of New York, 1851-1853. He once owned Hunter Island. Below Gun Hill Road was the ancient holding (1720s) of Reverend John Bartow that remained in the family for three generations. Above Gun Hill Road, Kingsland Avenue was originally Pell property which passed to Bartow and, in 1794, to the Givan family. The Givan mansion was located near Kingsland and Arnow Avenues. Earlier names of this avenue were Cedar Street and Tieman Avenue.

KINGSLAND PLACE. An 1865 map shows this area as being owned by G L. and A. C. Kingsland, the latter being a former Mayor of New York and owner of considerable property in all parts of the borough. The Place is really a flight of steps off West Tremont Avenue.

KINNEAR PLACE. This small Place is located on the pre-Revolutionary holdings of the influential Seabury family. At its southern end, Cebrie (Seabury) Creek once flowed into Westchester Creek. An 1864 deed mentions a John and Sarah Kinnier [spelled this way] who purchased 12 acres along Cebrie Creek from the Ferris family.

KINSELLA STREET. Early records show this to be the southern extent of W.B. Downing's land in 1838. The stream that flowed through the settlement of Van Nest was called Downing's Brook. The street honors Father Jeremiah A. Kinsella, who founded St. Raymond's parochial school in 1868.

KIP ROAD. This Riverdale road is laid out across the tract once owned by the Delafield family. The Kips stem from an old Dutch family and this particular landowner ran a famous roadside tavern at Southern Boulevard and Fordham Road.

KIRBY STREET. John E. Kerby [spelled this way], Spanish-American War Colonel, lived on City Island for many years, and was instrumental in having gas brought out from the mainland. He was an architect and designer, as well as a real estate appraiser, and built many churches and schools in The Bronx. He had been born in Tremont in 1859. His father, John Kerby, came from Ireland in the 1840s, and the name was so spelled on the Castle Gardens immigration records, whereas a brother, Michael, found his name had been spelled "Kirby." Both brothers held to their individual spellings, and their sons and daughters continued the practice. Ironically enough, when the City Islander John Kerby died, the street that honored his name was misspelled "Kirby!"

KNAPP STREET. This small street was, a century or more ago, part of the Palmer estate-the Palmers being descendants of Robert Givan, the Scotch miller who bought the land from the Bartows. Earliest reference to anyone of this name is in 1815, when Sheubel Knapp, farmer of Eastchester, purchased lands near the Hutchinson River. Later, in 1857, Alfred and Wright Knapp were listed as small property owners in Eastchester, so it can be assumed they were sons of Sheubel. It is not known if Halsey Knapp is of this farming family, but he was Pastor of the West Farms Baptist church around 1860, when baptisms were held in the Bronx River.

KNOLLS CRESCENT. This seems to be a descriptive name, alluding to the hilltop (knoll) and the curving road around Spuyten Duyvil hill. A Knowles family, long-time residents of Kingsbridge, disclaim any former ownership, and point out there was an estate "Cedar Knolls" located further north, belonging to the Barney family.

KNOLLWOOD AVENUE. This is a roadway running through the southeastern section of Woodlawn Cemetery. This is simply a descriptive name. A wooded knoll was once situated in that sector.

KNOX PLACE. This was cut across the edge of the Dickenson estate of the 19th century. At its southern end was the spring that formed School (Schuil) Brook that flowed along present-day Mosholu Parkway to Webster Avenue and eventually became the Mill Brook.

Avenues in this region carry names of Revolutionary War personalities, and Henry Knox (1750-1806) was a general of artillery. He was one of ten sons, owned a Boston bookstore and studied military science. General George Washington gave him a chance to test his theories at Ticonderoga. He founded the Order of the Cincinnatti. In his fifties, he weighed 280 pounds, and choked to death on a chicken bone.

KOLTOVICH PARK. See: William Koltovich Park.

KORONY SQUARE. This triangular patch is located outside the gates of the Park of Edgewater, bounded by Miles and Pennyfield Avenues. It was named in October, 1963. It honors a Summer resident of what had been Edgewater Camp, Theodore Burton Korony, who was inducted in August 1917 to serve in the Infantry overseas. As a Corporal, he died in front of the Hindenburg Line on September 29, 1919 in Ronssoy, France. His father had been a Civil War veteran, a Count Korony (according to The Bronx *Home News)* of Hungarian nobility. The land originally was a swamp, alongside Adee's Lane of the 1870s from which a spring fed into Weir Creek. Old-timers ice-skated there when the waters froze. Partially cleared in the 1920s, a local man named Barney Gallagher operated a refreshment stand there. It was drained and cleared just prior to World War II, and the first Theodore Korony Legion Post was built there. It (the clubhouse) was demolished in 1960 to make way for the Throgs Neck Expressway.

KOSSUTH AVENUE. At first this was named Kossuth Place, but changed to a full fledged avenue in May, 1916. Louis Kossuth (1802-1894) was a Hungarian fighter for freedom and came to America in 1849, to be greeted as a popular hero. People staged Hungarian balls in his honor and wore Hungarian dress, etc. He addressed Congress in Washington during his tour, having taught himself English while in political prison. Adjoining avenues honor foreign-born Revolutionary War generals, DeKalb, Rochambeau and Steuben, but Kossuth is not in this category. So credence must be given to an old story that this avenue was slated to honor foreign-born commander Kosciusko, but "Kossuth" was easier to spell!

LA GUARDIA AVENUE. This is a curving road leading through the hospital grounds on Riker's Island, honoring the memory of former Mayor Fiorello La Guardia.

LA SALLE AVENUE. In the 1820s, this was the northern bounds of a large Throggs Neck tract belonging to Elbert Anderson. Anderson sold individual farms to different owners in the 1830s and 1840 - and this became known as the Walsh farm. When the village of Schuylerville grew up, this was plotted as 2nd Street, but, in 1921, was given the name of LaSalle Avenue. No reason has ever been found to account for the name.

LACOMBE AVENUE. The avenue had been laid out across Pugsley Creek from Clason Point to Castle Hill Point since 1916, and was named in honor of Judge Emile Henry Lacombe, Corporation Counsel 1884-1887 and Judge in the U.S. Circuit Court in 1888. The junction of Lacombe and Leland Avenues is over the site of an Indian village called Snakapins. Near the junction of Lacombe and Castle Hill Avenues was the former mansion of Gouverneur Morris Wilkins.

LACONIA AVENUE. Laconia Park was a speculative holding laid out from 1888 to 1893 between Boston Road and Wakefield. It had been subdivided from the Blodgett Tilden estate. This avenue was called Tilden Avenue from Gun Hill Road to East 222nd Street.

LADD ROAD. This is a private road above West 254th Street in Riverdale, leading east from. Palisades Avenue. According to a 1908 Beers' *Atlas*, Dr. William Sergeant Ladd was the landowner at the time.

LAFAYETTE AVENUE. The lower end of Lafayette Avenue in the Hunts Point section was once called Lafayette Lane in honor of the French general who briefly traversed it during his triumphal tour in August, 1824. Men took the horses from the shafts of the general's carriage and pulled it themselves. From Hunts Point to the Bronx River, it was mapped in 1879 as Dickey Street as it ran across the Dickey estate of those days. In its middle portion (Bronx River to Westchester Creek), the avenue ran through what was once the Black Rock farm of the Ludlow family. It was crossed by Pugsley Creek where James Monroe Houses stand. In 1887, a Lucas Van Boskerck owned a farm at Beach Avenue. In Unionport, the avenue constituted the beginning of Gouverneur Morris Wilkins' property which extended from there to Castle Hill Point. An Indian trail crossed Lafayette Avenue at Castle Hill Avenue. On the far side of Westchester Creek, outside the new St. Raymond's cemetery, what is now Lafayette Avenue was once a winding creek up which it was possible to bring shallow-draft sloops. On this Baxter Creek (named for an early settler), a nineteenth-century landowner, Henry C. Overing, maintained a small dock. It was approximately in the school yard of Junior High School 101. The hill leading to East Tremont Avenue was much steeper, and on its crest once stood the original Baxter homestead of the 1660s. Subsequent owners built bigger houses until, in the 1890s, a man named Brinsmade was the occupant. When Lafayette Avenue was cut through around 1925, the mansion was due to be razed as it was squarely in the center of the proposed avenue. The mansion was purchased by a Swedish religious society and was moved 50 yards south to a corner plot. From East Tremont Avenue down to Weir Creek (now the Throggs Neck Expressway), Lafayette Avenue was the southern end of the Coster farm, and was first laid out as Gridley Avenue. An Indian trail crossed the creek at this junction. From the Throggs Neck Expressway to Long Island Sound, Lafayette Avenue runs through the former Century Golf links of the early 1900s. Prior to that, it had been the Ferris holdings, and the northern extent of the Bruce Brown estate.

LAFONTAINE AVENUE. This avenue runs through the nineteenth-century lands of the Ryer family, who were very early settlers. In 1897, this avenue was mapped as Lafayette Place. In March, 1900, the first subway was built, with the initial excavations being begun at City Hall. An impressive fountain had to be dismantled and removed, and it was brought up to Crotona Park. Newly opened Lafayette Place was on a straight line with this fountain, so the name was changed to Lafontaine.

LAKE AGASSIZ. Professor Louis Agassiz was a Swiss-American naturalist (1807-1873) whose name is given to a 5 1/2-acre pond formed by the Bronx River directly south of Pelham Parkway-Fordham Road.

LAKE AVENUE. This cemetery lane runs in the vicinity of Woodlawn Lake in the northeastern section of that burial ground. The tomb of Publisher Joseph Pulitzer is seen from Lake Avenue.

LAKE VIEW PLACE. This Place on the slope of "Upper Cortlandt" of Revolutionary times has a view of Van Cortlandt Lake.

LAKEWOOD PLACE. The short street is named after Lakewood, N.J., a noted Summer resort at the turn of the century. Around 1903-1905, Westchester Heights was developed as a wealthy residential section, and all streets were given the names of well-known Summer resorts, such as Narragansett Bay, Newport, Niagara, Saratoga and Ballston Spa.

LAMPORT PLACE. Nathan Lamport, a Jewish philanthropist and head of the Lamport Realty Corporation that developed "Tremont Heights" in 1905, is remembered in this Place. It represents the northern end of the extensive estate of John A. Morris, and was originally an orchard. For many years, the stretch of stone wail on East Tremont Avenue and Lamport Place was called "the haunted wall." See: Haunted Wall, The.

LANDING ROAD. Formerly known as Old Fordham Road, it was renamed on May 2, 1916 for the most celebrated landing on the Harlem River. A pier at that location served side-wheel passenger steamships, thus making that spot the Fordham landing.

LANE AVENUE. This street was once the boundary of John T. Adee's property. It was acquired in 1909 by the City, and laid out and paved a year later. James T. Lane had a grocery store and owned a large building on the south side of Westchester Square. The Lane family plot is in nearby St. Peter's churchyard.

LANZILLI WALK. In October 1989 Frank S. Lanzilli, who died in 1980, was honored for his active civic life. He founded a heating company on Williamsbridge Road near Jarrett Place, and it is in this area the street signs were erected.

LATKIN SQUARE. It was named in honor of David Latkin of 1280 Stebbins Avenue, who died of wounds in October 10, 1918. He served as a Pfc. in Co. K of the 28th Infantry in World War I, and was aged 27. The Square was named on December 7, 1926 and a news item credited Latkin with being the first Jewish soldier in Bronx County to lose his life in that war.

LATTING STREET. This street runs through what was the Seton estate alongside Westchester Creek (maps of 1850 to 1885). The street was named for John Jordan Latting, a noted genealogist and historian of Civil War times. He was a lawyer specializing in real estate, one such venture being the Seton Homestead Land Company. Edwards, Balcom and Vreeland were other Directors of the company and adjoining streets bear their names.

LAUREL AVENUE. This is a road in the southwestern section of Woodlawn Cemetery where laurel can still be found.

LAURIE AVENUE. Only 100 feet long, this is scarcely a street. Ordinarily it would be classified as a Place. Long-time residents of Middletown stated a Mildred (Paul) Laurie was a landowner at that junction alongside the Pelham Road (now Westchester Avenue) having inherited the land from her grandparents. The Paul farm was noted in military dispatches in the Revolutionary War when General George Washington and his staff stopped there. See: Mildred Place.

LAWN AVENUE. This is another Woodlawn Cemetery road, but in the northeastern section. The name evidently stems from the fact that this area was once a broad expanse of lawn. The tomb of John Haffen, the Bronx brewer, is alongside this road.

Also, the eight graves of the Barney family plot are there. The Barneys were landowners of Riverdale in the late 1890s, Tibbett Avenue being formerly known as Barney Avenue.

LAWTON AVENUE. Proceeding from west to east. Lawton Avenue was once a carriage road for the landed gentry from Balcom to East Tremont Avenues. From 1860 to 1920, it was known as Morris Lane. Lawton Avenue from East Tremont to approximately Kearney Avenues was the property of a John B. Stevens until 1877, when Frederick C. Havemeyer bought the land. Kearney Avenue and Lawton Avenue was once the southern extent of Weir Creek and the intersection was excavated as an Indian burial ground. From Meagher Avenue to Long Island Sound, the avenue crosses the former Pennyfield estate of Captain J. T. Wright (1840s), Herman LeRoy Newbold (1850s) and Francis DeRuyter Wissmann (1890s). In 1825, a Charles Lawton owned a small track in the area of East Tremont Avenue, but this is not likely the man for whom the avenue is named. The neighborhood is crisscrossed with Civil War and Spanish-American War heroes' names, and while there was a Civil War Colonel R. B. Lawton, this writer is inclined to favor a Spanish-American War general who was killed in action at Manila Bay, Henry Ware Lawton.

LAYTON AVENUE. Maps of Civil War vintage locate the Layton estate as being bounded by (present-day) Bruckner Boulevard, Rawlins and Layton Avenues and Eastchester Bay-52 acres in all. Mr. William C. Layton's mansion was situated at about Wilcox and Fairmount Avenues. In the 1890s to the 1920s, the land was a famous picnic grove, beach resort and entertainment center known as Lohbauer's. Layton Avenue was the original Town Dock Road of Schuylerville.

LEBANON STREET. Two overlapping stories might account for the naming of this short street off West Farms Square. The former owners, the Devoe family, once called their small estate "Lebanon" after the Biblical spot. The other story is that the Devoes transplanted some trees, and one of them - a Cedar of Lebanon - had to be sacrificed when the street was cut through.

LEE STREET. An 1868 map shows this street on the land of Mrs. Benson, but it was only in the late 1890s that it was laid out and named. It formed the western end of a district called "Tremont Terrace." It is only a possibility it is named for Gideon Lee, Mayor of New York 1833-1834, for the cluster of avenues honoring past Mayors is not in this neighborhood. In October, 1776, the British and their Hessian allies clashed with the Americans near Westchester Square. Prior to the battle, General Washington had visited the area of Lee Street on his way from the Paul farm. See: St. Paul Avenue. General Charles Lee, learning Washington was over at Throggs Neck, rode up the Pelham Road and joined him. It may have been at this point that the two men met.

LEGGETT AVENUE. Originally part of the Twelve Farms in the Debatable Lands, Leggett Avenue overlays the disputed boundary brook that was called Wigwam Brook, or Bound Brook, or Leggett's Creek. Gabriel Leggett married Elizabeth Richardson in 1661, she being coheiress of the original patent of West Farms. Major Abraham Leggett was a distinguished officer in the Continental army who saw service at the battles of Brooklyn, Harlem Heights and White Plains. A narrative of his Revolutionary career, written by himself, has been published. Samuel Leggett of West

Farms tried to furnish the City with fresh water from the Bronx River, but Croton River water was accepted instead. He founded the New York Gas Light Company. His sister, Charlotte, married William Fox. Still another Leggett, William (son of the Major in the Continental army), had many sea stories published, among them *Leisure Hours at Sea* and *Leggett's Naval Stories*.

LELAND AVENUE. Aaron and Submit Leland are mentioned in the town records of Westchester of 1810-1873. A Charles Leland owned 37 acres of land on Clason Point in 1883, and the family plot is off Beech Avenue in Woodlawn Cemetery. When the avenue was cut through the Mapes farm below Parkchester, it was called Saxe Avenue, after a Simon P. Saxe who owned land there. An Indian village was situated at the junction of Leland and Lacombe Avenues called Snakapins. The aboriginal wording was Sean-auke-pe-ing with two different translations: Place of Ground Nuts, and River-land-water Place.

LESTER STREET. At first, tiny Lester Street was called Wilson Place. It was laid out on part of Pierre Lorillard's extensive grounds at the time of the Civil War. In the 1890s, the tract was subdivided into building lots and acquired by the township of Olinville. It was called Lester Park, presumably after Charles Edward Lester, historian and friend of the Lorillard family.

LEYDEN PLACE. Adriaen van der Donck was the first of the Dutch patroons to hold possession of this land in the early 1660s. He studied law at Leyden, Holland.

LIBBY PLACE. This small street, once part of the original Pelham Road, was mapped in 1892 as part of the Benson estate. On yet another map, it is spelled Liberty Place, and was officially opened in 1934. The name might refer to Libby (Elizabeth) Cornell, a former landowner, whose property was bounded by this Place.

LIBRARY AVENUE. This avenue is laid out through the nineteenth-century Lorillard Spencer estate, which was once the eighteenth-century lands of the Ferris family, and the seventeenth-century grant of Nathaniel Underhill. It was named on October 20, 1922, and is a reminder of the world famous library of Archer Milton Huntington with its priceless collection of Hispanic art and literature. The northern end of Library Avenue pointed directly at the Huntington library in what is now Pelham Bay Park.

LIEBIG AVENUE. This Riverdale avenue skirts the "Pigeon Hill" of earlier times. In line with naming streets after scientists like Tyndall and Huxley, who were also once houseguests of William H. Appleton, owner of nearby "Wave Hill," this avenue honors Baron Justus von Liebig (1830-1873) acknowledged as the founder of the first chemical research laboratory for students. He developed the chemical process of coating a glass surface with metallic silver in 1835, thus greatly improving mirrors for telescopes and other devices. He advanced the theory, in 1844, that the use of soap was the true measure of a nation's civilization. He made many discoveries in many areas, including, metabolism, organic chemistry, nutrition. He invented manmade fertilizer, which he later regretted for all the damage it caused. "Nature herself... points out to man the proper course of proceeding for keeping up the productiveness of the land." He advocated natural methods in *The Natural Laws of Husbandry*. See: Wave Hill.

LIGHT STREET. There seems to be no reason for this name other than to offset the adjoining Dark Street. This thoroughfare was once the boundary of John H. Eden's estate, "Edenwald."

LINCOLN AVENUE. This avenue runs the length of Jonas Bronck's farm of 1641, and "Emmaus," the first European homestead, stood at the foot of this Avenue and East 132nd Street. Afterward, it was part of Colonel Lewis Morris' Long Neck farm of 313 acres, mapped in 1847. In Civil War times, this area was known as Old Morrisania, later North New York. The avenue was surveyed in 1883 and opened in 1890. It was a busy and well-populated avenue due to its proximity to the railroad and docks. It may well be that the avenue honors the Great Emancipator, Abraham Lincoln, but-by coincidence-there was a Brigadier General Benjamin Lincoln of the Revolutionary War who took part in some action in this very district.

LINCOLN SQUARE. It was named on May 23rd, 1899 by a vote of the Board of Aldermen, and approved by the Mayor to honor President Abraham Lincoln. At the time, East 137th Street and Lincoln Avenue was a lively district.

LINDEN AVENUE. This road is in the southwestern section of Woodlawn Cemetery, where a few linden trees are still found.

LINDEN DRIVE. A privately maintained street inside Parkchester, named for the tree.

LINDEN PLACE. This is a short thoroughfare in Silver Beach Gardens. See: Acorn Place.

LISBON PLACE. In 1884, George Opdyke was listed as the owner of this tract, which subsequently was divided into Cadiz, Grenada and Lisbon Places - all connected with the battles of St. George against the Saracens. St. George's Crescent is adjacent, and may reflect Mr. Opdyke's deep interest in the subject. Directly below this Place, Schuil Brook once flowed. The name was later anglicized to School Brook. An Indian trail passed on its northeastern corner.

LITTLE HOG ISLAND. On many maps this tiny isle is shown joined to Hog Island--but there is a chest deep channel between them, and Little Hog (.25 acre) is definitely a separate unit. It is located off the north shore of Hunter Island and can be reached by waders at low tide.

LITTLE LEAGUE PLACE. Near Westchester Square, this Place overlays a stretch of Westchester Creek. It runs alongside a little League ball field and received its name in November 1984, displacing the older name of Tan Place, but was not officially dedicated until June, 1985.

LIVINGSTON AVENUE. It was known in the 1890s as Indian Road as it passed Indian Pond. Later, it became Seminole Avenue, but in May of 1916, received its present name, which honors a family long resident in the area. John, James and Margarette Livingston are buried in the West Farms Soldiers' Cemetery as civilians.

LOCUST AVENUE. This was once farmlands of Jonas Bronck's bondsman, Pieter Andriessen, according to a map of 1639. It was later on Stony Island off the Manor of Morrisania and, when it was filled in around 1848, it became Stony Point. When Port

Morris was then laid out, this was 3rd Avenue, but shortly thereafter, took the name of Locust Avenue referring to the type of tree growing there.

LOCUST POINT. This was an Indian fishing camp which became part of the pre-Revolutionary farm of the Stephenson family. In the early 1800s, it became the property of James Drake, who sold it to John T. Wright, shipmaster of San Francisco, in August, 1848. It had gone by the names of Horseneck, Locust Island, and now became Wright's Island. It passed to George S. and Cordelia Wright in 1860. A nephew, George Foster, sold the peninsula around 1910, and it was surveyed and subdivided. Locust does not refer to the species of grasshopper, but to the trees that grew there. Locust wood was esteemed by the farmers for its toughness and natural resistance to rot.

LOCUST POINT DRIVE. The main street of this community was known for almost 50 years as East 177th Street. President James McQuade of the Locust Point Civic Association proposed, in October 1986, that it be given a more identifiable name, and in February 1987, new street signs went up. Happily, the name has historical value, as well. See: Locust Point.

LODOVICK AVENUE. This avenue was first called Valentine Avenue as it ran through the 28-acre Valentine estate along the Eastchester Road. In the mid 1800s, it was part of the Watson and Bruner lands. Before that, it was on the vast Givan farm which had been purchased in 1795 from the Bartow family. The first Bartow in 1725 called the land "Scabby Indian," and he purchased it from the influential Pell family. Givan's Creek once crossed Lodovick Avenue at its intersection with Mace Avenue. Incidentally, the Valentine house stands at Woodhull Avenue and Fielding Street. Of English descent, Charles Lodovick was a Lieutenant Colonel in early New York, where his father had been a trader there under the Dutch. Charles served as Mayor in 1694-1695. He died in England.

LOGAN AVENUE. Major General John Alexander Logan of the Civil War was honored in 1918.. In 1868, he issued an order fixing May 30th as Memorial Day. During the Revolutionary War (October, 1776), four companies of Hessian Grenadiers were stationed along present-day Logan Avenue to guard both the Throggs Neck Road (East Tremont Avenue) and to overlook Weir Creek, which was a water route in those days. Weir Creek has since been displaced by the Throggs Neck Expressway. From Dewey Avenue north to Lafayette Avenue, this was the estate of John David Wolfe (1820s), his daughter Catherine Lorillard Wolfe (1860s) and the Bruce Brown estate (1900s). From Lafayette Avenue north to Bruckner Boulevard, the avenue closely follows the line of trees in Coster's orchards (1850-1910).

LOHENGRIN PLACE. This land was owned by the Underhill family in 1685, by the Ferris family during the Revolutionary War, by J. Van Antwerp around 1865, then the Westchester Polo Club and a Dr. Ellis. The Ferris mansion stood on Rawlins Avenue and Lohengrin Place until 1961 when it was bulldozed to rubble overnight. Along with Parsifal and Siegfried Places, this street is a reminder of a Wagnerian opera. Eugene Rosenquist, President of the Westchester Electric Light Co., lived here, and is credited with naming them.

LONGFELLOW AVENUE. All streets in the vicinity carry names of poets, and this honors Henry Wadsworth Longfellow. It was, at first, called Division Street from East 176th Street to Boston Road as it divided the estates of Minford and Woodruff; and another short-lived name was Boone Street from Westchester Avenue to East 165th Street. Proceeding south from Westchester Avenue, this street traverses the former lands of Colonel Hoe, Faile and Fox to end at the site of what had been an Indian village.

LONGSTREET AVENUE. It is laid out across the original Pennyfield, later the Wright property, the Newbold estate and finally the Wissmann estate. A Locust Point map of 1923 shows an earlier name of Hooker Avenue, honoring a Civil War general. Why this avenue was named for a Confederate general has intrigued many historians. General James Longstreet was known as "the Dutchman" as he stemmed from an old Dutch family, Langstraat. The family migrated from the Netherlands in 1658 and settled in Manasquan, N.J. For some reason, Longstreet cast his lot in with the South. This writer believes the last property owner, Francis De Ruyter Wissmann, was instrumental in having the street named for "the Dutchman" as he too claimed Dutch ancestry.

LONGVIEW PLACE. This short Place is well named, for it overlooks the historic valley route of the Americans, British, Hessians, French and Indians through what is now Van Cortlandt Park. Longview Place is situated on the slope of the Cortlandt Ridge, or "Upper Cortlandt," purchased by Captain Joseph Delafield in 1829. It was sold by lots in 1873 to become part of Frederick Goodridge's estate.

LONGWOOD AVENUE. It was once called Lane Avenue. This avenue is a reminder of a large tract called "Longwood Park" which was owned in the 1870s by S. B. White. It was bounded by Westchester, Prospect, Leggett and Intervale Avenues. In those days, however, the boundaries were called Westchester Turnpike, East Morrisania, Leggett's Lane and Sacrahung Brook. The name "Longwood" was most popular during the 19th century, as the Emperor Napoleon was confined to a Longwood Park on the island of St. Helena.

LOOMIS STREET. This was Ash Street off Eastchester Road before 1900. Although Horace Loomis was a surveyor in the vicinity in 1904, the naming of this street goes back to landowners Charles Loomis (1797-1870) and his son John (1835-1889) who farmed the land, as did another son, Russel (1840-1906). Their family plot in Woodlawn Cemetery is shared with the Sacketts of the adjoining farmland along Eastchester Road. See: Sackett Avenue.

LORETO (Alfred) PARK. This large park at Tomlinson and Van Nest, Haight and Morris Park Avenues was formerly part of the Pearsall estate-Macomb's land - Morris Park race track - and the lawn of the Woodmansten Inn. Alfred Loreto was an off duty policeman who was killed in July 1950 by escaping gunmen. He lived in the immediate neighborhood, and the residents requested the park be named in his honor when it was opened in 1951.

LORILLARD FALLS. This is a waterfall of the Bronx River above the (restored) Lorillard Snuff Mill in the Botanical Garden. It was named after Pierre (Peter) Lorillard, who was the founder of the tobacco fortune. The term "millionaire" was not coined until he died in 1843. At the time, an obituary writer, plagued by a deadline, minted the phrase and it passed into the language.

LORILLARD PLACE. Emmet Powell, William Bayard and Jacob Lorillard purchased a tract of land – practically all of Belmont-in 1851. Other members of the Lorillard family owned property which is today the Bronx Botanical Garden and also the Bronx Zoo.

LORING PLACE NORTH. It is laid out through the former Andrews estate and named on March 3rd, 1936 by the Board of Aldermen. William Loring Andrews was a trustee and honorary librarian of the New York's Museum of Art. At one time, the Place was called Dayton Place. See: Andrews Avenue.

LORING PLACE SOUTH. See: Loring Place North.

LOU AUGER SQUARE. Thirty years of volunteer work with the little League and equal time with the 50th Police Precinct was recognized in 1988 by friends and neighbors when this square was named, at Kingsbridge Avenue and West 236th Street. Lou Auger died in December, 1986.

LOU GEHRIG PLAZA. This concrete island at East 161st Street off Walton Avenue honors the famous Yankee ballplayer who starred so spectacularly in nearby Yankee Stadium. The land was once part of the Gerard Morris estate, "The Cedars."

LOUGHRAN PLACE. George A. Loughran was a member of the taxpayers' committee that developed the Edenwald section in the 1920s. He was a life member of the New York Principals Association and taught history and civics at Evander Childs High School in The Bronx. He died in 1959, at the age of 75, and the City recognized his services by naming Loughran Place (Boston Road and Pratt Avenue) in his honor.

LOUIS NINE BOULEVARD. The naming of this road meant the disappearance of Wilkins Avenue on July 25th 1985. The change was in memory of the State legislator who represented the 79th Assembly District for 12 years in Albany. He was known for his special concern for housing low- and middle-income Bronxites.

LOWELL STREET. This very short street off Whitlock Avenue is on the former estate, "Brightside," that belonged to Colonel Richard Morris Hoe. It was named in February, 1911 and who it honors is not absolutely clear. The street probably honors James Russell Lowell, an American poet-just as adjoining streets were named after Longfellow, Drake and Whittier. However, there is a suspicion this street might honor a historian, Dr. Edward J. Lowell, noted for his research in the Revolutionary War. This possibility is voiced because nearby East 165th Street was once Bancroft Street, and George Bancroft was another historian of the Revolutionary War.

LOWERRE PLACE. This high spot was first known as Prospect Terrace for its view of the Bronx River at East 227th Street. The Lowerres were landowners in this region dating back to 1817, and were related to the Bell and Baldwin families, who also have streets named after them. An Edgar Baldwin Lowerre died in World War I.

LOZADA PLAYGROUND. The site, on the north side of East 135th Street between Alexander and Willis Avenues in the Mott Haven section of the Bronx, was acquired in 1938 and constructed in 1939. On November 21, 1987 the playground was named for Private First Class Carlos F. Lozada, who was posthumously awarded the Congressional Medal of Honor for his selfless actions on Hill 875 in Day To region of Vietnam three miles from the Cambodian border.

LUCERNE STREET. This street runs across the ancient manor lands of the Underhill family who migrated from England in the 1600s. There, their lands were known as "Kenilworth," and they gave the same name to this area. In 1776, General Howe's troops assembled nearby to embark in small boats for the opposite shore, Rodman's Neck. In the 19th century, the land retained its aspects of an estate belonging at different times to landowners named Edgar, Lorillard and Spencer. At the waterfront, the end of Lucerne Street was once the southern rim of Palmer Cove. Lucerne is another name for alfalfa, and the meadows here served as pasturage for the polo ponies and thoroughbreds. The street was named on October 20, 1922 by resolution passed by the Board of Estimate.

LUDLOW SQUARE. This junction of Bruckner Boulevard and Soundview Avenue was once the original Indian path that gave way to the Revolutionary-times road to Cornell's Neck (Clason Point). The Ludlow family acquired considerable property on Clason Point around 1830-one estate being the Black Rock farm, and the other called "Woodlawn." A Stephen Ludlow was a City surveyor. Ludlow Island, in the Bronx River, is now part of Soundview Park while Ludlow Avenue was an earlier name of Bruckner Boulevard.

LURTING AVENUE. This avenue bisects the former Morris Park race track that eventually became a field for early airplanes and balloon ascents. Thomas Haddon's Saw Mill Lane of the 1720s cut across this avenue at Allerton Avenue. The avenue honors Robert Lurting, Mayor of New York for nine successive terms, 1726 to 1735. He was born in England, and died in office.

LUSTRE STREET. This short street went by the name of Dark Street after John H. Eden's "Edenwald" was subdivided. On February 26, 1926, an ordinance of the Board of Aldermen changed it to Lustre Street. It is surmised that the home owners objected to "dark" and requested its opposite, lustre-brilliancy, shine, or sheen.

LYDIG AVENUE. Philip Leidig [spelled this way] was a ship's baker who migrated to America in 1750 and supplied sea biscuits to the merchant marine. His son, David, expanded into milling of flour and, in 1830, purchased Delancey's mills near West Farms. He died in 1842. Philip's grandson, also called Philip, who anglicized the name, inherited a half million dollars, and later sold the land which is now Bronx Park. One daughter married Judge William V. Brady, and the other wed Judge Charles P. Daly.

LYMAN PLACE. This short Place, opened in 1903, was named for Lyman Tiffany. Originally, it was at the extreme end of the Morris Manor lands, which were bounded by the Sacrahung Brook (Intervale Avenue). In the 19th century, it was part of the Tiffany estate which had belonged to Mrs. Tiffany's father, William W. Fox.

LYNCH (Edward P.) TRIANGLE. On December 11, 1937, off duty Patrolman Edward P. Lynch was killed near Gun Hill Road and Eastchester Road (Givan Square) after attending Mass at the church of the Holy Rosary. Local Law 378, September 21st, 1943, was passed designating the upper angle of the Square "Lynch Square" at the instigation of the N.Y. Police Department, the V.F.W. and the parishioners of the local church. Apparently, signs were never erected nor the Local Law followed out. In October, 1967, the three organizations again petitioned Councilman Merola to file a bill, this time using the word "Triangle." In May of 1968, Mayor Lindsay signed the bill. Patrolman Lynch was killed while in civilian clothes, pursuing a prowler.

LYON AVENUE. This was on the former Haight estate of 1868 called "Hopedale." An 1894 map shows a Dore Lyon as owner of "Lyon Court," a small estate which was crossed by Indian Brook, or Seabury Creek, midway between Doris Street and Zerega Avenue. On some maps, Mr. Lyon's first name is given as Darius, and this might be the origin of Doris Street, which runs into Lyon Avenue. A few homes on this avenue have stone lions on their lawns and stoops.

LYONS SQUARE. Located at Bruckner Boulevard, Longfellow and Bryant Avenues, it is on the former estate of Colonel Richard Hoe. It honors a nineteenth-century resident and small landowner who lived at Forest Avenue by the name of James L. Lyons. He was a sergeant in the Civil War, and a charter member of the Bronx Oldtimers. Later, he moved to Home Street, but at his time it was called Lyons Street.

LYVERE STREET. The Henry Lyvere farm of the 1850s and 1860s, where Green Lane (now Zerega Avenue) ran into Bear Swamp Road (now Bronxdale Avenue). The Lyvere family is buried in the graveyard of St. Peter's church near Westchester Square. In 1914, the City acquired title to the land wide enough to cut through a street, which was then called Lyvere Street. It had been "on paper" for many years before that.

MABLE WAYNE PLACE. East 158th Street, between Walton Avenue and the Grand Concourse, was given the name of a court buff who 'adopted' the County Courthouse, the officers and judges for 35 years. She carried the badge of an Honorary Supreme Court officer, and became a legend when she was murdered at age 84. The naming ceremony took place on April 18, 1987 and was attended by hundreds of policemen, court attendants and neighbors.

MAC CRACKEN AVENUE. This street in Roberto Clemente State Park honors Henry Mitchell MacCracken, Chancellor of New York University, who conceived the nearby Hall of Fame, and who was active in promoting parks in The Bronx. The name was suggested by the Bronx County Historical Society.

MACE AVENUE. Levi Hamilton Mace was born in Rye in 1825 and died at his home in Williamsbridge in 1896. For more than 32 years, he was a resident there, and for 26 years of this period, he was President of the Board of Education of District No. 2. His refrigerator factory was located at East 150th Street and the Harlem River. Louis Pollack wrote this in 1901: "In the course of my real estate business I was obliged to see Mrs. Mace in Williamsbridge. The country surrounding the Mace estate was one large swamp and the water was three or four feet deep. As the White Plains trolley came to a stop, there was a rowboat waiting to take me to high ground." The western end of Mace Avenue was once owned (1844) by John D. Wolfe, a son-in-law of the Lorillards. From Boston Road to Williamsbridge Road, the avenue ran across the Hitchcock estate. Springs at Bronxwood Avenue formed the headwaters of Downing's Brook, which eventually ran into the Bronx River. From Williamsbridge Road to Eastchester Road (1864) it crossed the lands of Denton Pearsall. Stoney Brook crossed at Seymour Avenue. From there to the Hutchinson River, this was the land of John Bartow (1727), Furman, Stillwell and Bayard (1820), and finally John Hunter III (1880s). At one tune, this avenue was called Ferris Avenue.

MACLAY AVENUE. It was laid out across the Adee estate, off Westchester Square, in the 1890s. Isaac Walker Maclay was a surveyor for the Department of Parks of the 23rd and 24th Wards. His wife was a Havemeyer. He was present at Lincoln's assassination, and helped to carry the President's body from the theatre. The Maclays are buried in the circular Havemeyer plot in Woodlawn Cemetery, Beech sector.

MACOMB'S DAM BRIDGE. In 1813, Robert Macomb was granted authority to build a bridge with a draw, but instead he had it erected without a draw and dammed the Harlem River at approximately East 160th Street. A number of Westchester farmers finally tore down the dam in 1836 and reopened the river to navigation. Later, a wooden bridge, this time with a swing draw, was built on the same spot and it was known as Macomb's Dam Bridge. It was replaced by the steel structure around 1895.

MACOMB'S DAM PARK. A 50-acre tract along the Harlem River that was opened in 1899, it was named after the Macomb family of millers who once operated a dam and mill in the 1800s. Before World War II, the athletic field and quarter-mile track was located due west of Yankee Stadium. It has since been relocated to the portion due north of the Stadium, and named (in 1962) Schoolman Athletic Field. See: Babe Ruth Field.

MACOMB'S ROAD. At one time, in the 1880s, it was known as Highwood Road. It was named for Alexander Macomb, who owned a dam across the Harlem River erected in 1813-1814.

MACY PLACE. The Macy family owned a small estate, flanked by the Westchester Turnpike and Leggett's Lane (now part of Longwood Avenue) according to a map of 1881. Josiah and Sylvanus Macy share a family plot with the Leggetts on Magnolia Avenue in Woodlawn Cemetery.

MADISON AVENUE BRIDGE. When this bridge over the Harlem River was opened in 1910, there was no question that Madison Avenue in Manhattan was better known and far more important than East 138th Street in The Bronx.

MAGENTA STREET. It was once known as Juliana Street west of White Plains Road. Prior to 1900, there was a small French colony of weavers in the neighborhood. A Rudolph Rosa was a surveyor in this sector of Williamsbridge in the 1880s, and the story is that he was allowed to name this street. His name meant "rose-colored" in German, so he chose a shade of red. A similar story is told of Amethyst Street when Van Nest was surveyed in 1892. However, the authentic reason for naming the street most probably involves a battle. In 1859, the French emperor, Napoleon III, led an allied force of French and Italians against the Austrians at the Italian town of Magenta and won a great victory. So impressive was this feat that French designers invented a new color and named it after the battle. In The Bronx, with the small colony of French weavers living in a neighborhood to which a significant number of Italian immigrants had located, Magenta was a logical name for a street to celebrate Franco-Italian amity and a historic battle in the long struggle for the unification of Italy.

MAGNOLIA AVENUE. This road is located in the northeastern section of Woodlawn Cemetery, passing the family plot of the Leggett and Macy families, as well as the grave of Augustus Van Cortlandt. The magnolia trees are common to the cemetery.

MAGNOLIA PLACE. This is a road in Silver Beach Gardens that was the carriage path leading to the Hammond mansion. The Place is named after the magnolia tree, and the tree honors Pierre Magnol (1638-1715), who was a French botanist. See: Acorn Place.

MAHAN AVENUE. Ownership of this land was at times in the hands of the Ferris family, Buff family, and Baxter family. It was named in honor of Captain Alfred Thayer Mahan who served 40 years in the Navy, including action in the Spanish-American War. He is the author of the book *The Influence of Sea Power upon History.*

MAIN STREET. The unofficial name of main road in the Park of Edgewater. The name is not on City maps, but is so listed on layouts of the Park, which is private property.

MAITLAND AVENUE. Although it was only opened in 1929, this street honors old settlers of Middletown from the past century. Part of the Cornell farm "along the Old Road leading from the Pelham Bridge Road," this tract was sold by Samuel and Mary Cornell to Sarah Maitland in 1853. A Dr. Maitland was listed in the 1861 records of the Ferris Brothers' drugstore, and an Edward H. Maitland was found in the World War I roster of St. Peter's P.E. Church. Situated on the flank of what was called Crow Hill by the townspeople, but mapped as Bridge Hill during the Revolutionary War, this area figured briefly in the October, 1776, cannonading of Westchester village by the British troops on the hill.

MAJOR WILLIAM F. DEEGAN EXPRESSWAY. See: Deegan Expressway (Major William F.).

MANHATTAN COLLEGE PARKWAY. This winding roadway once linked a 19th-century settlement called Warnerville on Broadway and the crest of Riverdale. It carried an historic name, Spuyten Duyvil Parkway, until 1953. That year marked Manhattan College's centennial, and the Parkway was renamed in honor of the occasion. Manhattan College stood on a 12-acre plot, acquired in 1922 by the Brothers of the Christian Schools who had previously taught in Manhattanville, hence, its name.

MANIDA STREET. This street is on part of the original purchase by Richardson and Jessup in 1663. Gabriel Leggett married Elizabeth Richardson and owned the land west of the Indian trail that eventually became Hunts Point Road. The road, in its winding course, almost touched Manida Street at Spofford Avenue. Manida Street was laid out in 1897 across the former estate of the Fox family called "Springhurst." A lower section ran through a small tract owned by E.T. Young in 1865. No one can prove how the street got its name, so some folklore is worth repeating. One story is that a servant of the Fox family had her cottage demolished to make way for the street, and her name was Ida Mann. Another story is that John S. Coster, another landowner, named it after his former home on Minetta Lane in Greenwich Village. Mr. Wuttge, a long-time resident of Hunts Point, points out the archaic Spanish word, "Manida," is a home, or a shelter, and most likely refers to the Sevilla Home for Children. This orphanage and school for Spanish-speaking children was instituted early in the 1900s, being financed by a half-million dollar estate left by a Peruvian, Juan de Sevilla.

MANNING STREET. An 1870 map shows this short street as the western edge of an estate on the Westchester Turnpike called "Hopedale," owned by the Haight family. A Joseph Manning was a landowner listed a few years later, and a Richard Manning was so noted in 1901, but according to Percy Mallett, a lifelong resident of Westchester Square, the street honors a Roman Catholic bishop. A Jewish realtor had been given the opportunity of buying a tract of land from St. Raymond's R.C. Church, and he asked that the first street to be cut through be named in honor of the churchman. The street was opened in 1913. See: Tratman Avenue.

MANOR AVENUE. The William Watson manor, "Wilmount," stood at East 173rd Street off Manor Avenue in the vicinity of James Monroe High School. It was first named Chanute Street, but then received its present name on February 7, 1911.

MANSION DRIVE. This is a carriage driveway in Van Cortlandt Park, from Broadway, north of West 242nd Street, eastward to the Van Cortlandt Mansion. The official naming took place on March 11, 1987.

MANSION STREET. This short street is, situated on what were the ancient holdings of the Underhill family, which passed to that of the Hunt family prior to the American Revolution. A man named Guerlain owned a sizeable portion in the early 1800s, and then it became part of the extensive Mapes farm. The entire tract was converted into building lots around 1900 under the name of "Park Versailles." Legend has it that Mr. Guerlain built a picturesque chateau for his wife. No trace of it remains, unless the latter-day Mapes mansion was built upon its foundations, which was a frequent practice. The Mapes house stood near the intersection of present-day St. Lawrence Avenue and Mansion Street and may have been the mansion referred to.

MAPES AVENUE. It was first known as Johnson Avenue. No mention could be made of West Farms without the name of Mapes cropping up. Right after the close of the American Revolution, there was mention of a merchant Mapes who owned one of the three stores in West Farms. This store was built on the foundation of Colonel Delancey's blockhouse off Boston Road and East 179th Street. A century later, it was the Mapes Temperance Hotel. A Daniel Mapes (1850s) was victualer to Fort Schuyler's garrison, and other members of the family were in the coal, feed and grain business along the Bronx River. The Mapes family was related to the Archers, Bartows and Arnows.

MAPLE AVENUE. This is a private street in Silver Beach Gardens. Maple trees are still quite common there. See: Acorn Place.

MAPLE DRIVE (1). This is a privately maintained street inside Parkchester. Maple trees were found all over the Protectory grounds on which Parkchester is built.

MAPLE DRIVE (2). This is the main driveway from Riverdale Avenue west to the present site of the Elizabeth Seton residence hall of Mount St. Vincent. The name is an arboreal one.

MARAN PLACE. This was once part of the old village of Bronxdale. Ely Maran was the head of a corporation of small property owners along White Plains Road. In the mid-1920s, their land was sold to the City, a street was cut through to connect to Bronx Park East, and it was called Maran Place.

MARCY PLACE. This once constituted the northern borderland of the ancient Manor lands of Morrisania. Later, in the 1850s, it was on the Andrew Findlay estate. When the Place was first laid out, it was called Findlay Place to honor the old-time surveyor who had mapped so many parts of The Bronx. The present name might have some connection with McClellan Street, several blocks away. General Randolph P. Marcy was a Union Army Chief of Staff during the Civil War, while his son-in-law, General George McClellan, commanded the Army of the Potomac. One old story, political in nature, credits the name of the Place to Tammany Hall's notorious "Boss," William Marcy Tweed. Back around 1870, he was instrumental in acquiring real estate for a friend "far up in the country." Rather than call it Tweed Place, the new owner discreetly used the middle name of the "Boss."

MARINA DRIVE. A two-block stretch of Schurz Avenue was given this name in 1982. It was to recognize the civic charities and business of the Randazzo family whose Marina Del Rey restaurant faced the East River. Their establishment had been on the site of an Indian village, later the Throckmorton colony of 1642, and the landing site of British and Hessian troops in 1776.

MARINE STREET. On old maps, the eastern half of this street was called Franklin Street on City Island. In 1874-1883, B. Frank Wood & Son had a small boat yard at the foot of this street, their work consisting mostly of small craft. Later, the yard became famous for its Gardner-design yachts. This was once the property of Rochelle Horton, who was a marine pilot. See: Horton Street.

MARIO MEROLA GROVE. On May 18, 1988, Indian Field in Van Cortlandt Park had 8 white pines added to an existing grove in memory of a Bronx District Attorney (1973 - 1987) who also was an ardent defender of the Bronx parklands. He and his family who lived nearby often visited the grove which was given his name after his untimely death.

MARION AVENUE. From East 184th Street to Fordham Road, it was once known as Virginia Avenue, later to be changed to Bainbridge Street. From Fordham Road north to present-day Mosholu Parkway, it was called Hull Street. It finally was named in honor of Francis Marion, the Revolutionary War "swamp fox" who harried the British forces in North and South Carolina.

MARMION AVENUE. Local lore is that the original name was Marion Avenue, but was changed slightly because of Marion Avenue in nearby Fordham. It is laid out across the former settlement of Fairmount, and some researchers believe the name is taken from a romantic martial poem by Sir Walter Scott, whose hero was a Lord Marmion. Scott's literary works enjoyed great popularity in America, and no doubt influenced the former owners who were all Englishmen by birth or descent: Bathgate, Minford, Dunham and Cochran. The town fathers, too, may have had a classical education.

MAROLLA PLACE. See: Marolla Terrace.

MAROLLA TERRACE. A bill was passed in December 1968 by Councilman Mario Merola to honor the late Edward A. Marolla. The terrace fronts on the Bronxwood Little League baseball field which Mr. Marolla helped to organize. He was also a Son of Italy, member of the Bronx Grand Jurors' Association and an official of the Sinclair Oil Co. He died in May, 1968. The former name had been Eden Terrace.

MARSHALL AVENUE. General Elisha Marshall [Civil War] has his name affixed to this Throggs Neck avenue. To date, the street has not been opened into Silver Beach Gardens.

MARTHA AVENUE. Both the Varian and Huestis families owned considerable property in the North Bronx, and this avenue in Woodlawn Heights is thought to hark back to Martha (Huestis) Varian. In prior years, it had been called Sparrow Avenue.

MARTIN LUTHER KING BOULEVARD. The stretch of University Avenue from Edward L. Grant Highway to West Fordham Road has had two street signs since November 1988. University Avenue recalls New York University's earlier site, while the second is a tribute to Dr. King. See: Martin Luther King Memorial Park.

MARTIN LUTHER KING MEMORIAL PARK. Named for the noted Black leader and national figure who was a Nobel Prize recipient. The park, at Shakespeare Avenue and West 168th Street, was dedicated in the early 1970s. See: Shakespeare Avenue.

MARVIN PLACE. This was once part of the property belonging to John T. Adee in the village of Westchester. Dr. Marvin D. Hubbell was prominent in real estate transactions before World War I. See: Hubell Street.

MATILDA AVENUE. Matilda Feth was a 1900 landowner with property fronting on this avenue at East 241st Street. Folklore has it that Henry Haffen, Commissioner of Streets, named it for his wife Matilda (Stoller). In the 1850s, Adelina Patti, a famous operatic star, lived in the village of Washingtonville on what is now Matilda Avenue.

MATTHEWS AVENUE. North of Allerton Avenue, where this avenue runs through the former Adee estate, it was first called Adee Avenue. Near Gun Hill Road, it was called Rose Street. David Matthews was a former Mayor of New York, serving from 1776 to 1784 during the occupation by the British forces. He was born in England and was sympathetic to the British cause.

MATTHEWSON ROAD. This Road in Roberto Clemente State Park honors Douglas Matthewson, President of the Borough of The Bronx, 1915-1918, who once lived nearby. The name was suggested by The Bronx County Historical Society.

MAYFLOWER AVENUE. This was Mapes Avenue until 1898, a reminder of the Mapes farm in this area. Streets in the neighborhood carry related names: Pilgrim, Puritan and Plymouth Avenues.

MAYFLOWER PLAYGROUND. See: Florence Colucci Playground.

MAZZEI PARK. A small park and playground at Mace Avenue and Williamsbridge Road was dedicated to the memory of Filippo Mazzei who, historians concede, influenced Thomas Jefferson in his composition of the Declaration of Independence. Mazzei's political essays in Italian were translated by Jefferson who used some of the ideas and phraseology. The bill, naming the park, was signed by Mayor Edward Koch in November 1980.

McCLELLAN STREET. This was part of the Morris estate, which later was subdivided into town lots and called West Morrisania. When the street was opened in 1879, it was called Maillard Place but later was changed to Endrow Place in 1897. It presently honors George B. McClellan of the Civil War.

McCORMICK LANE. See: John McCormick Lane.

McDONALD STREET. Originally swamplands drained by Stoney Brook, maps show it at various times belonging to the Timpson family (1870s), the Ferris family (1880s), and John Hunter at the turn of the century. In 1776, Washington and his staff crossed at this point on his way to Throggs Neck and, later, a body of American troops prevented a scouting party of British from crossing the creek. A C. H. McDonald was a surveyor in the early 1900s, and he may have gotten his name affixed to this street; but there is an off chance Assemblyman Thomas J. McDonald was able to get himself immortalized back in 1911 when the street opened.

McDONOUGH PLACE. This Place is laid out on the land grant of the Underhill family of 1685, which later became the Bayard farm, which, still later, became the William Spencer estate. Lorillard Spencer was a later proprietor before selling some land to James J. McDonough [spelled this way] in 1922. That same year the Place was officially opened and named.

McGOVERN FIELD. This is a football field on the grounds of Mount St. Michael Academy at Murdock Avenue on the city line. It was named after a benefactor who was a contractor in the 1920s and who helped to lay out and fill the grounds to make the field possible. Before that time, it had been a poorly drained field known as Bathgate Oval, as it was a part of the Bathgate farm.

McGOWAN STREET. Commander McGowan has his name affixed to a road running across the campus of the New York State Maritime College. He was Superintendent from 1891 to 1894.

McGRAW AVENUE. Today, this street is only a reminder of the extensive McGraw farm that is now incorporated into Parkchester. John McGraw was listed as a landowner in 1868, and the avenue was merely mapped as "the Old Road."

McKINLEY SQUARE. Once an important traffic hub and commercial center of Morrisania, it was officially named in 1902, honoring the President of the United States who held office from 1897 to 1901. Dr. Condon lived nearby in the 1880s. His daughter, Kate Condon, was a well-known schoolteacher, as was a son who gained fame as "Jafsie" in the Lindbergh kidnap case of 1934-1935.

McLAUGHLIN PARK. This small park at Greystone Avenue and West 238th Street was dedicated in 1971. Ruth McLaughlin had been a community relations assistant in the Bronx Borough President's office.

McLEAN AVENUE. This northernmost avenue of Woodlawn Heights acts as a common boundary line for New York City and Yonkers. James M. McLean was a property owner listed in 1867 and 1881, with 26 acres in The Bronx and 76 in Yonkers. McLean's Brook rose in a pond on his property and flowed south into Woodlawn Cemetery. Its course roughly followed Katonah Avenue.

McNAMARA SQUARE. See: John McNamara Square.

McOWEN AVENUE. This avenue overlays part of the ancient Huguenot lands of pre-Revolutionary times. It was the 1776 scene of Colonel Glover's retreat before the advancing British. The City took title in 1935. Anthony McOwen Sr. was born in County Wicklow, Ireland, in March, 1842, and came to the United States at the age of

16. He learned carpentry and later enlisted in the 99th Regiment of New York Volunteers, serving in the Civil War. He was Clerk of the 10th District Court and later Deputy Tax Commissioner. He was also the first Coroner appointed to The Bronx. His son, Anthony, Jr., was a builder and contractor of the early 1900s.

MEAD STREET. Van Nest streets carried names of our Presidents and this street was originally called Grant Street. Sometime around 1908-1910, the name was changed. It is almost certain the short street at Unionport Road owes its name to one of Captain Honeywell's volunteer horsemen during the Revolutionary War, Calvin Mead. He was still alive in 1846, when Judge MacDonald and Andrew Corsa - last of the Westchester Guides - interviewed him on his part in the Revolution. He was living "near the Hunt farm," and Van Nest had been the property of the Hunt and Fowler families back then.

MEADE (George) PLAZA. The triangular plot in front of the 42nd precinct police station house at East 160th Street and Washington Avenue was named in 1981 to honor a black patrolman who died in 1973 during a hold-up.

MEAGHER AVENUE. This Throggs Neck avenue was named in 1888 for General Thomas Francis Meagher (1823-1867), who was born in Ireland and exiled to Tasmania for treason. He escaped to America and became a politician, lawyer and editor of the *Irish News*. He organized the New York Zouaves in the Civil War, and became a Brigadier General. He was appointed Governor of Montana, and drowned in the Missouri river. Meagher Avenue, from Schurz to Harding Avenues, was opened in 1930 having been part of farmlands belonging to Abijah Hammond (1805), William Whitehead (1870s) and Frederick Havemeyer (1890s). From Harding Avenue to the Cross-Bronx Expressway, it was a very ancient lane, called Pennyfield Lane, on old maps dating back to Revolutionary times. From 1850 to 1910, it was called Adee's Lane, as it led to the Adee estate (now Park of Edgewater). St. Frances DeChantal's R.C. church was first organized in a tent on Meagher Avenue and Lawton Avenue. The correct pronunciation of this name is Marr.

MEANY (George) SQUARE. In the Spring of 1981, a triangle, formed by Westchester, Wilkinson and Continental Avenues, was formally dedicated to a local resident who had risen from plumber's helper to national stature as President of the AFL-CIO. George Meany held that office from 1955 to 1979. A year later, he died at the age of 85.

MELROSE AVENUE. Andrew Findlay surveyed this district in the 1850s and he was of Scottish origin, so he is credited with naming Melrose. Sir Walter Scott immortalized it in *Melrose Abbey* and the name was quite popular in the middle of the 19th century. Melrose Avenue was opened in 1891.

MELROSE PARK. This one-acre park at the northern end of Courtlandt Avenue, East 161st Street and East 162nd Street, was opened in 1902, and for many years was simply called "One Acre Park." Another nickname was "Baby Park," which might have alluded to its small size, or because it was well patronized by mothers and babies. On March 16, 1920, Borough President Henry Bruckner formally named it.

MELVILLE ABRAMS BALL FIELD. Assemblyman Abrams, who died on October 10, 1966, was a lifelong resident of the district he represented. A local law, September 14, 1967, designated this ball field in Soundview Park in his honor.

MELVILLE STREET. This was originally Hancock Street in Van Nest. It was, at first, only a short lane extending west from the railroad, but, in 1891, it was widened and paved, then lengthened to reach Morris Park Avenue. In that same year, Herman Melville died and was buried in Woodlawn Cemetery (24, Catalpa) and this gave rise to the contention the street honors this author of sea stories, including *Moby Dick*. Melville had lived in obscurity at 104 East 26th Street for many years, so the connection is tenuous at best.

MEMORIAL PARK. A small plot at the junction of City Island Avenue and Hawkins Street, formerly occupied by the police station. A war memorial tablet is in the park.

MERMAID COURT. The history of this street is the same as Fleet Court.

MEROLA GROVE. See: Mario Merola Grove.

MERRIAM AVENUE. This was the Indian land called "Nuasin" (the land between), an allusion to the land between the Harlem River and an estuary now filled in by Jerome Avenue. In 1671, it was mapped as Turneur's Land, and, in the 1700s, it was known as Devoe's Neck. A brook at West 168th Street and Merriam Avenue formed the boundary of the Manor of Fordham and the Manor lands of Morrisania. Around 1848, this land was the upper limits of Highbridgeville. Francis W. Meriam [spelled this way] was listed as a landowner in the 1850s, but when the avenue was opened in 1902, an extra 'r' was added.

MERRILL STREET. Three streets in this section, called Park Versailles, honors Electors of the ancient Town of Westchester, 1775: Guion, Odell and Merrill. Thomas Merrill is most likely honored by this street, which crosses the Mapes farm of the early 1800s.

MERRITT AVENUE. From the Boston Post Road north to the city line, this avenue runs through the former Louis B. Halsey estate. All the streets in the vicinity are named after early New York Mayors, so this avenue honors William Merritt (sometimes Merrett), born in England, and a mariner-turned-merchant in Little Old New York. He was elected Mayor for three successive terms from 1695 to 1698.

MERRY AVENUE. Thomas W. Merry, according to an 1811 deed, was a landowner of Middletown. In 1821 he was listed on a roll call of the Westchester Agricultural Society. Other family members, Robert, Henry C. and Marcus T. Merry, are all buried in St. Peter's P.E. churchyard. The avenue itself was originally called Vincent Avenue and formed the eastern boundary of Josiah Quinby's "Quaker Tract," which he deeded to the Society of Friends in 1808.

METCALF AVENUE. Charles and Hester Metcalf were listed as landowners in 1842 and 1846 deeds. The avenue runs across what was once the Ludlow family's "Black Rock farm," and the William Watson estate of 1868. Pine Rock Creek once wound in through the swamps from the Bronx River where it reached approximately to the corner of Story and Metcalf Avenues.

METROPOLITAN AVENUE. This is a privately maintained thoroughfare inside Parkchester, which was opened in 1939 to service the huge residential project. This development was owned by the Metropolitan Life Insurance Company, hence the name.

MEYERS STREET. Sebastian F. Meyers, physician, purchased land in 1836 along Throggs Neck Road (East Tremont Avenue) and Willow Lane (Bruckner Boulevard). In the 1850s, a small village sprung up because of the construction of Fort Schuyler and employment at the various estates. This settlement received the name of Schuylerville.

MICHAEL DEMARCO SPORTS COMPLEX. In June 1990 this longtime Bronx councilman had his name affixed to a Little League ball field and adjoining facilities, situated at Little League Place (formerly Tan Place) and Westchester Avenue. His efforts in City Hall were appreciated by the local citizenry.

MICHEL SQUARE. This odd-shaped plot at Claremont Parkway and Clay Avenue was acquired in 1897 by the Department of Parks. On May 22nd, 1940 (local law 62), it was named in honor of a World War I casualty who had served in the Field Artillery.

MICKLE AVENUE. Most of this avenue runs through the former lands of Robert Givan, who purchased a vast tract overlooking the Hutchinson River in 1795. The Givan home still stands on Mickle Avenue (2910 Mickle Avenue), although very much changed. At one time, the avenue was called Birch Street. Andrew Mickle was Mayor of New York from 1846 to 1847. Most avenues in the neighborhood carry names of New York Mayors.

MIDDLE REEF. This is a mussel-covered reef in Long Island Sound next to the boundary of Bronx and Westchester waters in the vicinity of the Nonations. Large boulders on its western end give it the appearance of a wall. Evidently, the reef got its name by being in the middle ground between the nearest Bronx and Westchester islands. See: East Nonations.

MIDDLE ROCK. According to the U.S. Geological Survey, this is the rocky ledge due south of Pelham Bridge, in Eastchester Bay. This rock is almost in midstream, which would account for the name.

MIDDLETOWN ROAD. This was once Main Street in the village of Middletown. In Revolutionary War times, this was the route of General Lord Howe's march to Eastchester Bay in October, 1776. In the early 1800s, old-time settlers named Baxter owned the triangle of land at Middletown Road and present-day Bruckner Boulevard occupied by the Indian Museum. The stretch from Bruckner Boulevard to Eastchester Bay was once called Waterbury's Lane, as it led to the Waterbury estate, now incorporated into Pelham Bay Park. Oddly, Middletown was midway to no two other towns. Perhaps it got its name from being midway from Westchester village to the Pelham Bridge.

MIELE PARK. See: George Miele Park.

MILDRED PLACE. This is a short street on which was once the Paul farm, where it sloped westward to the swamps of Westchester Creek. Military records tell of General George Washington meeting the Widow Paul during an inspection visit to Crow Hill (Dudley Avenue). The Pauls owned considerable farmland west of the Pelham Road (Westchester Avenue), and this Place is believed to hark back to a Mildred Laurie, a granddaughter of the original farmers.

MILES AVENUE. Miles Avenue, west of East Tremont Avenue was not opened until the 1920s, and ran through what was once the extensive John A. Morris estate. The portion from East Tremont Avenue east to Meagher Avenue was opened around 1903, and was then called Pennyfield Avenue as it led to the area bearing that name. An ice pond was located at Miles Avenue and Throggs Neck Boulevard, and once was stocked with carp. It carried the alternate names of Jackson's Pond and Soldier Pond. The oldest section of the avenue runs from the Throggs Neck Expressway to Long Island Sound and was a long-standing division lane between the estates of Herman LeRoy Newbold's "Pennyfield" and George T. Adee's "Edgewater." In the 1870s, it was mapped as Greene Avenue after a George Greene who owned property there. An ice pond was once located by Korony Square. An 1897 map shows the name Ocean Avenue, but, shortly thereafter, it acquired the name of Miles Avenue to honor Nelson A. Miles who captured Geronimo in 1887, and who became a general in the Spanish-American War.

MILL RIVER ROAD. This is a private road on the western side of the Fordham University campus paralleling the railroad tracks. In earlier times, the Mill Brook ran alongside this road.

MILTON PLACE.It is a reminder of Archer Milton Huntington (son of Collis P. Huntington), who was born in 1870 and educated in Spain. He was the author of many books on Spanish history, and the founder of the Hispanic Society of America. He was also cofounder of the Museum of the American Indian. Oddly enough, many Indian relics were dug up at Milton Place off Hollywood Avenue. George Younkheere, an expert historian, discovered an almost-intact Siwanoy earthen pot which predated the coming of the white men.

MINERVA PLACE. This Place was opened in 1913 across the former estate of Leonard W. Jerome, a wealthy financier and horse fancier of Civil War times. Its name has a doubtful history. One version is that it commemorates a classical statue of Minerva that graced the lawn of Jerome's villa, and which had to be removed when the Place was surveyed. The other tale is that Minerva was a famous race horse of the Jerome Park race track.

MINFORD PLACE. This is a rare example wherein an estate's name, "Minford Place," became the name of a short street. The Minford estate ran from Crotona Park to the Bronx River in the vicinity of the Cross-Bronx Expressway. A spring-fed pond was once located at nearby Suburban Place, and the brook ran along the present-day Minford Place to drain into the Bronx River. Thomas and Abigail Minford are buried in Woodlawn Cemetery.

MINNEFORD AVENUE. This is a reminder of one of the earlier names by which City Island is known. It was sometimes called Great Minneford, and is thought to be named after an Indian chief. In 1906, a real estate venture called "Minneford Park" was laid out east of City Island Avenue, through which this avenue runs.

MINNEFORD PLACE. See: Minneford Avenue.

MIRIAM STREET. The land was purchased by Isaac Varian in 1792, and was kept in the family for three generations. A grandson employed a William Rosa to survey and subdivided the tract for city lots. According to a lifelong resident of Fordham, Mrs. Dunn, this short street had the first apartment house to be built in that section, and the owner's daughter was named Miriam.

MITCHELL PLACE. This Place overlays ancient Hammond Cove which once made Locust Point an island at high tides. The street was named for Supreme Court Justice Richard A. Mitchell in 1920. He was contemporary with Justices Mullan, Tierney, Giegerich, Glennon and Hatting. See: Giegerich Place.

MOHEGAN AVENUE. Formerly known as Grant Avenue, its name was changed to commemorate our Indian predecessors. This tribal division included the sub-tribes that roamed over the region we call The Bronx.

MONROE AVENUE. It was officially named on May 2nd, 1916 for James Monroe, fifth President of the United States. This hillside, now divided by the Cross-Bronx Expressway, was once the Fisher farm of the early 1800s, which later was worked unsuccessfully by a man named Gamey. Around Civil War times, the County seized the farm on foreclosure, and an Elisha Monroe became the owner. In 1871, it was mapped as part of the Zborowski estate, which eventually became Claremont Park.

MONSIGNOR HALPIN PLACE. See: Halpin (Monsignor John) Place.

MONSIGNOR JOHN A. STELTZ SQUARE. The intersection of East 166th Street and Prospect Avenue received this name in June, 1990 to honor a priest who had served 33 years in St. Athanasius R.C. Church. Father Steltz, aged 67, died in 1989.

MONSIGNOR McCARTHY PLAZA. Since June 1990 the intersection of East 207th Street and Perry Avenue has been given the name of Msgr. John C. McCarthy as a tribute to his zeal in St. Brendan's R.C. School.

MONSIGNOR VOIGHT GREEN. See: Voight (Monsignor John) Green.

MONTEREY AVENUE. Pre-Revolutionary settlers in this area were the Ryers. One, Samuel Ryer, was an officer in the Mexican War, and he named his holdings above the Bathgate farm (now Crotona Park) "Monterey" to remind him of his campaigns in Mexico. His estate is shown on an 1866 map. Monterey Avenue was opened in 1905. East 180th Street was first called Samuel Street, and Belmont Avenue at East 181st Street was Ryer Place.

MONTGOMERY AVENUE. This was called Andrews Avenue, but in May, 1889, the name of a former resident was given to the avenue, that of Brigadier General Richard Montgomery. He was born in Ireland, served in the British army and then resigned to work a 75-acre farm in the area of Giles Place in the upper Bronx. It was from this farm that Montgomery left on the expedition to Quebec, where he was killed on New Year's Eve, 1775. It was not until 1818 that his body was brought down to New York on a funeral barge along the Hudson river and reinterred in the wall of St. Paul's Chapel.

MONTGOMERY PLACE. An 1868 deed dealing with property lines on Walker Avenue (East Tremont Avenue) has mention of a Montgomery Roselle. The Place was opened in 1919. On the opposite side of East Tremont Avenue, Roselle Street marks the boundary of his tract of land.

MONTICELLO AVENUE. An earlier name of this street was South 13th Avenue when it was part of Mount Vernon. It was later named Monticello after the estate of Thomas

Jefferson, just as Mount Vernon was the estate of George Washington. Furthermore, this avenue crosses Edenwald Avenue, which originally was called Jefferson Avenue.

MOORE PLAZA. At the junction of Barnes, Bronxwood and Pitman Avenues at East 236th Street, the plaza was never officially commemorated, but, in 1961, Msgr. McGuire finally had street signs erected. Reverend Francis P. Moore (1859-1928) was born in Ireland, but was brought to New York as an infant. He was ordained in 1884, and assigned to Wakefield in 1897 to establish a parish. It was through his zeal that St. Francis of Rome church and the parochial school became realities. He was a forceful speaker, linguist, an excellent artist and an expert in Colonial history. He was an authority on New York City topography, as well. He was often called "The Vicar of Wakefield," and literally burnt himself out in his efforts to complete the church and school.

MORGAN AVENUE. All the avenues in this vicinity bear the names of Governors, so it is safe to assume this avenue honors Governor E. D. Morgan. See: Fenton Avenue.

MORRIS AVENUE. This lengthy avenue takes it name from the Morris family that owned most of the territory through which the avenue runs. At the lower end of The Bronx, it was part of the original Lewis Morris estate during the Revolutionary War, and, in the 1840s, was mapped as the Long Neck farm. From East 138th Street to East 148th Street, it runs across Jordan Mott's purchase of "Mott Haven." At East 149th Street and Morris Avenue was the notorious Frog Hollow of the 1860s to 1880s. From East 163rd Street to East 167th Street, it runs across what had been the Fleetwood race track. It was 2nd Avenue from Claremont Park to the Concourse, Avenue A from East 181st Street to Fordham Road and Kirkside Avenue from Fordham Road to St. James's Park. It was not until 1906 that the name Morris Avenue was given to the entire length from Old Morrisania up to Bedford Park.

MORRIS PARK AVENUE. In the 1860s, it was part of the Hatfield estate, "Glenn Dale." "Round Meadow," a nineteenth-century farm belonging to John DeLancey Neill, was located around Morris Park Avenue and Bronxdale Avenue. In 1888, Eliza Macomb conveyed the tract of 152 acres to John A. Morris of Throggs Neck, who laid out an oval racetrack. It was called Morris Park. The track closed down in 1904, and the land was sold and earmarked for homes, but a lease was obtained on the land up to 1909 by an Aeronautical Society. The stalls were converted into workshops and hangars for the balloons and airplanes, and the world's first air meet was held there. Morris Park Avenue cuts through the former race track. The stretch from Williamsbridge Road to Eastchester Road was once called Saratoga Avenue. The section in Middletown (Hutchinson River Parkway to Westchester Avenue) was originally Liberty Street in 1892, renamed Alice Avenue in 1910, then took the name of Morris Park Avenue until 1968. That year, the avenue was renamed St. Theresa Avenue in honor of the Catholic church that is situated on it.

MORRISON AVENUE. This avenue runs across part of the Black Rock farm, mentioned in old land grants of the 1700s. In 1868, it was on property known as the Watson estate. Barrett's Creek crossed this avenue at Watson Avenue, and another small tidal stream, known as Pine Rock Creek, briefly touched Morrison Avenue at Seward Avenue. S. P. Morrison was a City engineer and a surveyor listed in 1874. His name was given to this avenue. See: Croes Avenue.

MORSE WALK. This is a footpath laid out across the campus of Bronx Community College (formerly New York University), and honors Professor Samuel F.B. Morse who occupied the chair of Literature, Arts and Design until his death in 1872. He constructed his first telegraph in 1835 and, as he was also an artist, painted several pictures of New York University.

MORTON PLACE. The western end of this short Place touches what was an important Indian trail long since obliterated by University Avenue. In the 19th century, part of the Benjamin Berrian farm was purchased by Thomas Morton in 1855.

MOSHOLU AVENUE. This Riverdale avenue bears the Indian term for Tippett's (Tibbett's) Brook, thought to mean "smooth stones" or "small stones" such as are still found in Tippett's Brook. It leads down to Broadway, on which the nineteenth-century settlement of Mosholu was situated at about West 254th Street.

MOSHOLU PARKWAY. This important highway was laid out as a link between Van Cortlandt Park and Bronx Park, just as Pelham Parkway was a further link between Bronx Park and Pelham Bay Park. The Parks Department designed this boulevard in the European manner. The name is an Indian one, referring to the small stream called by them "Mosholu," but known today as Tibbett's Brook. Professor Tooker, an authority on the tribal tongue, believed it meant "smooth stones" or "small stones." See: Mosholu Parkway North; Mosholu Parkway South.

MOSHOLU PARKWAY NORTH. The Parkway was laid out in 1884, and led to the settlement of Mosholu over on Broadway at Van Cortlandt Park.

MOSHOLU PARKWAY SOUTH. According to Robinson's *Atlas* of 1888, this thoroughfare was called Middlebrook Road. This avenue was formerly the course of Schuil Brook of the early 1700s. It was anglicized to School Brook, and ran southeast to join other tributaries at Webster Avenue to form the Mill Brook. This stream was sometimes called Middlebrook. At one time in the Revolutionary War, it was the source of water for the Negro Fort. See: St. George's Crescent.

MOUNT EDEN AVENUE. It was once called Wolf Street from Jerome to Walton Avenues, East Belmont Street from Townsend Avenue to Walton Avenue, Jane Street from Walton Avenue to the Concourse, and Walnut Street from the Concourse down to Webster Avenue. Mount Eden was a name in use before 1850, and harks back to Rachel Eden, who bought the hilly eastern portion of the Anderson farm in 1820. When Upper Morrisania was renamed Tremont, it was because of the three Mounts inside the town limits: Mount Eden, Mount Hope and Fairmont.

MOUNT EDEN PARKWAY. See: Mount Eden Avenue.

MOUNT HOPE PLACE. This hill was once called the Western Reserve (of Upper Morrisania) on an 1868 map. In 1889, this street was called Popham Street, as it was cut through the lower end of the Popham estate. The owner was a grandson of Richard Morris, who had owned Mount Fordham (Bronx Community College). For a short tune this street was called Washington Place. Mount Hope was not mapped as such, so it could have been a local name for the elevation, never committed to paper until the Place was cut through.

MULFORD AVENUE. The name is thought to be a combination of two realtors, Mullins and Colford. It was originally laid out as Pier Avenue, which was thought to refer to a pier on the headwaters of Westchester Creek (now covered by the State Mental Hospital grounds). More likely it referred to Emma Peere Miles, who, in 1889, lived at the junction of Libby Place and Mulford Avenue. The avenue was cut through in 1919.

MULINER AVENUE. The 1683 records list Thomas Mullinex as one of the early settlers. Subsequent spellings were Mallener and Molenaer. This avenue overlays, in part, a small stream known as Downing's Brook at about Bronxdale Avenue. This land once belonged to the Fowler family at the time of the American Revolution.

MULLALY PARK. John Mullaly was Secretary of the New Parks Commission (1888), and advocated the acquisition of more park lands. This park is along Jerome Avenue between East 162nd Street north to McClellan Street. The park was named in his honor in 1932. It was laid out over former Cromwell's Creek, which was the dividing line between the Morrisania Manor lands and the Turneur/Devoe grant.

MULLAN PLACE. This street touches the ancient road that led to Frog's Point (now Fort Schuyler). In 1793, it was mapped as Willett Leacroft's land, which was purchased by Abijah Hammond in 1810, who gave his name to Hammond's Cove. In the early 1900s, this street was given the name of Supreme Court Justice George V. Mullan.

MULLER PARK. The land was acquired at East Fordham Road and Creston Avenue by the City in December, 1897. It carried a local nickname of Ginko Square because of the ginko trees there, but, in 1930, it was named (Maurice) Muller Park after a prominent merchant and civic leader.

MULLIGAN COURTS. See: Robert "Mugs" Mulligan Courts.

MULVEY AVENUE. Vesting of title (acquisition by the City) took place in July, 1941. Prior to that year, the land had been a neglected tract once belonging to John Mulvey, according to an 1896 map. Some of his acreage was in adjacent Mount Vernon.

MUNDY'S LANE. Half of this street is in The Bronx, while the northern side is in Mount Vernon. It is the very ancient road leading from the Mile Square in Yonkers to Eastchester. In 1776, three companies of recruits were raised in Westchester to form one Corps under Captain Micah Townsend, and a Stephen Munday [spelled this way] is listed on the rolls. Throughout the 19th century, the Mundy family was prominent in the area. Bernard F. Mundy, a funeral director of the early 1900s, is believed to be the man whose name is on the Lane.

MUNSON WAY. See: Thurmond Munson Way.

MURDOCK AVENUE. Murdock Campbell was a Wakefield realtor listed in 1889. This was part of the Kenneth Cranford estate in 1931. Desbrow Place was the name of the stretch of Murdock Avenue from East 242nd Street to the city line. It was South 15th Avenue when the area was part of South Mount Vernon.

MURPHY (Arthur) SQUARE. Arthur Murphy was a former political leader of The Bronx. In fact, he was the first chairman of the Democratic Party in Bronx County. The Square was acquired by the City in 1901. Before that, it had been part of the Lorillard lands, and, in Revolutionary times, was on the Oakley farm.

MURRAY COURT. Clason Point has a few private walks and Murray Court leads in from Soundview Avenue, opposite Woodrow Wilson Square. It is 99 feet long and has ten homes facing each other in two rows. Tom Murray owned this parcel in the 1920s and he and his brother were well-known builders of row houses.

MYOSOTIS AVENUE. A roadway in the southwestern part of Woodlawn Cemetery. One of the graves nearby is that of Frank Munn, "Golden Voice of Radio." Myosotis is a botanical term for the herbage that includes the forget-me-not.

NACLERIO MEMORIAL COMPLEX. This is a Little League baseball field between Westchester and East Tremont Avenues on land that filled in Westchester Creek. Salvatore Naclerio was a Bronx contractor who devoted himself and his finances to youth activities.

NACLERIO PLAZA. Salvatore Naclerio's contracting firm was situated at Connor Street and Boston Road and, after his death in 1982, the intersection was given his name. Somehow, a small dead-end street - Clementine Street - was included and lost its identity. This occurred in March 1983.

NAPIER AVENUE. In this area, Chief Nimham and his Stockbridge Indians, who were allies of the colonists, were surrounded and slain by British cavalrymen in 1778. During the early 1800s, it was part of the Hyatt farmlands, the southern portion being sold to Reverend A. Peters just before the Civil War for a cemetery. In 1889, the remainder was surveyed for a development called "Woodlawn Heights," and this avenue was named for a resident, Milton Napier, architect and president of the Tidewater Building Company.

NAPLES TERRACE. This Terrace lies across the ancient island of Paparinemo of the aboriginal inhabitants. Colonial records refer to it as Hummock, or Humack, Island, which became the "Island Farm" of Alexander Macomb in 1794. It became the Godwin estate after 1853, and then was incorporated into the township of Kingsbridge. The director of the Naples Holding Corporation that sold the land in 1925 was Edgar H. Napolis, real estate and insurance broker and first Special Deputy Sheriff in 1914. He was born in Naples in 1881, and served in the Royal Corps of the Italian army from 1899 to 1904.

NARRAGANSETT AVENUE. What had been John Bartow's lands of the 1720s passed into the possession of the Ferris family during the next century, and was subdivided among the Pinchot, Abbot and Pearsall families from 1860 to 1880. Around 1900, a real estate development was planned called "Westchester Heights," and all avenues were given names of fashionable resorts. This avenue was to remind the residents of Narragansett Bay, Rhode Island.

NARROWS (The). This is a geodetic term still used by the Coast Guard to designate the narrow stretch of water separating City Island from the mainland. Also, it was charted as such in 1911 and 1917. A 1753 conveyance uses the term when Amos and Sarah Dodge sold the island to Samuel Rodman of Rodman's Neck.

NEEDHAM AVENUE. This Edenwald avenue harks back to a former landowner, Thomas B. Needham. Its junction with Eastchester Road was the start of Corsa's Lane, an early route to White Plains Road. From East 222nd Street northward, Needham Avenue was part of the Schieffelin estate of the 1860s. Its upper end was once called Beech Avenue. Ground was broken for Cardinal Spellman High School in December, 1959.

NEILL AVENUE. The Samuel Neal [spelled this way] estate was noted on an 1868 map as being located in the village of Bronxdale, southeast from "the Bleach," the local nickname of the Bolton Bleachery, the major firm in the village. Other deeds render the name, "Neill," and when the street was opened, this spelling was used, conforming to that on the family vault in St. Peter's churchyard. Originally, it ran across the Eliza Macomb property that became the Morris Park race track, and, where it continued eastward across Williamsbridge Road, the avenue was first called Niagara Avenue, running alongside the Bronx Municipal Hospital Center. It was originally part of the Westchester Heights development (1892) that gave all its streets the names of vacation spots. See: Narragansett Avenue.

NELSON AVENUE. In Beers' *Atlas of Westchester County* (1867), the I. Nelson house and farm is shown at a point above Highbridgeville. An earlier Nelson was listed in 1821 as a member of the Westchester Agricultural Society. The avenue runs from the historic Morris lands (below West 168th Street) up into the equally historic Manor lands of Fordham that were owned by the Archer family.

NEPTUNE COURT. The Shorehaven Beach Club locker rooms and a flower-lined concourse occupied this area. In the early 1900s, the main gate to the Clason Point Amusement Park - a 30-foot archway - stood there.

NEPTUNE LANE. In the years of the Shorehaven Beach Club, this area was part of its picnic area, a floral concourse and locker rooms.

NEREID AVENUE. No record of a N.E. Reid, either as landowner or surveyor, has ever been found, even though some residents of Wakefield repeat this bit of folklore. Rather, it is a remembrance of a long-ago firehouse of a Volunteer Hose Company once situated off Richardson Avenue before East 238th Street was cut through. Old-time pumping wagons carried "watery" names like Oceanus, Neptune, Poseidon and Pluvius - and this early Washingtonville fire fighting unit named their wagon "Nereid," which means a water nymph. Nereids were regarded as marine nymphs of the Mediterranean, in distinction from the Oceanids, or nymphs of the Atlantic Ocean, and from the Naiads, the nymphs of fresh water. The street was opened from the Bronx River to White Plains Road so that the firehouse had to be demolished in its path. The avenue was then extended to the Mount Vernon line in 1930, first being called West 5th Street.

NETANYAHU (Lt. Col. Yehonatan) LANE. See: Yehonatan (Lt. Col.) Netanyahu Lane.

NETHERLAND AVENUE. It begins on historic Spuyten Duyvil hill and was first mapped as Berrian Avenue, as the area was known as Berrian's Neck. Off West 231st Street was located the Revolutionary Fort No. 3 that was in use from 1776 to 1781. In 1889, the street was named Babcock Avenue where it crossed the lands of S.D. Babcock from West 252nd Street to 256th Street. It was officially opened in 1915, and its entire length named in honor of the Nieuw Netherlands Colony, of which it was a part.

NEUMANN-GOLDMAN MEMORIAL PLAZA. In Revolutionary War times, this land was part of the Valentine farmlands, which passed into the possession of the Varian family in the next century. Around the years immediately after the Civil War, the area formed a section of the Peter and John Bussing farm. Around 1900, a reservoir was constructed, and the concourse around it was descriptively named Reservoir Oval East and Reservoir Oval West. In November, 1959, part of Reservoir Oval West was re-named Neumann-Goldman Plaza to honor Corporal Jerome Neumann and Lieutenant George Goldman, paratroopers who lost their lives in World War II. On December 22nd 1962 the word "Memorial" was inserted.

NEW ENGLAND BALLFIELD. Its proximity to the New England Thruway was most likely the reason for naming the field.

NEW ENGLAND THRUWAY. This modern highway leads present-day Bronxites to the border of Connecticut.

NEW ENGLAND THRUWAY BRIDGE. This bridge takes its name from the highway it carries over the Hutchinson River. This wide thoroughfare spanning the Hutchinson River above Co-Op City almost obliterated a Bronx landmark when it was built in 1962. At its eastern end, in Pelham Bay Park, it narrowly missed demolishing the Split Rock in its path. See: Split Rock, The.

NEWBOLD AVENUE. Herman LeRoy Newbold did not live in the village of Unionport, but on a Throggs Neck estate called "Pennyfield," where his mansion stood at Miles and Longstreet Avenues. His grandfather was Herman LeRoy, a wealthy merchant of Colonial New York. Mr. Newbold was a vestryman of St. Peter's Episcopal church of Westchester, and, in 1848, he was one of three churchmen who purchased property along present-day Glebe and Starling Avenues, and Manning Street and conveyed it to the church. This avenue was originally laid out as 14th Street, but when it conflicted with the numbering system of New York City, Squire Newbold's generosity was remembered and his name was affixed.

NEWELL STREET. Leading from Gun Hill Road to the Bronx River Parkway, this street has not a store nor a house on its length. The *original* Newell Street was further east until it was obliterated by the southbound lane of the Parkway. In "the good old days," that earlier Newell Street led past L'Hermitage Restaurant, Voelker's Schutzenpark, and to the cut-off for "French Charley's." In the 1890s, Edward J. Newell lived there. He was a director of numerous corporations, and president of the Osage Oil Company. He was the author of a historical book called *Our Police*, and was the Democratic candidate for State Assembly in 1890, but was defeated.

NEWMAN AVENUE. Thomas Newman is mentioned in March, 1668, in a deed of the Town of Westchester: "setueate between the broke [brook] and the hieways which lieth to the north of the home lot which was formorly thomos neumans...." In later times, Elias Newman served as Sheriff of Westchester under the State Constitution of 1777. Newman Avenue overlays Pugsley Creek on Clason Point, a region noted for its muskrats as late as 1918.

NEWPORT AVENUE. A real estate development was opened in 1917, and it was called Pelham Heights to complement adjoining Westchester Heights. Streets were named after wealthy Summer resorts, and who hasn't heard of Newport, Rhode Island? More than a century before, the land was part of the Denton Pearsall estate that he called "Woodmansten." This portion, containing Newport Avenue and present-day Albert Einstein College, was purchased by a man named Trumbull.

NINE BOULEVARD. See: Louis Nine Boulevard.

NINTH AVENUE. This is a narrow road in the Park of Edgewater, ninth in order from the waterfront. See: Park of Edgewater.

NOBLE AVENUE. In the early 1700s, Robert Noble owned land in the vicinity of East 174th Street, which he allotted to Nathaniel Underhill in 1724. It passed to the Hunt and Pugsley families in 1777, and, a century later, it was the property of the Mapes family. South of Westchester Avenue, Noble Avenue runs across the former lands of Isaac Clason (1793), which became the William Watson property in Civil War times. There is a possibility the avenue does not honor the earliest settler at all, but a city engineer (*circa* 1885) named Alfred Noble, for five or six avenues next to Noble Avenue were given names of city engineers. See: Croes Avenue and Fteley Avenue.

NOELL AVENUE. The avenue is laid across part of the historic Pell Purchase, which later was subdivided among the Eastchester settlers. It adjoins the ancient landing-road on the Hutchinson river, and was included in the Conners holdings of the 1850s. Avenues in this sector carry names of former Mayors of New York. Thomas Noell was born in England and became a wealthy merchant in early New York. He served as Mayor, 1701-1702, and died in office, a victim of a smallpox epidemic.

NORTH BORDER AVENUE. As its name implies, this road runs across the northern section of Woodlawn Cemetery. The Oliver Tilden Post has a plot, featuring a Civil War soldier's statue, nearby. At its western end, the Bruckner family vault can be seen.

NORTH BROTHER ISLAND. On his voyage to the new world, Adriaen Block noted these islands on his charts in 1614, along with a surrounding cluster of islets which he called "de Gesellen," a term meaning Brethren. This was not in the family sense, but meant the commercial brothers (associates) of the Dutch West Indies Company. Around 1708 when the English occupied New York, the Governor of the Province of New York patented to William Bond the two islands, which he anglicized, "known as The Brothers." Due to its geographical location, it was designated as North. Both were also known as Two Brothers, on an 1874 map of Morrisania. Over the centuries North Brother Island has served many civic and federal purposes: hospital, army barracks, prison and quarantine station. The island attained world-wide attention on June 15, 1904, a weekday, when the excursion steamer, *General Slocum*, beached on its shore in sheets of flame in which over one thousand people, mostly women and children, died. See: South Brother Island, Pattry's Hook.

NORTH CEDAR AVENUE. This road is located in the northeastern section of Woodlawn Cemetery. The name serves a dual purpose by referring to the location of the avenue and to the cedars that are in the area.

NORTH OAK DRIVE. In 1885, a real estate development, "Bronxwood Park," was laid out, and the winding roads were named for trees. At first, this Drive was called Shimmer Street, as the land had belonged to Charles and Lily Shimmer.

NORTH STREET. In 1878, it was mapped as the north end of William and Samuel Archer's lands. Robinson's *Atlas* of 1888 shows it as the northernmost street on the Partridge & Craighead Plot of University Heights.

NORTON AVENUE. This avenue is laid out on the original Indian encampment that became part of the Cromwell Grant. It passed to the Wilkins family, and eventually, to John Screvin, who had married a Wilkins daughter. Delia Norton was a small property owner on Castle Hill Point listed in 1911.

NURSERY LANE. This is an unofficial name for an unpaved road leading into Van Cortlandt Park from the west side of Jerome Avenue near West 233rd Street to a row of potting sheds and nurseries.

NUVERN AVENUE. This is a hilly avenue that was cut through in 1958 into Mount Vernon. In short, the name combines *New* York and Mount *Vernon*.

O'BRIEN AVENUE. This Clason Point avenue honors Morgan Joseph O'Brien of the Supreme Court 1887 to 1896. A Siwanoy Indian trail cut across the avenue at approximately Beach Avenue, running due north and south. The land was part of the original grant to Thomas Cornell in 1646. Dairy Creek entered the peninsula from the Bronx river at the foot of O'Brien Avenue.

O'BRIEN SQUARE. The Tremont plot was acquired on March 10, 1896, for park purposes. It honors Captain Thomas A. O'Brien, whose family were long-time residents of 280 Burnside Avenue. Aged 34, he was killed in action during World War I on October 12, 1918.

O'NEILL PLACE. This Baychester Place was originally and correctly spelled "O'Neil" when proposed to the Board of Aldermen. The Honorable Thomas H. O'Neil had been appointed Justice of the Peace in 1895 when Westchester County relinquished its lower portion to The Bronx, and he went on to organize the Chippewa Democratic Club in 1898. In 1920 he served as Sheriff of Bronx County, and was the District Leader of the 6th Assembly, a division that embraced the entire borough east of the Bronx River. In 1931, as sometimes happens, the street sign appeared with two "L"s and was never rectified.

O'NEILL SQUARE. This small plot was acquired as park land on July 11, 1898 at Washington Avenue and East 161st Street. In 1931 and 1932 many small squares were given names of World War I soldiers and sailors who were killed in action, and Corporal Alfred H. O'Neil [spelled this way] of the 69th Regiment was remembered. He died on July 28, 1918 and had lived off nearby Webster Avenue.

OAK HILL AVENUE. A road located in the center of Woodlawn Cemetery, running past the ornate mausoleum of financier G. P. Morosini. The Ebling family plot is on this avenue and sons-in-law of this famous brewer are also buried there: Peter Doelger and Louis J. Heintz. The name is very descriptive of the terrain.

OAK LANE. It is named after the oak trees that originally formed its borders. See: Edgemere Street.

OAK POINT AVENUE. The original Landing Road of the early days on Hunts Point, leading to Leggett's Creek. On some maps, it was called Sacrahong Creek. Oak Point later became a well-known amusement park and resort of the 1890s, noted for its oak grove.

OAK STREET. This is one of Silver Beach Gardens' widest streets as it is a continuation of Throggs Neck Boulevard. See: Acorn Place.

OAK TERRACE. These were the uplands of the original Bronck Grant of 1642, and may well be the "Bronck's Hill" described in a later deed. In the Revolutionary War, it figured mainly as a vantage point from which to watch British forces on Randall's Island. Around the 1860s, it was part of the residential section called Wilton that was favored by theatrical folk and artists of that era. This terrace was named for the oak trees on its slope.

OAK TREE PLACE. This small Place figured in many ancient deeds dividing the farms that existed there before the American Revolution. Evidently, the original oak tree was "blazed," or marked, and must have been a conspicuous landmark, for it was mentioned in the 1849 deed - of the Lorillard and Ryer properties. It was originally surveyed in 1853, but it was not until 1902 that the Place was vested and opened.

OAKLAND AVENUE. This is a road in Woodlawn Cemetery, where oak trees have always been plentiful.

OAKLAND PLACE. This Belmont street was opened in 1903, and the name goes back to pre-Revolutionary times when the area was part of Miles Oakley's, "Oakley Grove."

OAKLEY STREET. This short street in Williamsbridge was laid out and named in December, 1913, after a Resolution by the Board of Aldermen. It was previously mapped as Ash Avenue. The Oakley family figured in the War of 1776, one of them serving as a Westchester Guide attached to the French allies. (Other guides were Andrew Corsa, the Dyckman brothers, and a man named Pine.) Isaac Oakley was a member of the Westchester Agricultural Society in 1820, giving his address as simply, Williamsbridge.

OBSERVATORY AVENUE. A graveled thoroughfare in the northeastern section of Woodlawn Cemetery. A 1904 photograph shows a six-sided gazebo literally draped in ferns and hanging plants. It was called "the Observatory," although how anything could have been observed of the celestial bodies is a mystery. The so-called observatory was later razed and in its stead was erected the impressive Amsinck mausoleum.

ODELL STREET. The original Episcopal church of Westchester was abandoned during the Revolution, and the townspeople decided to replace the dilapidated building with a new one. On January 26, 1789, the Church trustees of St. Peter's made an agreement with John Odell, carpenter, to build the second church for the sum of 336 pounds. Thus, the Odells can claim to be a very old family in the neighborhood. John Odell was one of the three Electors whose names were given to adjoining streets, Merrill and Guion being the other two Electors. Odell Street was first called Jackson Street, and in the early 1900s, Sehring's Washington Park, an amusement center and picnic grounds, was located at this street and Starling Avenue. See: Purdy Street.

OGDEN AVENUE. This thoroughfare follows a very old road once called Highbridge Avenue of the 1850s. The present name was adopted in 1876 in honor of William B. Ogden, who lived on a nearby estate, "Villa Boscobel." Joseph P. Day was the auctioneer when this estate was sold in 1,500 building lots-a four days' sale that was the largest partition sale in the history of the city up to that time.

OHIO FIELD. In April 1893, the former Loring Andrews property was put up for sale. It adjoined New York University (now Bronx Community College), and Chancellor MacCracken persuaded the Ohio Society of New York to purchase the property. The Society did so, in addition to raising a fund of $18,000 to give to the university. In gratitude, the athletic field on the new campus was named Ohio Field.

OHM AVENUE. Originally, this land was known as "Kenilworth" and was owned by a family named Underhill, who lived on the land in the late 1600s. In the 1800s, it was on the Furman estate, which later was the Lorillard Spencer property. Ohm Avenue ends at Rice Stadium, which was donated to the city by Isaac Leopold Rice, president of the Electric Storage Battery Company. As a result, the nearby streets have "electrical" names, such as Ohm, Watt and Radio. These names were approved by the Board of Estimate in October, 1922, and this particular street honors a German physicist who did research in electricity.

OLD ELM ROAD. This is a private lane on the western side of Fordham University grounds. There are many trees flanking its length, but few elms are left.

OLD FORT NO. 4 PARK. South of Reservoir Avenue near Sedgwick Avenue is this small park that was opened in 1914. It recalls a Revolutionary War fort that was built nearby.

OLD KINGSBRIDGE ROAD. This is the last surviving stretch of the original Indian trail that was widened to the Revolutionary road that was called Kingsbridge Road. It led from the Harlem River to what is now Bronx Park. This one-block unpaved road faces Southern Boulevard at about Grote Street. See: Kingsbridge Avenue.

OLINVILLE AVENUE. This street is named after the settlements called Olinville No. 1 and Olinville No. 2 that were incorporated in 1852-1854. The townships were named in honor of Bishop Stephen Olin of the Methodist church. From Gun Hill Road north to East 219th Street, it was once known as Pleasant Avenue.

OLIVER PLACE. It once formed part of the Briggs farm. In 1868, an avenue was plotted across Webster Avenue into the Lorillard estate, which later became the Botanical Garden. It was finally opened in 1893 as Oliver Street, but then demoted to a Place when it was blocked off and shortened. The Place is named after the Oliver family that lived on Webster Avenue, one member being Judge Frank Oliver, who was prominent in Bronx politics.

OLIVER TILDEN TRIANGLE. On December 5, 1974, the junction of Third Avenue, East 161st Street and St. Ann's Avenue was officially given this name to honor the first Morrisania man killed in the Civil War. Oliver Tilden was a carpenter, and his home and shop was located nearby at East 162nd Street and Eagle Avenue. At age 33, he was mustered in as a Captain, and saw considerable action until he was killed on September 1, 1862, at Chantilly, Virginia. His body was buried in the Bensonia (Morrisania)

Cemetery on St. Ann's Avenue, but, in 1878, the remains were transferred to the Sycamore plot in Woodlawn Cemetery. That the Triangle honors Oliver Tilden is due to the efforts of Berthold Sack, a member of The Bronx County Historical Society, and a grandson of two Civil War veterans.

OLMSTEAD AVENUE. The Olmsteads are old settlers of Westchester village dating back to Revolutionary times. Their gravestones are found in the oldest section of St. Peter's P.E. churchyard. Hiram Olmstead was Town Clerk in 1865 and a charter member (1859) of Wyoming Lodge 492, Masonic Order of Westchester. When the streets of Unionport were laid out, this was Avenue D, but it was officially changed in 1915, when L. .J. Olmstead was a surveyor.

ONEIDA AVENUE. This was formerly Devoe's Lane that figured prominently in a Revolutionary War battle. In 1778, Chief Nimham and his Indians fought Colonel Simcoe and his Queens Rangers across Devoe's Lane. The avenue was known as 4th Street in Woodlawn Heights until 1896, when it received its present name. The Indians were not of the Oneida tribe, so there may some credence to the story the avenue was named to honor a Civil War gunboat commanded by Commodore Farragut in his capture of New Orleans. A vague reference is made to a resident there who had served in that action aboard the gunboat, Oneida.

ORCHARD BEACH. This beach occupies what was once Pelham Bay, a shallow body of water south of Hunter Island. Up to 1929 or 1930 there had been a tent-and-bungalow colony opposite City Island by the name of Orchard Beach, where private individuals rented the waterfront land from the city. After a bitter fight with Robert Moses and the Parks Commission, the Summer campers were evicted to make way for a longer, larger beach and parking lot. Sand was floated up from Sandy Hook, and major landfill operations completely covered Pelham Bay, and joined Hunter Island to the mainland.

ORLOFF AVENUE. This was first mapped as Beebe Avenue when Oloff Park [spelled this way], a real estate development of 100 acres, was surveyed in 1869. Orloff, or Oloff, Van Cortlandt was the first of his name in the New World, and this area was part of his domains. Orloff Avenue was not officially opened until 1937.

OSBORNE PLACE. The Place was surveyed and cut through in 1902 across the ancient manor lands of the Archer family, which later sold the land to the Morris family. It was also the historic site of Fort Number 8 of Revolutionary times. Just before its subdivision into city streets, it was on the estate of Henry J. Cammann, who developed the modern binaural stethoscope, and a director of the Home for Incurables. E. Osborne Smith was a Fordham realtor from 1887 to the turn of the century, and it is believed he had a "modest" interest in the naming of this Place.

OSCAR COMRAS MALL. Born on June 10, 1915, Oscar Comras was brought to the Van Nest section as a child and lived to adulthood there. He was an energetic community worker, a talented drummer and worked hard to have band concerts in Bronx Park. He died in April, 1967, and on petition of the Pelham Parkway Citizen's Committee, in February, 1968, the lawn area in Bronx Park, where Summer concerts are held, was named in his honor.

OSGOOD STREET. Once known as Pell Street, it is hard by the City line at Mount Vernon. This had been named for S. Osgood Pell, a realtor active around 1890 to 1905 in that area, but when this portion of The Bronx was annexed to New York City (1898), there was already a Pell Street in lower Manhattan. This resulted in a change to Osgood Street, which occurred around 1900.

OSMAN PLACE. This short Place was once Chester Street, and the dividing line between the Penfield and Pitman estates, and either landowner may have had a hand in the present name. Frederic Penfield, whose estate ran north from Osman Place was listed in *Who's Who* as an author, diplomat and traveler. In 1909, he received the grand commander degree, Order of Osmanieh from the Khedive of Egypt. On the other hand, Oscar V. Pitman in the 1890s owned the tract of land from Osman Place south to East 217th Street. Letters sent to six families named Osman received no replies, so an "educated guess" would favor Oscar Pitman as being remembered by the Place.

OTIS AVENUE. In October 1776, Hessian Grenadiers were stationed along Logan and Otis Avenues to guard both the main road (East Tremont Avenue) and Weir Creek, which was a water-route. This creek is now overlaid by the Throgs Neck Expressway. In the 1850s, Otis Avenue was part of the Daniel J. Coster farm, which remained in existence until World War I. It was officially designated a public street by a City ordinance of 1929, and was named to honor General Elwell Otis, whose 40 years of military service included the Civil War, the Indian wars of Montana and the Spanish-American War.

OUR LADY OF GRACE AVENUE. This is a private walk circling Fonthill Castle on the grounds of Mount St. Vincent, overlooking the Hudson River. Our Lady of Grace is a term used by Catholics to denote St. Mary, Mother of God. This unwieldy name is used in the nuns' publications, but it is more popularly called the Cardinals' Walk.

OUR LADY OF SOLACE PLACE. On October 28, 1977, Mayor Abraham Beame signed into law a bill designating this Place at Holland and Morris Park Avenues to mark the 75th anniversary of the Catholic church. In the following week, a street sign was put up amid ceremonies attended by the clergy, parishioners and schoolchildren.

OUR LADY OF SOLACE PLAZA. November 2, 1977, saw a slightly different street sign put up in front of the church itself. See: above.

OUTLOOK AVENUE. It is so named as it affords an outlook over Eastchester Bay. The curve of this avenue follows the former Palmer Cove, as shown on early Bronx maps, when it was part of the Lorillard Spencer estate of the 1880s. It is believed in the neighborhood that Wilma was the name of a yacht that belonged to the Spencers. Noted on maps of 1908 and 1912, as Wilma Point, this promontory on Eastchester Bay formed the northern rim of Palmer's Cove. The City acquired title to this stretch of shoreline in 1930. See: Jaxon Point Division.

OVERING STREET. Laid out across the William Adee lands of Westchester village, it was originally called Washington Avenue. Upon annexation to New York City in 1898, the name was changed to Overing Street to avoid confusion with another Washington Avenue in Melrose and Morrisania. Around World War I, the Cherry Field baseball field was located at Overing Street and Frisby Avenue, behind Westchester Square.

References to games held on this field were noted from 1900 to about 1920. Henry C. Overing was a merchant and prominent member of nearby St. Peter's P.E. church. On May 4, 1805, he and his wife, Charlotte, purchased the Hobart property on Throggs Neck- 80 acres that extended from present-day East Tremont Avenue to what is now Ferry Point Park. In 1840, he built a 21-room Colonial mansion, which later became a Swedish Old People's Home (1920-1965) on Lafayette Avenue. The mansion was razed in January, 1965. See: Lafayette Avenue.

OWEN F. DOLEN PARK. See: Dolen Park.

OXFORD AVENUE. This avenue on Spuyten Duyvil hill was once part of the John Ewen estate of Civil War times. Opened in 1931, it was named for the English university town, as is the adjoining Cambridge Avenue.

PAINE STREET. This was originally Indian territory, then the Dutch outpost called Vriedelandt (Land of Peace), which became English, and, finally, after 1784, American. In 1862, this was the boundary line between the Seton estate and the Walsh farm. As other avenues in the neighborhood honor the Bay State, this short street is evidently named for Robert Treat Paine who signed the Declaration of Independence for the Massachusetts Bay Colony.

PALISADE AVENUE. In the 1840s this was a riverside road known as Quarry Road. The quarry itself was located near West 247th Street, and its granite went into the construction of many Riverdale mansions. Then it was called Bettner Lane, as it led through the Bettner property. The avenue is named for the majestic palisades directly opposite, one of the most inspiring stretches of landscape in the East.

PALISADE PLACE. From this high Place, which was opened in 1906, the New Jersey palisades could be seen-hence its name. Due to its elevation, it was a favorite vantage place for the soldiers of the Revolutionary War. This was the original seat of the Archer family who founded Fordham. In the early 1800s, it was part of the Lewis Morris estate, which passed to a grandson named Popham.

PALMER AVENUE. In 1692, a treaty was made with the Indians by the English colonists, and Samuel Palmer gave a gun valued at one pound. Since then, the Palmers have figured in all events through the Revolution to the 1900s. One, Benjamin Palmer, married Sarah Pell and bought City Island in 1761. He also helped build the Free Bridge over the Harlem River at West 225th Street and broke the monopoly of Colonel Philipse's toll bridge at Kingsbridge. A descendant, George Palmer, married into the Givan family in 1813, and when Robert Givan's lands were sold and subdivided, all the avenues were given names of his family. Palmer Avenue was almost blotted out by Freedomland, followed by Co-Op City, but a short stretch remains near Boston Road. This was once St. Agnes Avenue below Co-Op City, from the Hutchinson River Parkway to Stillwell Avenue, a distance of only two blocks.

PALMER COVE. This is an inlet on Eastchester Bay, with Radio Drive on the south, and Griswold Avenue on the north. It formerly extended further inland to Kennelworth Place. John Palmer was the proprietor of this tract in the 1740s, and he was still there during the American Revolution, for a 1777 military map showed his home due west of the Cove, near today's Ohm Avenue. The Cove was mentioned in nineteenth-century deeds as a dividing line, and its pier was used by the Furman, Campbell, Spencer and Waterbury families at various times. In the 1920s, the inlet was blocked at Ohm Avenue, and modernized with sea walls in 1944.

PANSY AVENUE. This is a roadway in the northwestern section of Woodlawn Cemetery, doubtlessly named for the flower so common to The Bronx.

PARK AVENUE (1). It was mapped as Terrace Place in Old Morrisania of the 1850s, but soon was renamed Railroad Avenue when the New York Central railroad was extended up from New York City. As the work progressed up into Morrisania, the avenue was given the name of Vanderbilt Avenue, as he was the man most responsible for the venture. In 1896, the entire stretch was christened Park Avenue as it was regarded as an extension of Park Avenue in Manhattan.

PARK AVENUE (2). This is an important thoroughfare in Woodlawn Cemetery, running from its southeastern corner to the northwestern end. The graves of Sam Harris, George M. Cohan and William Rhinelander are located along Park Avenue.

PARK DRIVE. This is a seldom-visited driveway along the northern boundary of Pelham Bay Park, serving six or seven lanes in Pelham Manor. See: Edgemere Street.

PARK OF EDGEWATER. This is not a park, but a privately-owned tract of about 40 acres on Eastchester Bay, containing approximately 500 bungalows. Indian arrowheads have been found along its shoreline, pointing to its use by the Siwanoys as a fishing camp before the coming of the white man. A deed, dated 1792, shows this promontory belonging to an Edward Stephenson, whose farm and woodlands took in most of Throggs Neck. The land passed through several hands until, in 1851, George T. Adee purchased the property for $20,000. It was his wife who is credited with called the estate "Edgewater," and the mansion, built in 1856, was luxuriously furnished. Around 1910, the sons of Squire Adee sold out to a Richard Shaw, who ran a stock farm there, and rented out camping sites for Summertime vacationers, using the name "Park of Edgewater" to make it sound attractive. The tents were replaced by bungalows, and, in time, these flimsy dwellings were built into year-round residences.

PARKCHESTER ROAD. This is a privately maintained thoroughfare inside Parkchester. An important Indian trail once crossed its junction with Unionport Road. The name was chosen by the Metropolitan Life Insurance Company sometime before 1938. The word "Park" has always been a favorite one with real estate operators to conjure up a mental picture of trees and lawns. The "chester" preserves part of the name of the ancient town of Westchester, in which the real estate development is situated.

PARKER STREET. At Westchester Avenue, it overlays the ancient Indian Brook (Seabury Brook) that emptied into Westchester Creek. Old-timers aver that the curve of Parker Street at Castle Hill Avenue is the bend of that creek. The street was part of the Haight estate of Civil War days, called "Hopedale." James Parker was Justice of the Peace in Westchester village in the 1850s, and is the likeliest man to be honored.

PARKSIDE PLACE. Located just below Gun Hill Road and west of Webster Avenue, this hillside street, named prior to 1888, faces the Botanical Garden, the top of Bronx Park.

PARKVIEW AVENUE. This avenue has a view of Pelham Bay Park. Its former name was Gainsborg Avenue, after a South American financier who had real estate dealings there. The avenue runs across land that once belonged to the Baxters (1600s), Ferris (1700s) and Burrs (1800s).

PARKVIEW TERRACE. This is a wide part of Jerome Avenue with some tall apartment buildings facing it. The park is most probably the long-vanished Jerome Park which used to be on the Lehman College grounds. See: East 196th Street, East 198th Street, Jerome Avenue.

PARSIFAL PLACE. On October 22nd, 1922, this Place was laid out and named after a knight in an opera by Richard Wagner. See: Lohengrin Place.

PATTERSON AVENUE. An Indian trail crossed this avenue at the northeast corner of Thieriot Avenue running towards the Bronx River. A few Clason Point avenues carry names of Justices, and Edward P. Patterson was a Supreme Court Justice from 1886 to 1896.

PATTERSON STREET. In front of Fort Schuyler's drained moat is a road named for Captain Patterson who was Superintendent of the New York State Maritime College in 1898.

PAUL AVENUE. This avenue is laid out across the former Jerome Park racetrack, and once was named Navy Avenue. Its name was changed to honor Dr. Francis Paul, for many years the principal of DeWitt Clinton High School and the person who was largely responsible for moving Clinton to The Bronx. He died on September 8, 1929, a few hours before classes started there.

PAULDING AVENUE. Once called Forest Avenue, this street runs across what had been the Morris Park race track of the 1880s and 1890s. Former Mayors of New York are honored by avenues in this part of The Bronx, and William Paulding served from 1825 to 1826. His wife was a Rhinelander. He was a descendant of an old Dutch family named Pauldinck, and remarked on several occasions his grandparents were more at ease in Dutch than in English.

PAULIS PLACE. This is a private road leading from Hawkins Street, on City Island, north to Fordham Street across what was once the Barker property of the 1870s. Earlier, it was mapped as the Vail estate. A recent owner was Fred Paulis, a Greek-American, who sold the property in 1959.

PAWNEE PLACE. This was part of the Denton Pearsall estate, mapped in 1868. At the turn of the century, "Westchester Heights" was begun as a residential subdivision, and three of the streets were given names of Indian tribes. Choctaw Place and Seminole Avenue were the other two streets, and their naming can only be attributed to some whim of the developers, Pinchot and Morrell.

PEARSALL AVENUE. The earliest record of the Pearsalls was a will, dated 1723, wherein Thomas Pearsall left "a certain Island called Spectacle Island or Harts Island, situate in the Sound" to one son, Nicholas. Denton Pearsall is listed in the County books as owner of considerable property between Williamsbridge Road and Eastchester Road and from Van Nest Avenue north to Allerton Avenue. The section of land around Allerton Avenue was purchased from the Palmer family around 1852, and had been part of their grandfather's lands known as the Givan Homestead. See: Palmer Avenue. The Pearsall property was called "Woodmansten" and it was watered by Stoney [spelled this way] Brook. The Pearsall family vault is on East Border Avenue in Woodlawn Cemetery.

PEARTREE AVENUE. There was never a peartree orchard along this street. This area was the original land grant of 1662 wherein several families, at the behest of Thomas Pell, agreed to "sett down at Hutchinsons." It became known as The Long Reach-a tract of 3308 acres, and so named from its shape. In the "New Patent"of 1708, Colonel Peartree was granted land there. The region is thought to be the site of Ane Hutchinson's cabin, and an old story advises a sighting taken through the cleft of the Split Rock (on the opposite side of the Hutchinson River) points directly to the farm. Peartree Avenue would fit that description. In Civil War times, Peartree Avenue was part of the 14-acre estate of Thomas Galway or Galwey. Below the New England Thruway, the avenue runs through the former G.S. Codling estate, which was turned into an amusement park called "Eagle Grove"(1868). William Peartree was born in the West Indies, prospered in New York and was elected Mayor for four successive terms (1703-1707).

PELHAM BAY PARK. This enormous tract was vested October 23rd, 1888, when the area was still part of Westchester County, but already earmarked for parks by the Parks Commission of New York. Louis Haffen, among others, was largely responsible for the surveying and later acquisition of this park, which was formed from estates belonging to the Hunter, Furman, Edgar, Lorillard, Morris, Stinard, Marshall, LeRoy and Delancey families, as well as that of the Roosevelts and Boltons.

PELHAM BRIDGE. The handsome stone bridge over the mouth of the Hutchinson River was built in 1908, replacing an earlier wooden span. Its name harks back to the former Pell lands which became Pelham.

PELHAM PARKWAY. It was once called Pelham Avenue and Fordham Road. In the early 1900s, it served as a trotting course for some of the landed gentry, and was popular in the 1920s as a bicycle race course. The official name is the Bronx and Pelham Parkway, as it connects Bronx and Pelham Bay Parks.

PELHAM PARKWAY NORTH. See: Bronx and Pelham Parkway.

PELHAM PARKWAY SOUTH. See: Bronx and Pelham Parkway.

PELL PLACE. This City Island street was originally called Pell Street, but the designation was changed to "Place"to avoid confusion with the Chinatown street. Thomas Pell was the first European owner of the island in 1654, and Pells have been living there ever since. Most likely, this short Place honors Henry Scofield Pell, who was born on the island in 1856. He was Clerk of the New York State Assembly, and a real estate dealer in later life.

PENFIELD STREET. The C.J. Penfield estate was partly in Bronx County and partly in Mount Vernon. It had been surveyed in 1889 by R.W. Burrowes, with a later (1906) survey by J.E. Crawford. The Penfield mansion was destroyed by fire on May 13,

1912, and the site was taken over for apartment houses. The Penfields were related to the DeMilt family and East 242nd Street was once called DeMilt Avenue.

PENNYFIELD AVENUE. This was originally the Throggs Neck Road mentioned in records of 1686 and 1723. After 1850, when Fort Schuyler was constructed, it gained the name of The Fort Road. According to George Ferriera, late historian of the Neck, Pennyfield got its name during the English occupation, when the fields were sold by the Indians for one English penny-not for its value, but for its metal. The fields were roughly bounded by Meagher, Miles, Harding Avenues, and Long Island Sound. The notation, "Pennyfield," is mentioned in a 1793 deed. During the American Revolution, General George Washington and some of his staff were surprised by a British warship as they cantered through what is now Silver Beach Gardens. The warship opened fire at the group of horsemen, who had to gallop madly along Pennyfield Avenue to escape the cannonade.

PEROT STREET. An 1853 Dripps' map shows a 63-acre estate belonging to Ellison Perot, located on the Tetard farm of Revolutionary War days (Devoe Terrace today), when it belonged to Dominie Jean Pierre Tetard. The Perots were related to the Tetards.

PERRY AVENUE. This avenue, laid out across the former Valentine and Briggs farms, was first named Glencoe Street. To keep the flavor of the War of 1812 (see Bainbridge and Hull Avenues), it now honors Captain Oliver Hazard Perry, who also participated in naval encounters in that war.

PETERS PLACE. During the Revolutionary War in 1778, the Queens Rangers under Colonel Simcoe assembled at approximately this spot and charged westward to surprise Chief Nimham and his Indians, around present-day Oneida Avenue. In the 1860s, Reverend Absalom Peters inspected the farmlands of the Valentine and Devoe families with the idea of establishing a cemetery "far out in the country." Heading a church group, he sponsored the Rural Cemetery Plan which resulted in the incorporation of Woodlawn Cemetery in December, 1863.

PHELAN PLACE. This Place is laid across the original holdings of the Archer family of Fordham Manor. In later years, it was the Mount Fordham estate of the influential Morris family. Around 1907, the tract was developed by Phelan & Billingsley, realtors, and sold into building lots. Thomas A. Phelan was listed as a business associate of the Schwab family who had owned an estate, now the Bronx Community College. When the short street was laid out, it was first called Winik Place, but in 1925 it received its present name.

PHILIP AVENUE. The western end of this Throggs Neck avenue was strictly "on paper" for many years, as it was a saltwater bog drained by Baxter Creek. Near East Tremont Avenue, it formed the 1870 boundary line of the DeEscoriaza farm, and, proceeding east to Long Island Sound, the limit of John David Wolfe's lands. Avenues that were surveyed on Throggs Neck around 1900 received names of Spanish-American War admirals and generals, and this particular avenue honors Commodore John W. Philip, who commanded the *U.S.S. Texas*.

PHLOX AVENUE. This is a curved road in the southeastern section of Woodlawn Cemetery named after a North American flower.

PIERCE AVENUE. In pre-Civil War days, this was part of the Sackett property, which, later, passed to the Hatfield and Pierce families. Many Van Nest avenues honor our Presidents, and this particular thoroughfare is thought by some to be a reminder of President Franklin Pierce (1853-1857), but this is not the case. Elisha and Maria F. Pierce were the landowners in the 1880s just before it was incorporated into the Morris Park race track. This enterprise was in operation from 1890 to 1904, and part of the track itself is under Pierce Avenue. Reverend John Bartow, who died in 1727, was buried in a family plot once located at Pierce and Tomlinson Avenues, but the exact location has been lost.

PILGRIM AVENUE. This avenue runs across the former Watson Ferris estate at Middletown Road. Along with Mayflower, Puritan and Bradford Avenues, it honors the settlers of Plymouth colony.

PILOT STREET. City Island was noted for its marine pilots-and it is a bit of local folklore that Rochelle Horton had a hand in naming the four streets on his property: Rochelle, Horton, Marine, Pilot. In 1874, the shipyard of David Carll spanned the foot of Pilot Street. Mr. Carll specialized in three-masted schooners, which were used in the yellow pine lumber trade from Georgia. Eventually, the shipyard was taken over by Henry Piepgras, born in Copenhagen, and, after many successful years, it became the Jacob Shipyard, specializing in yacht work, finally to become the Consolidated Company.

PINCHOT PLACE. This was once on the Denton Pearsall estate, "Woodmansten," a part of which was sold to James and Mary Pinchot in 1881. Later on, the firm of Pinchot & Morrell subdivided 70 acres to form "Westchester Heights," a real estate development.

PINE AVENUE. This is a road across the middle of Woodlawn Cemetery.

PINE DRIVE. This is a private street inside Parkchester.

PINKNEY AVENUE. Philip Pinckney [spelled this way] was one of the ten families who, in 1662, agreed to "sett down at Hutchinsons." His salt meadows are now occupied by Co-Op City. His daughter, Rachel, married Sir John Pell in 1673 or 1674. Its geographical background is identical with that of Peartree Avenue.

PITMAN AVENUE. This runs athwart the nineteenth-century Bussing farm at its lower end. The upper end of Pitman Avenue runs through what was once the 133-acre property of the Bathgate family, and on that 1905 map, a plot marked "Pitman" is shown in the vicinity of Barnes and Pitman Avenues. Oscar V. Pitman was a resident there in 1892. See: Osman Place. This portion of The Bronx was known at that time as South Mount Vernon, and the avenue was called West 6th Street. The City gained title to the street in 1937.

PLAZA BORIQUENA. This plaza at East 138th Street and Willis Avenue emphasizes the Puerto Rican influence in the South Bronx. The Plaza was named in the early 1970s.

PLIMPTON AVENUE. This was considered part of the village of Highbridgeville that was settled around 1851 by laborers on the Croton Aqueduct system. At West 169th Street, this avenue represents the southern bounds of the ancient Manor lands of Fordham, property of the Archer family. George A. Plimpton, publisher of educational tracts and treasurer of Barnard College, had a small estate there.

PLOUGHMAN'S BUSH. This is a small private community of twelve homes off West 246th Street and Independence Avenue. A Robert M. Field subdivided his Riverdale property around 1930, and gave it this fanciful name. His son stated that, so far as he knew, it has no historical or family connection.

PLYMOUTH AVENUE. This avenue is laid out across the former Mapes farm mapped in 1852. It was first named Bradford Avenue, after the first Governor of the Plymouth Colony. From Middletown Road to Roberts Avenue, it carried the name of Waldo Place. Finally, in May, 1926, the entire stretch was renamed Plymouth Avenue.

POE PARK. Created in 1902, on the site of an apple orchard located across the street from Poe Cottage, which was later moved into the park.

POE PLACE. Proximity to Edgar Allan Poe's cottage inspired the Board of Aldermen to name this narrow lane, which is on private land.

POLITE (Reverend James) BOULEVARD. In 1981 a short stretch of Stebbins Avenue was renamed in honor of the late pastor of the Thessalonia Baptist Church.

POLO PLACE. This is the only reminder that once a polo field belonging to the Westchester Country Club was laid out there. It was officially declared a public street in 1939. See: Campbell Drive.

POND PLACE. This was formerly known as 3rd Avenue in Bedford Park, being laid across a filled-in pond on the Briggs farm. Later, it was mapped as Ursula Place as the Ursuline Academy was close by, but finally it was changed to Pond Place. The overflow from this pond flowed downhill, eastward, to present-day Webster Avenue to join the Mill Brook.

PONTIAC PLACE. This was part of Gouverneur Morris II's lands that were conveyed to Henry Purdy in 1847. It passed into individual ownerships, including the Howes and Haffens, and the Place had to be hewn out of solid rock to reach the mansions at the top. In the 1940s, this rocky plateau was levelled, and is now occupied by St. Mary's Park Houses. A descendant of the Haffen family is sure the Place was named after a Democratic Club of the 1880s, and not after Pontiac, the Indian chieftain.

PONTON AVENUE. Richard Ponton was one of the early settlers of Oostdorp (which later became Westchester village) and who was forced to swear allegiance to the Dutch authorities. See: Fink Avenue. Captain Ponton later took part in the Indian massacre on the Mianus river in 1670. In the latter half of the 1800s, Ponton Avenue was the dividing line between the lands of Mitchell Valentine and William Cooper's property.

POPHAM AVENUE. Major William Popham was born in Ireland in 1752, and was the last of Washington's officers to survive the Revolutionary War. He died at the age of 95. Popham married Mary Morris, daughter of Richard Morris, and their son, Charles, inherited the extensive Morris estate in the West Bronx called Morris Heights. When this avenue was surveyed and laid out, it was first called Ridge Street for obvious reasons.

POPLAR AVENUE. This is a roadway located in the northeastern section of Woodlawn Cemetery. The Honeywell family plot is located alongside this avenue, as is one of the

Barney family plots. Barney Avenue was the former name of Tibbett Avenue. Poplar trees are plentiful in this beautiful cemetery.

POPLAR STREET (1). Named after the poplar tree, this is a private road in Silver Beach Gardens. See: Acorn Place.

POPLAR STREET (2). This street runs across the original Cornell farm of the 1790s, which was then transferred to the Arnow family. Andrew Arnow was listed as owner in 1807. It was called Arnow Place from Williamsbridge Road to Blondell Avenue. This same area was the scene of a "Holy Roller" convention in 1919, when Aimee Semple MacPherson conducted services in a huge tent there. Poplar trees are still evident in yards along this street. On some early maps, it was called Barlow Street.

POST ROAD. For centuries, this was the main road to Albany over which the mail, or post, was carried, hence its name. It was the scene of many events during Revolutionary times. During the 1870s, it carried the name of Courtlandt Avenue after the early landowners of that area, but, in 1900, it was given the name of Newton to coincide with other avenues honoring scientists. With due respect to Sir Isaac Newton (1642-1727), the Board of Aldermen was persuaded by Borough President Miller to restore the older, historical name. See: Cyrus Place.

POWELL AVENUE. This street was once 11th Street in the village of Unionport. Reverend William Powell was Rector of St. Peter's P.E. church at Westchester Square (1830-1849) and lived at the rectory, the site of which is now occupied by the Bronx Y.M.C.A. off Castle Hill Avenue. He married the widow Bayard and inherited the Bayard farm in the village of Belmont. The clergyman was born in Dublin in 1788.

POWERS AVENUE. This steep street on former Wilton Hill below St. Mary's Park has nothing left of its nineteenth-century charm. John and Mary Powers wer the owners of a summer villa in Civil War times. See: Beech Terrace.

PRATT AVENUE. On maps of the early 1800s, this was the eastern boundary of the Seton lands, and in the 1860s, it was charted as the George estate and the Halsey property. Since the avenue was laid out and named, no one is sure which Pratt is remembered for the choice is a wide one. Benjamin and Calvin Pratt were Colonial Judges of the Supreme Court. C. R. Pratt was an Assemblyman who voted "Yea" on the controversial Parks Acts in 1884. See: Pelham Bay Park. There was Colonel C. E. Pratt, Civil War commander of the New York Montezuma regiment, and George D. Pratt, benefactor who helped finance the Bronx Zoological Society, and Sereno Pratt, who was head of the New York Chamber of Commerce, not to omit Zadock Pratt, who was a Legislator of New York State. An "educated guess" would favor the Assemblyman who voted "yea," as Pratt Avenue runs alongside Seton Falls Park.

PRENTISS AVENUE. This street honors General Benjamin Mayberry Prentiss, the hero of Shiloh, who distinguished himself at the Civil War battle. He died in 1901, at the age of 82. The street is laid out across the region known, since Colonial days, as "Pennyfield." In the early 1800s, it was part of Captain Wright's land, then passed to Ogden Hammond, to Herman LeRoy Newbold and, eventually, to Francis DeRuyter Wissmann. The American Legion Post on Prentiss Avenue was once the palatial Wissmann mansion.

PROSPECT AVENUE (1). This was formerly the fashionable thoroughfare of Woodstock village. From Southern Boulevard to Westchester Avenue, it once had a center mall with trees and shrubs, and was called Eastern Avenue as it ran through East Morrisania. At its upper end, in the village of Belmont, it was called Taylor Avenue. It was finally named for the fine prospect (view) it afforded of the East River, and of the excursion steamers that sailed past.

PROSPECT AVENUE (2). This road with an extensive prospect (view) is located at about the approximate center of Woodlawn Cemetery.

PROSPECT PLACE. It is considered to be on Mount Hope, one of the three mounts that gave Tremont its name. The prospect (view) is of the Mill Brook valley, or Webster Avenue of today.

PROVOST AVENUE. This avenue, like others in the area, is named for a former Mayor of New York, David Provost, who served in 1699. By mere coincidence, a family named Prevost lived in this vicinity around 1850. Two of the Prevosts were English army officers. The former name of Provost Avenue was White Plains Road, a drovers' road that forked north from the Boston Post Road. Livestock was driven on it to get to market. John Adams, President of the United States, lived at the Vincent-Halsey house, 3701 Provost Avenue, for two months in 1797 to escape the yellow fever epidemic in Philadelphia. The house has been destroyed.

PUGSLEY AVENUE. The name is that of an old-time family prominent in this rural section of a century ago. In 1794, Talman Pugsley was listed as owning 200 acres of Cow Neck. One parcel of the farm is now incorporated in Parkchester.

PUGSLEY CREEK. This was the original Indian waterway leading past the village of Snakapins (Lacombe and Leland Avenues) to approximately Watson Avenue. The stream once served the Pugsley family as their access to the East river, and sloops moored at the Pugsley dock, now covered by White Plains Road. In the 1940s, the creek was terminated at Lacombe Avenue, and, around 1970, only a short stretch remained. In other centuries it was Cromwell's Creek, Wilkins' Creek, Clason's Creek and West Creek.

PUGSLEY CREEK PARK. In March of 1989, an army of Parks Department workmen began a clean-up of this 74-acre wetland preserve. Two weeks later, neighborhood volunteers, scout troops and schoolchildren completed the transformation from an eyesore to a park.

PULASKI PARK. This neglected park, overshadowed by Expressways, is situated on the former shoreline of the Indian encampment called Wana-Qua, Ranchaqua and other mutations. Later, it was on the Bronck farm, which became the Manor lands of the Morris family. In the 1890s, a well-known resort named Brommer's Park was located at about this spot, and it was a busy harbor for excursion boats heading up the Harlem river and Long Island Sound. Officially named in 1930, it honors the Polish patriot (1748-1791) who served in the American Revolution. His first name was Casimir.

PURDY STREET. The Purdy family is of Huguenot extraction, and were settled in New York City before the Revolution. Many Purdy men and women resided in what is The

Bronx in the 1700s. Samuel Purdy was a Justice of the Peace in the 1850s, and it would appear that it was this man that gave his name to the street, as he bought many plots during the 1860s, as old deeds attest. His grandson, Sylvanus, was listed as a 1902 property owner in the same neighborhood. This street was known as Washington Street until 1897. In the late 1890s and early 1900s Sehring's "Washington Park" picnic grounds was situated there, which, according to a 1902 map, ran through to Odell Street along Starling Avenue. See: Odell Street.

PURITAN AVENUE. Along with Bradford, Pilgrim, Mayflower and Plymouth Avenues, this avenue recalls the Bay State of Massachusetts. It runs across the former village of Middletown, the Cornell and Mapes farms, to end at the former Walsh farm (1873) at East Tremont Avenue.

PUTNAM AVENUE WEST. The avenue is prosaically named after the railroad (Putnam division) alongside it.

PUTNAM PLACE. This short Place is located on the nineteenth-century Valentine farmlands that, along with a part of the Bussing farm, was sold to Reverend Peters to form Woodlawn Cemetery. Its first name was Oxfield Street, but, later, it was changed to conform to the Revolutionary War pattern. It now honors General Israel Putnam, who was active in this area.

PYTHIAN CIRCLE. At the entrance to the New York State Maritime College (Fort Schuyler) is this Circle, named for Commander Pythian who was Superintendent from 1874 to 1878.

QUARRY ROAD. Originally, this road forked off from Webster Avenue at East 178th Street going in a northeasterly direction across Third Avenue to the grounds of the Home for Incurables. The quarry, it is surmised, became the cellars of the Lorillard mansion before that family donated it to be a hospital. An alternate name was Marble Avenue. By 1890, most of the road had been discontinued and built upon, with the exception of the short stretch in existence today.

QUARRY ROAD RECREATION FIELD. Dedication ceremonies took place on August 30, 1984 at Belmont Avenue and Oak Tree Place when this soccer, football, and softball field was sponsored by the Belmont-Arthur Avenue Development Corporation.

QUIMBY AVENUE. When Unionport was laid out, this avenue was 4th Street. Upon annexation to New York City in 1895, there was postal confusion between the numbered streets of lower Manhattan, so all the Unionport streets were given names of early settlers, thanks to the diligent work of County Clerks who searched through old land grants and deeds. One of the five patentees of Westchester was a Quaker named Quimby, who was listed in 1667. At times, the name was spelled Quinby and Quimbe.

QUINCY AVENUE. At its southern end, Quincy Avenue was once an Indian settlement, and later, it was the site of John Throckmorton's colony. It became, in turn, the Colonial estate of the Livingstons, and the nineteenth-century squires named Mitchell, Post, Havemeyer and Huntington. From Lawton Avenue to the Cross-Bronx Expressway, it was on the John Morris estate (1860s to 1890), and was cut through in 1937. A small truck farm survived near the junction of Quincy and Dewey Avenues until the 1930s, and folklore has it that it had been a thriving pig farm at the turn of the century. Proceeding northward, the avenue was cut through in 1932, bisecting the former DeEscoriaza farm at Philip Avenue and the Brinsmade and Overing properties at Lafayette Avenue. Proceeding to Bruckner Boulevard, the avenue overlays what was the 80-acre farm of Elijah Ferris (1815), which passed to Jonathan Ferris (1850), and, by 1864, was part of the Stillwell estate. Hoffmann's picnic grounds occupied this tract before World War I. The avenue was named for Josiah Quincy, patriot and orator and associate of Paul Revere.

RADCLIFF AVENUE. Starting at its southern end, this avenue runs across the Sackett farm of the 1860s, the Lott Hunt lands, which became Morris Park, the Waring estate, and above Boston Road, the Adee estate and the Tilden property of the 1880s. The names of former Mayors were given to avenues in this area, and Jacob Radcliff, a lawyer from Poughkeepsie, served as Mayor from 1810 to 1811.

RADIO DRIVE. This Drive follows the curve of what was once Palmer Cove. It is on the ancient manor lands of the Underhills of the 1680s, which became the property of the Ferris family before and during the Revolution. This became the fine estates of the Furman and Spencer families of the 1890s. At that time, the adjoining land was taken for park purposes and a philanthropist, Isaac L. Rice, donated a stadium and playing fields, swimming pool, etc. He was president of an electric storage battery company, so nearby streets were named Ohm, Watt, Radio and Ampere.

RAE STREET. Legend has it that this site was once an Indian burial ground. It became part of the Morris Manor lands, and was sold to John Rae in April, 1852. A larger tract was purchased by Benjamin Benson, so the entire area became known as Bensonia. Rae Street was the southern end of the Bensonia Cemetery. This burial place was cut through by newly-opened St. Ann's Avenue in 1868, and, eventually, the entire cemetery was evacuated when the Board of Education received permission to erect P.S. 38. The Rae family was related by marriage to the Mowats, who owned the land now occupied by Fordham University.

RANDALL AVENUE. As the district abounds in Civil War names, this avenue commemorates Brigadier General George W. Randall. It is just sheer coincidence that a Jonathan Randall is named in two deeds, dated 1794 and 1799, as owner of property in this same vicinity. The western end of Randall Avenue was marshland and winding creeks until the 1930s. From East Tremont Avenue to Long Island Sound, it was surveyed and laid out across the estate of Catherine Lorillard Wolfe (1850-1880) that became the Bruce Brown estate of the 1890s. Although it was originally Evans Avenue it was renamed to conform to the "Civil War pattern."

RANDOLPH AVENUE. In July 1983, a stretch of Crawford Avenue in the Wakefield area was renamed to honor A. Philip Randolph, a civil rights and labor leader of the Brotherhood of Sleeping-Car Porters.

RANDOLPH PLACE. This was once part of the 1680 holdings of the Underhill family, called "Kenilworth," which became the Bayard farm, the John Hunter III property and the estate of John C. Furman. Edmund Randolph bought the land in June, 1875. The short street was opened in 1960, filling in the inner shore of Palmer Cove.

RAOUL WALLENBERG LANE. This Pelham Parkway crosswalk was dedicated on June 29, 1986. The Lane connects Muliner and Bronxwood Avenues. Raoul Wallenberg, a Swedish diplomat in Hungary during the Second World War, risked his life many times to save Jews from the Nazis. He distributed certificates of protection, commonly called "Wallenberg passports" enabling them to migrate to Sweden and other countries around the world. He mysteriously disappeared after the war ended.

RAT ISLAND. This small rocky island was mapped as early as 1798. It was included in the original Pell purchase from the Indians, and it passed to the Delancey and Hunter families in the 19th century. It was owned in 1910 by a Dr. H.A. Parmentier. A natural channel cuts across the south end of the island, and is navigable at high tides to small craft. No information as to the origin of the name is available, except that it is a very old name.

RAVINE AVENUE. This is a well named road in the eastern part of Woodlawn Cemetery. Admiral Farragut is buried near this road.

RAWLINS AVENUE. Originally it was called the South Road, and is one of the oldest lanes in this section, appearing on Revolutionary maps. It led to the Ferris mansion, which stood at Rawlins Avenue and Lohengrin Place until it was razed in 1961. The road serviced the seventeenth-century lands of the Underhill family, and the eighteenth-century Ferris property. In 1860, John Van Antwerp became the owner, and the South Road became the boundary of the Westchester Country Club. Nathalie W. Rawlins was the last owner of record in 1916, and it is her family name that is preserved.

RED OAK DRIVE. This is a private street inside Parkchester. The name is descriptive of the horticulture once found there in profusion.

REED PLACE. In October, 1922, this short street was laid out "on the Lorillard Spencer Farm at Throggs Neck." The Country Club Association purchased land from the Waterbury and Spencer estates in 1890, and, during that same year, Albert S. Reed purchased a small tract from the Association. In 1913, a S.A. Reed was listed as proprietor of a parcel of land along Stadium Avenue.

REED'S MILL LANE. This lane dates back to the 1600s, when it led from Boston Road to a mill on the Hutchinson river. This was the first tidal mill to be erected on Eastchester Creek, (later, the Hutchinson River). The year was 1739. The mill was operated in succession by Thomas Shute, Joseph Stanton, John Bartow and (in 1790) John Reid was the miller. His son, Robert, continued on until the 1850s. In the ensuing century, the name was rendered "Reed." After the Civil War, it was abandoned and stood forlornly on the salt meadows for decades, finally to be blown down in a storm in 1900. In 1958, its site was blotted out by Freedomland, an amusement center, and later its site would be roughly the center of Co-Op City. The mill lane was asphalted in 1967 and incorporated into the Boston-Secor Houses.

REEDER STREET. The New York State Maritime College has a short road next to Fort Schuyler. Lieutenant Commander Reeder is remembered for his tenure (1897-1898) as the College Superintendent.

REGINA PLACE. A Regina Groshon owned a small tract of land alongside the Boston Road before 1900, at its junction with White Plains Road (now Provost Avenue). Back in the 1820s, both Henry M. and John P. Groshon were members of the Westchester Agricultural Society.

REISS PLACE. This is a reminder of George Reiss, landowner in 1893, whose property was on the eastern edge of the Botanical Garden. Reiss' Pond, with plenty of catfish, was used for swimming in the Summer and ice-skating in Winter by local boys. It is now filled in and covered by a playground at Bronx Park East north of Pelham Parkway. Crossing exactly in the center of the Place was once the *original* White Plains Road, which ran northeasterly to join the present-day White Plains Road.

RESEARCH AVENUE. It is laid out across the ancient lands of the Underhill family, which they called "Kenilworth" after their original estate in England. In the 1880s, the southern end of Research Avenue was the Lorillard dock on Palmer Cove. This avenue was "From Watt Ave. to Griswold Ave., laid out and named on map of Lorillard Spencer Farm at Throggs Neck ... approved by the Board of Estimate, October 20, 1922." The avenue honors the research done by George Ohm, John Watt, Thomas Edison and Andre Ampere.

RESERVOIR AVENUE. On a high ridge that once overlooked the long departed Jerome Park race track, this avenue has had a change of name to conform to its location outside Jerome Park Reservoir. When it was part of the Claflin estate (1810-1880), it was called Claflin Terrace.

RESERVOIR OVAL EAST. Along with Reservoir Oval West, this was the high-lying Peter and John Bussing farm around Civil War times. Later on, the Bronx River Distributing Reservoir was built there, and was so mapped in 1888. When the reservoir was installed, this name was descriptive. Today, it is a park, but the name is a reminder, and remainder, of its former name and purpose. Reservoir Oval West has undergone some changes. See: Reservoir Oval West.

RESERVOIR OVAL WEST. While its origins are the same as Reservoir Oval East, the name of that part of the Oval West, next to Bainbridge Avenue, was changed to Neumann-Goldman Plaza, in 1959, to honor two World War II casualties. It was changed again later on. See: Neumann-Goldman Memorial Plaza.

RESERVOIR PLACE. See: Reservoir Oval East.

REVERE AVENUE. The avenue is named in honor of Paul Revere, whose partly successful ride was made famous by an 1860 poem of Henry Wadsworth Longfellow. The foot of Revere Avenue at the East River shows indications of an Indian settlement overlaid by Colonial foundations. Military buttons from General Howe's troops were found there as well. John Throckmorton (or Throgmorten) established his colony of English-speaking settlers there in 1642, with the permission of the Dutch authorities who exercised sovereignty over this territory called "Vriedelandt." From the East River to Lawton Avenue, the avenue runs across the Colonial lands of Philip Livingston, and the nineteenth-century estates of the Ash, Post, Mitchell and Van Schaick families. An 1868 map shows the John A. Morris estate from Lawton Avenue to Dewey Avenue. On October 12 to 17, 1776, a British brigade was stationed at Revere and Dewey

Avenues to guard the Throggs Neck Road (East Tremont Avenue). From Schley to Randall Avenues, this street was Rosedale Lane. From Bruckner Boulevard to St. Raymond's Cemetery it was known as Harriet Place. There is a strong possibility it was named for Harriet Flanagan, whose father owned the property there prior to 1900. The name was changed in 1916. See: Calhoun Avenue.

REVEREND EDLER G. HAWKINS SQUARE. The intersection of East 165th Street and Prospect Avenue, since May of 1990, has been so designated, to honor the long-time pastor of St. Augustine Presbyterian Church for many years.

REVEREND JAMES POLITE BOULEVARD. See: Polite (Reverend James) Boulevard.

REVEREND PATRICK DeSOUZA WALKER SQUARE. See: Walker Square.

REVIEW PLACE. As it faces north to the Parade Grounds of Van Cortlandt Park, this short street off Broadway may hark back to some long-forgotten military review.

REVILLE STREET. The original name of this City Island street was Queen Place. Patrick J. Reville was a stonemason and general contractor, who later became Superintendent of the Bureau of Buildings. He was a Bronx Democratic leader who lived on Beach Street around World War I.

REYNOLD STREET. On the City Island Association roster of 1905, the name of Dr. George A. Reynold, dentist, appears. His son, Samuel, married Harriet Horton in 1904. The street marks the northern boundary of the Horton estate of the mid-1800s.

REYNOLDS AVENUE. Along with other avenues in the area, this street honors a Civil War general, John A. Reynolds, a West Pointer, who was killed at Gettysburg. This avenue is laid out across the land known since earliest days as "Pennyfield." From the bridge approach north to Wissman Avenue, it was once the Ogden Hammond lands, then the estate of Herman LeRoy Newbold and finally that of Francis DeRuyter Wissmann [spelled this way]. Beautiful gardens were once at the southeast corner of Reynolds and Wissman Avenues extending almost to Blair Avenue. The Wissmann coach house, a sturdy brick building, stood at this corner. Reynolds Avenue was extended north to the Park of Edgewater in 1959, across the former estate of Samuel Fox, and the smaller tract belonging to a Francis Albury. This area was described in an 1854 deed as the Pennyfield Orchards. See: Wissman Street.

RHINELANDER AVENUE. Laid out across the defunct Morris Park race track, it was known as Mianna Street. This can be traced back to a Mianna Hunt who died in 1802. Her nephew, Lott G. Hunt was the last owner before the City acquired the property. The Rhinelanders were of an old New York family, called "Dutch aristocracy." A daughter married Mayor William Paulding, but the naming of the avenue is connected with a relative who had real estate dealings in adjoining Van Nest.

RICHARDSON AVENUE. Fulton Street was its former name when this area was known as Washingtonville. The volunteer firemen had a Hall on this avenue and East 240th Street calling themselves the Empire Engine Company, until they were replaced in February, 1896, by the paid New York City firemen. A year later, Fulton Street became Richardson Avenue, but it is not clear which person it honors. John Richardson was one of the original patentees in 1663, and a William Richardson was one of the first settlers on Rattlesnake Creek in Eastchester a few years later. An "educated guess" would be Benjamin Richardson, who purchased land in 1882 "near the White Plains Road."

RICHMAN PARK. Located at Tremont, Valentine and Burnside Avenues, for over 75 years, it was called Echo Park. It was renamed Julius J. Richman Park in the Summer of 1973. Mr. Richman was a chairman of the Twin Parks Association, chairman of the Urban Action Task Force, and Assistant Administrator of the City's Finance Committee.

RIDER AVENUE. William E. Rider and Theodore Conkling were business partners who purchased 600 lots in Mott Haven from Jordan L. Mott in the 1870s. These partners completed the Mott Haven Canal that had been started in the 1850s by Jordan L. Mott. See: Canal Place.

RIDGE PLACE. It is a topographical name. See: Edgemere Street.

RIKER'S ISLAND. First sighted by Adriaen Block in 1614, the island belonged to the Dutch West India Company until 1664, when Abram Rycher petitioned and received from the Dutch authorities a patent to it to be used as a farm. It totaled 87 acres. Throughout the 1700s, the Riker family (who anglicized the name) maintained their ownership, although a 1777 military map shows it as Hulith Island. Sometime around 1850, the Rikers sold the island to the City, and, during the Civil War, it was used as a training grounds. In 1862, the first Negro regiment was organized and trained for combat there. At times, it was described as Hewlitt's Island. In 1884, it was incorporated in the city limits, and, around 1900, was used as a collection point for ashes, cinders and garbage by the Department of Street Cleaning (D.S.C.). In a short time, the ground was fertile enough to support a penitentiary. In 1936, a seven-story prison hospital was built, and, since then, 50 buildings have been added to house the prison population. A half-million shrubs and trees grow in the nursery and are used in city parks.

RISSE STREET. This is a short street at the end of the Grand Concourse that received its name on January 25, 1968 after submission by Councilman Bertram Gelfand and George Zoebelein of The Bronx County Historical Society. Louis Aloys Risse, French-born engineer and draftsman, designed the Grand Boulevard and Concourse under the direction of Louis Heintz, who was Commissioner of Street Improvements in the 1890s. Louis Haffen, who became first Borough President, encouraged the engineer, who also laid out Glen Island, Fleetwood, and the railroad plan of Port Morris.

RITTER PLACE. Formerly called Washington Place, it was renamed to serve as a reminder of the Ritter family, who lived on an estate in the vicinity. Charles H. Ritter was a registered voter of the 23rd Ward (1877) with a Boston Road address. He was an editor, and was on the Citizens' Committee that investigated the *General Slocum* steamboat disaster.

RIVER AVENUE. This avenue traverses what was once the Morris estate "The Cedars" (1840 map). In Indian days this was swampland, and part of a creek called Mentipathe, a watercourse later to be called Cromwell's Creek by the English settlers.

RIVER CREST ROAD. On a crest overlooking the river is a private road running parallel to the Hudson River and servicing four homes in the vicinity of West 254th Street and Independence Avenue.

RIVERDALE AVENUE. This is the principal road of the mid-nineteenth-century realty venture undertaken by the Spaulding, Dodge, Goodrich and other families according to an 1856 map depicting "Riverdale Park." The road itself was surveyed and vested in 1884, and was first called Westchester Avenue, as it led to Westchester County.

RIVERDALE PARK. This park flanks the Hudson River from West 232nd Street to West 254th Street, and borrows the name from the region behind it.

ROBERT "MUGS" MULLIGAN COURTS. "Mugs," a star athlete, fell to his death in a tragic accident at the age of 21. The Van Cortlandt Park basketball courts were named in remembrance of this fine athlete.

ROBERTO CLEMENTE PLAZA. See: Clemente (Roberto) Plaza.

ROBERTO CLEMENTE STATE PARK. In September, 1974, this park was dedicated to the memory of a Puerto Rican baseball player, star of the Pittsburgh Pirates. He was flying to aid earthquake victims in Nicaragua when his plane crashed into the Caribbean sea and he was killed. Formerly called Harlem River State Park, it is laid out alongside the Harlem River above Washington Bridge extending almost to Kingsbridge Road. Officially opened in 1970, it covers former railroad property, World War I shipyards and turn-of-the-century boating clubhouses. See: Clemente (Roberto) Plaza.

ROBERTS AVENUE. Valentine Street was the former name of Roberts Avenue from Westchester Square to Blondell Avenue when it ran through the property of Judge Mitchell Valentine in the 1890s. His mansion was on the corner of Main Street (Williamsbridge Road) and Valentine Street (Roberts Avenue) at Westchester Square, where the courthouse is today. Roberts Avenue in the village of Middletown was part of the Watson Ferris lands in the 1840s. Later, when the avenues were cut through, this street was given the name of Emily Street (1890s), which was renamed Tremont Road when a real estate development, "Tremont Terrace," was begun. A Roberts tombstone in the tiny cemetery behind the First Presbyterian church of Throggs Neck is the sole clue to the name the avenue bears today. See: Emily Street, Tremont Road.

ROBERTSON PLACE. The Place was named after a landowner, Philip Wilkins Robertson, who is buried in St. Peter's P.E. churchyard since 1850. For the history of the surrounding area, see Dwight Place.

ROBERTSON STREET. This short street lies athwart the former Penfield estate at the Mount Vernon boundary. Judge William H. Robertson was an early (1846) representative in Congress from that district in Westchester County.

ROBIN AVENUE. This is a road in the southeastern section of Woodlawn Cemetery. It is named for the bird, which is a departure in Woodlawn Cemetery nomenclature.

ROBINSON AVENUE. Colonel William Robinson of Westford, a town close to Concord, was at the North bridge on that day in history "when the shot was fired that was heard around the world." See: Buttrick Avenue.

ROCHAMBEAU AVENUE. This area was the scene of some Revolutionary action, but not necessarily involving the French officer whose name is attached to the avenue. It follows the "1776 pattern" of surrounding avenues, such as Wayne, DeKalb and Gates. Lt. General Rochambeau commanded 6,000 French troops during the American Revolution.

ROCHELLE STREET. This street on City Island was once called South Elizabeth Street (the land was then owned by Elizabeth Pell) to distinguish it from another Elizabeth Street on the island (land owned by Elizabeth King). Sometime around 1912, it was renamed to honor Rochelle Horton, whose name appears on lists of Hell-Gate Pilots, and whose property was laid out on the south end of the island. See: Horton Street.

ROCKWOOD STREET. Local residents believed the name was descriptive of the area, which is on a rocky, once-wooded slope off the Concourse. In 1906, Randall Comfort, the eminent historian, wrote a book called *History of Bronx Borough*, and George C. Rockwood was the photographer. He was also a writer and photographer-journalist for *Harper's Bazaar*. The Rockwoods were one of the early families of Kingsbridge and vicinity that were buried in the Dyckman-Nagel burial ground, but then removed and reinterred (1926-1927) in Woodlawn Cemetery. The short street was vested in 1901 through what had been the Findlay property of Civil War times.

RODMAN PLACE. This Place was formerly called Cross Street, as it connected West Farms Road to Boston Road. On it, in the 1850s, resided the Reverend Washington Rodman, rector of Grace Episcopal church of West Farms. When Rodman was appointed to his post, he was only 25, and his wife, Henrietta Blackwell Rodman, 23. He was Chaplain of Co. "K," 6th New York Infantry, 1862-1865, organized in West Farms. In 1866, he founded the Home for Incurables in what had been the Jacob Lorillard home on Quarry Road. Some accounts list the *original* Home for Incurables as the former Mapes' Temperance Hall at East 179th Street and Boston Road. Both Rodmans are buried in Flushing Cemetery, plot #1.

RODMAN'S NECK. This is part of the original Pell Purchase, called Pell Point (and, sometimes, Tom's Point) at the time of the Revolutionary War. Other names were Anne's Hoeck, Pell's Neck, Asumsowis and Pelham Point. It was then owned by Samuel Rodman, who married Mary Pell, and who once was the sole owner of City Island for 2,300 English pounds. He was mentioned in 1763, when a ferry system was established from his land across the Narrows to City Island. In the mid-1800s, this was the "Hawkswood" estate of L.R. Marshall. Acquired in 1888 by the Parks Department, it was turned over to the U.S. Government to be used as a Naval training camp in World War I. It then reverted to its wild state for many years, until, in 1950, the portion facing City Island was drained and used by the U.S. Army. In 1959, the barracks were transferred to the Police Department, and a firing range was established in 1960.

ROEBLING AVENUE. It was first known as Green Avenue in the village of Middletown. Then, in 1900, the name was changed to Roebling Street, which had no connection with John A. Roebling who designed the Brooklyn Bridge, nor his son, Washington A. Roebling, who actually built it. The street honors a local home owner named Frederick W. Roebling, who died in 1891. His son continued to live on the street until his own death in 1963.

ROGERS PLACE. The Rogers family owned a large estate on Boston Road before Jackson Avenue was cut through. It consisted of barns, stables, an orchard, and gardens that surrounded a large, Victorian mansion. Morris High School stands on the site of the Rogers mansion. Thomas Rogers married Emily Cauldwell, daughter of the Senator, and their wedding was the social highlight of the 1880s.

ROHR PLACE. This was originally a summertime retreat of 25 acres belonging to a family named Wenner who would come up to Ferry Point by ship from lower Manhattan. The Wenner Realty Company owned considerable property in The Bronx from 1840 onward. Around 1870, a Unionport resident, Michael Rohr, married one of the Wenner girls, but it is not clear if the couple resided on the land that faced Westchester Creek.

ROMBOUTS AVENUE. This avenue is in the original Pell Purchase, which, later became the township of Eastchester. Streets in the area honor former Mayors of New York, and Rombouts served in that capacity. Francois Rombouts was a Frenchman by birth, but his parents migrated to Holland. He came to America in 1658, married Helena Teller, of an old Dutch family, and became Mayor of the city in 1679.

ROOSEVELT AVENUE. This short street honors Colonel Theodore Roosevelt of the Spanish-American War, who later became President. In the early 1800s this was part of Captain Post's lands. In or around 1900, a real estate venture called "Tremont Heights" was headed by Benjamin Lamport, who purchased the tract from Collis P. Huntington. A few years before, that portion of Throggs Neck had been surveyed and the avenues given names of Spanish-American War generals and commodores.

ROOSEVELT PLACE. The history and location of this short Throggs Neck Place is identical to that of Roosevelt Avenue.

ROOT WALK. A footpath on the campus of Bronx Community College, formerly New York University, honoring Elihu Root who received his Law Degree in 1867. At the time the University was established in The Bronx (1894), he was Vice President of the Alumni Association. He was later Secretary of War, Secretary of State, United States Senator, President of the Hague Tribunal of Arbitration in 1913, and was on the committee of experts set up by the League of Nations to revise the World Court statutes in 1929.

ROPES AVENUE. This is on land, originally part of the Pell Grant of 1654 which, in turn, was deeded to "the Twelve Families." An 1886 deed lists Charles H. Ropes as owner of a tract bordering the Turnpike (Boston Road) at this point. It was part of the Peter Prove estate in 1905. The street was opened in 1924.

ROSE FEISS BOULEVARD. In June 1987 Murray Feiss, a manufacturer of lamps and electrical equipment, memorialized his mother who had organized a few housewives in the 1940s and taught them to sew lamp shades at home. What started as a modest enterprise grew into a successful business needing far larger quarters, and it culminated in a six-storied factory on Walnut Avenue, employing hundreds of local residents. Walnut Avenue carries the additional street sign of Rose Feiss Boulevard.

ROSE HILL AVENUE. There are two conjectures on the naming of this Woodlawn Cemetery road. One is that there may have been wild roses in the vicinity at the time the road was laid out. The second theory is that people who had lived on the "Rose Hill" estate (now Fordham University) are buried in the area. A search through the tombstones for such names and comparisons with resident lists, did not bear out this theory.

ROSE HILL PARK. It measures 0.72 acres and is located at the northeastern side of Webster Avenue at Fordham Road commemorating a once extensive estate, "Rose Hill," now occupied by Fordham University. Until 1945, the small park was called Fordham Square, but then regained the older, and now current, name.

ROSEDALE AVENUE. This avenue carries the name of a former estate, "Rosedale," that was owned by Hudson P. Rose in the vicinity of St. Lawrence Avenue.

ROSELLE STREET. The 1804-1805 records of the town of Westchester located this street on the 5-acre farm of the Cornell brothers, lying between Williamsbridge Road and Bear Swamp Road (now East Tremont Avenue at this point). The farm was owned by Ezra Cornell, but his brother, Elijah, and family lived there, too. The homestead was sold in 1808 to Andrew Arnow, and it was passed to Mattson Arnow around Civil War days. A landowner of the 1890s was a Mr. Lavin, mortician, whose wife's name, Rose, was incorporated into the newly-opened street: Rose L. This bit of local folklore is challenged by the fact a man named Montgomery Roselle figured in real estate deals in the year 1868, and Montgomery Place is a short distance away. A search through the Lavin biography did not disclose any member of the family named Rose, so Mr. Montgomery Roselle is remembered.

ROSEWOOD STREET. Although it is a short street, it has had four different names, the other three being Elizabeth Street, Locust Avenue and Post Street. As to its naming, one tale has it that two old-time Williamsbridge families, the Roses and the Woods, lived at opposite ends of the street when it was first cut through. The Engineering Department maps show this street as the extreme edge of the Lorillard property when surveying was done prior to 1913. A stand of imported Brazilian rosewood trees had to be cut down to make way for the street, which then received the name. This was an occasional practice, as witness Minerva Place, Cottage Place and Nereid Avenue.

ROTA SQUARE. See: Tony Rota Square.

ROWE STREET. It is laid out on the original Colonial homestead of the Seabury family next to Seabury creek. The Rowe families were early settlers, there being a Benjamin Rowe listed as pewholder in St. Peter's P.E. church in 1790. This particular street is evidently named for Richard M. Rowe, who, in 1854, owned property in the vicinity.

ROWLAND STREET. Horace Rowland, mentioned in 1834 deeds, is buried in the West Farms Soldiers' Cemetery. He purchased land from Darius Lyon in 1853. See: Lyon Avenue. The street was opened in 1912.

RUDD PLACE. This was formerly the carriage road that led to the Rudd mansion on Clason Point, it being located near O'Brien and Newman Avenues. Colonel A.G. Rudd and his wife, Susan (Husson) Rudd inherited the mansion and part of the estate, called "Sans Souci," when the earlier owner, Joseph Husson, died. His son, Matthew, was executor. The Hussons and Rudds are buried in Woodlawn Cemetery.

RUPPERT PLACE. Colonel Jacob Ruppert, brewer and long-time owner of the New York Yankee ball team had his name given to this thoroughfare in 1933, displacing the older name of Doughty Street. The Yankee Stadium owes its existence to this man, and to "Babe" Ruth, the ballplayer. The Place overlays Doughty's Brook, or Cromwell's Creek of the 19th century ancient Maenippis Kill of the Dutch days, and Mentipathe of the India times.

RUTGERS AVENUE. This road is located in the northeastern corner of Woodlawn Cemetery. All the tombstones, flat on the ground, were transferred from the Rutgers Street Church in downtown New York City (along with the bodies) thus accounting for the name.

RUTH PLACE. Located on the grounds of the Westchester Country Club on the shores of Eastchester Bay, this short street possibly honors a wife or daughter of some long-departed landowner. For a history of this area, see Campbell Drive.

RYAN RECREATIONAL COMPLEX. See: Aileen B. Ryan.

RYAN SQUARE. This small plot at East 143rd Street and Morris Avenue, Mott Haven, was acquired by the city in 1900. It was named in honor of Corporal George P. Ryan, who was killed in action in France on September 29, 1918.

RYAWA AVENUE. This avenue is laid out across the filled-in bay noted on 18th- and 19th- century maps of Hunts Point. Shortly after the Civil War, General E.L. Viele planned a magnificent port here catering to Barge Canal traffic and oceangoing liners, and this odd name might be the abbreviation of his Railway and Water Association. There is also the possibility that, on the blueprints, this was to be the Railroad Yard and Warehouse Area, also abbreviated to R.Y.A.W.A. The nineteenth-century historian, Bolton, noted the aboriginal words "rekawi" (sandy) "ani" (path) would describe the trail leading to the Indian encampment out on Hunt's Point.

RYER AVENUE. It was called Bassford Place from Burnside Avenue to East 180th Street. The Ryer family is an old one in Bronx annals, for Ryer Micheilszen's name appears on a 1736 deed of the Manor of Fordham, although at other times it is anglicized to Roger Michelsen. After the Revolution, some of the descendants adopted the name Reyer and Ryer. One Ryer farm was located off present-day Kingsbridge Road west of the Concourse. Another Ryer estate belonged to two brothers, veterans of the Mexican War, who named their land "Monterey" as a reminder of their adventure. Samuel Ryer was remembered-for a time-when East 180th Street was called Samuel Street. James Cardinal, Assistant District Attorney of Westchester County, gives a fascinating account of a notorious John Ryer of Fordham, who, in 1792, killed a sheriff in a tavern run by Levi Hunt. See: Bronx Park South. Ryer escaped to Canada and worked as a chainbearer for a group of surveyors in the wilds of Quebec. He was recognized by a man from Westchester County, arrested and extradited to White Plains jail to await trial. On October 2nd, 1793, John Ryer was hanged - and the final irony is that on Ryer Avenue there is located the 46th Precinct station house of the Police Department.

SACKETT AVENUE. The Sackett farm, extending from Williamsbridge Road to Eastchester Road, was noted back in Revolutionary times, and there was a Captain Sackett listed in 1781 in a surgeon's report. Later on, the farmlands were enlarged to reach Bronxdale Avenue as noted on 1854, 1868 and 1891 maps with the Sackett house being located at approximately Radcliff and Pierce Avenues. The thoroughfare was first called Hilton Avenue. North on the Sackett farm was the Givan estate, and a Givan granddaughter, Harriett Palmer, married States Mean Sackett. They are buried in the Sackett family plot on Forest Avenue, Woodlawn Cemetery. Sharing this plot is the Loomis family, whose lands adjoined the Sackett property at Eastchester Road. See: Loomis Street.

SAGAMORE STREET. Local legend is that the patriotic founders of Van Nest, Messrs. Levy and Morris, wanted all their streets named after Presidents. Theodore Roosevelt was already honored by a land on Throggs Neck, so they adopted the name of Roosevelt's home on Long Island-Sagamore Hill.

SAGE PLACE. This name is a reminder of the Warren B. Sage estate, prior to 1900. His mansion overlooked Spuyten Duyvil Creek, and in the 20th century became Edgehill Inn, a famous resort.

ST. ANN'S AVENUE. From East 132nd Street to East 138th Street, this was the Cherry Lane of Gouverneur Morris' estate in the 1850s, when, according to all accounts, it was lined with cherry trees. From East 138th Street to East 149th Street, it was named in honor of St. Ann's Episcopal Church, which was dedicated to Gouverneur Morris' mother, Ann Randolph Morris. Oddly enough, she is not buried in that churchyard, but in a private graveyard between the Bowery and Second Avenue, 2nd and 3rd Streets. After the Civil War, the neighborhood was built up, and the graveyard hidden by tenements. According to an 1851 survey, St. Ann's Avenue, from Westchester Avenue to East 160th Street, was called Benson Avenue or Fordham Road. The former name referred to Benjamin Benson who purchased the land in 1853, and the latter name was sometimes in use as the road led to Fordham. Around 1875, the avenue was extended to East 161st Street, cutting the Bensonia Cemetery in half. The stretch from East 134th Street to its junction with Third Avenue was called Carr Street, after a William Carr who had purchased the tract from Gouverneur Morris. At East 149th Street, briefly during the Revolutionary War, this was a military camp for Loyalist officers. At East 156th Street, around Civil War times, Aurora Park was a German picnic park which was later occupied by Ebling's brewery and casino. At East 160th Street was another brewery owned by the Hupfel family. See: Ebling's listings.

ST. DOMINIC SQUARE. In 1966, 'the Five Corners" of Van Nest was renamed to honor a nearby Catholic Church, and its pastor the Right Reverend Monsignor Fiorentino. Councilman Mario Merola initiated the bill in the City Council.

ST. GEORGE'S CRESCENT. This is the Negro Hill of the Revolutionary War maps, for a small fort on its summit was garrisoned by the Negroes from Virginia. In the 19th century, it was on land belonging to Leonard W. Jerome, which passed to John Corsa and, eventually, to George Opdyke. See: Lisbon Place.

ST. JAMES PARK. This park, measuring 11.83 acres, adjoining the church of the same name, was opened in 1901. St. James' Church is a nineteenth-century institution, predating the park by at least 45 years.

ST. JOSEPH'S WAY. Until November 1987 this curving road, leading from the northbound Bruckner Boulevard service road to St. Joseph's School for the Deaf was unnamed. It was officially opened with ceremonies on March 18, 1988 - one day before St. Joseph's Day. All speeches were translated into sign-language.

ST. LAWRENCE AVENUE. The Siwanoy Indian trail that led from the upper Bronx to Clason Point crosses this avenue twice: midway between East 172nd Street and Westchester Avenue, and again at the corner of Story Avenue, where it runs diagonally from the northwest to the southeast corner. From Westchester Avenue to East Tremont Avenue, it was in the estate of Lewis Guerlain (1804), which later became the Mapes farm of 1869, and finally the real estate development known as Park Versailles. One of the backers of this plan was the banker, Cyrus Jay Lawrence, but no official records were found to prove he is the man so honored. Early records of the "Burrow Town of Westchester" frequently mention the Lawrence (Larrance) family and, at one time, they owned lands in the Clason Point area. One story is that the street was to be called Lawrence Avenue, but there being an avenue of that name in Highbridgeville, the "Saint" was added to avoid confusion. Frank Wuttge, a fellow Names enthusiast, pointed out the Park was laid out in gridiron pattern - and this "grid" is the symbol by which St. Lawrence is known, having met his death on a torture-grid. If so, the planners were most erudite.

ST. MARY'S PARK. Plotted and mapped in 1868 as simply Mary's Park, this tract was then a cluster of private estates. A few wealthy families lived on the hilltops, with fine lawns and gardens extending to St. Ann's Avenue. It was opened in 1883-1888 as part of the New Parks Act. In 1904, stables were acquired and asphalt walks were added in 1905. The hills were opened to the children in 1906, and ice-skating in Winter and band concerts in Summer became popular features, thanks to Parks Commissioner Haffen. The first playground in the Bronx was built in the park in 1914, along with the first Christmas tree that same year. On the site of the pond that was once the ice-skating rink was built a community recreational center in 1951. A guidebook of The Bronx states the park was named for St. Mary's church (Protestant Episcopal) that stood on Alexander Avenue and East 142nd Street until its demolition in 1959.

ST. MARY'S STREET. This street was laid out shortly after St. Mary's Park was opened in 1888 and was a fashionable promenade for a time. See: Beech Terrace.

ST. OUEN STREET. This small street is laid out across the former Penfield estate near the Mount Vernon line. None of the residents knows the origin of the name, so there is a possibility this is a reminder of a small city in France visited by Frederic Penfield. St. Ouen is now a northern suburb of Paris. This is borne out by the fact the Penfield mansion was built to resemble the St. Ouen manor house on the island of Jersey. See: East 231st Street.

ST. PAUL AVENUE. There is no explanation of the "Saint" being added, as the avenue is laid out across the Paul farm, of which mention is made during General Washington's inspection of Throggs Neck. The widow Paul welcomed the officers on their arrival from Eastchester Road. An 1811 deed mentions a Captain Simon and Mary Paul. Another home was situated on the Old Pelham Road (now Westchester Avenue near Wilkinson Avenue).

ST. PAUL'S PLACE. This is the northern boundary line of the Morris manor lands, and the southern bounds of the Bathgate farmlands. The Place is named after the Protestant church of St. Paul's of Morrisania village, which was begun as a chapel on July 8, 1849. A few years later, it severed its dependence upon the mother church of St. Ann's. See: St. Ann's Avenue.

ST. PETER'S AVENUE. This was once the western boundary of the Adee estate in the village of Westchester. On a map of 1868, it was Union Avenue. It was then renamed in honor of St. Peter's Episcopal church to which it leads.

ST. RAYMOND'S AVENUE. Its present name honors St. Raymond's Catholic church at the lower end of this avenue. For a time, it was 3rd Street in the village of Westchester from Zerega Avenue to East Tremont Avenue, and Evadna Street from Williamsbridge Road to Blondell Avenue. Evadna Arnow was a daughter of the nineteenth-century landowner and married Daniel Mapes.

ST. THERESA AVENUE. This five-block stretch from the Hutchinson River Parkway to Westchester Avenue, formerly Morris Park Avenue, was renamed St. Theresa Avenue in December, 1968. It runs past the Catholic church of St. Theresa of the Infant Jesus. Former names of this same thoroughfare were Liberty Street (1892 map) and Alice Avenue (1910).

SALTERS SQUARE. Facing Boston Road adjacent to Charlotte Street, the intersection's street sign recalls June Salters, a longtime community leader, who died in 1985.

SAMPSON AVENUE. It was named for Admiral William T. Sampson, Commander-in-Chief of the North Atlantic Squadron during the Spanish-American War, and who once bombarded Matanzas, Cuba. From Ferry Point Park eastward to East Tremont Avenue, this was once a sharp hill rising from the swamps. A small creek called Boundary Creek wound in from Morris Cove (now covered by the park). It was part of the estate owned by John A. and Cora Morris, who raised racehorses. He died in 1895 and she died in 1922. The Morris mansion-a 40-room showplace, which was built in the 1860s-was bought around 1924 by Najeeb Kiamie, a Lebanese millionaire, who had to move it at the expense of $150,000 as it was situated partly on Sampson, Avenue. Nothing remains of this mansion, but at 2850 Sampson Avenue a garden wall is built of what had been an ornate fishpond built of lava imported from Vesuvius. Eastward from East Tremont Avenue, Sampson Avenue was part of the original Baxter (1680) homestead which became the Ludlam/Brown property of the 1840s. The Jackson farm was mapped there, and remained until the early 1900s. The eastern end of this avenue, at Hollywood Avenue, held the remnant of an orchard until 1958-then the entire area was demolished to make way for the Throgs Neck bridge approach.

SANDS PLACE. The earliest settlers who owned land in the vicinity of Pelham Bay Park were the Sands. Ferris Sands was a Revolutionary soldier, as was Comfort Sands, who is buried in an Eastchester churchyard. A Mary Sands married John Pell. An 1873 map charts this plot as "Lands of William Bowne Sands."

SANDY COURT. See: Dune Court.

SANTA MARIA AVENUE. The Roman Catholic church of Santa Maria is located on this short street off Zerega Avenue. It was formerly part of St. Raymond Avenue.

SANTO DONATO PLACE. At the behest of local residents in 1979, a short street, Mildred Place, was renamed to honor a saint in Italy. According to legend, the holy man averted a drought in the region from whence many of the residents had migrated. An early reference to this area was a military report that General Washington visited the Paul farm during his survey of Throggs Neck.

SAXON AVENUE. This was once part of Orloff Van Cortlandt's land. The avenue was laid out in 1897 through what had been the John Dickinson estate. Evidently, Mr. Dickinson was interested in history for the streets laid out recalled Harald, the Saxon king who fought the Normans. The names were given in 1890, and there may have been some confusion because a 1916 map shows Saxon as a prior name for the adjoining Hillman Avenue as well. The warrior's names did not survive the years. See: Dickinson Avenue, Hillman Avenue.

SCANLAN SQUARE. A triangular plot of land in front of St. Helena's R.C. church at Olmstead and Benedict Avenues. It was named in June, 1965, in honor of Monsignor Arthur J. Scanlan, pastor of the church.

SCENIC PLACE. This is a private road, overlooking the Hudson River at West 232nd Street on the crest of historic Berrian's Neck, later known as Spuyten Duyvil Hill. The scenic view is best in months when the trees are bare.

SCHIEFFELIN AVENUE. This avenue honors a wealthy family who owned land in Edenwald from Eastchester Road to Rattlesnake Creek (ancient boundary between Eastchester and Westchester) near Ely Avenue. A Schieffelin's Lane crossed Baychester Avenue in the vicinity of Crawford Avenue. The Schieffelins were prominent merchants, specializing in pharmaceuticals, and were listed in "American Millionaires, 1892." Eugene Schieffelin is not fondly remembered for he released the first starlings in America in 1890 to exterminate caterpillars.

SCHIEFFELIN PLACE. Another reminder of this family who had previously owned considerable property on Clason Point in the 1870s. This tract was sold to the Christian Brothers who ran a well known military academy there for many years, while the Schieffelins moved up to this Baychester region. See: Schieffelin Avenue.

SCHLEY AVENUE. This was the northern limit of the John A. Morris estate on Throggs Neck, west of East Tremont Avenue. When first laid out, it was mapped as Burdett Avenue, but later honored Admiral Winfield Scott Schley of the Spanish-American War. From East Tremont Avenue down to Long Island Sound, it was once the lands of the Baxter family, then Drake, John D. Wolfe (1830-1870) Montgomery family and the Bruce Brown estate. The hill was called Wolfe's Hill and Brown's Hill. At the foot of Schley Avenue was an important fishing station. Mummies, artifacts and other Indian relics have been found there, and are on display in the Museum of the American Indian.

SCHNEIDER-SAMPSON PARK. Hard by the Bruckner Boulevard curve at Baisley Avenue, this was once the northeast corner of the Henderson mill yards, next to Willow Lane. The triangle honors two Schuylerville residents, victims of World War I. It was placed under the jurisdiction of the Department of Parks on July 1, 1929. George Schneider was killed in action on October 16, 1918, at the age of 23. He was a star of the Franklin Athletic Club on Blondell Avenue, behind Westchester Square. Corporal William P. Sampson, a machine-gunner, was gassed in August 1918 but survived the war. He died some 11 years later of pneumonia, at the age of 33.

SCHOFIELD STREET. Along with the Bownes, Hortons and Fordhams, the Schofields were counted among the earliest settlers of City Island. They owned considerable property across the waist of the island, through which the street runs.

SCHOOLMAN ATHLETIC FIELD. The running track and stadium, located at the southeast corner of Macombs Dam Park at East 161st Street and River Avenue, was named on February 6, 1962. Irving M. Schoolman was an athlete (New York University, '28) who organized numerous Schoolboy Meets - was a manager of the 1956 Olympic team - a track coach at many athletic meets and who received countless civic awards. He died in 1961.

SCHORR PLACE. This short street was once part of the Givan estate, and overlays the original driveway. Minnie Schorr's name appears in the 1920s, but in 1940 the name was changed to Maple Place. Neighborhood old-timers stated the Givan driveway was lined with maple trees, which prompted them to request the Place be named for the trees. At a later date, the Place once more regained its present name.

SCHULTHEISS PLAZA. This is a triangular plot named in honor of Rev. Gustav R. Schultheiss, pastor of St. Helena's Catholic church, and Dean of the Bronx Catholic Clergy in 1969. The triangle is on Metropolitan Avenue, between Castle Hill Avenue and Purdy Street. The bill was signed by Mayor Lindsay after sponsorship by Councilwoman-at-Large Aileen B. Ryan.

SCHURZ AVENUE. This riverside avenue is laid out on historic land, as it is the site of a Siwanoy Indian settlement, the place of the Throgmorton colony, and the landing of General Howe's army in 1776. See: Calhoun Avenue. It runs through the former lands of Philip Livingston, Ash, Mitchell, Van Schaick, Huntington, Havemeyer and Morris. Names of Civil War generals abound in this Throggs Neck section, and Karl Schurz served in that conflict. He was born in Germany in 1829, was an editor of a German-language newspaper, and was appointed (after the War) Secretary of Indian Affairs. He died in 1906.

SCHUYLER PLACE. This short street has no connection to Fort Schuyler, which honors Philip Schuyler of Revolutionary fame. It harks back to a Throggs Neck surveyor in the years 1872-1874 named John Schuyler.

SCHUYLER TERRACE. This is a small bungalow colony on the narrow neck separating the East river from Long Island Sound. The homes had originally been situated on nearby Schuyler Hill, but moved to this spot around 1930. Schuyler Hill took its name from nearby Fort Schuyler, which was constructed in the years 1848-1851. The land was once on the 1790 Stephenson property, which was purchased by Abijah Hammond around 1810.

SCOTT PLACE. This Place was most likely a meadow alongside the ancient road to "Frogges Point" in records of the 1600s. It was part of Thomas Baxter's 365-acre homestead of those days which, in the mid-1800s, was part of the Coster farm. It was the Bruce Brown estate in the 1920s when it was laid out and cut through. This short street was first named Douglass Place (1916) before the actual cutting through, but then was named in honor of General Winfield Scott of Mexican War and Civil War fame.

SCREVIN AVENUE. John Screvin was the nineteenth-century owner of Castle Hill Point. On some early Bronx maps "Screvin's Point" was used. He inherited the property, extending to Lafayette Avenue, from his father-in-law, Gouverneur Morris Wilkins. Title went back to Isaac Wilkins, who was once rector of St. Peter's P.E. church. The Screvin family is said to have been of Scottish descent, emigrating from a town called Torry or Torrey in Scotland.

SCRIBNER AVENUE. The Scribners were landowners whose gravestones can be seen in St. Peter's Episcopal churchyard off Westchester Square. Scribner Avenue was once part of the Ferris farmlands, then the Taber farm and finally the Stillwell estate. Stephen and Deborah Scribner were landowners off the Fort Schuyler Road in 1847, and bought more property in 1855.

SEABURY AVENUE. This avenue behind St. Peter's P.E. church at Westchester Square was named in memory of its fourth rector, Dr. Samuel Seabury III. He was born in Groton, Conn., in 1710, was a graduate of Yale and was consecrated a Bishop in Scotland in 1784. Why in Scotland? Because the American Church, until the end of the Revolution, was under the jurisdiction of the Bishop of London. All the colonists had to go to England for Ordination, but, after the Revolution, none could go to England for it was a requirement that every Bishop must take an oath of allegiance to the reigning monarch. Therefore, it was to Scotland Dr. Seabury went, where, upon assurance of his loyalty to his faith (and not to the King), he was consecrated.

SEABURY PLACE. This short street, laid out across the former estate of Thomas Minford, recalls Amelia Seabury who married into the Minford family. See: Miniford Place.

SECOND STREET (1). This is a paved road that runs from east to west across Hart Island- the second of five numbered streets.

SECOND STREET (2). This is a footpath in the Park of Edgewater, two rows in from the waterfront. See: Park of Edgewater.

SECOR AVENUE. This name is intimately connected with both the region and the Boston Road. The Sicard [spelled this way] family were early Huguenot settlers who arrived there in the 1670s. Since then, the name has been rendered Seacord, Secor, etc. The avenue runs through the former Darke estate of the 1880s. When first surveyed and laid out, it was given the name Johnson Street.

SEDDON STREET. This short street was planned to be called Tryon Row when the tract was sold by the Seddon Realty Corporation in 1910. The name, Seddon, is not encountered in old deeds of Westchester, so there may be some truth in the story that Hans Truelson, who was president of the corporation, played a joke on the townspeople: he said the name was a concocted one, whereas there was a James A.

Seddon, *Confederate* Secretary of War. Civil War animosity was still extant in 1910, the 50th anniversary of that North-South conflict.

SEDGWICK AVENUE. It is named after Major General John Sedgwick of the Civil War who was killed at Spottsylvania. The avenue received its name in 1886, displacing Emmerick Place where it crossed Kingsbridge Road. The original name was a reminder of the Hessian officer Emmerich [the correct spelling] who was so active against the colonists in the Revolutionary War. The avenue itself runs from the former village of Highbridgeville through the vanished estates of the Ogdens, Pophams, Morris, Cammanns, Andrews, Claflins and Jeromes. Some Revolutionary forts were once located along this ancient road. See: Emmerick Place.

SELWYN AVENUE. It was named on May 2nd, 1916, in honor of one of the early owners of Fordham Manor. In 1683, John Archer lost the Manor of Fordham by foreclosure to the wealthy Steenwyck family. Later, the widow Steenwyck married Dominie (Reverend) Henricus Selwyns, and they endowed the land to the Dutch Reformed Church. Moreover, they generously gave Archer's descendants an opportunity to buy 300 acres at a reduced price. The minister's name appears as Sellinus (1692), Selyns (1700) and also Selwyn. O'Callaghan, a famous historian who translated the Dutch records of Nieuw Amsterdam into English, stated that many Irish were part of the Dutch colony-among them, the Sullivans whose name became Swelwin and Selwyn in later generations. Selwyn Avenue's lower end marks the boundary of the Manor of Fordham. It was known as Eden Avenue from Morris Avenue to Mount Eden Avenue and it was also called 3rd Avenue from Mt. Eden Avenue to the Concourse.

SEMINOLE AVENUE. As far back as 1868, maps show this as part of Denton Pearsall's estate, "Woodmansten." In 1881, James Pinchot purchased acreage across which land Seminole Avenue now runs. When this real estate was developed as Westchester Heights, in 1917, the streets were named after wealthy resorts, such as Newport, Narragansett Bay and Saratoga. The original name of Seminole Avenue was Saratoga Avenue. The present name honors the Indian tribe of Florida, but for no apparent reason.

SEMINOLE STREET. This street is laid out across what were once swamps that were owned by Reverend John Bartow in the 1720s, figured briefly in October 1776 in a Revolutionary skirmish, and then were in the possession of the Ferris family (1840s), Timpson (1860s), John Hunter (1890s) before being vested by New York City upon the annexation of The Bronx in 1895. Drainage took place around 1920. The swamps might have reminded the surveyors of the Everglades, and casting about for an Indian name, the Seminole tribe would be a natural choice.

SENECA AVENUE. It was known as Irvine Street until 1907. This avenue may have some connection with nearby Lafayette Avenue. Red Jacket, a famous Seneca chief and friend of Lafayette, accompanied the Frenchman on part of his triumphal tour in 1824-1825, a part of this tour including Hunts Point. It is on the Indian lands called Quinnahung, "The Planting Neck," which became the lands of Richardson and Jessup in 1663. The Hunt and Jessup families were principal proprietors until the Revolutionary War. In the 19th century, various rich men developed estates, and this particular avenue was the dividing line between the lands of W. W. Gilbert and E. G. Faile.

SENGER PLACE. This is on historic Ferris Point, which was settled in the 1640s by Englishmen named Spicer and Brockett. The land passed into the possession of the Hunt family (1686), and later, the Ferris families. In 1840, the tract was mapped as farmlands belonging to Charlton Ferris and, twenty years later, to J. H. Ferris. The Senger family arrived around 1840 from Germany, settled in downtown New York and later purchased some land along Westchester Creek from a man named Lawton, who had bought it from J. H. Ferris.

SETON AVENUE. When this territory was South Mount Vernon, the avenue was South 12th Avenue. The Setons were land owners of the past century, and prominent in affairs of Eastchester. Rattlesnake Creek was dammed as it flowed through their 51-acre estate and formed a waterfall that gave its name to Seton Falls Park. The estate was called "Cragdon." Mother Seton, who founded the Sisters of Charity in 1812, was the daughter of a Protestant minister, and yet another son, Robert, was the first Catholic priest in America to become a Monsignor.

SETON FALLS PARK. This is 29.25 acres of park land which remain of the 51-acre estate of the Seton family of Eastchester. The park was acquired by the City in two parcels from the Seton Falls Realty Corporation. On June 10, 1914, 29 acres were purchased for a hospital site, but was assigned for public park use on June 11, 1930. On March 2, 1932, an additional seven lots were condemned to add to the park. The falls, once a beauty spot of the Seton estate, has been allowed to degenerate to a sluice. The stream flowing through the park is called Rattlesnake Brook (or Creek) and was already noted in records of the 1600s.

SEVEN BROTHERS SQUARE. On June 10, 1981, the seven Santini brothers were honored when the area of West 170th Street and Jerome Avenue was so designated. The brothers were Pasquale, Pietro, Paride, Rinaldo, August, Godfrey and Martin who came from Lucca, Italy, and started a carting business in 1905. Over decades the firm grew to a huge enterprise with 1000 workers, a fleet of trucks and a string of warehouses. The last of the original founders, Rinaldo, died in 1980 but the business continues into the fourth generation.

SEVENTH AVENUE. This is a narrow road in the Park of Edgewater, seventh in order from the waterfront. See: Park of Edgewater.

SEWARD AVENUE. Barrett's Creek formerly came out into the Bronx River at the foot of Seward Avenue. Another creek, called Pine Rock Creek cut across this avenue at Morrison Avenue, and Pugsley Creek crossed this avenue at approximately Pugsley Avenue. The Indian path of the Siwanoys crossed at Leland Avenue. Judges' names are on adjoining avenues, so most likely this avenue honors Samuel S. Seward, who was Judge of the New York Circuit Court.

SEXTON PLACE. In 1913, the Eastchester Syndicate Company assembled several estates, bisected by the New York, Westchester and Boston Railway, and sold almost 1,500 lots at auction-stressing the railroad transportation as a selling point. One of the tracts was owned by Harvard University, a bequest of a former director of the Bronx Zoo, Lawrence E. Sexton. Regretfully the author had to rule out James and Freelove Sexton who bought land near the Boston Post Road in 1819, almost a century before the Place was named. Stony Brook crossed Sexton Place at Allerton Avenue.

SEYMOUR AVENUE. It honors Horatio Seymour, Governor of New York State during the Civil War. On an 1868 map, the street was almost entirely on Denton Pearsall's land, but by 1905, it was partly on the J. H. Watson estate and the S. Sullivan property. At its junction with DeWitt Place, it was crossed by Thomas Haddon's sawmill lane that led from Williamsbridge Road to Eastchester Road and down to the Hutchinson River. Stony Brook crossed this avenue at Mace Avenue and again at Astor Avenue.

SHAKESPEARE AVENUE. From Jerome Avenue to West 168th Street, it was called Judge Smith's Hill because of a roadhouse at the foot, on Jerome Avenue, known as "Judge Smith's." In Indian days, this avenue was on the flank of a steep hill known as Nuasin that overlooked the creek called Mentipathe (now Jerome Avenue). Above West 168th Street, it was called Marcher Avenue until 1912. This harks back to a deed, dated 1865, when Rebecca Marcher purchased an estate which she called "Rockycliff." According to a Highbridgeville old-timer who once worked as their stableboy, the Marchers had statues on their lawn depicting Shakespearian characters-and even named their horses and dogs in the same fashion: Thisbe, Puck, Othello and so on.

SHANDLER RECREATION AREA. See: Allen Shandler Recreation Area.

SHCHARANSKY SQUARE. Wide attention was generated when this Soviet Jew differed with the Russians and sought to emigrate to Israel in the 1970s. Borough President Stanley Simon proclaimed the intersection of West 225th Street and Mosholu Avenue "Shcharansky Square" to embarrass the Russian Mission there. Finally able to migrate, Anatoly Shcharansky did visit "his" Square in May, 1986.

SHEPARD AVENUE. On the East River side of the New York State Maritime College is a curving road named for Commander Shepard who was Superintendent from 1883 to 1886.

SHERIDAN AVENUE. It is named after General Phil Sheridan of the Civil War. It is laid out across part of the Manor of Morrisania, specifically belonging to General Staats Morris in pre-Revolutionary times, which he used as a racecourse. In the 19th century, it was changed to the Fleetwood Race Track, and some of the stables were located on Sheridan Avenue at East 166th Street. From East 167th Street to Mount Eden Avenue, this thoroughfare was once called Main Avenue.

SHERIDAN EXPRESSWAY. It was officially named in 1952 for Arthur V. Sheridan, a highway engineer, who was killed in an auto accident.

SHERIDAN PLAZA. Located at Broadway and Mosholu Avenue, this plaza honors David Sheridan, once a resident of nearby Post Avenue. His was an old Riverdale family. Sheridan enlisted in 1940 as a private, and was a second lieutenant at his death in 1945 during a drive on the Rhine.

SHERMAN AVENUE. This was Indian territory, which became a Dutch possession, eventually to be a part of General Staats Morris' racecourse of the 18th century. As part of the (1870) Fleetwood Race Track, the track was underneath Sherman Avenue from East 165th to East 166th Streets. The clubhouse and grandstand were at this avenue and

East 166th Street until the enterprise closed down in 1898. When avenues were cut through, this particular street was named in honor of William Tecumseh Sherman (1820-1891) of Civil War fame.

SHOELACE PARK. On July 7, 1973, a section of Bronx River Road from East 211th to East 227th Street was set aside for a recreational park. Due to its long, narrow, shape, the name seemed appropriate. Mayor Lindsay, Parks Commissioner Clurman and members of the Wakefield-Edenwald Office of Neighborhood Government were present at the official opening.

SHORE FRONT PARK. Dedicated in the early 1970s, this park is on land once part of the T.H. Edsall estate. It is in the shadow of the Henry Hudson Bridge over the Harlem River in Spuyten Duyvil.

SHORE ROAD (1). This is a circumferential road around Riker's Island.

SHORE ROAD (2). On Throggs Neck, it runs through the former Wissmann and Adee estates at the very edge of Eastchester Bay. In the Pelham Bay sector, it is one of the oldest routes in our borough in that it evolved from the Indian trail of long ago. This was used by both the British and Colonial troops during the Revolutionary War, and, following that, was the carriage path of the nineteenth-century squires whose estates lined the shoreline of Eastchester Bay.

SHOREFRONT PARK. This undeveloped parcel (also called Riverdale Park parallels the Hudson River from West 232nd Street to West 254th Street. It was formed from the properties of two prominent families named Dodge and Perkins.

SHOREHAVEN. At the end of Clason Point, overlooking the East River, the Bronx River and Pugsley Creek, Shorehaven is a nearly 1200-unit condominium complex. It was begun in 1988 on the cleared site of Shorehaven Beach Club - a Summer resort that had flourished from 1947 to 1986. Local Historian Arthur Seifert noted the same 48-acre tract had once been Kane's Park and the Clason Point Amusement Park, popular places from the late 1890s to sometime in the 1940s.

SHRADY PLACE. This is on the Continental farm of Reverend Tetard, and considered part of the village of Kingsbridge. In 1860, Charles Darke conveyed 12 acres of land to Dr. George Shrady, editor of the *New York Medical Record*, and his mother, Maria. Dr. Shrady was a Civil War veteran, and his father, John, had been a soldier in the War of 1812, and a grandfather had served in the Revolutionary War as Private. John J. Schreder, 3rd Regiment of the New York Line under Captain DeWitt.

SID AUGARTEN BALLFIELD. In July 1989 a Little League ball field at West 254th Street and Riverdale Avenue was opened and dedicated to a tireless volunteer.

SIEGFRIED PLACE. See: Lohengrin Place.

SIGEL (Franz) PARK. See: Franz Sigel Park.

SIGMA BROOK. It originates in the vicinity of Sigma Place and Palisade Avenue, makes an S-turn, runs south then west into the Hudson River. Riverdale residents have no name for it, but Sigma Brook it was on a real estate (unofficial) map.

SIGMA PLACE. Sigma is the Greek letter "S," which is roughly the contour of this curving Place. This land was once part of Phillipsborough (Frederick Flypse's land in pre-Revolutionary times) and the scene of some encounters from 1778 to 1784 when it was called Upper Cortlandt. The Place was included in Lower Yonkers until 1874.

SILVER BEACH GARDENS. This is a 350-home community near the tip of Throggs Neck. Its high bluffs were an excellent lookout for the Siwanoy Indians, and later for the adversaries in the Revolutionary War. It was part of the Stephenson farm of the 1790s. A daughter, Glorianna, sold it to Abijah Hammond in 1795, who built a mansion there between 1805 and 1810. It became the Whitehead lands of the 1850s, Havemeyer property in 1869 and eventually the joint real estate venture of the Peters and Sorgenfrei families. From a Summer bungalow colony of the 1920s, Silver Beach Gardens became a permanent settlement, with residents purchasing the property and forming a corporation in 1972. The name takes certain poetic license with the sandy, slightly-curving beach that at times has a silvery appearance. No local folklore has ever been heard regarding buried silver!

SILVER STREET. It was formerly Silver Lane. It was platted by Ezra Cornell in 1807 to connect Bear Swamp Road (East Tremont Avenue) with Eastchester and Williamsbridge Roads. The Cornell farmhouse stood on the northwest corner of Silver Street and Williamsbridge Road. Silver Street was the boundary of a 5-acre homestead, later sold by Cornell to Andrew Arnow, whose son, Matson Arnow, sold to George F. Jarrett. The reason for the naming of Silver Lane has not survived the centuries, so the supposition of an earlier historian, George Ferriera, is advanced. Most transactions of those post-Revolutionary times were accomplished by the barter system, so if land was gotten by silver (a term for currency in those days), it was worth recording for posterity.

SIMEONE SQUARE. The intersection of East 186th Street and Hughes Avenue, since November 1989, has been named for Frank Simeone, an active resident of Belmont for 75 years.

SIMPSON STREET. An 1865 map shows J. B. Simpson's house on the Westchester Turnpike, and his estate, "Ambleside," was bounded by Bound Brook (Intervale Avenue), Southern Boulevard and Westchester Avenue. A generation later, William Simpson owned the adjoining 120-acre "Foxhurst" estate, and was famous for his string of trotting horses. The Simpsons, natives of England, were related to the Woodruff family whose property ran roughly from Crotona Park to the Bronx River.

SIWANOY TRAIL, The. Siwanoy Indians once inhabited the eastern part of The Bronx, so a 1.8 mile nature trail in Pelham Bay Park was given this name. During the winter of 1988 Appalachian Mountain Club volunteers cleared the path and on March 20th, the trail - beginning on the Shore Road south of the Pell-Bartow Mansion - was officially opened to the public.

SIXTH STREET. A footpath in the Park of Edgewater, sixth in order from the waterfront. See: Park of Edgewater.

SLATTERY PARK. Located at East 183rd Street and Valentine Avenue, it honors Robert Augustine Slattery who was the first New York policeman to die in World War I. He lived on Tinton Avenue. Entering the army in June of 1918, he sailed for France in

July, was in the front lines by August, and was killed in action on September 14, 1918. He was a corporal.

SOLIMINE FIELD. A premier umpire on local baseball diamonds for many years, Dan Solimine's name was given to a sports field between Unionport Road and Bronx River Parkway in 1980.

SOMMER PLACE. This Place is laid across the Coster farm of the late 1800s, and when the land was subdivided around 1920, John Sommer, a real estate developer *(circa 1900-1911)*, was remembered.

SOUNDVIEW AVENUE. It was Clason Point Road until May 2nd, 1916, when it received its present name by a Board of Aldermen Ordinance. It is part of the Colonial road that led to Cornell's Neck (earlier name of Clason Point) and overlays some of the Indian trail leading to a village at Leland and Lacombe Avenues. As late as 1870, the East River at this point was mapped as Long Island Sound-hence the name.

SOUNDVIEW PARK. The City took title to 93 acres of swamplands, creeks and low-lying land along the left bank of the Bronx River in 1937. This area took in the former "Woodlawn" estate of the Schieffelin family, the Dominick Lynch lands, and the extensive property of the Ludlow family. Included were Dairy Creek, Barrett's Creek and Cedar Brook, as well as the historic Black Rock which figured in the original Cornell deed of 1643. It is local lore that the original name was Lafayette Park, given by Commissioner Robert Moses. Later, he acceded to the wishes of the residents who had asked to have the park given a "neighborhood" name. See: Soundview Avenue.

SOUNDVIEW TERRACE. This is a hillside road at the lower end of Silver Beach Gardens laid out over a much earlier road leading from the Hammond estate to the Bayard farm (now Fort Schuyler). From this road, one has a view of the Sound as well as the East River.

SOUTH BROTHER ISLAND. Formerly named Gesellen (Brethren) along with North Brother Island, it acquired its Dutch name in the 1640s and was Anglicized to South Brother a century later. In 1708, the English governor patented to William Bond the islands known as "The Brothers." Daniel Ludlow was the owner in the later 1700s, Edward Acheson in 1802, and Jacob Eaton in 1805. Colonel Jacob Ruppert bought it in 1894 and erected a Summer house there, which he occupied until 1907. Yorkville youths would sometimes swim to Rat Island, or Ruppert's Island (both unofficial names). The brewer organized many clambakes and venison roasts on his island. The mansion burned down in 1909, but it was not until 1944 that the Colonel's estate sold the island to John Gerosa. It was listed as the property of the Manhattan Sand Company in 1958. Although the 11-acre island is definitely in Bronx waters, most charts showed it to be Queens territory. The Bronx County Engineers' Office amended the county lines in 1940 to include the island, but many subsequent maps have not reflected the change. See: North Brother Island.

SOUTH NONATIONS. This reef lies in Long Island Sound, north and east of Hart Island, and slightly to the south of East Nonations. The name is thought to be a corruption of "No nation." Unwilling to fight over such a tiny piece of land, the early

English and Dutch conceded it belonged to no nation. Another school of thought is of the opinion the real meaning is "Notations," i.e. abbreviations.

SOUTH OAK DRIVE. "Bronxwood Park" was an 1880 real estate development and the streets were named after various trees. Originally this street was called South Chestnut Drive. Evidently, in the 1880s, chestnut trees were predominant. A casual survey in 1970 showed more maple trees than either chestnut or oak trees.

SOUTHERN BOULEVARD. Like the Grand Boulevard and Concourse, this wide and important boulevard did not evolve from an animal run, through an Indian trail to a Colonial footpath to a carriage road to a paved highway. Southern Boulevard was a drawing board creation envisioned by the Engineering Department of the Annexed District sometime in the 1870s to be a grandiose thoroughfare sweeping up from East 133rd Street and Third Avenue, cutting through the wide estates of the eastern Bronx, bypassing Crotona Park and Bronx Park and terminating at the Botanical Garden. A latter-day engineer said its name was first suggested when the countryside was still under the jurisdiction of Westchester County and therefore the proposed boulevard was to be the southernmost in the county. Upon annexation in 1874 the name was still applicable to the boulevard originated in the South Bronx. It was laid out across the original Jonas Bronck farmlands which became the holdings of the Morris family, on through the lands of the Leggett homesteaders, the "Longwood Park" estate of S. B. White and the Fox, Simpson and Tiffany properties. It was the boundary line of the Minford, Woodruff, and Lydig estates, and cut off part of the academy grounds of St. John's (now Fordham University). The Lower end of Southern Boulevard was settled first and became industrialized early in the 20th century, whereas the upper end, near Hunts Point and running to Bronx Park, was a more attractive residential area. In 1981, the upper end of this Boulevard from Fordham Road to Allerton Avenue was renamed Dr. Theodore Kazimiroff Boulevard.

SPENCER AVENUE. William Appleton, publisher, once lived at nearby "Wave Hill" (West 248th Street) and had as his houseguests many distinguished visitors. One was the English philosopher, Herbert Spencer, whose name was given to this avenue. Other honored guests had been John Tyndall and Baron von Liebig and their names, too, are on adjoining avenues.

SPENCER DRIVE. This is a reminder of the bayside estate of the Spencer family. William Spencer, husband of Elenore Lorillard, was a benefactor of the New York Public Library, and his son, Lorillard Spencer, was the publisher of *The Illustrated American*. The drive was originally a carriage path along the north end of a polo field, that terminated at a small dock on Palmer Cove.

SPENCER PLACE. See: Spencer Avenue.

SPENCER TERRACE. See: Spencer Avenue.

SPLIT ROCK ROAD. Once this was one of the most important roads in the region and was the colonists' improvement on an earlier Indian trail that ran from City Island to Pelham. The colonial lane crossed the lands of the Pell family, and passed a landmark known as the Split Rock overlooking the Hutchinson River. During the Revolutionary War, Colonel John Glover's small forces took advantage of the stone walls that lined the sides of the road. From their positions, the Americans slowed down the advance of the British and Hessian troops in October, 1776. The road is now inside the Split Rock golf course of Pelham Bay Park, and is not used for vehicular traffic. Sections of the stone walls can still be seen at its western terminal, near the Split Rock. See: Glover's Rock, and Split Rock, The.

SPLIT ROCK, The. This is a large glacial boulder, a landmark in use since the coming of the European settlers in the 1600s. It is situated on the edge of Pelham Bay Park, overlooking the Hutchinson River, just south of New England Thruway bridge. Legend has it that, sighting through the wide cleft that splits the boulder into almost equal halves, one can see the site of Anne Hutchinson's farm on the opposite bank. Some historians believe her farm was located near the rock itself at the side of Split Rock Road. This pioneer settler was murdered by Indians in 1643. In 1962, when the New England Thruway was extended across the Hutchinson River, it was at first planned to blast away part of the boulder to make room for the super-highway. Edgar Brown, Pelham historian, and Dr. Theodore Kazimiroff, Bronx historian, succeeded in having the line for the new road moved a few feet north and the ancient landmark was spared. See: Split Rock Road.

SPOFFORD AVENUE. An 1868 designation of Hunts Point was Spofford's Point when Paul N. Spofford maintained his estate, "Elmwood," there. He was a director in banks, railroads, insurance companies and a sugar refinery. His name is found on the membership rolls of the exclusive Union Club from 1870 to 1887.

SPRUCE AVENUE. This is a road located in the northwestern section of Woodlawn Cemetery. The family plots of two related City Islanders, William Bowne and Elnathan Hawkins, are there as well as the mausoleum of the Hupfel family, Bronx brewers. The name is in keeping with the early arboreal aspect of the former farmlands.

STADIUM AVENUE. Originally laid out as East Road, as it bordered a former polo field, from Rawlins Avenue to Country Club Road. In 1922, East Road was renamed in honor of Rice Stadium. See: Rice Stadium.

STANTON COURT. This is a Shorehaven street named for Michael Stanton, the corporation that promoted the housing complex. See: Dune Court.

STARBOARD COURT. See: Dune Court.

STARLING AVENUE. At one time this was called Railroad Avenue. An ancient Indian trail crossed it at its junction with Unionport Road. At the upper end, Indian Brook (Seabury Brook) crossed it at Glebe Avenue. The origin of this name is uncertain. One tale is that the surveyors facetiously dubbed in the name after being annoyed by starlings. Another bit of folklore is that a nearby club, the Starling Athletic Club, numbered in its membership some influential politicians of Unionport who saw to it that their name was "immortalized." Sehring's "Washington Park," a picnic resort, was located on this avenue. See: Purdy Street.

STEARNS STREET. Once called Rose Place, it was officially opened in 1914. At its northern end once flowed a brook that originated in the vicinity of Bronxdale Avenue and emptied into Westchester Creek. It was called Seabury Creek, Indian Brook and

Barnes Creek at various times. T. Gerald Stearns, active in Bronx construction and real estate, was a Spanish-American War veteran who received a royal welcome upon his safe return. He was active in politics, and was part-owner of the Suburban Baseball Club, located near present-day Parkchester. His father was J. Thomas Stearns, real estate broker of the 23rd and 24th Wards, listed in 1885 publications.

STEBBINS AVENUE. It was the northern extremity of the Manor of Morrisania, which later became the William W. Fox estate. A further division created "Pioneer Park," a picnic grounds at Home Street and Stebbins Avenue. The Stebbins family were wealthy landowners near the present Concourse and not in this area at all, so it is believed the avenue honors Henry Stebbins, who was (in 1876) president of the Board of Commissioners of the Department of Parks. A Reverend George Stebbins was once minister of the West Farms Presbyterian church (1828-1835), but it is doubtful if this man of God is remembered. The name is pure Saxon: Styb (stump) and Ing (meadow).

STEDMAN PLACE. Ernest C. Stedman owned a tract of 36 acres in this neighborhood, and the Place is laid out across the northern border of his land. His property was approximately bounded by the Saw Mill Lane (Allerton Avenue), Mace Avenue, Laconia Avenue to a point just west of Wilson Avenue.

STEENWICK AVENUE. Cornelius Steenwick (or Steenwyck, as his signature attests) was listed as a Trader in the city's records of 1653. He became a wealthy merchant and held a mortgage of 10,000 guilders on Jan Arcier's (John Archer's) Manor of Fordham. Steenwick became the owner in 1683 by foreclosure. He died in 1684, leaving his last will and testament in the Dutch language. His wife was Margaretta DeReimer, who later married the Reverend Henricus Selwyns. They endowed the land to the Dutch Reformed Church, whose elders sold it to the Archer descendants, and to the Varians, Valentines and others. See: Selwyn Avenue. The reason Steenwick's name is not in the Fordham section is that he was a Mayor of New York in 1667, and in this Baychester area, all the avenues carry names of former Mayors, including DeReimer. Steenwick took his name from the village of that name in the District of Drenthe, the Netherlands.

STELL PLACE. This is laid out across farmlands that bordered the historic Williamsbridge Road. It is named for Andrew Stelle Hammersley, son of Thomas *(obit.* 1831), whose estate extended up to Allerton Avenue. On some maps, the name is spelled Hamersley.

STELTZ SQUARE. See: Monsignor John A. Steltz Square.

STEP STREET. This is a flight of steps leading down from Knolls Crescent to Kappock Street. It is noted on commercial maps, and listed in the Street Name Index of the Department of City Planning.

STEPHENS AVENUE. Clinton Stephens bought the end of Clason Point in the late 1890s, and moved into the historic house that had been owned by Isaac Clason, which incorporated parts of the older Willett house of the 1700s, and that of Thomas Cornell of the 1600s. Mr. Stephens did considerable restoration, and had a plaque on the wall attesting to the previous occupations. Born in 1834, Stephens was a contractor and engineer whose contracts included the Erie Canal, the San Hilario tunnel in Mexico, the Erie Railroad and the New York State Aqueducts. Mr. Stephens developed his

Clason Point holdings into a well-known amusement center, and ran his own boats from the foot of East 138th Street and the East River to the Clason Point Amusement Park. It was also serviced by trolley cars and by ferryboats. The Park went out of business in the late 1920s. The tip of land at the foot of White Plains Avenue on the East River was known as "Stephens' Point."

STEUBEN AVENUE. This street was opened in 1907, and honors Frederick Wilhelm von Steuben who was born in Magdeburg in 1730. As Inspector General of the American army during the Revolution, he transformed the ragged troops into a well-disciplined army. He was granted a 16,000-acre tract in New York State in 1786 by a grateful Congress.

STEVENSON PLACE. It is laid out over the original Van Cortlandt patent, which overlooked-in the 1860s-famed Jerome Park racetrack. Orloff Park, a real estate venture, incorporated the name of Oloff Stevenson Van Cortlandt, Burgomeister of Nieuw Amsterdam from 1654 to 1664. It was the John Dickinson property of the early 1900s.

STICKBALL BOULEVARD. In the Summer of 1987 a reunion was held on Clason Point with a nostalgic turnout of former teenagers of the 1950s and 1960s. A marathon stickball tournament was held on a two block stretch, from Lafayette to Randall Avenues, one block east of White Plains Road. A bill designating that stretch as Stickball Boulevard was passed and five street signs were duly posted.

STICKNEY PLACE. A reminder of the 51-acre estate owned by Joseph Stickney that extended from Bronxwood to Laconia Avenues, East 217th to East 224th Streets. East of Paulding Avenue stood the Stickney mansion at approximately East 219th Street, but it was in the middle of the street and had to be razed. The land had been purchased from Oscar V. Pitman in 1896.

STILLWELL AVENUE. In the area between this avenue and the State Mental Hospital grounds was once the upper reaches of Westchester Creek. A minor skirmish took place in October, 1776, between Americans on the western slope and British and Hessian scouts advancing from Throggs Neck. Although there was a prominent Stillwell family in lower Westchester in the 19th century, this avenue might well honor State Senator Stephen Stillwell (1906), who was prominent in Williamsbridge Democratic circles. He sponsored a bill which eventually made The Bronx a County in 1914.

STORY AVENUE. As there are other magistrates honored by neighboring streets, this avenue could be named for Supreme Court Justice Joseph Story, 1779-1845. A local candidate for the honor might be William H. Story, who was, at the turn of the century, principal of P.S. 100, a school that formerly stood at Westchester and St. Lawrence Avenues. At that time, the Watson property was being subdivided, but it is unlikely an avenue would be named for a living person. Turtle Creek once wound in from the Bronx River at Story Avenue. In the mouth of this creek (foot of Story Avenue), a small island, called Ludlow's Island, was situated; on some maps it was simply Sedge Island. Proceeding eastward, the street was on part of the Ludlow "Black Rock" farm. An Indian trail crossed Story Avenue at about Fteley Avenue, which eventually became the Colonial road that was replaced by Soundview Avenue. Between Soundview Avenue and White Plains Road, it was part of the Bradish Johnson estate (1851). It was crossed by Pugsley Creek at Beach Avenue, where an 1887 map lists a Van Boskerck farm. Beyond this, Story Avenue was once on the ancient sheep pasturage (1600s) that became the township of Unionport.

STRANG AVENUE. This Edenwald avenue was once called Randall Avenue, after a landowner Leslie F. Randall, of Yonkers. Daniel L'Estrange was one of the first Huguenot settlers that arrived in 1683. He was called Streing in the Rye Grant in April, 1705. During the Revolution, some of the family were Loyalists and migrated to Canada. This avenue is named for Feltus Strang, landowner of Williamsbridge, who was on the 24th Ward voting rolls in 1877.

STRATFORD AVENUE. Laid out across the former William Watson estate, this avenue may have no connection with Shakespeare's Stratford-on-Avon, although Mr. Watson was an Englishman. Stratford is a town in the Commonwealth of Virginia, and all three names are on avenues in that vicinity. See: Commonwealth Avenue.

STRONG STREET. It was once part of the northern end of Jan Arcier's (John Archer's) land, which was called the Manor of Fordham. Through foreclosure, the land became the property of the Steenwick family, then of the Dutch Reformed Church, whose elders sold the tracts by huge parcels. In Civil War times, this was part of the H. B. Claflin estate, and the subdividing of his property took place at the time The Bronx east of the Bronx River was annexed to New York City. William L. Strong was Mayor of the city at that same time, from 1895-1897, and the one most likely to have a short street given his name. He is one of the few Mayors buried in Woodlawn Cemetery.

SUBURBAN PLACE. This was once on the extreme eastern boundary of the extensive Bathgate holdings of the early 1800s. By mid-century it was on the estate of Thomas Minford. When mapped in 1887, a spring-fed pond lay at the junction of Suburban Place and Boston Road, its overflow running down to the Bronx River. The name was descriptive in its day, when the neighborhood had a suburban aspect.

SULLIVAN PLACE. This short Place is laid out across the Baxter lands of pre-Revolutionary times-and the Ellison Anderson farm of the early 1800s, which became the Coster farm after the Civil War. Names harking back to the Revolutionary War and Civil War abound on Throggs Neck, so two Sullivans are involved: Brigadier General John Sullivan, whose Continental army crossed upper New York State to finish the Indian threat, and General Jeremiah Sullivan of the Shenandoah and Vicksburg campaigns. As adjoining Scott Place is of Civil War vintage, Sullivan Place is most likely of the same.

SUMMIT AVENUE. One of the few streets in the Ebling area to retain its original name for more than a century. It appears on a map dated 1851, and refers to its location.

SUMMIT PLACE. This is a Fordham street that was officially opened in 1901, and well-named it is, being at the very summit of what was Tetard's Hill of Revolutionary times. In the 19th century, this was a favorite picnic spot for the residents of Fordham and Kingsbridge, who called it "The Tee-Taw."

SUNNYSIDE AVENUE. This road in the northeast section of Woodlawn Cemetery catches early morning sunlight.

SUNSET BOULEVARD. One of a dozen private roads in Shorehaven. When the Shorehaven Beach Club existed the children's pool and an outdoor theatre occupied this space. A Midway and a section of a rollercoaster track were crossed by this Boulevard when the Clason Point Amusement Park was in operation (1890s - 1945).

SUNSET CIRCLE. This turnabout in the Shorehaven community was once part of Higgs' farmland of the 19th century. William Higgs rented out campsites to Summertime residents and this enterprise was the forerunner of nearby Harding Park.

SUNSET TRAIL. This is a scenic walk in Silver Beach Gardens in private ownership, doubtlessly used by the Indians in their day as a look-out, as it commands a fine view of the East River. It was briefly occupied by British Rangers during the Revolutionary War. As the trail is on the western boundary of the community, the name "Sunset" is very appropriate.

SURF CIRCLE. This turnabout overlooking the East River on the end of Clason Point exists on landfill covering what had been extensive wetlands. White's Boatyard was situated there from approximately 1914 to the 1950s. The Shorehaven Beach Club's marina then took its place.

SURF DRIVE. Shorehaven's serpentine road follows, more or less, an earlier road that passed the picnic area and baseball field of the Shorehaven Beach Club. When the Clason Point Amusement Park was in operation, launches and excursion boats docked there, and a million gallon outdoor swimming pool, billed as "the nation's largest" was in the vicinity of Beacon Circle and Surf Drive. Near the southwestern end of Surf Drive once stood a Ferris Wheel that was blown over by a sudden Summer squall in 1922, wherein seven riders were killed.

SUTHERLAND STREET. A David Southerland [spelled this way] was an early settler, mentioned in the Westchester Deeds of 1702. A latter-day Edward G. Sutherland was Chairman of the Westchester County Democratic Committee. Upon City Island's annexation to New York City, he was appointed Sheriff until 1902.

SWINTON AVENUE. From the East River to Lawton Avenue, this was the ancient settlement of John Throckmorton (1641)-later the estate of Philip Livingston (1780-1811) and the nineteenth-century Hammond, Ash, Van Schaick, Havemeyer and Huntington families. From Lawton Avenue to the Throggs Neck Houses, it was on the lands of John A. Morris (1860s-1890s), a noted turfman. Proceeding northward, it was on the slope of estates belonging to Captain Wright Post, Henry C. Overing and the Stillwells and Crosbys. Its extreme northern end was part of the 80-acre farm of Elijah Ferris, who sold it to Jonathan Ferris in 1815, who called it "the Swanton farm," as he and his wife, Ursula, hailed from Swanton, VT. If Swinton Avenue does not derive from this farm, there are two more possibilities. On the 1686 Grove Farm patent to Thomas Hunt (present-day Ferry Point), two signatures are found, these being Governor Thomas Dongan and J. S. Swinton. Another 1686 Westchester Patent also is signed J. S. Swinton without further identification. Not acceptable, but amusing, is a bit of folklore that a small pig farm was located at the junction of Dewey and Swinton Avenues where the streets were being surveyed. Its local nickname was "Swine Town" and the surveyors, mindful of the nickname, submitted the name in slightly changed form. An 1888 Robinson map shows the avenue laid out from Bruckner Boulevard to Lafayette Avenue as Putname Avenue.

SYCAMORE AVENUE. At different times, Mark Twain and Theodore Roosevelt occupied the mansion known as "Wave Hill" on Sycamore Avenue and West 252nd Street, when it was on the Appleton estate. It was first called Seward Place, but on May 2nd, 1916, by Board of Aldermen Ordinance, it received its present name, due to the sycamore trees then in abundant evidence.

SYCAMORE DRIVE. This is a private street inside Parkchester.

SYLVAN AVENUE. Riverdale was once a sylvan area, which prompted the naming of this street.

TAPPAN WALK. This is a footpath on the campus of Bronx Community College that honors a Professor Henry P. Tappan who taught Philosophy and Belles-Lettres (1832-1838) in New York University, the predecessor of Bronx Community College that now occupies the property.

TAYLOR AVENUE. Although there was a Moses Taylor homesteading in present-day Van Nest in 1808, it is doubtful if the avenue honors him. More likely, it is named for the 12th President of the United States Zachary Taylor, as other streets in Van Nest honor Presidents. At first, it was called Harrison Street. When the Leonard Mapes farm (bounded by Westchester and East Tremont Avenues) was sold and subdivided into lots-"Park Versailles"-Taylor Avenue was extended across the tract where old-timers recall it as a simple boardwalk called Devine's Lane. On Westchester Avenue, a tavern keeper named Devine catered to thirsty travelers along the Turnpike. An Indian trail crossed Taylor Avenue at McGraw Avenue, and further south, again at Lafayette Avenue, on its way to the settlement on Clason Point. At about Story Avenue-now covered by the James Monroe Houses - an 1887 map shows this avenue was part of a farm owned by a Lucas and Katherine Van Boskerck.

TEASDALE PLACE. The Place itself was opened in 1888, and received the name of 1st Place in the town of Morrisania. The Teasdale family were among the early settlers of the town,-a George R. Teasdale being listed as a property owner in 1853, serving as a Commissioner in 1861. John R. Teasdale in 1868 was on the Town Board, and William Teasdale that same year owned lots at East 167th Street and Fulton Avenue.

TELLER AVENUE. It was originally mapped as Fleetwood Avenue, named after the Fleetwood Race Track that flourished there for many years. The name Teller goes back to the Dutch origins of Nieuw Amsterdam as it is in the church rolls of 1680. A Helena Teller was the wife of Francois Rombouts, a Mayor of New York in 1679. A William Teller was a Colonial surveyor (1740) who worked on Lewis Morris' lands. In the Revolutionary War, there is mention of a Captain James Teller of Westchester (now The Bronx) and in the Civil War there was a Colonel C. D. Teller. This avenue was named for Richard H. Teller, one of the Commissioners who signed the approved map of Morrisania in 1871. Other signers were Jordan L. Mott, Gouverneur Morris II and William Cauldwell.

TENBROECK AVENUE. Surrounding avenues honor State Governors and some (Yates, Haring and Woodhull) had been Presidents of the Provincial Congress. Abraham TenBroeck was President of the Provincial Congress in 1777. North of Allerton Avenue, Tenbroeck Avenue was part of the Hamersley estate, as spelled on an 1850 map. South of Allerton Avenue, it was on the lands of John Bartow (1700 map) and Denton Pearsall (1800s). Thomas Haddon's pre-Revolutionary saw mill lane crossed Tenbroeck Avenue at its southwestern corner of Allerton Avenue, and was still discernible in the 1960s.

TENNY PLACE. This street was opened in 1925, but no previous owner of that name can be found. A Daniel Denny was listed as a registered voter in 1877 at Lind Avenue (now University Avenue) and, in 1892, Levi S. Tenney acted as Administrator in some property transactions of the West Bronx. Levi is the better of the two candidates to be remembered by a Place.

TENTH STREET. This is a footpath in the Park of Edgewater. See: Park of Edgewater.

TERRACE PLACE. Some 20 yards of this Place is in The Bronx, above Pelham Bay Park, and was formerly terraced. See: Edgemere Street.

TERRACE POINT. The official name of the northern tip of City Island, once owned by Henry D. Carey. It is unofficially known as "Carey's Point." See: Terrace Street.

TERRACE STREET. This is the northernmost street on City Island, and may have been terraced when Judge Carey's estate was situated there.

THEODORE KAZIMIROFF BOULEVARD. See: Kazimiroff (Dr. Theodore) Boulevard.

THIERIOT AVENUE. The old road that led past the McGraw farm to Clason Point crossed this avenue at about McGraw Avenue. The avenue runs across the former estate, "Rosedale," that was owned by Hudson P. Rose. At about Watson Avenue, this street was, in the 1700s, part of Cow Neck farm (Stephen Leggett, prop.), who sold it to the Pugsley family (1794-1854), whence it became the Cobb farm. At the junction of Thieriot, Leland and Lacombe Avenues was the large Indian village known as Snakapins. Many artifacts, even mummies, were found by field teams from the Museum of the American Indian Heye Foundation. When "Park Versailles" was begun as a real estate development, John S. Mapes sold to Albert Thieriot, of Brooklyn, several lots on April 8, 1893. He may have been related to one of the three financiers behind the "Park Versailles" venture: James Boorman Colgate, Cyrus Jay Lawrence and Ferdinand M. Thieriot.

THIRD AVENUE (1). This important artery has had a number of names and nicknames along its route. Beginning in Old Morrisania as Morris Avenue (1848), becoming the Boston Road in the village of Melrose (1850s), it was also Carr Street at East 160th Street, and Morse Avenue for a three-block stretch, not to forget its popular nickname of "Huckleberry Road." It was the main thoroughfare of Morrisania (East 163rd Street to East 170th Street) and Tremont (East 171st Street to East 180th Street) where it carried the name of Fordham Avenue. It was an important brewing center, shopping area, theatrical district, and a concentration of shops, tobacco factories, social halls and other business establishments from the 1850s to World War I. The 1871-1872 Director of Morrisania and Tremont calls it 3rd Avenue, although the numbered avenue was not officially used until 1891. It carried the name from Manhattan's Third Avenue because of the extension of that borough's Third Avenue elevated line. From East 183rd Street up to Fordham Road, it overlays an ancient Indian path and was called Kingsbridge Road until 1891.

THIRD AVENUE (2). This is a narrow road in the Park of Edgewater, third in order from the waterfront. See: Park of Edgewater.

THIRD AVENUE BRIDGE. This bridge was opened in 1901 at the foot of Third Avenue to carry traffic from Manhattan over the Harlem River. At first, it was called the Second Harlem Bridge, as it took the place of an earlier structure that had been in operation since 1860.

THIRD STREET. This is the third of five numbered streets on Hart Island.

THOMAS BROWN MALL. A stretch of White Plains Road, from Brady Avenue to Pelham Parkway South, was named in June 1989 to honor the late District Manager of Community Board #11.

THOMAS PELL WILDLIFE REFUGE AND SANCTUARY. The Committee on Parks approved a local law on September 14, 1967, setting aside parts of Pelham Bay Park as a wildlife refuge. It encompasses an area adjacent to the Split Rock golf course, the Hutchinson River and the New England Expressway. Thomas Pell was made first "Lord of the Manor" in 1666, and his manor lands included Pelham Bay Park. The area was saved due to the efforts of Dr. Kazimiroff, Bronx historian, naturalist John Kieran, and representatives of The Bronx County Historical Society, the Audubon Society, Linnaean Society and the New York Zoological Society, not to omit U.S. Interior Secretary, Stewart Udall.

THOMPSON PLAZA, PLAYGROUND, PAVILION. This street in Roberto Clemente State Park had honored Julius J. Richman. Dennis A. Thompson, a 20-year-old local resident died when he and 240 U.S. Marines were killed by a terrorist's bomb explosion in Lebanon. Friends and neighbors requested the young man be remembered. On Veterans' Day 1984, residents of River Park Towers on the Harlem River, honored him.

THORNE LANE. In the Fieldston section of West 259th Street this three-block-long, unpaved lane runs uphill from Huxley to Tyndall Avenue, where foundations of a house could still be seen in 1989. Wealthy merchants named Thomas and William Thorn [spelled this way] had grain and feed warehouse on Spuyten Duyvil Creek in the 1880s, and the lane may be a reminder of those times. See: Thorn's Basin in the Old Bronx Names.

THROGMORTON AVENUE. It was first called Valentine Avenue after a Judge Valentine of Westchester Village, but the name was changed in 1916. It now commemorates the pioneer of the region known as Throggs Neck. John Throckmorton, eighth in direct descent of Lord John Throckmorton of Worcester, England, was given permission by the Dutch to settle on the extreme end of the Dutch province, called Vriedelandt (Land of Peace). In the Dutch deed, dated 1642, he is called "Jan Trockmorton." To escape an Indian massacre, he left and later settled in New Jersey. The name is derived from the Saxon Thurrock-mere-tun referring to a town (tun) by a drained lake. Present day members of the family believe it means the Rock-Moor Town.

THROGS NECK BOULEVARD. At its lower end, it was mainly low lying swamps of Weir Creek's southern arm, and the site of an Indian burial ground. Along its upper end, the Boulevard was also swamplands of the Wolfe, Brown and Coster properties of the 19th century. See: Throgmorten Avenue.

THROGS NECK BRIDGE. This bridge leaves The Bronx at Locust Point on Throggs Neck, crosses over Fort Schuyler and the East River, to terminate in Queens. The span was opened in January, 1961, and the first two people to cycle over were 11-year-old John McNamara and his father, the author of this book. As the bridge was closed to cyclists, the ride was made, unofficially, before dawn of the opening day. The spelling of "Throgs" and "Throggs" is derived from the ancient name of Throckmorton, the first settler on the Bronx peninsula.

THROGS NECK EXPRESSWAY. This wide boulevard overlays the once-important waterway known as Weir Creek that flowed into Eastchester Bay, at Weir Creek Park. See: Throgmorton Avenue.

THROGS NECK PARK. It is located at the end of Myers Street, but has no park equipment. It was formerly known as Schuylerville Park and carried the local nickname of "Bowling Green" or "Bolen Green" as the Bolen family owned adjoining property . It was acquired as a public place in 1836.

THROOP AVENUE. The avenue, as other avenues in the neighborhood do, honors a Governor of New York State. In this case it is Enos Throop who served from 1829 to 1833. At its lower end, the avenue runs through the former lands of Reverend John Bartow (1727), Watson (1840s) and Denton Pearsall (1860s). Above Allerton Avenue, it runs through the Kidd and Hamersley [spelled this way] estates. Thomas Haddon's saw mill lane crossed at the northwest corner of Allerton Avenue and, although in use before 1725, was still discernible in 1960.

THURMOND MUNSON WAY. On December 13,1979, Mayor Koch signed a bill authorizing a stretch of East 156th Street in Concourse Village be given this name. Thurmond Munson was a Yankee all-star catcher, who had died in an Ohio plane crash three months earlier while at the controls.

THWAITES PLACE. The Thwaites family lived on this small street for three generations, and their property is listed in Beers' 1868 *Atlas*, and Bromley's 1891 *Atlas*. Mention of this family is found in very early records, dating back to 1703 when the name of Dan Twaites [spelled this way] is found on a list of schoolmasters.

TIBBETT AVENUE. Once the extreme western rim of the island of Paparinemin which was bounded by Tibbett's Brook-a name which is a corruption of George Tippett who acquired land there in 1668-the avenue follows, more or less, the former course of the brook, and, at one time or another (1848, 1868 maps), carried the name of River Street. Its nickname was "Long Branch" after the fashionable New Jersey seaside resort-for the brook frequently overflowed onto the front lawns of the residents, so that they had swimming right at their doorsteps in much the same fashion as the vacationers at Long

Branch. Tibbett Avenue resumes existence above West 244th Street opposite Van Cortlandt Park, and once this stretch was known as Barney Avenue. Hiram Barney was a landowner of an estate called "Cedar Knolls" and he was a good friend of the poet, Edgar Allan Poe. Barney died in 1895 and is buried in Woodlawn Cemetery.

TIEBOUT AVENUE. A very ancient Dutch family of Huguenot extraction, the Tiebout name is found in the 1689 and 1703 Lists of Inhabitants, with sometimes the spelling "Thibaud." James Tiebout purchased land from Peter Valentine in 1851, embracing this area. The avenue was opened in 1897 from East 180th Street to East 183rd Street. From there up to Fordham Road, it was opened in 1911. This last portion was laid out through the former farm, "The Elms," owned by the Keary family.

TIEMANN AVENUE. This avenue is laid out across what was land belonging to the Givan family. Givan's Drive crossed Tiemann Avenue at about its junction with Arnow Avenue. Along with many other avenues in this neighborhood, the street honors a former Mayor of New York. Daniel Fawcett Tiemann, a wealthy paint manufacturer, who served as Mayor from 1858 to 1860 and is credited with having street signs attached to lamp posts for better visibility.

TIER STREET. Daniel Tier was one of the original petitioners of Benjamin Palmer to settle on New City Island in 1784. Early 1800 records list a Jeremiah Tier, landowner, and, in 1890, a Jennie L. Tier bought land on City Island from the Schofield family. Tier Street was originally called Cemetery Lane as it led to the burial grounds on City Island.

TIERNEY PLACE. This waterfront street honors Supreme Court Judge John M. Tierney, contemporary of Justices Hatting, Glennon and Giegerich. It overlays the southern shoreline of what was once Wright's Island, and records mention the stand of locust trees there. See: Giegerich Place.

TIFFANY PLAZA. In June 1981, Terence Cardinal Cooke celebrated a Mass of Thanksgiving as part of the dedication of this Plaza located between Fox Street and Southern Boulevard. It faces St. Athanasius R.C. Church. See: Tiffany Street.

TIFFANY STREET. It runs through the nineteenth-century estates known as Longwood Park, the Rose Bank estate of the Leggett family, and the former properties of the Whitlock and Casanova families. The street is a reminder of the landowners named Tiffany who owned "Foxhurst" on Hunts Point. A daughter of W illiam Fox married H. D. Tiffany. The street was laid out in 1906 across what was once the Debatable Lands between the West Farms Grant and the Morris manor lands. Tiffany Street overlays what was Wigwam Brook (so called by the Hunt family) or Bound Brook (insisted upon by the Morris clan) and which eventually became Leggett's Creek. At one time, a short stretch of this street was Blythe Avenue, named after an estate. The intersection of Tiffany Street and East Bay Avenue marks the location of a buried island, Long Rock or Duck Island, that once lay in the mouth of Leggett's Creek.

TILDEN STREET. This street does not honor Democratic Governor Samuel J. Tilden (1874), nor Oliver Tilden who was the first Civil War casualty buried in Woodlawn Cemetery. This short street is all that remains as a reminder of the Blodgett-Tilden estate. William Blodgett and William Tilden purchased this property from John and George Palmer, who were grandchildren of Robert Givan. See: Givan Square.

TILDEN TRIANGLE. See: Oliver Tilden Triangle.

TILLOTSON AVENUE. It was once known as Cornell Avenue from Eastchester Road to Black Dog Brook (now covered by Bruner Avenue), but around 1897, was renamed to honor Gouverneur Tillotson, Commissioner of Deeds and a local landowner who had been active in Eastchester affairs in Civil War times.

TIM HENDRICK PLACE. In July, 1977, ceremonies were held at the easternmost section of West 236th Street which became Tim Hendrick Place, honoring a well-known community leader of Kingsbridge.

TIMPSON PLACE. This short street in East Morrisania is, like adjoining Austin Place, named after a family who purchased property in the 1870s. Earliest mention of the Timpson family is found in a 1741 list of inhabitants of New York. The firm of Timpson, Adee & Co. was well-known in the 1870s. This street most likely honors Thomas W. Timpson, one of the original trustees of Morrisania. Other Timpsons, Edward (1802-1870) and William (1818-1893), owned land along Eastchester Road and Pelham Parkway.

TINA ZAFFUTO SQUARE. On October 13, 1989, Throggs Neck residents and officials paid tribute to a crossing guard who had died in 1986. Her post was at East Tremont and Barkley Avenues, where she protected her charges and made a lasting impression on everyone she met.

TINTON AVENUE. The name is a corruption of Tintern, the original country seat of the Morris family in Wales. In the late 1600s, Colonel Lewis Morris owned most of the lower Bronx, across which this avenue runs. In 1673, he bought property in New Jersey which he called "Tintern Manor" and today is known as Tinton Falls, N.J. The avenue itself runs across the ancient manor lands of Morrisania which, below Westchester Avenue, became the Dater estates. Above Westchester Avenue, the land traversed by Tinton Avenue was once called the Shingle Plain (1700s) and, in the late 1800s, Woodstock or Deckerville. A famous political-social club, was once situated on the corner of Tinton Avenue and East 161st Street. See: Tallapoosa Club.

TOMLINSON AVENUE. It was once Munroe Avenue from the railroad to Morris Park Avenue and overlays the former Morris Park racetrack that became a flying field in the early 1900s. Somewhere near the junction of Pierce and Tomlinson Avenues was the ancient burial grounds of the Bartow family, but the exact location has been lost to history. John C. Tomlinson owned numerous parcels of land on Haight and Lydig Avenues, and on Williamsbridge Road and was president of the Williamsbridge Taxpayers' Association.

TONY ROTA SQUARE. In 1933 Rota's Restaurant was established where Bronxdale and East Tremont Avenues meet. A triangular concrete traffic island at that junction was named in April 1986 for the restaurateur. He closed it in 1990, when he retired.

TOPPING AVENUE. It was opened in 1897 with the earlier name of Lafayette Avenue after it was cut through the "Claremont" estate of the Zborowski family. Ira Topping bought land in October, 1853, in what was then called South Fordham. His nephew,

William A. Topping, was a Tammany politician and auctioneer (1888) who helped sell the land.

TORRY AVENUE. On or about this place on Castle Hill Point was the fortified Indian village which was sighted by the first European explorers sailing up the East River. In 1685, this was Cromwell's Neck; then, three generations of Wilkins' property from 1770 to late 1880s, when it was inherited by a son-in-law, John Screvin. The birthplace of the Screvin family in Scotland was a small town named Torry-now a suburb of Aberdeen.

TOWNSEND AVENUE. One story is that this street marked the town's end, in this case the town of West Morrisania. More likely is it that the street is a reminder of the Townsend family who settled there in Revolutionary times. In 1853, a Townsend Poole was a landowner along Featherbed Lane.

TRAFALGAR PLACE. This Place is laid out on what was once the Fairmount estate of Thomas Minford. It was opened in 1909 and commemorates the British naval victory by Admiral Horatio Nelson over the French in 1805. An English-born Colonel Dunham, who once owned this land, was thought to have asked the Board of Aldermen to so name the Place, but this is a doubtful story, as Dunham was the landowner some 50 years prior to the time of street naming. More likely, it would be the Bathgates or the Minfords, both of English ancestry, too, who lived on the land around the 1900s.

TRASK AVENUE. Benjamin Trask was a wealthy ship owner who had an 80-acre estate off Pugsley Creek, according to an 1886 map. One of his packet ships was the *Virginia* and Virginia Avenue forms an angle of his former estate. Squire Trask is buried in Woodlawn Cemetery.

TRATMAN AVENUE. This was part of the Adee estate of the 1890s. When sold in 1910 and cut through in 1912, it was 1st Street in the village of Westchester. A George M. Troutman's name appears in deeds dated 1843 and 1849, but b 1905 the property was charted as "Tratmann" (two 'n's). A bit of local lore credits Mr. Tratman with the naming of Manning Street. See: Manning Street.

TREMONT AVENUE. See: East Tremont Avenue, West Tremont Avenue.

TRIBOROUGH BRIDGE. As the name implies, three boroughs (Manhattan, Queens and The Bronx) are served by this bridge. One span enters The Bronx over the Bronx Kill at Port Morris.

TRINITY AVENUE. This was originally laid out through St. Mary's Park where it was called Passage Avenue on 1857 maps of Wilton, Port Morris and East Morrisania. It was known as Cypress Avenue where it crossed present-day East 149th Street and also Avenue C from East 158th Street to East 161st Street. Progressing northward, it was once the farmland of H. P. DeGraaf. It was called Delmonico Street in 1875 from East 161st Street to East 166th Street in the village of Forest Grove, and was also known as Grove Avenue around the time of the Civil War. The avenue supposedly was named for a small Lutheran (Evangelical) Trinity church off Pontiac Place that was razed many years back.

TRUXTON STREET. It is located on what was the "Woodside" estate of E. G. Faile, and honors the American naval officer, Thomas T. Truxton (1755-1822) who fought in the

Tripolitan War of 1802-1804 against the Algerian pirates. At one time it was called Garrison Square, reputedly after an encampment of a Civil War garrison, but more likely after the Garrison family who had lived in the vicinity.

TRYON AVENUE. This is an abbreviated street running from Gun Hill Road to Woodlawn Cemetery and formerly part of the Revolutionary farm of the Valentine family. General William Tryon's career does coincide with those of DeKalb, Rochambeau and Steuben, whose names are on adjoining avenues, but he was a Loyalist Governor of New York during the Revolution and quartered his troops in New York City during the British occupation. His headquarters were in King's College (which became Columbia University) and the street next to Tryon Avenue is King's College Place.

TUDOR PLACE. Opened in 1897, this short Place was part of the vast William Morris holdings that once reached from Cromwell's Creek (Jerome Avenue) to the Mill Brook (Brook Avenue). An oft-repeated tale to explain its name concerns a Tudor cottage at that point that had to be demolished when the Grand Concourse was laid out.

TULFAN TERRACE. Some 50 yards long, and considerably uphill, Tulfan Terrace supports eight homes, and one of them has a sign: "Tulfan's Peak. 246 ft. above sea level." It is on the historic slope overlooking the former island of Paparinemin (Kingsbridge), but is not on Bronx maps as it is a private road. Located off Oxford Avenue and West 236th Street, it owes its name to two contractors of the 1920s, Tully & Fanning.

TURNBULL AVENUE. This avenue is laid out across the ancient sheep pasture of the town of Westchester. Later, it was the property of Gouverneur Morris Wilkins, and then the Screvin family. Catherine V. R. Turnbull was a landowner of 51 acres of uplands and marsh, from this avenue to approximately Lacombe Avenue, and from Howe Avenue to Westchester Creek. She is buried in St. Peter's P.E. churchyard.

TURNEUR AVENUE. The original proprietors of this Castle Hill peninsula were the Cromwells. Perregreene and Neomi Turner were related to John and Elizabeth Cromwell, according to the records of 1687. Elsewhere, the name appears as Turnure and Turneur. The "Indian castle" is believed to have stood on the slope of Turneur Avenue, and the latter-day Wilkins mansion was also located in the same area. See: Turnbull Avenue.

TURTLE COVE. A shallow inlet on the southwest side of Rodman's Neck that figured in one Revolutionary War adventure. A British picket ship was moored in Eastchester Bay to control shipping coming down the Hutchinson River. A band of daring Yankees sailed a whaleboat down from Connecticut, traversed the inlet behind Hunter Island and then portaged their boat overland to Turtle Cove. From there they rowed out to a market sloop, hid aboard and then boarded the British ship, "Schuldam," by a ruse, and captured it. Turtles have always been plentiful in Eastchester Bay in former times, and the shallow cove was a natural attraction for them.

TWIN LAKES. These are small ponds formed by the relocation of the Bronx River at Allerton Avenue roadway, inside the Botanical Garden. An earlier (1905) name was Blue Bridge Lake, well-known to Gun Hill ice-skaters of that era.

TWINS, The. See: East Twin Island and West Twin Island.

TYNDALL AVENUE. This Riverdale street was originally named Hill Street when it was opened in 1914, but was later renamed Tyndall Avenue in honor of the British physicist, John Tyndall (1820-1893), who once had been a house guest at nearby "Wave Hill" (West 248th Street). See: Liebig Avenue.

UNDERCLIFF AVENUE. This was part of the Ogden estate overlooking the Harlem River until 1884, when Marianna A. Ogden sold the property. George C. Hollerith surveyed the tract in 1896. The name is obviously a topographical name as it runs below the steep slopes of the West Bronx.

UNDERHILL AVENUE. This was formerly mapped as Cottage Grove Street. An early settler was Captain John Underhill (1597-1672) who served under the Dutch and led the Indian massacre at Mianus River in 1637. He later retired to Oyster Bay. Nathaniel, a son-born in 1633-married May Ferris, and owned a tract of land on Eastchester Bay. See: Kennelworth Place. This avenue evidently honors a grandson, Nathaniel (1690-1771), whose holdings included most of Parkchester. His son, Israel, was a vestryman of St. Peter's P.E. church from 1788 to 1795.

UNION AVENUE. This was the eastern edge of the Shingle Plain mentioned in old deeds of the Manor lands of Morrisania. See: Tinton Avenue. From East 147th Street to East 157th Street, it became the Dater estates which were subdivided into the village of East Morrisania. From Westchester Avenue to East 165th Street, Union Avenue was included in the village of Woodstock, and from there to Boston Road was considered part of Eltona. This avenue was named directly after the Civil War with an idea of honoring the Union.

UNIONPORT BRIDGE. This span over Westchester Creek has been replaced a few times since the days when Unionport was a village with well-defined limits. The present lower-level span was built in 1952, but then was doubled in width about 15 years later.

UNIONPORT ROAD. This was originally an Indian path that led out to Castle Hill Point. In the 1700s, it was known as Sheep Pasture Road as it led to the Commons (now Unionport). It was Parsonage Road of the 1800s, referring to St. Peter's parsonage at Westchester Avenue and Castle Hill Avenue, then known as Centerville. In 1851, the village of Unionport was established, and the road leading to it received its name shortly thereafter. From the 1880s to 1940, Unionport Road ran through the grounds of the Catholic Protectory, which was transformed into Parkchester thereafter.

UNITED STATES SHIP CANAL. This is the official name of the "cut" extending from Spuyten Duyvil to the Harlem River at West 225th Street. The Army Engineers so designate it on their official maps, although most New Yorkers refer to it as the Harlem Ship Canal or just the Harlem River. The first section of the artificial channel was completed on June 17, 1895, cutting Marble Hill off from the rest of Manhattan island. See: Marble Hill.

UNIVERSITY AVENUE. It was named in honor of New York University, whose campus is now used by the Bronx Community College. Until 1917, the stretch below Sedgwick Avenue was known as Lind Avenue, possibly in honor of Jenny Lind the Swedish

Nightingale whose visit to America in 1850 coincided with the naming of this avenue. The section north of Washington Bridge to Fordham Road carried the name of Aqueduct Avenue from 1886 to 1913.

UNIVERSITY HEIGHTS BRIDGE. This bridge over the Harlem River leads from University Heights in The Bronx, which accounts for its name. It was opened in 1908.

UNIVERSITY PARK. Measuring 2.75 acres, it was opened in 1901 and took its name from New York University, which has since been displaced by the Bronx Community College. It was the scene of some Revolutionary activities as its slope led up to the British-held Fort No. 8.

UNKNOWN SOLDIER PLAZA. Situated at University and West Tremont Avenues, this space was named in 1954 to honor all American servicemen who died, unidentified, in all our wars.

VAIL WALK. This walk, on the campus of Bronx Community College, honors Alfred Vail, a student when the campus and buildings belonged to New York University. He assisted Professor Morse in the early experiments with the telegraph, and it is believed by many that the dot-and-dash code was actually devised by Vail, although it is commonly called the Morse code. The land crossed by Vail Walk was once the ancestral domain of the Archer family and, during the Revolutionary War, was dominated by Fort No. 8.

VALENTINE AVENUE. It was mapped as Carlin Avenue north of Fordham Road, and locally, in the 1890s, was known as Lovers' Lane. Dennis Valentine Sr. and Jr. both gave land at different times to the Dutch Reformed Church. This avenue runs through the former Valentine estate on which was once located one of the prettiest springs in the city. It was 50 feet east of Valentine Avenue and 300 feet north of East 194th Street.

VALHALLA PLACE. The land was owned by the Underhill family in 1685, by James and Charity Ferris during the Revolutionary War, and in 1868 by J. Van Antwerp before becoming part of the Westchester Polo Club. See: Lohengrin Place.

VALLES AVENUE. When this Riverdale street was first surveyed, it was given the name Park View Avenue as it overlooked Van Cortlandt Park. On May 2nd, 1916, it was named for the librarian of the Corporation Counsel, first name not given.

VAN BUREN STREET. This street was opened in 1914 and followed the Van Nest pattern of being named after a President of the United States. It constituted part of the widespread Hunt farmlands that had flourished there for centuries. Martin Van Buren, who served from 1837 to 1841, has been credited with originating the expression "O.K." as he belonged to the "Old Kinderhook Party" in politics. More factual was his sojourn while President at the mansion of John Hunter, on Hunter Island.

VAN CORTLANDT AVENUE WEST. A short hilly street on land once owned by the powerful and influential Van Cortlandt family, and in the 1800s, the Dickinson estate.

VAN CORTLANDT AVENUE EAST. This is one of the oldest roads in The Bronx, having been an Indian trail that subsequently became part of the original road to Boston in pre-Revolutionary times.

VAN CORTLANDT PARK. Second in size to Pelham Bay Park, but equally important in Bronx history, this park is over 1,100 acres in size. It was the lower end of the immense Philipse tract of the 1600s which was bought by son-in-law Jacobus Van Cortlandt in 1699-and over the decades and through the following generations, the Van Cortlandt holdings grew to a vast estate. Revolutionary War history was made there, and to this day the park is an important section of the upper Bronx.

VAN CORTLANDT PARK EAST. In Revolutionary days it was the famed Miles Square Road. Later on, in the 1800s, it acquired the name of Mount Vernon Avenue, but in 1894, a City ordinance changed it to its present descriptive one.

VAN CORTLANDT PARK SOUTH. This street is on land once owned by the Van Cortlandts, and the name is a descriptive one.

VAN CORTLANDT VILLAGE SQUARE. In 1976, the City Council designated by law the intersection of Van Cortlandt Avenue West, Sedgwick Avenue, Gouverneur Avenue and Stevenson Place as Van Cortlandt Village Square. The designation was sought by the Van Cortlandt Village Council to identify the area in which the members lived and worked-lands once owned by the Van Cortlandt family, south of the park of the same name.

VAN HOESEN AVENUE. The name does not appear on any list of local property owners, so it is a guess that the short street off Pelham Parkway preserves the memory of Justice George M. Van Hoesen, Provost Marshall in the Civil War, and a prominent Democrat in Bronx politics. He originated the bill for Elevated Railways in New York City.

VAN NEST AVENUE. This avenue is laid out across the Hunt farmlands of the 18th century and part of the Morris Park racetrack of the 19th century, but was originally called Columbus Avenue. The land was developed in 1892-1896 by the Van Nest Land & Improvement Company, Ephraim B. Levy, Director. Dave Hennen Morris and Alfred Hennen Morris were listed as secretary and treasurer, respectively. They were sons of John A. Morris, who owned the nearby racetrack. Van Nest was subdivided into 1700 lots from property that had been Neill, Delancey and Sackett farmlands. The development took its name from the Van Nest railroad station that served the racetrack-an odd case of a depot being built before there was a town. Pieter Pietersen Van Neste, common ancestor of the family, came from the Netherlands in 1647, and the name was rendered also as Van Ness and Van Neste. The name appears in church records of 1667 and 1687. Descendants are found in New Jersey as well as in New York. Reynier Van Nest was a wealthy saddlery owner of the early 1800s and his son, Abraham R., became a director of some railroad companies, including the Suburban line of the New York, New Haven & Hartford Railroad. It was at the son's request that one of the depots was named in honor of his father (1771-1859). Abraham (1810-1888) was a wealthy blacksmith with a town house at West 35th Street and a Summer home in Tuxedo, N.Y. His son, Alexander T. Van Nest (1844-1896) died in Prussia, but was buried in Woodlawn Cemetery along with his ancestors. The Van Nests never lived in The Bronx, asserts Nicholas DiBrino who has become the local historian, but most of them are buried in the Lake plot of Woodlawn.

VAN NEST MEMORIAL PARK. It is located on the ancient road from Bronxdale to the Westchester Turnpike, on the farmlands of the Hunt family. It was named in 1926 at the junction of White Plains Road and Unionport Road to honor the men of Van Nest who gave their lives in World War I.

VANCE STREET. These were the Bassett lands of the 1690s, and part of John Bartow's purchase which he mentioned, in 1727, as "Scabby Indian." Swamps here were a connecting link in the Indian waterways of upper Westchester Creek and Givan's Creek. In this area honoring former Mayors, this street is a reminder of Samuel B. H. Vance, who served temporarily as Mayor of New York upon the death of Mayor Havemeyer.

VARIAN HOUSE PARK. This park adjoining Williamsbridge Oval and Bainbridge Avenue between Van Cortlandt Avenue East and East 208th Street was created in 1965 as the site for the historic Valentine-Varian House. It is named for Isaac Varian, Jr., who grew up in the house, and became Mayor of New York in 1839.

VAULT HILL. It derives its name from the ancient burial place of the Van Cortlandts. It was in this vault that Augustus Van Cortlandt, then City Clerk, hid the records of the City of New York when the city was evacuated in 1776. It is 149 feet above sea level.

VERVEELEN PLACE. This Place off Broadway, opened in 1922, immortalizes the ferryman, Johannes, and his son, Daniel, whose home and inn were approximately on Godwin Terrace from 1669 to 1693. This became a tavern during the Revolutionary War. It next served as the foundations for the Macomb house, and finally the Godwin mansion, which was demolished in 1920. The Verveelens operated a primitive, but serviceable, ferry system over Spuyten Duyvil Creek.

VETERANS' GATE. This is the entrance to Fordham University at Third Avenue and East Fordham Road to memorialize the servicemen of World War I.

VETERANS' MEMORIAL DRIVE. This stretch of the Throgs Neck Expressway service road from Longstreet to Schley Avenues passes the Bicentennial Veterans' Park (dedicated 1976). The Drive partially covers Weir Creek. In 1986, veterans Pat Devine, Tom Hansen and others succeeded in having this name officially adopted.

VETERANS' MEMORIAL PARK. On May 20, 1990 this small park in Co-Op City was dedicated with impressive ceremonies by civic leaders and many veterans organizations. It is not to be confused with Bicentennial Veterans Memorial Park, on Throggs Neck.

VETERANS MEMORIAL PLAZA. Situated at the important crossing of the Grand Concourse and Fordham Road, it can be assumed local Posts were instrumental in naming it in 1958 or 1959.

VICTOR STREET. Downing's Brook once crossed Victor Street at its junction with Morris Park Avenue. The small cemetery of the Hunt family was formerly located at Unionport Road, Van Nest Avenue and Victor Street, as this was part of the Hunt farmlands. Victor Street was opened in 1915 and is thought to honor Orville J. Victor, innovator of the "dime novel." His wife, Matta Fuller Victor, was an authoress, and their son, Orville, was a newspaperman. This son served in the Spanish-American War, 8th N.Y. Infantry, and later was caretaker of the Poe Cottage. He died in 1917 and lies buried in the West Farms Soldiers' Cemetery. A perhaps facetious story is that the first Italian families that moved into the newly-opened street petitioned the Board of Aldermen to name it after King Victor Emmanuel of Italy.

VIELE AVENUE. General Egbert Lodovicus Viele, Class of 1845 of West Point, was a descendant of an old New York Dutch family. In a list of "Chirugeons practicing in the City-1695" was Dr. Cornelius Viele. As an Army Engineer, he laid out Central Park, following the plan of Olmsted and Vaux. He also planned to make Hunts Point a vast port and commercial center, under a project called the East Bay Corporation. He designed workmen's homes, recreation centers, work areas and railroad spurs-but East Bay never materialized. General Viele and his wife are buried in a West Point mausoleum, flanked by stone sphinxes and bearing an inscription in ancient Etruscan on the lintel. Dr. Ray Kelly translated it as follows: "The body of Egbert Lodovicus Viele is in this tomb with Juliette his wife and companion."

VILLA AVENUE. A map dating back to the Civil War shows this land as the property of Leonard W. Jerome, who organized the American Jockey Club. A home is clearly indicated south of the junction of Villa Avenue and Bedford Park Boulevard and may be the villa mentioned. The Villa Site Improvement Company purchased the land in 1893, and the avenue was named in 1900. It must be pointed out that the wealthy Jerome family owned more than one villa in what is today The Bronx.

VIMOUNT ROAD. See: Vinmont Road.

VINCENT AVENUE. This avenue runs through what was once the Bruce Brown estate, the Dr. Turnbull property and the Century golf course of the early 1900s, alongside Eastchester Bay. As so many avenues in this area honor Civil War generals, this avenue might honor Brigadier General Thomas McCurdy Vincent (West Point class of 1853) who fought at the battle of Bull Run. Residents of this avenue might be intrigued to know Vincent Avenue overlays an Indian trail from Philip Avenue, leading to a fishing camp at the foot of Schley Avenue.

VINEYARD PLACE. At one time this was part of the Bathgate farm, and later the property of Thomas Minford. The name is evidently derived from the fact a vineyard once flourished on the spot.

VINMONT ROAD. This is a semi-circular lane, listed sometimes as Vimount Road in the Postal Zone Guide, that services a private development of brick homes. It is a private road, maintained by the owners of the property, a family named Weinberg. A son, Robert C. Weinberg, was the architect. Although the road points towards Mount St. Vincent, this is not the reason for its name: Weinberg (vineyard) if changed to French, becomes Vinmont!

VINMONT VETERANS PARK. Originally opened as a playing field in 1951 for the pupils of P.S. 81, this name was proposed in October 1986 by residents of the neighborhood. It was officially put under the maintenance of the Parks Department in 1987.

VIREO AVENUE. During the Revolutionary War in 1778, the Queen's Rangers led by Simcoe, crossed this present-day avenue westward to cut off Chief Nimham and his Indians who were gathered at Devoe's Lane. See: Oneida Avenue. Some of the

Woodlawn Heights avenues had names such as Sparrow Avenue and Quail Avenue, and a vireo is a small woodland bird. In 1897, the other avenues had their names changed.

VIRGIL PLACE. Publius Vergilius Maro was born near Mantua in 70 B.C. and is known for his poems eulogizing husbandry and nature. His name was given to this Castle Hill street by Solon Frank, a realtor who was -active in this area around 1900. Perhaps influenced by his own first name, Frank was partial to Greek and Roman names. Virgil Place is just inside the property line of the former holdings of Gouverneur Morris Wilkins, and that of a son-in-law, John Screvin.

VIRGINIA AVENUE. It was known as Gray Avenue until December 1924. This avenue was laid out through the Cobb farm of the 1890s-and skirted the edge of the Cow Neck farm of 1805. The avenue might possibly be named after a packet ship, Virginia, owned by Benjamin Trask, a landowner in the area. Or else, a surveyor wished to commemorate his birthplace, Stratford in the Commonwealth of Virginia-for the first two names are in adjacent avenues. Local lore tells of Emily Virginia Sullivan, a lifelong resident on Virginia Avenue. She was a family friend of J. J. Gleason, who owned the property around 1890, and her parents requested that one street honor the little girl. Sullivan's Hill was the rocky outcropping at Virginia Avenue and Hugh Grant Circle that has been demolished to street level.

VOIGHT (Monsignor John) GREEN. In May 1983, the islands in the center of Metropolitan Avenue, Parkchester, from McGraw Avenue to the Oval were given this name in memory of a former pastor of St. Helena's R.C. Church, noted for his tireless work in the church and school, Little League and senior citizen centers.

VREELAND AVENUE. By a strange coincidence, this avenue is the one bordering Westchester Creek-the historic boundary of Vriedelandt (Land of Peace) during Dutch days. It was also written "Freedland" by A. van der Donck in 1656. In the 1870s, Alfred Seton sold his estate to the Seton Homestead Land Company, and one of the directors was Enoch Vreeland.

VYSE AVENUE. This avenue is laid out across the former village of Fairmount. The Vyse family owned an estate called "Rocklands" and, when the property was divided into city streets, this avenue was first called Chestnut Street as it was lined with trees of that sort. After 1905, the avenue became heavily populated when rows of apartment houses were built. A predominantly Jewish neighborhood up to World War II, it was temporary home to Leon Trotsky, Russian exile in 1917 when he and his family lived at 1522 Vyse Avenue, Apartment 5-D. However, some historians believe he lived on nearby Stebbins Avenue.

WADE SQUARE. Local Law 16, Council Procedures of 1940 gave the name of a World War I soldier to this square at Oak Tree Place and Quarry Road. Private. Edward Wade of nearby Valentine Avenue was killed in action in October, 1918.

WADHAMS STREET. Commander Wadhams is remembered by this short road running through the grounds of Fort Schuyler, from the East River to Long Island Sound. He was Superintendent of the New York State Maritime College from 1901 to 1902.

WAGNER AVENUE. Leading to the Administration Building on Riker's Island, this road is unique in that it was named after a living person, Mayor Robert Wagner.

WAKEFIELD. This was the name of a separate village, surveyed in 1855, and incorporated into New York City in 1895. Its original bounds were the Bronx River on the west, Laconia Avenue on the east, and from East 215th Street up to East 233rd Street. Upon its annexation to New York City in 1895, Wakefield was extended up to East 238th Street to absorb Jacksonville, and later up to East 243rd Street to absorb Washingtonville. See: Wakefield Square.

WAKEFIELD SQUARE. It was surveyed in 1853, and given the name of Washington's birthplace. Its location is at East 222nd Street at Bronxdale Avenue.

WALDO AVENUE. It was so named because it passes through the center of what had been the Waldo Hutchins estate. This property extended from Riverdale Avenue to Irwin Avenue, and from West 246th Street to Dash Place. It was opened in 1916.

WALDO SQUARE. This square was laid out on part of the original Waldo Hutchins estate in Riverdale. In 1940, the name was changed to Brust Square.

WALDO SQUARE NORTH. See: Waldo Avenue.

WALES AVENUE. This street is laid out through the former Manor lands of the Morris family, whose ancestral roots go back to Tintern, in Monmouthshire, Wales. In the mid-nineteenth-century, the land was sold to the Dater brothers and became known as East Morrisania. When the avenues were opened in 1896, this thoroughfare was first called Tinton Avenue, but then the name was shifted to an adjoining avenue. Both Tinton Avenue and Wales Avenue have a historic connotation with the Morris family.

WALKER SQUARE. The intersection of East 222nd Street and Barnes Avenue was named in March 1989 to honor Reverend Patrick DeSouza Walker, a longtime priest, in nearby St. Luke's Episcopal Church, noted for his civic and religious accomplishments.

WALLACE AVENUE. It was known as Jefferson Street in Van Nest, until 1922; and in the Williamsbridge sector, the original name was Hickory Street. For want of official records, two different magistrates could be candidates for the honor of having the avenue named for them: U.S. District Judge William J. Wallace, or Judge-Advocate General William Copeland Wallace.

WALLENBERG LANE. See: Raoul Wallenberg Lane and U. Thant Park.

WALNUT AVENUE (1). It is in the approximate center of Woodlawn Cemetery, where walnut trees are still to be seen.

WALNUT AVENUE (2). See: Rose Feiss Boulevard.

WALTER J. (Corporal) FUFIDIO SQUARE. See: Fufidio Square.

WALTON AVENUE. Mary Walton was the wife of Lewis Morris. Their son, Captain William Walton Morris, was an aide-de-camp to General Anthony Wayne, and died in 1832. The Morris holdings in the West Bronx were later subdivided: the lower end of Walton Avenue around East 164th Street was called Butternut Street, while the upper stretch from East 177th Street to Burnside Avenue was Punnet Street. In 1904, the land between Butternut Street and Punnet Street was surveyed and named Sylvan Avenue, but only for a short time before the entire length was renamed Walton Avenue. It would seem Gerard Walton Morris was the specific land owner the Commissioners had in mind when the avenue was named.

WARD AVENUE. Richard and William Ward were the earliest English settlers (1656) of Westchester who swore allegiance to the Dutch in order to remain on their lands. The first church built in what is today The Bronx was St. Peter's Episcopal church (Westchester Square), and a notation in 1700 reads "Richard Ward was to receive forty pounds for doing the work." Adjoining avenues were named for other colonists who likewise had to pledge loyalty: Thomas Wheeler and John Cloes (Close).

WARING AVENUE. In maps, dated 1859, 1868 and 1872, there was an estate belonging to Maria Waring in the vicinity of Williamsbridge Road. The western end of Waring Avenue was originally the carriage entrance to Pierre Lorillard's snuff mill (in the Bronx Botanical Garden). A later period (1844) shows this area as the holdings of John David Wolfe, who married Doriane Lorillard. At Barker Avenue, the original White Plains road that figured in Colonial history crossed from the southwestern corner to the northeastern corner. From Williamsbridge Road to Eastchester Road, it runs through what was the Denton Pearsall lands known as "Woodmansten." From Eastchester Road to the eastern terminus it was the early 1700s property of Reverend John Bartow-later, Furman, Stillwell, Bayard and Hunter families. The end of the avenue overlays ancient Givan's Creek (Black Dog Brook).

WASHINGTON AVENUE. Originally this fashionable street of the 1850s ran through the villages of Melrose, Morrisania and Tremont from East 153rd Street up to East 187th Street. The lower end, in Melrose, was later rechristened Elton Avenue. The remaining stretch was for many years the scene of patriotic parades. Andrew Cauldwell erected the first house in Morrisania at about East 169th Street in the Fall of 1848. Washington Avenue was famous for its churches, social halls, synagogues and fine homes and gardens. From Third Avenue north to Fordham Road, the avenue was laid out as Delancey Street in the village of Belmont. It was named after a former landowner, Delancey Powell, who had purchased the farmland in 1849. But, because the same DeLancey [spelled this way] had Tory connections in The Bronx past, the town fathers decided to extend the existing Washington Avenue, and absorb the Tory name with a more patriotic one.

WASHINGTON BRIDGE. Back in 1888, this majestic bridge was opened to the public. Although our first President comes to mind, the bridge's name is more likely a geographical term as it leads over the Harlem River to Washington Heights in Manhattan.

WASHINGTON BRIDGE PARK. This small park consists of nine acres of steep hillside overlooking the Harlem River at the north end of Washington Bridge. It was declared city property in 1899.

WATERBURY AVENUE. From Westchester Creek to East Tremont Avenue, this avenue was the southern boundary of the Seton estate. When the estate was sold into city lots by the Seton Homestead Land Company, the lane was designated as Marran Avenue on some maps and as Marian Avenue on others. Neither was correct if it was to honor Joseph J. Marrin, who was a director of the realty company. From East Tremont Avenue eastward to Bruckner Boulevard it was 1st Street in the village of Schuylerville. From the boulevard to Eastchester Bay it was on land belonging to James and Charity Ferris (1777), later the Van Antwerps, Laytons and others. Lawrence Waterbury (1812-1879) was a trustee of St. Peter's P.E. church on Westchester Square, and owned a considerable estate on Eastchester Bay that is now incorporated into Pelham Bay Park, called "Plaisance." It was located at the foot of Middletown Road which was called - around 1870 - Waterbury Lane. A son, James M. Waterbury sold the property to the Country Club Association in 1888.

WATERLOO PLACE. This street was on the north portion of the Bathgate farm, which became Crotona Park. Thomas Minford owned the land for a time. The Place was officially opened in 1908 and, along with nearby Trafalgar Place, owes its name to a British victory-although the Germans helped defeat Napoleon at Waterloo in 1815. Who was responsible for naming both Places is not known, but it was evidently one who admired the English, or was of English extraction-the Bathgates, the Dunhams, or the Minfords.

WATERS AVENUE. Along with adjacent Waters Place, this short L-shaped Street overlays what was once extensive swampland that was filled in during the 1940s. A bend in Westchester Creek is preserved in the I-shape of Waters Avenue.

WATERS PLACE. Edward Waters was one of the English settlers of East Town, which the Dutch called Oostdorp, who was compelled to swear allegiance to the Dutch authorities: "This first January A° 1657: in east towne in the N. Netherlands, Wee hose hands are vnder writen do promes to oune the gouernor of the Manatas as our gouernor..." Oostdorp/East Town is today Westchester Square. Edward Waters' name appears also in a witchcraft trial when one Katherine Harryson was tried. He and Thomas Hunt appeared against her, asking that the court order that she be removed for the town. "In dark & dubious matters" the convicted party could appeal to the Governor, which the woman did. She was released, on the condition she would go back to New England.

WATSON AVENUE. This street recalls William Watson who was born in Scarborough, England in 1849. A landowner of consequence in this country, his property ran from the Bronx River at West Farms Square, across Westchester Avenue to Bruckner Boulevard. His estate was called "Wilmount" and the Watson manor house was once located in front of James Monroe High School. Squire Watson formed a realty corporation and, when his estate was divided into lots and streets, Manor Avenue was named for his domicile, and Watson Avenue for the man himself. From Commonwealth Avenue to White Plains Road, the avenue was first laid out as Larkin Avenue, after a Francis Larkin, farmer, of 1854. From White Plains Road to Pugsley Avenue, the street was known as Walter Street and, in the village of Unionport, it was 9th Street. An Indian trail cut across this avenue at Croes Avenue, leading from the interior to Clason Point.

WATT AVENUE. James Watt was a Scottish physicist whose work in generative power was equal to that of Ohm, Ampere and Edison. See: Ohm Avenue.

"WAVE HILL." This is a residence of historical importance that is located at West 248th Street above the Hudson River. Built of fieldstone in 1830 by William Lewis Morris, it was later occupied by the publisher William H. Appleton from 1865 to 1903. It is now part of the New York City Park Department Arboretum.

WAYNE AVENUE. This was once the lower end of the Bussing farm, and when first mapped, was called Fort Street. The avenue honors General Wayne who captured Stony Point on July 15, 1779. He detested his nickname of "Mad Anthony."

WAYNE PLACE. See: Mable Wayne Place.

WEBB AVENUE. This avenue overlays historic land that figured prominently in Revolutionary War skirmishes. In the early 1800s, this was part of the Valentine farmlands, and in Civil War times, William H. Webb, a naval architect, erected an Academy & Home for Shipbuilders, which was replaced by the Fordham Hill Houses in 1949. A local nickname for this high street was Tee-Taw Avenue, which was derived from Tetard, a French-Swiss minister who once lived there.

WEBSTER AVENUE. Directly after the Civil War, this thoroughfare was surveyed from East 162nd Street to East 165th Street. In 1879, it was opened up to East 184th Street and three years later advanced north to Fordham Road, absorbing a small lane called Thomas Street. It had acquired the name of Webster Avenue and most people supposed it honored Daniel Webster, the great pre-Civil War orator and statesman, but more likely is it that the name is more localized. Albert L. Webster was an engineer in the Department of Public Works at the time the avenue was lengthened, and Joseph O. B. Webster was the surveyor-so either man (or both) might be remembered. The avenue, from East Fordham Road northward to Gun Hill Road, was at first called Berrian Avenue, after landowners along that stretch; and from Gun Hill Road to the city line it was Bronx River Road.

WEEKS AVENUE. Zeno and Johanna Weeks were listed as innkeepers in 1834, having purchased nine acres of land in Upper Morrisania. In 1854, sons Charles and William added some acreage which was eventually inherited by a nephew John Weeks in 1879. The farm was cut through by the City, and the street traversing the land was originally called Clinton Avenue. This was changed around 1898 to avoid confusion with another Clinton Avenue near Crotona Park.

WEIHER COURT. This is an oddly-shaped alley running from Washington Avenue to Third avenue above East 164th Street, and lined by miniature brick houses that are only 16 to 20 feet deep. It was mapped in 1898 and laid out in 1902-extending at that time only half-way from Washington Avenue to Third Avenue. In 1908, the Court was finally cut through to the avenue. The name, in German, means a fish-pond, but no earlier maps indicated a pond at that spot, so the Court is evidently a reminder of William H. Weiher, a builder and contractor of Morrisania.

WEISSMAN MEMORIAL PLAZA. This junction of Williamsbridge Road, Astor Avenue and the Esplanade was named in 1962 in honor of Private. Sidney Weissman.

WELLMAN AVENUE. It was originally called Cornell Avenue because it was laid out across the Cornell farm of the past century. The Wellman Finance & Realty

Corporation was located at Westchester Square before 1913, and Francis L. Wellman had a hand in the sale and platting of the former Cornell farm.

WENNER PLACE. This was originally a Summertime retreat of some 25 acres belonging to a large family named Wenner, who lived in downtown Manhattan around the 1840s. They journeyed up to Ferris Point (Ferry Point Park) in their own schooner and spent two or three months on the peninsula. One of the daughters, Magdalene, married a Unionport resident named Rohr. See: Rohr Place. Last owner of record was Anthony Wenner who formed the Wenner Realty Corporation. This Place is the southern boundary of the Ferris homestead, facing Westchester Creek.

WEST 161st STREET. The northern banks of Cromwell's Creek extended to this street which, in the 19th century, was swampland.

WEST 162nd STREET. It was mapped once as Cross Street. This street represents the southern extent of the nineteenth-century settlement of Highbridgeville that came into existence when the Croton Aqueduct was built.

WEST 163rd STREET. This street runs across what was once the Anderson farm. Later, the farm was sold to a family named Ketchum who named their estate "Woody Crest."

WEST 164th STREET. This street was on the Anderson farm, and their stone house was located on West 164th Street at Anderson Avenue. It was built of native stone, called gneiss, quarried within 200 feet of the site.

WEST 165th STREET. See: West 163rd Street.

WEST 166th STREET. It was once known as Devoe Street, according to an 1876 map, running from University Avenue to Anderson Avenue. From Jerome Avenue to Gerard Avenue, it was mapped as Endrow Place in 1879. The area facing the Harlem River was deeded to the Devoe family in 1676. In 1794, James Anderson was listed as owner.

WEST 167th STREET. It is likely that part of this street, from Sedgwick Avenue to Ogden Avenue was the very old Woolf Street which followed the bed of the brook dividing Daniel Turneur's property from that of John Archer's manor lands of Fordham. The Woolf family descended from a Hessian soldier who settled there, and the name appears as Woolf in the 1840s, but as Wolf in 1876. At one point in its history, West 167th Street was called Beach Street, harking back to W.B. Beach, a property owner on an 1884 map. From Ogden Avenue east to Jerome Avenue, West 167th Street was known as Union Street. Near Woodycrest Avenue there was once a spring-fed pond, belonging to Reverend Hasbrouck DuBois in the 1890s.

WEST 168th STREET. It was mapped as late as the 1890s as Birch Street from High Bridge down to Jerome Avenue. Later, it was known as Orchard Street, according to a 1901 *New York Eagle* ad, publicizing Patrick Clancy's Hotel in Highbridgeville.

WEST 169th STREET. Laid out across the old settlement of Highbridgeville, this street was on the Nelson farm of the 1870s. Down at Jerome Avenue, it was originally called Arcularius Street, after the proprietor of Jerome Park Hotel, listed in 1859.

WEST 170th STREET. From the Harlem River to Sedgwick Avenue, this was once a noted amusement center from approximately 1870 to 1900. It was called Kyle's Park and excursion boats from East 138th Street docked there. Mr. Kyle had a pedestrian bridge built from the pier to his Park over the N.Y. Central tracks. It is now overlaid by the Major Deegan Expressway. From University Avenue east to Edward L. Grant Highway, this street runs across the Nelson farm of the past century.

WEST 171st STREET. This short street marks the southern boundary line of William B. Ogden's estate, "Boscobel."

WEST 172nd STREET. Almost facing the eastern end of Washington Bridge was a double gate leading to the "Boscobel" estate. At Jerome Avenue, it was once mapped as Randolph Avenue.

WEST 174th STREET. This street runs athwart the former Nelson farm, which was absorbed by the "Rose Hill" estate of J. D. Poole.

WEST 175th STREET. A short street west of Grand Avenue was once part of J. D. Poole's "Rose Hill" estate. The Townsend Poole cottage, built in 1783, stood on Macombs Road near Featherbed Lane. A family of Eskimos, brought to The Bronx by Arctic explorer Admiral Robert E. Peary, lived in the cottage for a number of years before World War I. See: Peary Gate.

WEST 176th STREET. This thoroughfare runs across the historic manor lands of the Archer family that were purchased by the wealthy and influential Morris family of the 19th century. Major William Popham married Mary Morris, and their son inherited Morris Heights as it was then called.

WEST 177th STREET. This is a short street that marks the southeastern corner of the Popham estate. See: Mount Hope Place.

WEST 178th STREET. Once known as Powell Place from the Harlem River to Sedgwick Avenue in 1897, the street is almost obliterated by the Major Deegan Expressway.

WEST 179th STREET. From the Harlem River to University Avenue this street runs across the former estate of Henry Cammann, credited with developing the modern binaural stethoscope. The estate was mapped in 1868. The streets were laid out in this sector from 1891 to 1892 as part of Logan Billingsley's East Fordham real estate syndicate.

WEST 180th STREET. It is laid out over the ancient Fordham Manor that was owned by the Archer family before the American Revolution. This land was part of the Schwab property of the 19th century. At the western end of West 180th Street once was situated Fort No. 8 that saw much service during the Revolutionary War.

WEST 181st STREET. This street stretches across the William Loring Andrews estate, mapped in the 1870s and 1880s.

WEST 182nd STREET. The former name of this street, from Aqueduct Avenue down to Jerome Avenue, was Andrews Place, a reminder of the former landowner. See: West 181st Street.

WEST 183rd STREET. This street crosses part of the Jan Arcier (John Archer) property of 1683, which in the 19th century was part of the Morris holdings. From Loring Place to University Avenue this street was originally called Hampden Avenue. G. W. Hampden was a partner to Logan Billingsley of the East Fordham real estate syndicate in the 1890s.

WEST 184th STREET. This was once mapped as Cammann Street alongside the Harlem River, where was once the Cammann estate known as "Roseland." Dr. Henry Cammann helped found St. Barnabas Hospital in the village of West Farms before the Civil War. Part of the estate was owned by his brother, Oscar.

WEST 188th STREET. At first this street ran from Sedgwick Avenue to Grand Avenue, but now is but one block long. The western part, a curve, follows the course of an ancient brook that originated on the Devoe farm (Aqueduct Avenue and West 190th Street) and emptied into the Harlem River near Fordham Landing. This curve has been called Father Zeiser Place since March, 1953.

WEST 189th STREET. It was once called Academy Place, a reminder of Webb's Academy & Home for Aged Ship Builders that once overlooked the Harlem River.

WEST 190th STREET. Known as Croton Avenue when it was opened from University Avenue to Jerome Avenue, this street runs across part of the Valentine and Devoe farms of Civil War days. In 1884, when St. James Protestant Episcopal church was built, the entire stretch (West 190th Street and East 190th Street - which was Croton Avenue) was renamed St. James Street. A small pond was located alongside West 190th Street-the probable source of Valentine Brook that ran westward into the Harlem River. In 1894-95, the street was cut through, and was numbered shortly thereafter. See: East 190th Street.

WEST 192nd STREET. It was once mapped as Primrose Street, from Aqueduct Avenue over to Jerome Avenue.

WEST 193rd STREET. Before acquiring its number, the street was known as Knox Place, for a one block stretch at Bailey Avenue. It was also mapped as Vietor Place, according to an 1887 map, after a landowner George F. Vietor, importer.

WEST 195th STREET. This street is laid out across the former Claflin estate of the 19th century. Near its junction with Webb Avenue was once Fort No. 5 of Revolutionary times. It was built by Loyalists, and for a time was commanded by Colonel DeLancey. It had no cannon, nor did it figure prominently in any action. The fort was destroyed and abandoned by the British in 1779.

WEST 197th STREET. This was once part of the Claflin estate. See: Claflin Avenue.

WEST 205th STREET. When this area was the Jerome Park racetrack, West 205th Street was a lawn of Leonard Jerome's villa. Later, this was the edge of the Jerome Park reservoir, when that body of water extended almost to Jerome Avenue.

WEST 225th STREET. Originally, this street, running east from Broadway, was on opposite sides of the Harlem River (now covered by Exterior Street). In 1759, the Farmers' Bridge was built by Benjamin Palmer to break the toll-monopoly enjoyed by the Philipse family, who owned the King's Bridge farther upstream. This bridge was also called the Free Bridge, Hadley's Bridge, Dyckman's Bridge and (during the Revolution) the Queen's Bridge. Subsequent bridges served the area until 1911, when the Harlem River was diverted, and the original channel filled in. A plaque, describing the former bridge, was formally unveiled in June, 1959, by the Kingsbridge Historical Society. Old-timers recall with nostalgia the Velodrome that was situated on this street featuring all types of bicycle races in which national and international cyclists competed.

WEST 227th STREET. At Spuyten Duyvil, this street was originally mapped as Sidney Street. Two Revolutionary forts were once directly to the north. See: West 230th Street.

WEST 229th STREET. Between Heath and Bailey Avenues, this street was known as Reed Place until 1916.

WEST 230th STREET. Two Revolutionary forts were once built south of this street. Fort Swarthout was erected by American troops under Colonel Swarthout in 1776 near Arlington Avenue, but it was captured by the British and given a number, Fort No. 2. Fort No. 3 was located at Netherland Avenue, and was occupied by both sides from 1776 to 1781. Spuyten Duyvil Creek formerly flowed under this street from Tibbett Avenue to Broadway, where it was known as Depot Street. West 230th Street was the southern shore of the island of Paparinemin, later Hummock Island, Phillipsborough, "Island Farm" estate of Alexander Macomb, and finally, Kingsbridge. The creek was filled in, 1916-1917. The steep grade from Broadway up to Sedgwick Avenue was once known as Break Neck Hill.

WEST 231st STREET. In 1889, it was Morrison's Lane, off Riverdale Avenue. It was called Macomb's Street west of Broadway, for Alexander Macomb's mansion stood at Godwin Terrace. He had purchased the entire island of Kingsbridge around 1801 and called it "Island Farm." Tibbett's Brook crossed West 231st Street just east of Irwin Avenue. During a 1922 real estate venture, West 231st Street between Albany Crescent and Bailey Avenue was slated to be renamed Martin Terrace after a Michael J. Martin, but the plan was dropped.

WEST 232nd STREET. This street lies across the Morrison estate of the 1880s from Palisade Avenue up to approximately Fairfield Avenue. From Irwin Avenue to Broadway, it crosses the former island of Kingsbridge that was formed by the two branches of Tibbett's Brook and Spuyten Duyvil Creek.

WEST 233rd STREET. A few yards east of Broadway, this street was crossed by a branch of Tibbett Brook. The waterway was filled in around 1916.

WEST 234th STREET. Tibbett's Brook crossed West 234th Street at Irwin Avenue. At Kingsbridge Avenue, this street was crossed by the "Wickers creek" Indian trail. The Weckguasgeeck Indians lived in this area, and most likely were the tribesmen that met Henry Hudson. An early name of this street was Varian Street, after a pioneer family of Spuyten Duyvil.

WEST 235th STREET. This street runs across the high ridge called "Upper Cortlandt" in Revolutionary times, although it had been sold in 1761 to William Hadley by Jacobus Van Cortlandt. It was the 19th-century estate of John Warner.

WEST 236th STREET. From approximately Tibbett Avenue to 40 yards east of Broadway, this was the island of Paparinemin, later part of Macomb's "Island Farm," 1801. Tibbett's Brook flowed past, between Corlear and Tibbett Avenues. See: West 235th Street.

WEST 237th STREET. At its junction with Broadway, West 237th Street was once the northeastern corner of the island of Paparinemin, later called "Island Farm." A footbridge over a branch of Tibbett's Brook led from the island to the Van Cortlandt estate. See: West 235th Street.

WEST 238th STREET. From Riverdale Avenue down to Irwin Avenue, this street overlays the former Waldo Hutchins property. At Kingsbridge Avenue, Tibbett's Brook forked to form the island of Paparinemin. In the middle 1800s, this was the approximate southern limit of a settlement called Warnerville. Mr. Warner, land owner, lived on the site of Croke (Gaelic) Park. See: West 235th Street.

WEST 239th STREET. This street was once mapped as Northern Terrace, and marked the limit of William Hadley's purchase of 1761 from Jacobus Van Cortlandt. This street later became the southern boundary line of the Dodge estate in the 19th century. At West 239th Street is a Riverdale landmark, a tower that was built in 1922 by popular subscription. It houses a bell brought here by General Winfield Scott after the Mexican War, and which was first used by the Riverdale firehouse to serve as an alarm.

WEST 240th STREET. This street crosses the ancient manor lands of Frederick Philipse (1663-1785 were the years his family owned it) which was seized by the Commission of Forfeiture at the end of the Revolutionary War. It was sold to William Hadley, who owned property south of West 239th Street. It was originally mapped as Dash's Lane, after Bowie Dash who lived on Waldo Avenue, opposite Van Cortlandt Park.

WEST 242nd STREET. From Riverdale Avenue down to Irwin Avenue, this street runs through the former estate of Waldo Hutchins. At Broadway, on a map dated 1851, this was in a settlement called Warnerville.

WEST 244th STREET. This street crosses what was part of the extensive Dodge estate at its junction with the Waldo Hutchins property. The street was opened in 1916.

WEST 246th STREET. The street runs across the nineteenth-century estate of Cleveland Dodge and, from Arlington Avenue to Riverdale Avenue, forms the south boundary of "Dodgewood, Inc." It constituted part of the Albany Post Road to the Delafield estates. Scotch Hill was a small settlement of Scottish families on Old Post Road, now covered by Horace Mann school. See: West 240th Street.

WEST 247th STREET. Dogwood Brook crosses under West 247th Street east of Arlington Avenue as the street (once called Dodge Lane) bisects the nineteenth-century estate of W. E. Dodge and the twentieth-century lands of Cleveland Dodge. Their handsome country house was designed in 1863 by James Renwick, famed as the architect of St. Patrick's Cathedral. A granite quarry was located at this street near Riverdale Avenue, and the section was known as Quarrytop. It supplied the granite for many Riverdale mansions, including Font Hill (Mount St. Vincent).

WEST 248th STREET. From the Hudson River to Independence Avenue, this countrified lane has been the boundary between various estates, notably the Hadley holdings and the Dodge estate. Dogwood Brook (see: Alder Brook) runs parallel to this street.

WEST 249th STREET. It was once mapped as Spaulding Lane, as it was on the estate of the Spaulding family. A spring was situated at the corner of this street and Riverdale Avenue, one source of Dogwood Brook (Alder Brook). It formed a pond, which was drained and filled in, in 1961. The Riverdale Presbyterian church at Henry Hudson Parkway was built in 1863. The designer was James Renwick, and his motif was English Gothic Revival.

WEST 250th STREET. It runs across what had been "Upper Cortlandt" of Revolutionary days, which became the nineteenth-century estates of Colonel Delafield, Dr. Goodridge and the Livingston family. The street was opened in 1916. See: Tulfan Terrace.

WEST 251st STREET. This is a block-long street connecting Post Road and Broadway with a background identical to adjoining streets.

WEST 252nd STREET. It was once called South Street. At different times, Theodore Roosevelt and Mark Twain lived on Sycamore Avenue and West 252nd Street on the palatial Appleton estate. This 20-acre tract was given to New York City as an arboretum by Mrs. Freeman (nee Perkins) in 1960. Christ Church, a gem of Victorian Gothic structure, was designed by Richard Upjohn, and stands at the corner of West 252nd Street and Riverdale Avenue.

WEST 253rd STREET. From Riverdale Avenue down to Broadway, an 1889 map shows this street as Riverdale Lane. It was later called James' Lane, after J. B. James, landowner, whose mansion overlooked Van Cortlandt Park. The mansion was demolished to make way for Henry Hudson Parkway.

WEST 254th STREET. This was formerly called River Avenue as it ended at the Hudson River. At the top of the hill was once the Riverdale Inn, an 1890 hotel owned by William Olms who was called "Jumbo" because of his size. A localism for this intersection was "Jumbo's Corner."

WEST 255th STREET. This short street marks the northern limit of the J. B. James estate of the 1890s. See West 253rd Street. The property had previously belonged to Leemon Tripp, County Sheriff in 1861.

WEST 256th STREET. From Fieldston Road to Sylvan Avenue, this street was first known as St. Vincent Avenue. Proceeding downhill to Broadway, its name was Hawthorne Avenue.

WEST 257th STREET. Around 1900, the Chestnut Grove Picnic Grounds were located at West 257th Street and Riverdale Avenue. Before that, a small settlement, Irishtown, extended to West 259th Street.

WEST 259th STREET. It was originally named Rock Street, and was the northern limit of Irishtown, a small community along Riverdale Avenue.

WEST 260th STREET. It was known as Beech Street from Riverdale Avenue to Liebig Avenue. At West 260th Street and Liebig Avenue, the slope was known as Pigeon Hill. From Spencer Avenue down to Broadway this was the lower boundary of George H. Forster's estate of 1888.

WEST 261st STREET. This is a steep winding road that once formed the southern bounds of "Font Hill," the nineteenth-century estate of Edwin Forrest. From the Hudson River to Independence Avenue, it was once called Randolph Lane, after a former landowner. From Independence Avenue to Riverdale Avenue, it was called Cuthbert Lane after another landowner whose estate was below West 261st Street.

WEST 262nd STREET. This steep street was what had been the Lispenard Stewart estate boundary line, separating it from the George Forster property to the south.

WEST 263rd STREET. From Riverdale Avenue east to Broadway, this was once called Stuart's Lane-evidently a misspelling of the former landowner's name, Lispenard Stewart of the 1890s. An earlier proprietor of the steep property was Jonathan Odell, whose family antedated the Revolutionary War.

WEST AVENUE. This is a privately maintained street in the western sector of Parkchester that was opened in 1940 through what had been the Catholic Protectory grounds.

WEST BORDER AVENUE. As its name implies, this road runs on the west border of Woodlawn Cemetery, past the graves of sportsman O. P. H. Belmont and brewer George Ehret.

WEST BURNSIDE AVENUE. This avenue runs across a part of the historic Archer Manor, that eventually was owned by the Morris family. Its nineteenth-century name was Dashwood Avenue, after the last owner, G. L. Dashwood. It was renamed to commemorate the Civil War general, Ambrose E. Burnside.

WEST CLARKE PLACE. Maps & Profiles of the 23rd & 24th Wards (1895) lists a Clarke family in this lower section of Fordham.

WEST FARMS ROAD. This thoroughfare overlays a good deal of an Indian path that led from Bronx Park south to Hunts Point. The earliest colonists knew it as the Lower Road or the Back Road, and its official name "ye Queen's Road" was first mentioned in 1723. Another early (and unofficial) name was the Hive Town Road. It was in 1802 that the Westchester Highway Commission referred to the West Farms Turnpike. Although it had been in use for centuries, the road was only officially opened and paved throughout its entire length in 1893. It ran past the estates of the Fox and Tiffany families, the residence of Dr. Freeman, and "Minford Place," the palatial estate of the Minford family. See: West Farms Square.

WEST FARMS SQUARE. This was once the business and social center of the town of West Farms from the time it was a thriving river port in Civil War times to about the 1930s. West Farms received its name as it (the district) lay west of the Bronx River, the original name being the Ten Farms.

WEST FORDHAM ROAD. The original settlement of Fordham was begun in 1669 alongside the Harlem River near Broadway and grew into a prosperous farming community. The American Revolution brought down on the hamlet an occupying force of British and Hessian soldiers who pillaged the farms and used the houses for firewood during the Winters so that Fordham was completely devastated at war's end. The inhabitants never returned. It was not until the mid-1850s that people began to live there in any numbers. West Fordham Road overlays the ancient path down to Fordham Landing. In the mid-1800s, it was known as Highbridge Road, and later was given the name of Cammann Street as it was the northern boundary of the Cammann estate. Oswald and Henry Cammann owned an estate, "Roseland," that was bounded by the Harlem River to University Avenue, and from West 183rd Street up to this road. See: East Fordham Road.

WEST GUN HILL ROAD. See: Gun Hill Road.

WEST KINGSBRIDGE ROAD. See: Kingsbridge Road.

WEST MOSHOLU PARKWAY SOUTH. See: Mosholu Parkway South.

WEST MOSHOLU PARKWAY NORTH. See: Mosholu Parkway North.

WEST MOUNT EDEN AVENUE. See: Mount Eden Avenue.

WEST SHORE DRIVE. This is a driveway along the western shoreline of Hart Island.

WEST STREET. This short street is the most westerly street in Pelham Manor, and crosses over into Bronx County. It was opened in 1935, and named by the Village Clerk, Julius Dworschak. The street is on the original Pell Grant, and figured briefly in October, 1776, as the scene of Colonel Glover's orderly retreat before a larger British-Hessian force.

WEST TREMONT AVENUE. This cuts across the lower end of what had been Archer Manor, the estate of the first proprietors of Fordham. Later, it became the lands of Lewis G. Morris, "Mount Fordham," which was inherited by a grandson, Charles Popham. This avenue was a country road from the Aqueduct down to Jerome Avenue, curving northward to avoid the swampy end of Cromwell's Creek. A local name was Snake Hill. See: East Tremont Avenue.

WEST TWIN ISLAND. This was originally part of the Indian hunting and fishing preserves known as Laap-Haw-Wach-King, which were purchased by Thomas Pell. In 1713, John Pell described part of his land grant as including "one small Island known by the name of Twins." West Twin was joined to Hunter Island by a stone span, presumably built by John Hunter as he had spent $40,000 landscaping his island home. The narrow channel and the sandy beaches on both Hunter Island and West Twin Island were extremely popular with canoeists and campers once it was acquired by the Parks Department in 1888. West Twin was joined to the mainland when Orchard Beach was extended in 1947.

WESTCHESTER AVENUE BRIDGE. In 1938, this bridge replaced an earlier wooden structure. It takes its name from the avenue.

WESTCHESTER AVENUE. This is a very ancient road overlaying in part a still older Indian trail. It was used in pre-Revolutionary times to connect the Manor of Morrisania to the town of Westchester. In 1867, it was graded somewhat and called the Southern Westchester Turnpike. It was extended beyond Westchester Square in 1916 and 1917, replacing the old Pelham Road through Middletown as the preferred route from the old village of Westchester to Pelham.

WESTCHESTER ESPLANADE. It was given its name by the developers of the residential area below Pelham Parkway when this was still a part of Westchester County. Hardly anyone knows its full title.

WESTCHESTER SQUARE. This was the original Dutch outpost of the early 1600s which developed into a village called Oostdorp. The name meant East Village, and denoted its location east of the colony of Nieuw Amsterdam. It was later absorbed by the English expansion down from Connecticut and rechristened West Chester, as it was west of the New England colonies.

WESTERVELT AVENUE. In the early 1700s, this was part of Reverend John Bartow's holdings, which he described in his will as "Scabby Indian." Later it was owned by the Givan family and descendants, until around 1903-1906 when it was subdivided into city streets. Avenues were arbitrarily given the names of former Mayors of New York. Jacob Westervelt, Mayor from 1853 to 1855, stemmed from an old Dutch family who were wealthy shipbuilders. He served as a sailor on clipper ships and, in later life, himself became a prosperous shipbuilder.

WHALEN PARK. Henry A. Whalen, past commander of the United War Veterans Committee of The Bronx was, for many years, Chairman for the Annual Memorial Day parade on the Grand Concourse. In World War II, he saved the life of General (then Captain) Westmoreland, who later commanded the American troops in Viet Nam. The park, at East 205th Street and Perry Avenue, was dedicated on May 26, 1974.

WHALEN STREET. School trustee John Whalen was a lawyer of the New York City Corporation Counsel, a Tammany Democrat and a prominent man in the 23rd Assembly District. Whalen Street is partially cut through near West 261st Street, west of Broadway.

WHEELER AVENUE. All the avenues on this former Watson estate are (or were) named after the Westchester colonists who, in 1656, agreed to pledge allegiance to the New Netherlands so that they could remain on their homesteads. Thomas Wheeler was the Lieutenant of the small English colony that had established itself in what was then considered Dutch territory. His companions were William and Richard Ward, John Cloes (Close), John Genner (now Boynton Avenue), William Fenfell (now Evergreen Avenue) and Sherrood Damis (now Colgate Avenue).

WHITE OAK AVENUE. This road is located in the northeastern section of Woodlawn Cemetery, running past the Maple Plot where the graves of the Abram Bassford family are seen. Old-time Bronx brewers-Christian Kolb, the Hupfels and Johann Eichler - have adjoining family plots on this road. The name is descriptive of the trees that once grew in profusion in the area.

WHITE PLAINS ROAD. This is not the original road to White Plains that figured so frequently in Revolutionary annals, and which was a winding lane roughly paralleling the straight and wide thoroughfare we call White Plains Road or White Plains Avenue today. This modern thoroughfare was laid out in the year 1863 and widened in 1902-1908, passing over parts of the original lane. White Plains was so called from the dazzling fields of balsam flowers that were so abundant there. At one time, the road was called Washington Street in Van Nest, and 3rd Street in Williamsbridge.

WHITE WALK. According to Robert Farkas, student-historian of New York University, this walk crossing the campus of what is now Bronx Community College honors a former educator, Richard Grant White. It bisects a lawn, once the field surrounding Revolutionary Fort No. 8.

WHITEHALL PLACE. Local lore is that a former owner, Oscar V. Pitman, called his mansion "White Hall" and when his property was sold off, it was by the Whitehall Realty Corporation. A 1905 map shows the tract to be subdivided ran from East 234th Street to East 238th Street, between Old White Plains Road to about Barnes Avenue.

WHITEHEAD PLACE. This Place is laid out in mid-channel of Hammond Cove, between Locust Point and Silver Beach Gardens. William Whitehead was an owner of present-day Silver Beach Gardens, his property being mapped in 1872. He was listed as a vestryman of St. Peter's P.E. church in 1849, and is buried in the churchyard. He built the "newer" stone wharf on the East River, south of the village-owned Whitestone Ferry slip.

WHITEWOOD AVENUE. This principal roadway runs across Woodlawn Cemetery from west to east. Whitewood is the generic name for tulip, linden and basswood trees, all of which are native to this area.

WHITLOCK AVENUE. The avenue is laid out across the former Richard March Hoe estate of the 1860s. Benjamin Whitlock, a wealthy merchant of that era, was the owner of the Leggett lands, "Rose Banks," on the East River. He renovated the old mansion and rebuilt parts of it with stone imported from Caen, France. Whitlock was ruined when the South lost the Civil War, and the price of cotton tumbled. Innocencia Casanova bought the land from Whitlock's widow.

WHITTEMORE AVENUE. During the 18th century it was known as Bowne Road as the Bowne family had a farm on its north side at East Tremont Avenue. The farmhouse was demolished around 1950. The road led from Westchester village (now Westchester Square) to the Whitestone ferry (Ferry Point Park). The Whittemores were landowners dating back to the List of Freeholders of May, 1775.

WHITTIER STREET. This Hunts Point street runs across the former estates of Colonel Hoe, Squires Faile and Fox. At the foot of Whittier Street was once the important Weckguasgeeck Indian camp. The Bronx River divided this tribe from the Siwanoys. The street is named for John Greenleaf Whittier, Quaker poet who died in 1892.

WICKHAM AVENUE. The avenue is laid out across the ancient Pell lands which-from Stillwell to Hammersley Avenues-was purchased by Reverend John Bartow in the 1720s. After three generations, it passed to the Givan family and their descendants. It formed part of the Givan farm, north of Gun Hill Road. It closely parallels the ancient bed of Black Dog Brook, which was the boundary between Eastchester and Westchester, according to the records of 1667. In the Edenwald section, this avenue was first called Burke Avenue and, in the Wakefield area, it was known as Chestnut Street. In line with the policy of naming these avenues after former Mayors of New York, this street honors William H. Wickham, a diamond merchant and importer, who served from 1875 to 1876. He is buried in Woodlawn Cemetery.

WILCOX AVENUE. In part, the avenue overlays an Indian trail that led to an important fishing camp at the junction of Schley and Wilcox Avenues. Excavations before 1920 revealed burial pits, mummies, flint knives and some evidence of trade with English and Dutch settlers in the form of brass buttons and clay pipes. The avenue is named for Colonel Vincent M. Wilcox of the Civil War.

WILDER AVENUE. It was formerly known as Sound View Place because Long Island Sound could be seen from its elevation. Later it was known as South 16th Avenue when the area was a part of South Mount Vernon. At its upper end it was part of the Heintze property, and below that, it was called the Bathgate estate. The Sound Realty Corporation developed the property in 1904, and it is believed Wilder was one of the directors.

WILKINSON AVENUE. Robert and Ellen Wilkinson were the owners of 16 acres of land on the Pelham Road (now Westchester Avenue) and this avenue cuts directly across their former tract. An earlier name was Evelyn Place.

WILLETT AVENUE. The Willetts go back to the earliest roots of New York, Thomas Willett being the first Mayor in 1665. He was an Englishman, conversant with the Dutch language, and had originally settled in Plymouth. Since that time, the Willetts were prominent on Long Island and in The Bronx. Three families of that name were residents of Williamsbridge, listed in 1873, 1881 and 1900 as property owners.

WILLIAM AVENUE. Nicholas Haight in the 1820s was sole owner of City Island, with the exception of four small parcels. In 1826, William Schofield bought property, and his house still stands on the corner of William Avenue and Schofield Street.

WILLIAM C. BERGEN PARK. See: Bergen Park.

WILLIAM F. DEEGAN (Major) EXPRESSWAY. See: Deegan Expressway (Major William F.).

WILLIAM KOLTOVICH PARK. For over 40 years this Pelham Bay resident tended bocci courts and a garden plot on Jarvis Avenue on a voluntary basis. At age 68 he suddenly died, and the bocci players and neighbors saw to it that his faithful work would be remembered. The special ceremony took place on March 28, 1986.

WILLIAM PLACE. This short Place is situated on Bridge (Crow) Hill which figured in the Revolutionary War cannonading of Westchester village by British troops in 1776. It was the nineteenth-century estate of the Harrington family that, in the 1920s, was subdivided into building lots. George and William Jorgenson built one-family houses on this Place. See: George Street.

WILLIAMSBRIDGE ROAD. This was an Indian footpath for centuries before it was widened to accommodate the rude drags of the early settlers. A John Williams had a farm near the present Gun Hill Road station at White Plains Road, and the bridge over the Bronx River was called Williams' Bridge, later to become contracted. The road was an important artery during the Revolutionary War and continues in importance to this day. Ezra Cornell's home was on this road at Silver Street (1807) and, in the 1890s, the road flanked Morris Park racetrack.

WILLIAMSBRIDGE SQUARE. This open space at Gun Hill Road and White Plains Road was acquired as a public square in 1910, and placed under the jurisdiction of the Department of Parks in 1926. The name is a reminder of the village of Williamsbridge which was annexed to New York City in 1895. See: Williamsbridge Road.

WILLIS AVENUE. This was the main thoroughfare of what was called Old Morrisania, and then North New York. It was planned in 1876, confirmed in 1880, plotted in 1894 and opened in 1897. In 1901, it was widened. The name is an old one in Bronx history, a Samuel Willis having mapped the Manor of Fordham in 1756, and for the next century the name appears principally on deeds. The avenue's name harks back to Edward Willis, who owned property at East 143rd Street from Alexander Avenue to present-day Willis Avenue in the 1860s, and who was in the real estate business.

WILLIS AVENUE BRIDGE. Opened in 1901, this bridge over the Harlem River, was an important link between Willis Avenue in The Bronx, and First Avenue in Manhattan.

WILLOW AVENUE. The land across which this avenue runs was farmland leased by Jonas Bronck to Pieter Andriessen (Map of 1639). It was marshland surrounding Stony Island, but in the 1840s Gouverneur Morris II had the island joined to the mainland, and the area adopted the name of Port Morris. The avenue is named after the willow trees once so abundant in the neighborhood.

WILLOW LANE. Willow Lane formerly ran from the Unionport bridge to the Pelham bridge, but was renamed Eastern Boulevard and once again renamed Bruckner Boulevard. It was the most important road passing across Throggs Neck, but almost lost its identity entirely when this last remaining stretch was called Becker Street. Finally, in 1927, the name Willow Lane was restored near Westchester Avenue. It skirts the former modest estate of the Widow Arnow. According to old-timers, willow trees flanked the lane for most of its length.

WILSON AVENUE. In this area most of the streets honor former Mayors of New York, and there was a Mayor Ebenezer Wilson who served from 1707 to 1710. Beginning at its lower end, this avenue was once part of the Denton Pearsall lands (1860), which was later the Astor estate (1900). Around Allerton Avenue, it was crossed by Stony Brook, and was on the Stedman property. Proceeding north, it was once the western end of the Givan property (mid-1800s), which was the Kidd property that passed to Giles Bushnell after the Civil War. At Boston Road it was the extreme edge of the Joseph Stickney estate (1900).

WILSON SQUARE. At Soundview, Underhill and Patterson Avenues, this Clason Point square was acquired in October 1912, and later named in honor of President Woodrow Wilson. An ancient Indian path ran alongside Soundview Avenue at this point.

WINTER STREET. This is a very old name in the annals of Westchester County, of which it was a part. John Winter's signature is on the Patent for the Town of West Chester in 1686, and Gabriel Winter was a Commissioner in 1820. George Winter was a City Island resident in 1907.

WISSMAN AVENUE. A country lane in the 1840s, called Morgan Avenue after Edwin D. Morgan who bought the land in 1854 from Herman LeRoy Newbold. Next owner was Francis DeRuyter Wissmann [spelled this way] born in 1860, son of Frederic Wissmann and Celine Frances Bull Wissmann. He married Helen Adele Jones, who was born in Paris in 1866, the daughter of Lewis Colford Jones and Catherine Berryman Jones. Her father was the landed proprietor of Jones' Woods, which he developed into blocks of apartment houses below Yorkville. The Wissmanns purchased property south of the avenue in 1890-1891, and enlarged a small villa into an imposing mansion. Squire Wissmann was an eccentric man, always engaged in lawsuits over property and riparian rights, and used to patrol his sea wall with a spy-glass and warn off passing boats. He used a jaunting car out on the sandbars to reach his sloop. Every morning Wissman Lane was swept clean by his coachmen. The coachhouse, a sturdy brick building, burned down in February, 1953, having stood at the southwest corner of Blair and Wissman Avenues. The original church of St. Frances de Chantal was set up in a tent on the Wissmann lands at Meagher Avenue. Mrs. Wissmann died in December 1949, age 82, and her husband died a few months later at the age of 90. They are buried in Sleepy Hollow Cemetery in North Tarrytown.

WOOD AVENUE. An earlier name was Cornell Avenue. There is a strong likelihood this avenue overlaps an earlier wood road to the McGraw farm. There is also a slight possibility it was named for Fernando Wood, a controversial Mayor of New York in the 1850s who tried to secede New York City from the rest of the state. He was pro-South during the Civil War. Another controversial Mayor, Hugh Grant, had his name attached to a nearby Circle.

WOOD ROAD. This is a private street inside Parkchester that was opened in 1939. It overlays an ancient Indian path leading to Clason Point, and the general direction indicates it was part of the wood road to the McGraw farm.

WOODHULL AVENUE. The land across which this avenue runs was originally Pell property of the 1600s, the Bartow lands of the 1700s, and then part of the Givan farm which had been called "Ednam." Around Civil War times, it belonged to the Watson family, then to the Valentines. The Valentine mansion still stands at Fielding Street and Woodhull Avenue. Avenues in this neighborhood carry names of former Mayors of New York, and Caleb Woodhull served from 1849 to 1851.

WOODLAWN BROOK. It originates as a spring in the Lakeside sector of Woodlawn Cemetery, and flows southeasterly down to Webster Avenue. It is then led by conduit under the avenue and Bronx River Parkway to empty into the Bronx River at approximately East 225th Street.

WOODLAWN CEMETERY. It was incorporated in December, 1863, due to the efforts of Reverend Absalom Peters, a theologian, poet and a man of civic vision. He sponsored the "Rural Cemetery Movement" to insure uncongested burial spots outside the city. The first internment in the 310-acre tract was made in January, 1865-a few months before the end of the Civil War. The name conjures thoughts of woods and lawns in keeping with Reverend Peters' intention.

WOODLAWN LAKE. A private lake inside Woodlawn Cemetery, artificially formed sometime after 1865 when the burial ground was laid out. The runoff is led under Bronx River Parkway into the Bronx River.

WOODMANSTEN PLACE. Named for an estate owned by Denton Pearsall (1870's), the Place is not located on the estate itself, however, according to the old maps. See: Pearsall Avenue. The name supposedly was borrowed from a vast estate in England.

WOODROW WILSON SQUARE. See: Wilson Square.

WOODYCREST AVENUE. It is named after a former estate "Woody Crest," belonging to the Ketchum family. Mrs. Ketchum was Angelica Anderson, whose forebears farmed some 60 acres on the same tract. From 1850 to 1880, the avenue was plotted as Bremer Avenue after a small landowner, but regained its name when cut through. Near the junction of this avenue and West 167th Street was a pond belonging to a Reverend Hasbrouck DuBois, who was pastor of the Dutch Reformed Church of Mott Haven, 1866 to 1877.

WORTHEN STREET. It is considered to be a corruption of Worden, as streets in this area are named after naval heroes. Admiral John Lorimer Worden (181,8-1897) fits this category, for as a lieutenant he was first commander of the *Monitor* during the Civil War. In the course of the battle against the Merrimac, Worden was nearly blinded, and only partially recovered. He retired on full sea pay in 1886, and lived out his life in Washington, D.C.

WRIGHT AVENUE. The lower end of this avenue was eliminated by Co-Op City, but a two-block stretch remains off Boston Road. In the 1880s, it formed part of the Ruser estate, later the Hollers property, the ice pond being overlaid by this avenue. In a row of avenues honoring former New York Mayors, Silas Wright's name is an oddity, for he was a Governor who served in 1845-1846.

WYATT STREET. This is a street off West Farms Square that was opened in 1915 across the former property of the Devoe family. Colonel J. Milton Wyatt was a resident at about that spot when West Farms-in Civil War times-was a thriving river port and mercantile center. His father-in-law was John B. Haskin, who was a prominent nineteenth-century statesman of what is today Bronx County.

WYTHE PLACE. This short Place is laid out through the former Findlay property of the 1850s. A family of that name briefly held title to lots there in 1896 when the Grand Boulevard and Concourse was being surveyed.

YANKEE MALL. This Parkchester social- and shopping center was named in honor of the New York Yankee ball team on April 28, 1967, through the efforts of the Parkchester Merchants' Association. Colorful ceremonies were enacted that day on the short street, and were attended by sportsmen and political dignitaries. Prior to 1940, this land was part of the Catholic Protectory grounds.

YATES AVENUE. It crosses the lands of John Bartow, who owned it from 1702 to 1727. It was the Denton Pearsall's estate, "Woodmansten," of the 19th century. In 1892, Messrs. Pinchot and Morrell began a real estate development below Pelham Parkway which they called "Westchester Heights." Yates Avenue was extended into this district, and absorbed a Richfield Street that had already been laid out. The avenue is named in honor of Governor Joseph Yates, who served from 1822 to 1824.

YEHONATAN (Lt. Col.) NETANYAHU LANE. On February 10, 1977 Mayor Beame signed into law a bill giving this name to the sitting area on Pelham Parkway at Holland Avenue. It is named for a Bronx-born Israeli soldier who lost his life rescuing passengers on a hijacked airliner that had taken off from Tel Aviv and was diverted to Entebbe, Uganda. The dramatic rescue took place on July 4, 1976.

YEW AVENUE. This road is located in the southeastern section of Woodlawn Cemetery, and refers to the yew trees once prominent there.

YOUNG AVENUE. Avenues in the area are named for State Governors, and John Young was serving in that capacity in the years 1847-1848. The avenue is laid out across the Kidd estate as mapped in 1868, and its deep woods gave it a local nickname, "Tanglewood," that persisted into the 20th century.

YOUNG PARK (1). Located at East Tremont and Van Nest Avenues, and East 180th Street, this park was named on November 4, 1933. It honors a local resident, James A. Young, who fought in World War I.

YOUNG PARK (2). The margin of this narrow park was formerly the edge of the millpond of Westchester village. A local inhabitant, Samuel H. Young, is remembered, and the park was named on December 31, 1941.

YZNAGA PLACE. This Place is laid out on what was once Coney Island – a wooded 4 1/2-acre island on the salt marshes of Ferris Point (Ferry Point Park). It was on the Havemeyer property of the 1880s. Louisa L. Seaman was the landowner in 1890, and the Place was not mapped until 1898. There is a bare possibility that the Yznaga sugar importers of Cuba had business and social ties with the Havemeyers, who were known as "the Sugar Barons" with property on the peninsula. Frank W uttge, a fellow name-hunter wrote, "Since Yznaga Place first appeared on maps in 1898 at the time of the Spanish-American War, it is possible that someone serving in the Navy at the bombardment of Cienfuegos, Cuba, found the tower of the Yznaga Brothers' sugar plantation with its great sign on top YZNAGA, and wrote home about it." According to Ralph Fornes, a lineal descendant, the brothers Pedro and Pablo, competed in an unusual contest: one tried to build the tower higher than the other could excavate a deeper well. A Jose Aniceto Yznaga once went to Venezuela to meet Simon Bolivar and bring back an army to liberate Cuba, spending a vast fortune in this vain effort. He then lived as a political refugee in New York where he died. It is a question whether he, or the socialite Fernando Yznaga was remembered by Squire Havemeyer when he sold his property.

ZAFFUTO SQUARE. See: Tina Zaffuto Square.

ZEISER (Father) PLACE. This curving Place was once part of West 188th Street from Webb Avenue to Grand Avenue. The curve follows the course of vanished Valentine's Brook. It was named in March, 1953, after a former pastor of St. Nicholas of Tolentine R.C. church when it was a small wooden edifice. Blasius J. Zeiser was born in Mauch Chunk, Pa., in 1878 and attended Villanova College and was ordained to the priesthood in 1907. Father Zeiser was assigned to the new parish in Fordham from 1908 to 1912. After a stint in Philadelphia, he returned to St. Nicholas to serve as pastor from 1917 to 1946. During this time he built the present "Cathedral of The Bronx," added a section to the parochial grammar school, and finally opened a high school department. He pioneered in dramatics, chorals, socials and athletics and became one of the legends of Fordham Heights. Father Zeiser returned to the monastery at Villanova where he died on May 9, 1951.

ZEREGA AVENUE. Originally it was called Green Lane (later Green Avenue) and only ran from East Tremont Avenue to Westchester Avenue. Around 1851, the village of Unionport was established and the road from the Westchester Turnpike to the Unionport bridge over Westchester Creek was called Avenue A. About 1910, the avenue was extended southward toward Castle Hill Point. The street pointed directly at "Island Hall," the manor house of the diZerega family, across Westchester Creek in what is Ferry Point Park today. In short, the avenue never ran across the former estate. Augustus Zerega diZerega was a wealthy shipowner, born on the island of Martinique on December 4, 1803. His wife was a Danish baroness, Eliza von Bretton, who was born in the Danish West Indies on the island of St. Thomas, January 3rd, 1810. She died at the age of 99, having outlived her husband, some sons and even a grandson. Mr. diZerega, who migrated to New York from Venezuela with ten children and a retinue of slaves, was certainly an odd commercial genius. Transacting an immense business that ran into thousands of dollars, he kept no account books, and bills were immediately paid in cash. This honest old merchant depended entirely upon a prodigious memory in all his numerous transactions. His ships of the Red Z Line were captained by some of his sons and sons-in-law, and some of them saw subsequent service in the Civil War. Mr. diZerega paid $70,000 for the 114-acre estate on Ferris Point (most of which is now covered by the Bronx-Whitestone Bridge) and, in 1854, enlarged the mansion of the previous owner. The estate had 21/2 miles of waterfront from Baxter Creek to the East River to Westchester Creek. The diZerega mansion, "Island Hall," burned down in 1895, and a smaller one was rebuilt on the foundations during the following year. The property was sold by one surviving son around 1916 to a Catholic Order, which remained there until the late 1930s, when the City acquired the property for a park and the approach of the Bronx-Whitestone bridge.

ZIMMERMAN PARK. Louis Zimmerman was a New York City Policeman who served as an Army corporal in World War I. He was killed in the Argonne Forest on November 5, 1918. Although he lived on Webster Avenue, the park that honors his name is at Barker and Olinville Avenues and Britton Street. The site was acquired with funds collected by the Police Department as a memorial to policemen killed in the war. The park and playground were opened in 1934.

ZINNIA AVENUE. This road, in the southeastern section of Woodlawn Cemetery, memorializes a native flower.

ZULETTE AVENUE. In District School #3 of Westchester village, the student roll of 1853 carried the name of Mary L. Zulette, whose family were property owners on "Crow Hill." While still in the blueprint stage, this avenue was slated to become John Street, a reminder of John Cornell whose early farm was on that slope, but by 1922, the present name prevailed. From Mayflower to Gillespie Avenues, this avenue runs through the Cornell farm, and from Gillespie to Merry Avenues, this was the Quaker Tract of 1808-1900. A small brook called Middle Brook originated near Gillespie and Zulette Avenues, and flowed southward into Weir Creek, a tidal run. See: Bicentennial Veterans Memorial Park.

226th DRIVE. This is a street in the Edenwald Housing Project curving northward to connect East 225th Street with Schieffelin Avenue.

229th DRIVE NORTH. This is a bracket-shaped street, north of East 229th Street, in the Edenwald Housing Project.

229th DRIVE SOUTH. This is a semi-circular street, south of East 229th Street, in the Edenwald Housing Project.

McNAMARA'S GUIDE
TO THE OLD BRONX

This guide is a compendium of colorful, quaint and descriptive names by which our Indian, Dutch, Colonial and 19th-century predecessors identified the landmarks of their times.

The natural landmarks include farms and estates that have been subdivided, hills that have since been leveled, woods that have disappeared, brooks and ponds that were rifled in, and islands now engulfed by landfill.

The manmade edifices are old inns and taverns, bridges, picnic grounds and breweries, silent movies houses, lifeguard stations and volunteer firehouses and most of all - the street signs that marked an earlier Bronx.

A STREET. See: Hobart Avenue. (Current).

ABBOTT PLACE. It ran off Eastchester Road, above Morris Park Avenue and was mistakenly spelled "Abbatt Place." William Abbott was the proprietor of this 9-acre tract in the 1890s. In 1945, the property was incorporated into the grounds of Jacobi and Van Etten Hospitals.

ABBOTT'S BROOK. It ran through the present Bronx County Medical Center, under Eastchester Road, joining Westchester Creek in swamps now filled in to form the grounds of the Bronx Psychiatric Center. See: Stony Brook, Saw Mill Brook, Kidd Brook.

ABEL COURT. This pathway is shown on a few Bronx maps. Investigation in 1960 showed a gravel path, fenced at both ends, running from Independence to Netherland Avenues, above West 256th Street.

AC-QUE-HO-UNK. See: Hutchinson River (Current).

ACADEMY PLACE. An earlier name of West 189th Street, from Sedgwick Avenue to Devoe Terrace, after the Webb Academy & Home for Aged Ship Builders.

ACKERMANN STREET. See: Corlear Avenue (Current).

ACME CAFE. This was a small frame restaurant established in 1902 by Robert and Martha Mayer at East 233rd Street and the Bronx River. The roadhouse with German style cuisine prospered into Mayer's Parkway Restaurant that was inherited by two sons, William and George. The landmark was almost completely destroyed by fire in 1965, but was rebuilt in 1971 on a far larger scale.

ACORN STREET. This was a short street near Reed's Mill Lane that is now covered by the New England Thruway. See: Reed's Mill Lane.

AC-QUE-HO-UNK. This is the Hutchinson River. Translations differ. It could mean "High Bank" or "Red Cedar Tree." See: Aqueanouncke, Eastchester Creek, Pelham River.

ACQUEEGENOM. This was a Siwanoy Indian name for present-day Pelham Parkway at Bronx River meaning "Where the Path goes over."

"ACTORVILLE." This was a localism for the village of Wilton because so many actors and actresses made their home there. See: Wilton.

ADAM STREET. This is the former name of East 184th Street at Third Avenue.

ADAMS AVENUE. This is now Adams Place, from Grote Street to Crescent Avenue.

ADAMS STREET (1). See: Bay Street. (City Island).

ADAMS STREET (2). The original name of East 184th Street, from Park to Washington Avenues. It was on the former Bassford farm that was mapped as Adamsville (2).

ADAMSVILLE (1). This is the name of a former settlement, east of Fordham University, north of Fordham Road and west of Southern Boulevard. The land had belonged to the Reverend Carson Adams in 1876.

ADAMSVILLE (2). This tract was surveyed by Andrew Findlay in 1853. Part of the Bassford farm, it centered around Webster and Third Avenues below Fordham Road.

ADEE AVENUE. Matthews Avenue's original name, from Burke Avenue to Adee Avenue in Williamsbridge, formerly owned by John T. Adee. See: Rose Street, Wright Street.

ADEE ESTATE (1). This refers to the property of William T. Adee around 1894. The bounds were East Tremont Avenue to Benson Street to Westchester Avenue to St. Peter's Avenue. His mansion stood on the northern end of Benson Street.

ADEE ESTATE (2). The 1850-1915 estate of George T. Adee and his four sons, bounded by Weir Creek on the north and west, Long Island Sound on the east, and present-day Miles Avenues on the south. It measured 35 acres, and became the Park of Edgewater.

ADEE PARK. This 1865 tract belonging to John T. Adee was subdivided in and around 1890 by real estate developers. Its boundary lines were approximately White Plains Road on the west, Burke Avenue on the north, Bronxwood Avenue on the east, and Williamsbridge Road on the south.

ADEE POINT. This promontory was on the "Edgewater" estate of George T. Adee. Its present-day location would be Edgewater Park.

ADEE'S LANE. This is the former name of Meagher Avenue from Harding Avenue north to a short distance past Miles Avenue. It referred to the estate of George T. Adee (now Park of Edgewater) of the 1850s to early 1900s. See: Pennyfield Lane, Pennyfield Road.

ADEE'S POINT DIVISION. This Volunteer Station of the Life Saving Service of New York City was listed in 1916. Adee Point was an earlier name of Edgewater Camp, now Edgewater Park. The Volunteer crew consisted of three officers and eight surfmen.

ADRIATIC HOTEL. An 1869 notice of this hotel, F. Lamer, proprietor, located it on Locust Avenue near Broadway (East Tremont Avenue near Crotona Park). An 1878 map credits this hotel with 7 acres bounded by Vyse, East Tremont and Bryant Avenues and East 179th Street. The inn itself was located in the roadbed of East 178th Street between Vyse and Bryant Avenues.

ALAMO AVENUE. See: Goodridge Avenue.

ALBANY ROAD. This is a far older name of Bailey Avenue.

ALDINE STREET. This was a former name of Aldus Street, according to an 1879 atlas.

ALERT NO. 2 ENGINE COMPANY. An 1874 map of West Farms locates this firehouse on the east side of the junction of Boston Road and present-day East Tremont Avenue, about in the bed of West Farms Road. James Grayson was Foreman, and Henry Hooper, Assistant.

ALICE AVENUE. In 1910, this was an alternate name of Liberty Street, which is now St. Theresa Avenue, the 5-block stretch from Hutchinson River Parkway to Westchester Avenue. It is believed to be a reminder of Alice Haight (nee Stinard). See: Liberty Street, Morris Park Avenue.

"ALL BREEZE." This was the former estate of Colonel Jacob Lorillard on Ferris Point in the 1880s, which was occupied by Pierre Lorillard in 1908-1910. It is now incorporated into Ferry Point Park, facing Westchester Creek. See: Lorillard Point.

ALLAIRE AVENUE. A 1910 map shows this short avenue plotted south from Lafayette Avenue across the marshes to Westchester Creek, but it was never cut through. Captain Allaire was a resident of Unionport, and a Civil War veteran.

"AMBLESIDE." This was the former Simpson estate near Hunts Point, in 1868. It was bordered by Intervale Avenue, Southern Boulevard and Westchester Avenue.

AMERICAN BREWING COMPANY. Bromley's map of 1897 shows this brewery on the northeast corner of Third Avenue and East 168th Street. It was separated from the David Mayer brewery by a coal yard.

AMERICUS HOOK & LADDER COMPANY. It is listed in an 1899 souvenir booklet of Exempt Firemen, and was located on Third Avenue in the vicinity of East 166th Street. Members listed had Morrisania addresses, and a few were Civil War veterans.

AMSTERDAM AVENUE. See: Hobart Avenue. (Current).

ANDERSON FARM. An ancient deed of 1794 shows James Anderson in possession of a 60-acre farm now traversed by Woodycrest Avenue, Anderson Avenue and West 164th Street. It was known as "Woody Crest" due to its thick woods and height. After the Civil War, his granddaughter, Angelica, married Edgar Ketchum, and the estate retained its name. A secondary Anderson farm was located due east, near the Concourse, in 1854.

ANDREWS PLACE. This is a former name of West 182nd Street from Aqueduct Avenue to Jerome Avenue, and named after the Loring Andrews estate of 1889.

ANN STREET. This is an earlier name of East 181st Street from Daly Avenue to Bronx Park, in the settlement of Tremont. See: 5th Street, Clover Street, Irene Place, Irving Street, John Street and Ponus Street.

ANN'S NECK. This is the 17th-century designation of Rodman's Neck. It commemorated Anne Hutchinson, the early settler who was massacred by Indians in 1643. See: Anne's Hoeck, Asumsowis, Camp Mulrooney, Tom Pell's Point, Pell's Neck, Pelham Point, Pelham Bay Naval Training Center, Tom Pell's Point.

ANNA PLACE. This was the original name of Kindermann Place in Morrisania. The land had belonged to Elliott and Anna (Bathgate) Zborowski, and was sold in 1888. It was eliminated by the Butler Houses.

ANNES HOECK. Early Dutch name of Rodman's Neck, from its first white inhabitant, Anne Hutchinson, 1643. The Dutch word "Hoeck," later anglicized to Hook, means a neck of land or peninsula. See: Asumsowis, Ann's Neck, Camp Mulrooney, Pelham Point, Pelham Bay Naval Training Center, Pell's Neck, Tom Pell's Point.

"ANNESWOOD." This was the estate of John Hunter III, son of Elias Desbrosses Hunter and grandson of John Hunter who owned Hunter Island. This estate was formerly the

Bayard farm, and now is part of Pelham Bay Park in the vicinity of the Police Academy paddock. The grounds were used as anaval base in World War I. See: Belleau Wood.

ANNEXED DISTRICT. This was the name for that section of The Bronx, annexed on January 1, 1874, west of the Bronx River, including the villages of Mott Haven, Port Morris, Hunt's Point, West Farms, Highbridgeville, Tremont, Fordham, Woodlawn Heights, Kingsbridge and Spuyten Duyvil.

ANTHONY PLACE. This is a former name of the Grand Concourse, north of Kingsbridge Road, on maps of 1885 and 1889.

ANTHONY STREET. This is an earlier name of Clay Avenue, from East 172nd Street to East 174th Street. See: Crestline Street, Elliott Street.

ANTIN PLACE. This Bronxdale Place, named in 1918, came into existence when the Dyre Avenue Line was built. Before that, it was the back lot of a small tape factory in the Village of Bronxdale. It is named for State Senator Benjamin Antin of the 22nd District. A Mr. Dunn of Fordham asked the Senator how this was done, and Mr. Antin explained he and his associates had built a number of houses in the vicinity and the architects could not find any map or record indicating the name of a street that the last building fronted upon. One architect arbitrarily used the name "Antin Street" and asked the associates to submit the resolution to the Board of Aldermen. It was passed, but the Street was modestly reduced to a Place. See Mercy College Place.

APOLLO, The. This silent movie house at 747 East 180th Street, between Clinton and Prospect Avenues, was listed in an atlas of 1912. It had 433 seats, and went out of business in the late 1930s.

APPLEBY'S ISLAND. This is one of the many names by which Hunter Island was known. This was noted in 1771. See: Blagge's Island, Henderson's Island, Jesse Hunt's Island, Lapp-haw-wach-king, Pell's Island and Pelican Island.

APPLETON AVENUE. This avenue was once part of the Pelham Road when it wound its way from Westchester Square to Middletown Road. Possibly it was named for Henry Cozzins Appleton who owned property there. The avenue was absorbed by the Hutchinson River Parkway.

AQUAHONG. This Siwanoy Indian name for the Bronx River has been translated as "High Bluffs," possibly referring to Bronx Gorge. Tooker's *Amer-indian Names in Westchester County* can be cited. See: Bronck's River (1645), Broonks River (1778), and West Farms Creek (1834).

AQUEANOUNCKE. See: Hutchinson River (Current).

AQUEDUCT AVENUE. Named in February, 1886, it referred to the Croton Aqueduct. In 1913, the name was officially changed to University Avenue, referring to New York University (superseded by the Bronx Community College). See: Lind Avenue.

AQUEDUCT BRIDGE. The original name of High Bridge, begun in 1839 and completed in 1848, it was a favorite walk of Edgar Allan Poe, who was entranced with its view.

"ARCHER MANOR." This was the family homestead of the Archer family before and during the American Revolution, and now the grounds of Bronx Community College. In 1856, William Archer sold the land to the Schwab and Mali families. See: Fort No. 8.

ARCULARIUS PLACE. This is the former name of East 169th Street from Jerome Avenue to the Concourse in West Tremont. Charles P. Arcularius was the owner of the Jerome Park Hotel, listed in the 1871 Directory of Morrisania and Tremont. Charles Place became East 168th Street. See: 7th Street.

ARDEN ESTATES. This was a real estate development around 1902-1904, which was part of the extensive Givan lands, purchased in 1795. Miss Agnes Arden was a daughter of Robert Givan's daughter, Alison, and last owner of record when she died in 1881. The property was bounded by Burke Avenue, Eastchester Road, Boston Road, and Bruner Avenue. The original Givan lands were more extensive.

ARION HALL. A roof-garden restaurant and noted social center of North New York on Alexander Avenue near East 143rd Street. Adam Epple was proprietor from 1895 to 1903.

ARION LIEDERTAFEL HALL. On the west side of Courtlandt Avenue just north of East 154th Street, this was a well-known meeting hall for German singing societies from the 1870s to around 1910. Emil Hass was the manager in the early 1900s. Louis Haffen, first Borough President, was numbered among the members of the Arion Liedertafel (Singing Society) who often sang at the band concerts given in St. Mary's Park during the 1890s. The Hall later became a center of Polish-American activities. It was razed around 1953 to make room for the Melrose Houses project. See: Bronx Central Hall.

ARMY AVENUE. A wide boulevard that was planned to pass near the Jerome Park Reservoir, but was eliminated by Hunter College (present-day Lehman College). See: Navy Avenue, President Avenue.

ARNOLD AVENUE. This is a former name of Appleton Avenue, prior to 1910. Richard Arnold was a property owner in 1890, north of the mill pond behind Westchester Square. See: Appleton Avenue.

ARNOLD STREET. It was surveyed and mapped in 1890 from approximately Leggett Avenue to the East River in the vicinity of the Casanova Railroad station. It is now covered by the railroad tracks. B. G. Arnold once owned this particular tract of land.

ARNOLD'S POINT. This is an earlier name of Oak Point, and a reminder of B. G. Arnold who was a 19th-century landowner there. See: Jeffeard's Neck, Leggett Point, Oak Point, "Ranaque."

ARNOW AVENUE. This is the former name of Arnow Place, where a Mrs. Arnaux [spelled this way] was listed as a homeowner in 1868.

ARNOW PLACE. This was the former name of Poplar Street, from Williamsbridge Road to Blondell Avenue. It was originally part of the Cornell farm that was sold to Andrew Arnow in 1807, remaining in the Arnow family until 1877. During Civil War times, this was a small settlement around the railroad depot. The last remaining tavern on the corner of Blondell Avenue and Poplar Street was demolished in 1958. Arnaux was the original spelling of a French Huguenot family, pioneers of Westchester County.

ARROWHEAD INN, The. A famous Riverdale showplace and resort frequented by high society in the 1920s and 1930s. Broadway luminaries and topnotch orchestras were its attraction, and the location at West 246th Street and Riverdale Avenue gave it a stunning view of the Hudson River and the Palisades. Legend had it that Indian arrowheads had been found by the surveyors, hence the name. In the 1940s the Arrowhead Inn was gutted by fire and Ben Riley, debonair host and gourmet, died in the flames.

ART, The. This small silent movie house was located on Southern Boulevard between East 167th Street and Westchester Avenue. It had 600 seats.

ARTHUR STREET (1). This is an earlier name of East 213th Street, from White Plains Road to Barnes Avenue.

ARTHUR STREET (2). This is a former name of Arthur Avenue, from East 187th Street north to Fordham Road. See: Broad Street, Central Avenue.

ASH AVENUE. This was a 1900 name of Oakley Street in Williamsbridge.

ASH STREET. This is the former name of Bruner Avenue, from approximately Adee Avenue to Boston Road. This was the edge of the Givan farm that was subdivided around 1900 into the Arden Estates. See: Oakes Avenue, Willis Place.

ASH'S CORNER. This former junction of East Tremont and Lawton Avenues led west to the Thomas Ash estate, called "The Elms," according to an 1831 deed and an 1853 map. "The Elms" was a 60-acre tract bounded by the East River, Hollywood, Lawton and Swinton Avenues.

"ASHBURNE." This Clason Point estate was owned by E. Schieffelin in 1865, which had been the Dominick Lynch property of the 1790s.

ASHLEY STREET. East from Broadway at approximately West 228th Street, this short street was eliminated by Marble Hill Houses. See: Hyatt Street (2).

ASHY BOTTOM. According to Ken Raniere, the name applied to the area around Blondell Avenue, and Cooper Street, behind Westchester Square. The landfill used, in the early 1900s, to encompass the swamps of Westchester Creek was mainly ashes.

ASTOR PLACE. This short street on Locust Point was eliminated in 1958 by the Throgs Neck Bridge approach.

ASUMSOWIS. This was the site of a considerable Indian settlement, now known as Rodman's Neck. See: Anne's Hoeck.

ATHENEUM, The. This theatre was once located on Washington Avenue and East 167th Street. James S. Parshall, proprietor, was listed in the 1869 *Directory of Morrisania*.

AUGUSTA PLACE. See: Calhoun Avenue. (Current).

AURORA PARK. See: Ebling's Brewery.

AVALANCHE HOSE CO. NO. 7. Woodlawn Village had its volunteer firemen, but the exact location of their firehouse cannot be ascertained.

AVENUE A (1). This was the former name of Eagle Avenue from East 156th Street to East 163rd Street in the settlement of Bensonia, 1868. See: Westray Street.

AVENUE A (2). This was the earlier name of Morris Avenue, from East 181st Street up to Fordham Road. See: 2nd Avenue and Kirkside Street.

AVENUE A (3). Zerega Avenue went by this name, before 1900, from Westchester Avenue to the Unionport dock. See: Green Lane.

AVENUE A (4). This is the name first given to East 214th Street from White Plains Road to Barnes Avenue. See: Shiel Street.

AVENUE B (1). This is the name by which Cauldwell Avenue from East 156th Street to East 161st Street was once known when it constituted the margin of Aurora Park, a noted ballpark and picnic grounds of the 1870s. See: Park Street, Terrace Street.

AVENUE B (2). When Unionport was laid out, the avenues were lettered and the streets were numbered. This was Havemeyer Avenue. An even earlier name was Lowerre's Lane, as it led to Lowerre's farm.

AVENUE C (1). This is the former name of Trinity Avenue in 1889, from East 158th Street to East 161st Street. See: Cypress Avenue, Delmonico Street, Grove Avenue.

AVENUE C (2). In the village of Tremont, in 1886, this was the name of Anthony Avenue. See: Prospect Avenue.

AVENUE C (3). This was an earlier name of Castle Hill Avenue from Westchester Avenue to approximately Turnbull Avenue. See: Lafayette Avenue. (Current).

AVENUE D. This is the former name of Olmstead Avenue in the village of Unionport, i.e. from Westchester Avenue to Lafayette Avenue. See: Sand Street.

AVENUE E. Prior to 1905, this was the name of Pugsley Avenue in the village of Unionport.

AVENUE ST. JOHN. This is the former name of Hughes Avenue, and was so called as it led to St. John's College (now Fordham University). In 1895, the name was changed to Hughes Avenue to avoid confusion with Avenue St. John in East Morrisania. Reverend John Hughes was the priest who was instrumental in having the Catholic college located there. See: Frederick Street, St. John Avenue.

AVIATION FIELD (1). This was a local nickname for the former Morris Park racetrack after it had given up racing and had been changed to a landing field for airplanes and balloons. The world's first aviation meet was held there on November 3, 1908.

AVIATION FIELD (2). This large meadow, bounded by Newman, Bolton and Gildersleeve Avenues and White Plains Road on Clason Point, carried this nickname from World War I days. Neighborhood lore has it that the tract was used for balloon ascents and airplane flights by flying enthusiasts. Another tale is that a promoter solicited money from the Clason Pointers to build an airport there, but then absconded.

"AVYLON." This is the name of the former estate of the Morris family that extended along the East River from Ferry Point Park to Balcom Avenue. In the 1820s, it was the Bayard Clark homestead, which was sold to Francis Morris in the 1850s. It passed to his son, John A., grandson Alfred Hennen Morris and to a great-grandson John A. Morris. The mansion and grounds were sold in 1922. The colonnaded white mansion was demolished in the 1930s when Schurz Avenue was cut through.

B-B, The. Herman and John Bolte opened a small silent movie house on the west side of White Plains Avenue near East 221st Street in August, 1913. They had an awning and shades manufacturing shop known as Bolte Brothers, so they used these initials, and the theatre was called the B.B. It had a seating capacity of 599, and the neighborhood children irreverently called it "the Bed Bug." The little B.B. closed down in 1930, and its interior was cleared out to form part of a machine shop.

BABCOCK AVENUE. This is the former name of Netherland Avenue from West 252nd Street to West 256th Street, when it led through the 1897 estate of S.D. Babcock.

BABY PARK. See Melrose Park (Current).

BACK ROAD, The. This is a very early reference (1700s) to West Farms Road. See: "Hive Town Road," the Lower Road, the Queens Road.

BACK ROAD. This is the former name of Spuyten Duyvil Road to the depot.

BACON STREET. The original Hunts Point Road of Colonial days crossed Bacon Street at East Bay Avenue, running east to west. Although a Frank L. Bacon is listed as a resident in 1904, it is more likely that Francis Bacon, the poet and author, was honored, along with the other literary figures in the neighborhood such as Drake, Longfellow and Whittier. It was obliterated by the Hunts Point Market around 1970.

BAGLEY STREET. This short street on the blueprints for many years and laid out on former Ferris property was named for Lieutenant Bagley of Spanish-American War fame, according to the *New York Times* of December 8, 1918. The entrance to the Bronx-Whitestone Bridge cut off the street, and it was discontinued by the Board of Estimate in September, 1964.

BAILER'S HOTEL. This was an old-time tavern in the village of Unionport. An 1897 ad locates it at Avenue C and Ludlow Avenue, which, in present-day directions, would be Castle Hill Avenue and Bruckner Boulevard. It featured bicycle racks outside the barroom for thirsty cyclists.

BAIN'S CREEK. An 1804 deed lists this watercourse as bounding the property of Nicholas Haight in the vicinity of Zerega Avenue. See: Bame's Creek (1824), Barn's Creek (1800), Indian Brook, Little Creek (1727), Seabrey Creek (1850), Seberry Creek, and Zeabrey Creek (1868).

BAINBRIDGE STREET. This was the former name of Marion Avenue from East 184th Street to East 189th Street. See: Hull Street, Virginia Avenue.

BALCOM AVENUE. An 1858 map of Bensonia (Rae and Carr Streets) shows this as the original name of Hegney Place. See: German Place.

BALL'S POND. This was a small body of water near the mouth of Spuyten Duyvil Creek which had been created when the railroad tracks were laid. In Wintertime, it became a popular ice-skating rink for the youngsters (circa 1890). The Ball family lived nearby.

See: *Schools and School Days in Riverdale, Kingsbridge, Spuyten Duyvil* by Dr. William Tieck.

BALLSTON AVENUE. 1905 maps carry this street, but it was never cut through. A real estate development known as Westchester Heights, stressed names of famous vacation spots, such as Newport, Saratoga and Narragansett. Ballston Spa was an upstate fashion center. The hospital grounds of Jacobi Hospital cover the projected stretch of this avenue. See: Tuxedo Street, Lenox Place.

BAME'S CREEK. An 1824 deed lists this as flowing near the Friends' Meeting House, near today's Zerega and Westchester Avenues. See: Bain's Creek (1804).

BANCROFT STREET. This former name of East 165th Street from Westchester Avenue to Whitlock Avenue is believed to honor Hubert Bancroft, a historian. The name was changed by Municipal Ordinance, 1911. See: Guttenberg Street, Third Street, Wall Street.

BANDBOX, The. This silent movie house at 37 West Fordham Road, from the 1920s to the 1930s, had a seating capacity of 600.

BANTA LANE. Banta is an old City Island name, and this is a short lane that ran at right angles to Fordham Place.

BARBOUR STREET. This short street led off the early Indian path that eventually became Eastchester Road. Westchester County records carry the name of a James Barbour, active in politics, but do not list his residence. The dates are 1857 and 1860. In the early 1960s, this street was eliminated by the New York State Mental Hospital grounds.

BARETTO STREET. This is an old-time name of Fox Street.

BARLOW STREET. This is a former name of Poplar Street. See: Arnow Place.

BARN'S CREEK. This watercourse is described in an 1800 deed giving the bounds of St. Peter's Episcopal church. See: Bain's Creek.

BARNARD STREET. This street off Eastchester Road was named after a Quaker family that lived on that site around 1800. The Cornell brothers' mill was on Westchester Creek, behind Barnard Street on the area now covered by the State Hospital. Eunice Barnard married one of the brothers, who were also Quakers, and she became the mother of Ezra Cornell (1807), who later founded the upstate university. A later generation Horace Bernard married Louisa diZerega of the prominent Bronx shipping family. Today, the street is covered by the southern tip of the State Hospital grounds.

BARNETT PLACE. A narrow alley that ran from White Plains Road to Matthews Avenue, about midway between Morris Park Avenue and Rhinelander Avenue on a 1905 map.

BARNEY AVENUE. This was an earlier name of Tibbett Avenue, from West 246th Street to West 253rd Street. A 1905 Bronx land book lists Hiram Barney as a property owner in Mosholu. He was Collector of the Port of New York appointed by President Lincoln, whom he knew personally. Squire Barney (1811-1895) was an executive of an Anti-Slavery Society and a Temperance League.

BARNEY'S LANE. 1894 police blotters of Kingsbridge Precinct note the name in several instances, and refer to present-day Tibbett Avenue. Hiram Barney was also a friend of Edgar Allan Poe when the poet resided in Fordham. He is buried in Woodlawn Cemetery.

BARNUM PLACE. Jeremiah Barnum was a registered voter of the 23rd Ward in 1877, his address being Boston Road. Barnum Place ran west from present-day East 169th Street, crossing Tinton Avenue.

BARRETT'S CREEK. One of the original settlers of Clason Point was Samuel Barrett who came into the region in March 1656 from Connecticut. He married Hannah Betts, and his father-in-law deeded him 6 acres of meadow, and 20 more acres to John Barrett, his grandson. Barrett's Creek ran through Ludlow's "Black Rock farm"as a tidal wash that formed a swamp at the Westchester Turnpike. The creek followed Metcalf Avenue diagonally across to Seward Avenue. See: Ludlow Creek.

BARRICK, The. This is an odd-shaped projection from the Bronx shore into Spuyten Duyvil Creek in pre-Civil War days. The point of land was cut through by the Harlem River ship canal in 1895. See: United States Ship Canal.

BARROW CREEK. It was noted on U.S. Geodetic charts of the Hutchinson River and depicted as a small tidal creek partially bisecting Rose Island (now the southern end of Co-Op City) opposite Goose Island. In 1963, it was blotted out by landfill operations. One meaning of a barrow is a mound, or hummock.

BARRY STREET. A proposed name for Albany Crescent at West 231st Street, during a 1922 real estate development, but not adopted. See: Gallagher Avenue, Martin Terrace.

BARTOE'S CREEK. This name was mistakenly ascribed to Rattlesnake Creek on a British Headquarters map of 1777, possibly because the Bartows owned the land in that Baychester area from 1720 to 1795.

BARTOW. This village along the Shore Road, north of the Pelham bridge, from the 1820s to about 1900, was the terminus of the monorail from City Island in 1910. The town hall of Bartow remained until 1956, some 50 yards south of present-day Bartow Circle in Pelham Bay Park. See: Pelham Park & City Island Railroad.

BARTOW PARK. A 1915 map shows a Bartow Park west of City Island Avenue from Buckley to Pell Streets and fronting on Eastchester Bay. It was a real estate development.

BARTOW PLACE, The. This was a former estate of the Pell family on the mainland behind Hunter Island. It was purchased by John Bartow, a son-in-law of Joseph Pell II, in March of 1780. The Bartow-Pell mansion was built between 1836 and 1842.

BASEMENT CHURCH, The. This church at 2547 East Tremont Avenue is the oldest of the Methodist denomination in The Bronx, having been chartered in 1803. (Some annals list 1808). From 1913 to 1948, the congregation worshipped in the basement until the present (third) church was built. See: Eelpot Church.

BASS ROCK. This rock in Eastchester Bay is mentioned in an 1835 deed as a boundary marker of William Bayard's farm, now part of Pelham Bay Park. In earlier deeds it is referred to as Bess Rock. It lies at the foot of Watt Avenue.

BASSFORD PLACE. Abraham Bassford was listed as owner on 1872 maps. It is the former name of Ryer Avenue from Burnside Avenue to East 180th Street.

BASSFORD STREET. This was an earlier name of East 186th Street from Park to Washington Avenue. Once formed part of the Bassford property of the 1840s.

BATES STREET. Robert Bates was one of the original settlers of Westchester in 1668. Like Pell, Archer, Waters and others, he came from New England, specifically Stamford. His descendants included John S. Bates, District Attorney of Westchester County under the Act of 1818, and N. S. Bates, who was Clerk of Deeds, *circa* 1800. The street was eliminated by the State Hospital grounds in the early 1960s.

BATHGATE ESTATE. This 133-acre tract of the early 1900s was bounded by Grace Avenue, Barnes Avenue, Bissell Avenue to Mundy Lane, to Bussing Avenue.

BATHGATE FARM (1). This Morrisania tract was inherited by the three sons of Alexander Bathgate, who had been overseer of the Morris Manorlands. The farm was approximately 140 acres, and the farmhouse was located off Third Avenue and East 172nd Street. The lands were sold to the City around 1883, and became Crotona Park.

BATHGATE FARM (2). This farm was owned by James Bathgate, brother of Alexander, the overseer of the Morris Manorlands. In 1866, it was sold to the Jerome Park Racing Association and made into a race track. Today, the track is covered by the Jerome Park Reservoir, Harris Park and Lehman College.

BATHGATE FARM (3). See: Bathgate Estate.

BATHGATE OVAL. This baseball and football field was laid out across the Bathgate Estate off Bussing Avenue where Mount St. Michael School is now situated. It was in use during the 1920s.

BATHGATE PARK. This was the original name of Crotona Park. As it had been Bathgate property, this name was decided upon, but, following an argument between the surveyors and the family, Crotona Park was decided upon.

BATHGATE PLACE. See: Claremont Parkway.

BATTERY HILL. This is the Revolutionary name and early 1800 designation for the bluff overlooking the Harlem River on the site of Fort No. 8. The Hall of Fame covers this site.

BAXTER CREEK. Thomas Baxter was the original settler in the Throggs Neck region when it belonged to the Dutch in the mid-1600s. His home is believed to have been located on Lafayette Avenue about Quincy Avenue facing out over what had been marshlands and Baxter Creek. The tidal creek was famous for its fabulous yield of eels and blueclaw crabs, and wound its way inland past present-day St. Joseph's Home for the Deaf. Landfill for Ferry Point Park blotted out the creek around 1950.

BAYARD FARM (1). This 19th-century farm is now part of Pelham Bay Park, the last owner of record being John Hunter III. It ran from today's Bruckner Boulevard to Eastchester Bay, north of Middletown Road.

BAYARD FARM (2). This farm covered approximately the area between Third Avenue, East 188th Street, Fordham Road and Lorillard Place. William Bayard was listed as the owner in 1834. His widow married the Reverend William Powell, and the farm became known as the Cedar Hill farm. In Civil War times, the property passed to Delancey Powell. See: Cedar Hill farm.

BAYARD STREET (1). East 188th Street's former name, from Third Avenue to Beaumont Avenue. The Bayard farm was here, extending to Fordham Road in 1800. In 1840, it was inherited by Reverend Powell who married Bayard's widow. See: Walsh Street.

BAYARD STREET (2). This was the original name of Ely Avenue, north of Nereid Avenue. See: Doon Avenue, Hazel Street, South 21st Avenue.

BAYARD'S BROOK. It is mentioned in an 1837 deed of West Farms "next to the salt marshes belonging to Richard Hunt." See: Bayard Farm (2). *Robinson's Atlas* of 1888 shows a small stream running from the Bayard farm to the Bronx River.

BEACH AVENUE. This earlier name of Tinton Avenue is listed in the 1871 Directory as part of the Dater estate of the 23rd Ward, now partially covered by the Adams Houses.

BEACH STREET. This is the former name of West 167th Street from Ogden Avenue down to the Harlem River. W. B. Beach was the property owner, listed in 1884. See: Wolf Street (1876), Woolfe Street (1840s), Union Street (1890).

BEACH, The. This localism for Silver Beach Gardens is used on Throggs Neck. The shortening came into usage around 1920 when the bungalow colony came into existence out of the former Havemeyer estate.

BEACON AVENUE. This is the former name of East 174th Street adjacent to the Bronx River Houses. See: Spring Street, Twelfth Street.

BEACONSFIELD INN. This was a large hotel in South Fordham, on the west side of Jerome Avenue, between East 184th Street and Fordham Road. Its grounds extended westward to approximately Grand Avenue, according to an 1893 map.

BEAR SWAMP. It is mentioned in ancient deeds, and referred to a vast swamp that extended from Wallace Avenue on the west to Fowler Avenue on the east, from Bronxdale Avenue on the south to Lydig Avenue on the north.

BEAR SWAMP ROAD (1). This is the former name of East Tremont Avenue from Castle Hill Avenue to Williamsbridge Road, according to Cornell records of 1804.

BEAR SWAMP ROAD (2). This is the Colonial name for Bronxdale Avenue, leading from the Bronx River to Westchester Square, past the infamous Bear Swamp.

BEAVER SWAMP ROAD. This was a road through the former Lorillard estate, now the Botanical Garden, leading west from the Bronx River, above Lorillard Falls.

BECK STREET (1). This is the former name of East 151st Street from Jackson Avenue to Cauldwell Avenue, past Aurora Park. Anna Beck sold the property. See: Pontiac Street.

BECK STREET (2). See: East 156th Street.

BECK'S HOTEL. Originally a Temperance Hall built in 1872, the hotel featured a saloon, and catered to travellers en route to Albany. It was located on Broadway in Warnerville (West 240th Street) and was destroyed by fire in July, 1913. See: Tremper's Pond.

BECKER AVENUE. In the 1880s, it was East 241st Street from Bronx Park to White Plains Road. See: Hyatt Street.

BECKER STREET. This is the former name of Willow Lane, from Westchester Avenue to Parkview Place. The name was changed in 1927.

BEDFORD, The. A 1912 atlas lists this silent movie house on Webster Avenue between Oliver Street and East 199th Street. It went out of business in 1920.

BEEBE AVENUE. This is the former name of Orloff Avenue.

BEECH AVENUE. This is the original name of Needham Avenue.

BEECH STREET (1). This is the former name of Bolton Avenue on Clason Point.

BEECH STREET (2). This was once located off Steenwick Avenue, but it is now covered by the New England Thruway.

BEECH STREET (3). This is the 1899 name of West 260th Street, from Riverdale Avenue to Liebig Avenue.

BELDEN POINT AVENUE. This was the former name of Belden Street on City Island. William Belden was the owner of a waterfront resort. See: Windmill Street.

BELLE PAREE, The. This small "hole-in-the-wall" silent movie house was located on East 149th Street, west of Bergen Avenue. It was in existence from approximately 1920 to sometime around 1930. "Fat" Jacobi was the bouncer.

BELLEAU WOOD. This name gained nationwide prominence in World War I, representing a battlefield in France where American troops fought and died. To inspire recruitment, the War Department gave this name to a Naval Hospital Base on the shores of Eastchester Bay, partly on Rodman's Neck. In the years 1916-1919, it contained a fully equipped hospital and ambulance unit, a gym and parade grounds, barracks and even a "brig" (prison). See: "Anneswood," Camp Mulrooney .

BELLEVUE PARK. This small picnic park was mapped in 1870 on the east bank of the Bronx River below Gun Hill Road, at the foot of Magenta Street (then called Julianna Street). The official opening took place in May, 1874, and the proprietor was August Reidinger. Today, his dance hall and restaurant would be in the middle of the Bronx River Parkway.

BELMONT PLACE. This is the former name of East 184th Street in the village of Belmont. Its name was changed in 1886 to conform to the grid-pattern of numbered streets.

BELMONT STREET (1). This was an alternate name for Featherbed Lane, noted on 1884 maps. Both names were listed in parenthesis.

BELMONT STREET (2). This was the earlier name of Mt. Eden Parkway from Morris Avenue to Webster Avenue. See: Jane Street (1), Walnut Street (1).

"BELMONT." This former estate of Jacob Lorillard was located to the east of Third Avenue at approximately East 182nd Street. In the 1880s, it was subdivided into building lots, and the locality given the name of Belmont.

BENDER STREET. Listed on an 1883 map, this street in West Tremont was eliminated by a bend of Walton Avenue.

BENENSON, The. See: Fenway.This silent movie house, listed in 1922, was located at 1546 Washington Avenue, near Claremont Parkway, where 1,312 patrons could be accommodated. The theatre was in operation from 1921 to 1932. Benjamin Benenson was a real estate developer of the early 1900s.

BENSON AVENUE. An 1851 survey of Bensonia shows this name affixed to what is today St. Ann's Avenue from Westchester Avenue north to East 160th Street. Benjamin Benson was a landowner of the times. See: Fordham Road, St. Ann's Avenue.

BENSON STREET. This is the original name of East 149th Street in Melrose. Benjamin Benson purchased the land from Gouverneur Morris II in 1853.

BENSONIA. This tract was sold to Benjamin Benson, and administered by the Trustees of Morrisania. It was bounded by East 149th Street, Brook Avenue, East 156th Street, and Eagle Avenue.

BENSONIA CEMETERY. The land was purchased in 1853 by Robert H. Elton, and extended from Rae to Carr Streets east of the Mill Brook. It was once a picturesque plot, densely shaded by elms, poplars and evergreens. The Sons of Liberty interred 150 of its members, but after 1868, no more burials were permitted. When St. Ann's Avenue was extended, it cut the cemetery in two. Around 1890, what remains could be reinterred were moved to Woodlawn Cemetery (and Green-Wood Cemetery in Brooklyn, some say) and a public school was built upon the site.

BERKELEY OVAL. This was a large, clear area with an oval track on University Heights just south of P.S. 26 on Burnside Avenue. It was demolished when Andrews Avenue was cut through.

BERRIAN AVENUE (1). This former name of Webster Avenue above East 180th Street to Gun Hill Road, was called after an early family of landowners. See: Bronx River Road, Thomas Street, Worth Avenue.

BERRIAN AVENUE (2). This was the earlier name of Netherlands Avenue, when the region was known as Berrian's Neck, and sometimes, Spuyten Duyvil.

BERRIAN'S LANDING. This was the 18th-century name for Fordham Landing. The Berrians were a pre-Revolutionary family with extensive farms in various sections of The Bronx.

BERRIAN'S NECK. This was the 17th-century name of Spuyten Duyvil. See: Konstabelsche Hoek, Shorarakopkock, Tippett's Hill, Tippett's Neck.

BERRIAN-BASHFORD BURIAL GROUNDS. See: Kingsbridge Burial Grounds.

BERRY STREET. See: East 179th Street.

BESS ROCK. This rock in Eastchester Bay is mentioned in an 1817 and 1823 deed, describing it as a boundary marker of William Bayard's farm (now incorporated in Pelham Bay Park). Bess might refer to Elizabeth Bayard, his wife, or to Elizabeth (Bayard) Campbell, his daughter. See: Bass Rock (1835).

BETTNER'S LANE. This was the former name of Palisade Avenue along the Hudson River, where it ran through the Bettner estate.

BICKLEY PATENT. This was an 18th-century name for the high land near High Bridge. See: Dangerville, Devoe's Point, Highbridgeville, Nuasin, Turneur's Land.

BICKNELL'S LANE. Although noted in an 1877 firehouse report, the name is not found on Riverdale maps. The Bicknell mansion stood at West 253rd Street and Post Avenue.

BIGGART'S CANAL. In the 1880s, just south of High Bridge, this short channel extended in from the Harlem River to accommodate coal barges. Mr. Biggart, the owner of the coalyard, lived at Summit Avenue and West 165th Street. See: Crab Island.

BILIKIN, The. This small silent movie house, located in the vicinity of Southern Boulevard and East 167th Street, was in operation around 1912.

BILLAR PLACE. This was a small footpath on City Island that once ran from Fordham Street to Carroll Street, at approximately Minneford Avenue. W. and S. Billar ran the post office on City Island in the 1860s.

BILTMORE DRIVE. This was an extension of Pawnee Place that is now covered by Van Etten Hospital.

BIRCH STREET (1). This is one of the former names of West 168th Street from High Bridge down to Jerome Avenue. See: Orchard Street.

BIRCH STREET (2). This was a 19th-century name of Shakespeare Avenue, from West 168th Street to Jerome Avenue. See: Judge Smith's Hill, Marcher Avenue.

BIRCH STREET (3). This former name of Mickle Avenue, from Givan Avenue to Boston Road evidently was named for the trees so common in the area. See: Westervelt Avenue.

BISHOP AVENUE. This short street in the settlement of Bartow, facing the town hall, ran parallel to the railroad tracks, west of Bartow Circle in Pelham Bay Park, but all traces of the hamlet are gone. The avenue was named for Theodore Bishop, a Civil War sailor who had served aboard the U.S.S. San Jacinto.

BLACK DOG BROOK. This is the ancient water boundary between Eastchester and Westchester Townships that is mentioned in 1664, 1682 and 1689. The English settlers evidently brought the name with them from Suffolk, England, where the legend of the Black Dog is well-known. Bungay in Suffolk is an ancient market town and the coat-of-arms features as its crest "the notorious black dog." The motif is also to be seen on weather vanes in the town. On a Sunday night in August, 1577, the Black Dog "or the Divel in such a likeness" appeared in the church. Black Dog Brook originated in a cluster of springs west of Baychester Avenue and flowed southeastward to join the Hutchinson River. In the 1930s, it was cut off at Boston Road, near Ely Avenue. In the following years it was filled in entirely.

"BLACK ROCK FARM." This estate was owned by R. H. Ludlow, but is now covered by the Soundview Houses. The land was traversed by Barrett's Creek and included Ludlow's Island in Bronx River. Later, the tract was sold to the Schieffelin family who held it to the 1880s. See: Ludlow's Creek.

BLACK STREET. This Castle Hill street, laid out, but never cut through, ran from Zerega Avenue to Westchester Creek, between Virgil and Homer Places. See: Coppee Street.

"BLACK ROCK." This great boulder embedded in the salt marshes near the junction of Ludlow's Creek and the Bronx River was thought by the early inhabitants of Clason Point to be a meteorite. Thomas Cornell, first European settler in 1643, used this Black Rock in platting his holdings. The boulder is now inside Soundview Park.

BLACK SWAMP, The. East of the Zborowski lands (now Claremont Park) was an infamous Black Swamp, where cattle had been lost since the time of the Indians. For years, it defied the efforts of contractors to fill it up. Morris Avenue now crosses it.

BLAGGE'S ISLAND. This is one of the names used to describe Hunter Island. It was so mapped in 1791, when John Blagge sold it to Alexander Henderson. See: Appleby's Island.

BLAND AVENUE. This is the former name of Harper Avenue, from East 233rd Street to Light Street.

BLEACH, The. This is a localism for the bleach mill established on the Bronx River in 1818 by James Bolton of England. This enterprise gave impetus to the growth of the village of Bronxdale. It was sometimes referred to as Bolton's Bleach. The site of this mill and the dam are now part of Bronx Park at Lake Agassiz.

BLEEKER STREET. This was an earlier name of Allerton Avenue from Bronx Park to Cruger Avenue.

BLENHEIM, The. Built in the late 1920s, at 466 East 169th Street between Park and Washington Avenues, this silent movie house had a seating capacity of 1,847. It occupied the site of an earlier house called the Park Theatre, which is known to have been in operation in 1916. The Blenheim was razed in the 1940s to make way for a housing project.

BLIZZARD ISLAND. This former island in the mouth of the Hutchinson River was accessible at low tide. It once supported the eastern arch of the *original* Pelham bridge. 1868 Beers' *Atlas* lists David Blizzard as the proprietor of a small hotel on Tallapoosa Point who rented out fishing tackle and rowboats.

"BLOOD ALLEY." This was a nickname for Heath Avenue at Albany Crescent. This one-block lane received its nickname from a slaughterhouse in the vicinity of Summit Place and Kingsbridge Terrace. See: Darke Street.

BLUE BRIDGE LAKE. This is a lake formed by the Bronx River in the present-day Botanical Garden in the vicinity of the Allerton Avenue roadway. A wooden bridge, painted blue, spanned the lake in early 1900s. See: Twin Lakes.

BLUFF, The. This was an elevation of land on the Jerome Park Racetrack (1865-1894) on which the clubhouse was built. Randall Comfort, the historian, referred to the immense rock overlooking Jerome Park Reservoir, which was built on the former racetrack.

BLYTHE AVENUE. This was the earlier name of Tiffany Street from Spofford Avenue to Oak Point Avenue, evidently to save the name of Francis J. Barrette's estate, "Blythe."

"BLYTHE." This was the former estate on Hunts Point owned by Francis J. Barretto. The mansion was of Revolutionary times and, when its inside shutters were closed, it was a miniature fortress.

BOARDWALK, The. Prior to World War I, Downing's Brook, which meandered through the Van Nest area, was bridged over by a wooden boadwalk. In a 1914 souvenir journal of St. Martha's P.E. Chapel, the address of the manse was Cruger Avenue and the Boardwalk.

BOCKET'S COT. This was mentioned in an 1804 deed as Bocket's cot or landing place on Barretto Point. It is supposed the word "cot" meant "cove." In another deed of the same year it was Boescity's Cot. An 1805 indenture mentions Bockett's Cot, and one in 1806 speaks of Bocketz's Cot. This last one was signed by James Graham whose lands were located between present-day Port Morris and Hunts Point.

BOHEMIAN PARK. The Bohemian Hall and Park was located on the southeast corner of East 166th Street and Park Avenue. Among other groups, the Schweizer Turnverein (Swiss Gymnast Club) celebrated their festivals there from 1900 to 1911.

BOLEN GREEN. This was the town square of Schuylerville, northeast of the intersection of Bruckner Boulevard and East Tremont Avenue. It was once known as Schuylerville Park. John Bolen purchased some lots from Sebastian Meyers in 1843, and the square acquired the nickname of "Bolen Green." A later generation called it Bowling Green.

BOONE STREET (1). The first designation of Boone Avenue. The change occurred on March 14, 1904.

BOONE STREET (2). This was the former name of Longfellow Avenue from Westchester Avenue to East 165th Street across what was once the Hoe estate. See: Division Street.

BOOTH MOVIE, The. This was a silent movie house on Boston Road, between East 180th Street and East 181st Street, in operation before World War I. William H. Booth's name appears on West Farms records of the 1890s and 1900s, when part of his land was acquired in the widening of Boston Road.

BORO, The. Originally it was The Miracle. The Boro Theatre was a silent movie house on Melrose Avenue, east side, between East 156th and East 157th Streets. It ran from around 1911 to 1936, seating 559 patrons. The building was then converted to a post office near East 156th Street.

BOSCOBEL AVENUE. See: Grant (Edward L.) Highway.

BOSCOBEL ESTATE. See: Villa Boscobel, Boscobel Place.

BOSTON AVENUE. This was the 19th-century name of Albany Crescent.

BOSTON HILL. This was a local name for Albany Crescent which was part of the original Boston Post Road. It was in common usage around 1810 and survived into Civil War days. See: "Hut Hill."

BOSTON POST ROAD. (1.) See: Coles' Boston Road.

BOSTON ROAD. This was the earliest name for Third Avenue from East 129th Street to East 164th Street.

BOSTON TERRACE. This was a 19th-century name of Albany Crescent. See: Boston Hill, "Hut Hill."

BOUND BROOK. This is one of the names of Sacrahong Creek given to it by the West Farms disputants over the Debatable Lands. They claimed it marked the bounds of their territory, but this was contested by the Morris family. See: Bungay Brook, Sacrahong Street.

BOUNDARY CREEK. This creek that wound in from Morris Cove on the East River through the marshlands is now covered by the Throggs Neck Houses. It once flowed in the vicinity of Sampson Avenue, and around 1905 to 1920 was a favorite haunt of numerous Finns and Norwegians of the neighborhood.

BOWLING GREEN. See: Bolen Green.

BOWNE ROAD. This is the earlier name of Whittemore Avenue as it ran past the Bowne farm on present-day East Tremont Avenue next to St. Raymond's Cemetery. In the 1800s, this road led from Westchester village to Old Ferry Point (Ferry Point Park).

BOWNE STREET. This is the 19th-century name of Ferris Place off Westchester Square. Sidney Bowne was once the most important merchant in the village. See: Dock Street.

BRACKEN AVENUE. In the 1890s this avenue was laid out across the Edenwald sector and named for Henry Bracken, a prominent member of the Tammany Society of the Annexed Districts. It is now Edson Avenue. See: South 19th Avenue, Sycamore Street.

BRADFORD AVENUE. This is the earlier name of Plymouth Avenue. The change of name took place on May 2nd, 1916. See: Waldo Place.

BRANCH RAILROAD, The. When the towns of Wilton, Port Morris and East Morrisania were surveyed in 1851 by Andrew Findlay, this was the designation given to the railroad line running from Port Morris north through Morrisania. See: Old Pokey, Pocahontas Line, Spuyten Duyvil & Port Morris R. R.

BRAUN'S HALL. It was situated on the north side of East 143rd Street near Third Avenue in the 1870s. Henry Braun was a manufacturer of sashes and blinds, and rented out his adjoining Hall for weddings and theatricals. It was used by the Pocahontas Lodge No. 234 of the Temperance League-a name chosen to honor Gouverneur Morris' mother who claimed descent from Pocahontas.

BREAK NECK HILL. This is a Revolutionary War name for steep West 230th Street from Sedgwick Avenue down to Broadway. See: Tetard's Hill.

BREINLINGER'S OLD POINT COMFORT PARK. This resort and picnic park on Boston Road near Dyre Avenue, was popular for clambakes, dances, weddings and athletic events. Hollers Pond formed its western margin. Kilian Breinlinger owned the 4-acre Park from 1926 to 1957. The frame hotel, demolished in 1960, was over 125 years old-and was believed to be built over a Revolutionary inn. See: Dickert's, Old Point Comfort Park, Odell's Tavern.

BREMER AVENUE. This is the former name of Woodycrest Avenue from West 161st Street to about West 168th Street. The name appears on maps from 1851 to 1880, with an 1873 showing a Bremer estate. "Woody Crest" was a name in use when the land was farmed by the Anderson family in the 1790s.

BRENDAN HILL. This slope between Mosholu Parkway, Gun Hill Road, Jerome Avenue and Webster Avenue, was so named on May 3rd, 1910, by the Board of Aldermen. The tract was formerly known as North Bedford Park, Norwood Heights and other local names.

BRIDGE HILL. This is the site of a Revolutionary War cannonading of Westchester village in October, 1777, when British gun batteries were dragged to its summit. The summit is located at Dudley Avenue and Harrington Avenue. See: Crow Hill.

BRIDGE STREET (1). The former name of Wilgus Street, now covered by Parkside Houses.

BRIDGE STREET (2). This was a gravel path that ran to Pelham Bridge from the Huntington estate that is now part of Pelham Bay Park.

BRIGGS AVENUE. This was an early name of Gun Hill Road from Williamsbridge Road to Eastchester Road in use around 1895.

BRIGHTON AVENUE. The 1905 maps carry the name of this street in the Westchester Heights realty development, but it never was cut through. It is covered by the Van Etten Hospital grounds.

"BRIGHTSIDE." This was the name of Colonel Richard Hoe's estate on Hunts Point Road.

BRINSMADE ESTATE. The N. F. Brinsmade estate in 1868 was measured at 44 acres. Its approximate bounds were Fort Schuyler Road (now East Tremont Avenue) from Philip to Barkley Avenues, and extending west to Baxter Creek (now part of St. Raymond's new cemetery). The mansion later was sold to the Swedish Lutheran Society as a home for their aged.

BROAD STREET (1). The former name of Crotona Avenue. See: Broadway, Franklin Avenue, Lafayette Street.

BROAD STREET (2). This was an earlier name of Arthur Avenue, from East 175th Street to East Tremont Avenue. See: Arthur Street.

BROADWAY (1). This was the former name of Crosby Avenue, from Middletown Road to Westchester Avenue. See: Schuyler Street.

BROADWAY (2). This was another designation of Crotona Avenue from East 180th Street to Bronx Park. See: Broad Street, Lafayette Street, Franklin Avenue.

BROCKETT'S POINT. This was the ancient name of a tract along Westchester Creek, downstream from the Unionport bridge. In 1667, it was owned by a colonist named Brockett. Later, it was incorporated into the "Grove Farm" of Josiah Hunt in 1694, which passed to Josiah Cousten in 1760, and to John Ferris in 1775. See: Grove Farm, Grove 'Siah, Ferris Point, Spicer's Neck.

BROMMER'S PARK. This was an amusement park of the 1880s at the foot of Willis Avenue that was well patronized by residents of Old Morrisania, Wilton, Mott Haven and Harlem. See: Christ's Park, Harvey's Hill, "Scratch Park. "

BROMMER'S UNION PARK. Alois Brommer was the proprietor of this picnic park and amusement center located at East 147th Street and Southern Boulevard. An 1889 advertisement featured an Ox Roast & Pic Nic of the Columbian Club of the Annexed District. See: Union Park.

BRONCK'S RIVER. This is the earliest European name to be attached to this important stream, known to the Indians as Aqua-Hung, or Hocque-Unk. It was the boundary between the Weckquasgeeck Indians west of the river, and the Siwanoy Indians east of it. The stream was mapped as Broonks River in 1778, and West Farms Creek in 1834.

BRONCKS STREET. This is the former name of East 140th Street. Jonas Bronck bought the land from the Weckguasgeeck Indians in 1639, and his widow sold it to Arent Van Corlear in 1644.

BRONX CENTRAL HALL. It was situated on the west side of Courtlandt Avenue between East 154th Street and East 155th Street. It originated as a social hall for German societies in the 1870s and continued into the 20th century. During the 1920s, it was called the Bronx Central Hall, and was demolished to make way for Melrose Houses. See: Arion Liedertafel Hall, Central Hall.

BRONX CLUB, The. The clubhouse of this prominent social organization stood at 1261 Franklin Avenue in the early 1900s. William Ebling, a leading brewer, was president.

BRONX GARMENT CENTER, The. This was a huge factory located atop a rocky bluff overlooking Cauldwell and Westchester Avenues. It had been the Ursuline Convent (1854-1888) and Lebanon Hospital (1892-1943). In 1956, the block-long building was demolished to make way for St. Mary's Park Houses. The rocky bluff was dynamited and brought down to street level.

BRONX GIANTS' BALLFIELD. This was a baseball diamond with a wooden grandstand that was once located off Westchester Avenue, near Wheeler Avenue. It had been part of the Watson estate in the 1920s, and Watson's Lane ran behind the grandstand. The star of the Bronx Giants was Heinie Zimmerman.

BRONX HILLS, The. This range is now inside St. Mary's Park. The hills were referred to in an 1840 land book on Gouverneur Morris' holdings. An 1885 reference to the Bronx Hills was found in the Police blotters of the 33rd precinct of Morrisania.

BRONX OPERA HOUSE, The. This prominent showplace was located east of Third Avenue on East 149th Street from 1912 to about 1930. It featured very little opera, but excelled in its presentation of Broadway stars such as George M. Cohan, Pat Rooney and Marion Bent, Francis X. Bushman, the Barrymores, Weber and Fields and the Great Houdini. The site had been occupied by Schnaufer's stables and, prior to 1900, was the northern end of Karl's Germania Park, later called the 23rd Ward Park and Loeffler's Park. The course of the Mill Brook lay diagonally across (under) the lobby of the Bronx Opera House.

BRONX OVAL. This was located at the entrance to Hunts Point Road at Southern Boulevard, and was a large fenced-in field before 1900. A tanbark arena was used as a barn to house horses and ponies belonging to the Simpson family. About 1904, the arena was turned into a baseball field and was called the Bronx Oval. Heinie Zimmerman of the Chicago Cubs, Tim Jordan of Brooklyn, "Bugs" Raymond of the New York Giants, Paul Dietz and Jack Coffee played there.

BRONX PLACE (1). The former name of Crawford Avenue, four blocks to the city line.

BRONX PLACE (2). This was one short street now covered by Bronx Park East, just north of Pelham Parkway.

BRONX RIVER DISTRIBUTION RESERVOIR. This was the former designation of Reservoir Oval when it was in use according to Robinson's *Atlas* of 1888.

BRONX RIVER ROAD. This was a former name of Webster Avenue, from Gun Hill Road north to the city line. See: Berrian Avenue, Thomas Street, Worth Avenue.

BRONX STREET. This was the former name of East 140th Street, before 1888. See: Broncks Street, Second Street (Port Morris).

BRONX TERRACE. This is the original name of Bronx Boulevard in Wakefield from East 219th Street to East 226th Street. See: Marion Street, Second Avenue.

BRONX THEATRE, The. This was the first theatre in The Bronx. Percy Williams bought the land at E. 150th Street and Melrose Avenue and built the theatre which opened to great fanfare in 1908. It soon became a popular vaudeville house. Later, it was changed to Miner's in The Bronx, and became a burlesque house and then showed comedies and dramas. In the 1930s, it was used by traveling stock companies, followed by conversion to a movie theatre and closed down sometime around 1947.

"BRONX VIEW PARK." In 1895, Ephraim B. Levy purchased a tract of land for development. It ran from Unionport Road easterly to Barnes Avenue and from Morris Park Avenue to Bronxdale Avenue. Levy gave it this name to distinguish it from adjoining "Van Nest Park," but the designation never caught on, and residents considered themselves part of Van Nest.

"BRONXWOOD PARK." This was an early real estate development bounded by North Oak Drive and South Oak Drive, White Plains Road and Bronxwood Avenue.

BROOK STREET. This is a street that adjoined the Mill Brook from East 173rd Street to East 175th Street, east of Webster Avenue that was mapped around 1904. It is now a private road leading down to a winery and a railroad siding on Park Avenue.

BROOK, The. This was a small silent movie house that was once located at East 137th Street and Brook Avenue.

BROOKLINE STREET. This was the former name of East 193rd Street from Marion Avenue down to Webster Avenue. It followed the line of Briggs Brook that rose on the Valentine estate, near present-day East 194th Street and Valentine Avenue, and flowed due east into the Mill Brook (Webster Avenue).

BROONKS RIVER. On the "Map of New York with the adjacent Rocks and other remarkable Parts of Hell-Gate" by Thomas Kitchen, Sr., Hydrographer to his Majesty, 1778, is Broonks River.

BROTHERS, The. This is a 1708 reference to North Brother Island and South Brother Island, when the English Governor of the Province of New York patented to William Bond the two islands "known as The Brothers, which have become separated in the long march of history. See: North Brother Island, South Brother Island.

BROWN AVENUE. This was a former name of Cruger Avenue, from White Plains Road to Bronxdale Avenue. See: Timpson Avenue.

BROWN'S HILL (1). This is a late 19th-century name for Schley Avenue from East Tremont Avenue down to Weir Creek (now covered by the Throgs Neck Expressway). Bruce Brown was the owner of the property. See: Wolfe's Hill.

BROWN'S HILL (2). This is a hill overlooking the Harlem River in the vicinity of West 167th Street and University Avenue. In the 1960s, the Brown family still lived in the neighborhood, but their Hill is now occupied by Highbridge Houses, a New York City housing project.

BRUCKNER'S BREWERY. In 1869, John A. and George and Louis Bruckner were listed as brewers of Melrose. Their brewery was located at Washington Avenue and William Street (Elton Avenue at East 161st Street).

BRUCKNER'S HALL. The German Lutherans of Melrose established a temporary place of worship in this Hall in 1852. It was situated at present-day Elton Avenue and East 161st Street.

BUCK ROCK. This was an oblong rocky isle in Eastchester Bay off Tallapoosa Point near the Pelham bridge. It was so labeled on Bronx maps of 1924, 1931 and 1941. In 1963, a sea-wall was built out beyond the island, and by 1966 it was completely covered over by landfill and garbage.

BUCKBEE'S ROCK. In old deeds, dated 1719 and 1720, mention is made of a prominent rock on the salt marshes of what is now Clason Point. John Buckbee signed the grants, but elsewhere the name appears as Bugbee and Buggbee. The rock was used as a reference point, and was described as being at the head of a tidal creek (Pine Rock Creek) emptying into the Bronx River.

BUCKHOUT STREET. See: Echo Place.

BUENA RIDGE. It is listed in the 1871 Directory of Morrisania and Tremont as the area west of the Concourse below East 149th Street. See: Buena Vista Ridge Road, Grove Street.

BUENA VISTA RIDGE ROAD. This was the name of the Concourse from East 153rd Street to East 158th Street in 1900.

BUGBEE'S ROCK. See: Buckbee's Rock.

BUILDERS' FIELD. An undeveloped stretch of land north of East 233rd Street near Baychester Avenue that was used in the 1920s and 1930s as a baseball field and football field. Wakefield old-timers are vague on the origin of this name, but think some builders (carpenters, masons and plumbers) contributed fencing, rails and some grandstands.

BUNGAY CREEK. This is the 18th-century name of the Sacrahung (Bound Brook) that originated in Crotona Park's Indian Lake, flowed down present-day Intervale Avenue and emptied into the East River at East 149th Street. It is mentioned in a deed of 1835, but figured in 17th-century boundary disputes between the Morris family and the Leggett, Hunt, Richardson, Jessup and other families. Bungay is a market town in Suffolk, England, and in its records, a Francis Jessup [spelled this way] was mentioned in 1641 as an art agent. Lionel Throckmorton is another man mentioned in those ancient records. In England, Bungay's situation on a point of land gave it the name of Le Bon Eye, Bongeye, Bungia and Bungay. The similarity of the mouth of this creek in the New World likely influenced the early English settlers to give it the same name.

BUNGAY STREET. This is the 1870 name of East 149th Street from Southern Boulevard to the East River, as it overlaid Bungay Creek. Its name was changed to East 149th Street in 1883.

BURDETT AVENUE. The former name of Schley Avenue, from East 177th Street to East Tremont Avenue.

BURIAL GROUNDS, The. This was the local name for what had been the Bensonia or Morrisania Cemetery. All bodies had been removed around 1890, and the barren tract from Rae to Carr Streets, Hegney Place to St. Ann's Avenue was used as a ballfield, later to be the site of a public school.

"BURIAL POINT." This was a Siwanoy burying-ground on present-day Ferris Point (Ferry Point Park) located 1/4 miles northeast of the tollbooths of the Bronx-Whitestone bridge.

BURKE (James) PLAYGROUND. See: James Burke Playground.

BURKE AVENUE. This was the former name of Wickham Avenue. See: Chestnut Street.

BURLAND, The. It debuted in 1896 as an open-air theatre used only for summertime. In 1916, it was covered over. This silent movie house was listed in 1922 at 985 Prospect Avenue near East 163rd Street. It had a seating capacity of 283.

BURYING POINT. Drafted on a Revolutionary War map in 1778 by Major John Andre, it is obviously the latter-day Stephens' Point at the foot of White Plains Road where the Bronx River joins the East River. See: Stephens' Point.

BUSHNELL AVENUE. One avenue that proves the old saying that "the map is not the territory," this avenue was laid out across the marshes of Eastchester. Though it appeared on maps, the avenue was never constructed. The area was, at one time,

Freedomland-and is now Co-Op City. Giles F. Bushnell was an Eastchester landowner who claimed relationship with Cornelius Bushnell who helped finance Captain John Ericsson to build the *Monitor* in 100 working days. His father, David Bushnell, invented a hand-operated wooden submarine during the Revolutionary War and called it "the Turtle."

BUTTERNUT LANE. See: Butternut Street.

BUTTERNUT STREET. It was listed in 1872, running approximately four blocks, parallel to the Concourse from East 162nd Street to East 166th Street. The lower end is now Walton Avenue, and the upper end is a yard alongside All Hallows' School at East 164th Street.

BYLES' SHIPYARD. Around 1884 George and William Byles began business on City Island building small boats. Many racing shells used in intercollegiate races were built there.

BYRON STREET. This is an earlier name of Byron Avenue, in Wakefield.

CABOT STREET. Laid out in 1888 in what was called, in earlier times, "the Debatable Lands" as both the Morris family and the Hunt and Jessup families claimed the region, Cabot Street overlaid, in part, the Bungay Creek, which was the ancient boundary line. The proximity to Worthen, Truxton and Du Pont Streets (all named after Navy men) suggests a naval origin to Cabot Street, and history tells us the first American fleet (1775) included the brig Cabot commanded by Captain John Burroughs Hopkins. The street has been absorbed by the Hunts Point Market.

CADIZ PLACE. This is listed in 1884 as one of the bounds of George Opdyke's property. Other Places named were Potter Place (now East 204th Street), Lisbon Place, and Grenada Place (now East 206th Street). Cadiz, Grenada and Lisbon were all historically linked with the Saracen wars fought by St. George. St. George's Crescent is nearby. All these names were suggested by Mayor George Opdyke (1862-1864) who was interested in history, and who owned the property. Cadiz Place was absorbed by Mosholu Parkway South.

CALHOUN DIVISION. This was a Volunteer Station of the Life Saving Service of New York City that was constructed in 1932 at the foot of Calhoun Avenue on the East River. It was manned by three officers, ten surfmen and a weekend medical officer.

CALHOUN TERRACE. This was a short street in Belmont running from present-day Cambreleng Street to Belmont Avenue near its junction with Crescent Avenue. It was shown on an 1851 map.

CALLAHAN'S HILL. This Irish family lived on Elton Avenue (then known as Washington Avenue) in the Civil War times. When East 156th Street was cut through, down to Third Avenue, the Callahans moved their house a few yards, and remained atop 'their' hill. Miles Callahan joined the Bronx Oldtimers when they organized in 1911.

CAMAC PLACE. The Camac Street Company, Inc., of 12 West Fordham Road purchased land on Ferris Point in 1908 from the Lowenstein estate. Camac Place was mapped as running from Ferris Lane (now Hutchinson River Parkway) to Westchester Creek, south of the high schools, but was never cut through.

CAMBELLINA STREET. This is the former name of Hughes Avenue, from East 186th Street to Fordham Road. It was noted on an 1874 map of West Farms, and might be a fanciful spelling of Cambreleng, the name of a former landowner.

CAMBRELENG STREET. Stephen Cambreleng was a landowner below Fordham University in the 1850s. Later on, Cambreleng Street was renamed Belmont Avenue.

CAMEO, The. This silent movie house was on Third Avenue, between East 179th Street and East 180th Street.

CAMMANN STREET (1). The former name of West 184th Street from the Harlem River to Hampden place. It was the lower end of the Cammann estate that was there from 1850 to the late 1890s. See: "Roseland."

CAMMANN STREET (2). This was the first name of West Fordham Road, which was once the upper border of the Cammann Estate. Oswald Cammann was a wealthy landowner and vestryman of St. James P.E. church of Fordham. Henry J. Cammann was Director of the Home for Incurables, and is credited with developing the modern binaural stethoscope. The Cammanns intermarried with the Lorillard and Mali families.

CAMP MULROONEY. This was the summer home of the New York Police Academy from 1930 to 1936. Its headquarters building was the large drill hall left behind by the U.S. Navy on Rodman's Neck. The Police rookies wore shorts, athletic shirts and French berets. In 1962, the Neck was again taken over by the N.Y.C. Police Department which maintains the largest pistol range in America. See: Ann's Neck, Belleau Wood.

CAMP SEUSS. This was the unofficial name of the Army Base Hospital called Chateau Thierry by the War Department, from 1916 to 1918. It was located at Gun Hill Road and Bainbridge Avenue, on the open lots that then surrounded Montefiore Hospital owned by Columbia University. The medical corpsmen, trained in first aid and ambulance work, were quartered in barracks on the northeastern corner, and Commandant Jacob Tennessee Seuss was the officer so honored. See: Chateau Thierry.

CAMPBELL'S ISLAND. From about 1919 to 1930, this was a German and Czech resort in the swampy region north of Givan's Creek and west of the Hutchinson River. The island supported a Summer colony of some 60 tents, rented by Yorkville residents who had to bring in all food, fuel and liquid refreshments by rowboat from Baychester. Although owned by the Steers Sand & Gravel Company, the island was named for a year-round resident, a Mr. Campbell whose two daughters paddled their canoe daily to shore to take the railroad downtown. Today, the island would be a substrata of Co-Op City's southern edge at approximately Elgar Place.

CANAL AVENUE. This is the name of a former avenue flanking the canal that ran from Bronx Kill to East 144th Street, east of present-day Triborough Bridge approach. It appears on an 1865 map.

CANNON'S HOTEL. This was a popular inn at the junction of Bear Swamp Road, Walker Avenue and Lafayette Avenue around 1900. It was opposite St. Raymond's R.C. church. This location is now Bronxdale Avenue, East Tremont and Castle Hill Avenues. Later, the hotel became a training camp and gymnasium for prizefighters, and was known as Dal Hawkins'.

CANOE BEACH. This was a popular beach on the southwest side of West Twin Island, facing Hunter Island. Up to 1930, it was a favorite rendezvous for canoeists, but is now incorporated into Orchard Beach.

CAREY'S POINT. The northern tip of City Island is officially known as Terrace Point, but most islanders refer to it as Carey's Point. The name harks back to Judge Henry Carey, owner of the horsecar line that ran from Bartow station to City Island in the 1900s. His son, Harry Carey, gave up the study of law to become a movie actor, moved to Hollywood and became a cowboy film star.

CARLIN AVENUE. This is the former name of Valentine Avenue, north of Fordham Road.

CARLL'S SHIPYARD. David Carll established the first commercial shipbuilding yard on City Island after purchasing land from George Washington Horton in 1862, and from Richard W. Peary in 1864. The plant was located at the foot of Pilot Street, facing Hart Island. He sold the business to Henry Piepgras in August, 1886. See: Piepgras' Shipyard, Jacob's Shipyard.

CARR HILL. A part of Gouverneur Morris' lands, it was conveyed to William Carr in 1847. In 1853, it was recognized as a postal address. It was the hilltop at East 156th Street, Eagle and Cauldwell Avenues.

CARR STREET (1). This is the former name of St. Ann's Avenue from East 134th Street to its junction with Third Avenue. See: St. Ann's Avenue.

CARR STREET (2). From East 161st Street to Teasdale Place, this was the early name of Third Avenue. See: Fordham Avenue, Morris Avenue, Morse Avenue.

CARR STREET (3). Originally a small lane on the northern end of Bensonia Cemetery, this short street below East 156th Street and St. Ann's Avenue was eliminated in 1978. After Bronxchester Houses were built Carr Street became a grassy strip between the Project and South Bronx High School.

CARROLL LANE. The former name of Dorsey Street from Zerega Avenue to Seddon Street.

CASSEL'S HOTEL. In 1880, this hotel was on the southeastern corner of Webster Avenue and Claremont Parkway. The Mill Brook ran on its eastern side. The brick, 4-story building survived until the 1970s.

"CASTELLO DE CASANOVA." An estate on Hunts Point, owned in the late 1860s by Innocencio Casanova. Much of his wealth financed revolutions in his native Cuba. His mansion was used as a rendezvous for political exiles and a storeroom for munitions.

"CASTLE HILL." This was the former estate of Gouverneur Morris Wilkins at the end of Castle Hill Point. It consisted of 341 acres, mapped in 1868, with the entrance to the estate being located at present-day Cincinnatus Place. The Wilkins manor house stood at Castle Hill Avenue off Lacombe Avenue. See: Castle Neck, Cromwell's Neck, Screvin's Point.

CASTLE NECK. An ancient deed, dated 1685, mentions an exchange of acreage upon Castle Neck between Thomas Hunt and John and Elizabeth Cromwell. See: Cromwell's Neck, Screvin's Point.

CASTROP'S PAVILION. This was a popular picnic area and dance pavilion, near the Bartow railroad station, as advertised in 1891 and 1892 by the New York Eagle. Lieutenant Castrop of Station No. 7, Pelham Bay Division of the Volunteer Life Saving Corps, was frequently mentioned in the *North Side News* for his rescues (1900-1904).

CASWELL ACADEMY. In the 1870s, this academy was situated on Hunts Point, northwest of Hunts Point Avenue, facing the Bronx River.

CASWELL AVENUE. This was mapped as one block south of Ryawa Avenue, but the site is now occupied by the Hunts Point Sewer Disposal plant.

CATHERINE AVENUE. A former name of Daly Avenue, from East 179th Street to Bronx Park in 1897. Catherine (Lorillard) Cammann's father owned what is today the Bronx Botanical Garden. See: Elm Street.

CATHERINE STREET. This is an earlier name of Carpenter Avenue, and was thought to have been named by Louis Haffen, Commissioner of Street Improvements, for his mother Catherine (Hays) Haffen in 1896. See: 2nd Street.

CATHOLIC PROTECTORY. Its full title was the Society for the Protection of Destitute Roman Catholic Children in the City of New York, It occupied 150 acres of woodlands, farmlands, school buildings, churches and chapels and an athletic field. It had been made up from Leonard Mapes' farm of 20 acres, purchased in 1869, the Leggett farm of 112 acres in 1866, and an additional 7 acres of the Cow Neck farm in 1880. It was situated between White Plains Road, Westchester Avenue, Unionport Road and East Tremont Avenue, and was sold to the Metropolitan Life Insurance Company in 1938. The huge housing development of Parkchester now occupies the grounds.

CAUSEWAY PLACE. It was located next to the toll booths of the Bronx-Whitestone bridge at the entrance to Ferry Point Park. The name harks back to a causeway that crossed Coney Island Creek to reach the estates of the Lorillard and diZerega families. The Place was eliminated by the Board of Estimate in September, 1964.

CEBRIE PARK. This was an early real estate development bounded by Westchester Avenue, St. Peter's churchyard, Seabury Creek and Zerega Avenue. It was displayed on a 24th Ward Map of 1905. The name is a corruption of Seabury, a family of pre-Revolutionary prominence in the affairs of Westchester village.

CEDAR AVENUE. This is a former name of Sedgwick Avenue at West Tremont Avenue. See: Emmerick Place.

"CEDAR GROVE" (1). This was the former estate of Gerard Morris, mapped in 1859, which eventually became Cedar Park of 17.5 acres in 1880. Cedar Park was renamed

Franz Sigel Park some time after the General's death in 1902. See: Cedar Park, West Morrisania.

"CEDAR GROVE" (2). William B. Hoffman's farm, mapped in 1888, was south of St. John's College that is now Fordham University.

"CEDAR HILL." This is another name for the Powell farm when Delancey Powell lived there in the 1860s. It was bounded by Fordham Road, Third Avenue, East 180th Street and Arthur Avenue.

"CEDAR KNOLL." This is a wooded knoll, formerly owned by the Theodore Roosevelt family on the shore of Pelham Bay, at the city line. It was reputed to be haunted by a band of headless Siwanoy Indians. See: Haunted Cedar Knoll.

"CEDAR KNOLLS." This estate was mapped in 1856, extending from Riverdale Avenue east to Broadway, and from approximately West 246th Street to West 253rd Street. Hiram Barney was the landowner, and a good friend of Edgar Allan Poe when the poet resided in Fordham.

CEDAR OF LEBANON. This well-known tree once stood on the shore of the East River at the foot of East Tremont Avenue, and served as a navigational guide for mariners for over a century. It was planted in 1790 by Philip Livingston, and was alive until 1938. It was blown over in the hurricane of September 14, 1944, and the trunk remained on the southeastern corner until 1959, when sections were removed for display in the Bronx Botanical Garden Hall. The rest of the trunk was removed by The Bronx County Historical Society.

CEDAR PARK. This is the former name of Franz Sigel Park. Its western slope (Walton Avenue) was once an Indian path. General Washington and Andrew Corsa, his scout, watched British shipping from the summit of this park, according to accounts dealing with the Revolutionary War. See: "Cedar Grove" (1).

CEDAR PLACE. This is an earlier name of East 158th Street from Third Avenue to Westchester Avenue. See: Milton Street.

CEDAR STREET (1). On a map dated June 13, 1888, this was the name of what we call East 159th Street, from St. Ann's Avenue up to Eagle Avenue. See: John Street.

CEDAR STREET (2). This is the former name of East 161st Street from the Harlem River to the Concourse, evidently taking its name from the Morris estate, "Cedar Grove," or sometimes, "The Cedars." The name, Cedar Street was noted in the 1871 Directory of Morrisania.

CEDAR STREET (3). In the village of Belmont, this was the earliest name of East 178th Street from Third Avenue to Hughes Avenue. See: Elmwood Place, Mechanic Street.

CEDAR STREET (4). This is a former name of Barnes Avenue, from Burke Avenue to Gun Hill Road. See: 4th Street, Kingsbridge Road, Madison Street.

CEDAR STREET (5). In Baychester, this was the former name of Kingsland Avenue, from Adee Avenue to Givan Avenue. See: Tieman Avenue.

CEDARS PARK. This is a variant of Cedar Park. Although the name had been changed to Franz Sigel Park in 1902, many old-time residents of Melrose used the older name as late as 1920.

CEMETERY LANE. This is the original name of Tier Street, leading from City Island Avenue to the Pelham Cemetery.

CENTERVILLE. This small settlement was located around Castle Hill Avenue and Westchester Avenue in the 1840s to Civil War times.

CENTRAL AVENUE (1). This is the 1850 name of Jerome Avenue, so called because it was at the approach to the Central Bridge (Macomb's Dam Bridge).

CENTRAL AVENUE (2). This is a former name of Arthur Avenue, from Crotona Park to East 187th Street.

CENTRAL AVENUE (3). In Baychester, this was the earlier name of Stillwell Avenue, from Baychester Avenue to the Hutchinson River.

CENTRAL BRIDGE. This was the name, before the Civil War, for Macomb's Dam Bridge.

CENTRAL HALL. This gathering place of early Melrose was on the northeastern corner of Third Avenue and East 150th Street. It was operated by Peter Kirchof in the 1870s, whose brewery adjoined it. In 1900, it became Piser's furniture store. See: Bronx Central Hall.

CENTRAL STREET. See: East 179th Street.

CENTRE STREET. See: East 179th Street.

CENTREVILLE. An 1851 map in the New York Public Library and an 1850 deed spell it this way. See: Centerville.

CENTU FARM. This was a 26-acre farm on the east side of the Throggs Neck Road, approximately from Randall to Lafayette Avenues, and running east to Weir Creek (now the Throgs Neck Expressway). In 1829, D. Centu sold the farm to Elbert Anderson, who had purchased orchards, woodlots and farmlands extending from Dewey Avenue to LaSalle Avenue. See: Doric Farm, Hawthorn Farm, Willow Lane Farm.

CENTURY GOLF CLUB. It was originated in 1896 by Augustus G Miller, who later became Golf Supervisor of the Bronx parks in 1924. The clubhouse was the former Turnbull mansion, near Clarence and Lafayette Avenues, and the 9-hole golf course was in operation from 1898 to 1915. The links lay south of Layton Avenue, between Weir Creek (now the Throgs Neck Expressway) and Long Island Sound. Its name might have had some connection with the Centu Farm, a homestead on the opposite bank of Weir Creek in the 1850s.

CENTURY WOODS. This was a stand of fir, oak and maple trees, east of Weir Creek, belonging to the Century Golf Club (1896-1915). The woods, in the vicinity of Lafayette Avenue and Ellsworth Avenue, were cut down in 1959 when the Throgs Neck Expressway was built.

CHA-TI-E-MAC. See: Hudson River [Current].

CHALFIN TERRACE. This was a planned street in the Kingsbridge sector that never materialized. S. F. Chalfin was a surveyor in 1888 in the Spuyten Duyvil-Kingsbridge area.

CHANDLER'S POND. "Jan. 14, 1891. Minutes of the Board of Commissioners, Department of Public Parks: Request from John Deesser and John Young, applying for permission to cut ice on the Bronx River at the locality known as Chandler's Pond. Referred to the President for report." A search for this pond on maps of that era have been in vain.

CHANUTE STREET. This is a former name of Manor Avenue, which was changed on February 7, 1911.

CHAPEL FARM. This was an 18-acre tract in the Fieldston area with the highest elevation (248 feet above sea level) in The Bronx. It was once the property of Clement and Genevieve Griscom. He died in 1918, she in 1958, leaving the mansion and grounds to Manhattan College that annexed it in 1969.

CHARLES PLACE. This is the first name of East 168th Street from Gerard Avenue up to the Concourse. Charles Arcularius owned a hotel at the intersection of East 168th Street and Jerome Avenue in the 1880s. East 169th Street was formerly Arcularius Place. See: Glen Street, Sixth Street.

CHARLOTTE PLACE. It was named for Charlotte (Leggett) Fox, and is the former name of Jennings Street.

CHATEAU LAURIER. This was the residence of William Belden on the southern tip of City Island which later became the Monte Carlo Hotel. Since 1937, it has been the clubhouse of the Morris Yacht Club.

CHATEAU THIERRY. This name was given by the War Department to an Army Base Hospital that was located at Gun Hill Road and Bainbridge Avenue in the World War I years of 1916-1918. Prior to that time, there had been an athletic field and tennis courts on the land belonging to Columbia University. The name represented a battlefield in France where American troops had fought, and was given to this Bronx location to inspire recruitment. See: Camp Seuss.

CHAUNCY STREET. This is the former name of Paulding Avenue from East Tremont Avenue to the railroad. See: Forest Street, Sixth Street (4).

CHEEVER PLACE. This was the earlier name of East 140th Street, from the Harlem River to the Concourse where the land belonged to John H. Cheever, 1859. The name was changed in 1911.

CHERRY FIELD. (Baseball field) See: Overing Street.

CHERRY HILL. Once at East 132nd Street from St. Ann's Avenue to Cypress Avenue on the former lands of Gouverneur Morris, at one time it was a fair sized hill, but it was leveled for additional fill in the railroad yard.

CHERRY LANE. See: St. Ann's Avenue.

CHERRY STREET. See: East 136th Street.

CHERRY TREE POINT. A 1900-1925 designation of the promontory in the Country Club sector off Agar Place.

CHESTER AVENUE. This was the earlier name of Chester Street near Boston Road and Eastchester Road. It was originally planned as a main street of a development called Seneca Park around 1908.

CHESTNUT AVENUE. An 1898 map locates this short thoroughfare in the settlement of Bartow as west of the railroad depot, running north and south.

CHESTNUT GROVE PICNIC GROUNDS. This was once located at West 257th Street and Riverdale Avenue from 1904 to 1911.

CHESTNUT HILL. This is the local name (*circa* 1880-1930) of a hill outside Spuyten Duyvil village, east from Johnson Avenue near West 230th Street. It originally sloped into Spuyten Duyvil Creek.

CHESTNUT STREET (1). This is the former name of Vyse Avenue, from East Tremont Avenue to Bronx Park South. The Vyse estate was noted for its fine chestnut trees.

CHESTNUT STREET (2). This is an earlier name of Wickham Avenue, noted in 1888.

CHESTNUT STREET (3). This former street one block long from Reid's Mill Lane to Oak Street is now covered by the New England Thruway.

"THE CHESTNUTS." An 1868 map of South Yonkers (now The Bronx) shows this estate belonging to H. Ketchum located west of Broadway at about West 258th Street.

CHRIST'S HOTEL. The former manor house of Lewis Morris, demolished in 1891 by the New York, New Haven & Hartford Railroad. It was a hotel from roughly 1870 to 1890, and was believed to have been built on the site of Jonas Bronck's house. That would be East 132nd Street between Willis and Brook Avenues.

CHRIST'S PARK. An 1872 map shows this amusement park, picnic grove and dance pavilion, at the foot of Willis Avenue. See: Brommer's Park, Harvey's Hill, "Scratch Park."

CHURCH STREET. This is the former name of Kingsbridge Avenue from West 230th Street to West 234th Street. It was named for the Church of the Mediator, an Episcopal frame building built in 1855. In 1905, the frame church was replaced by a stone church.

CINDER ROAD, The. This was a local name for a footpath that led from East 177th Street north to Layton Avenue. It was laid out over the sewer line, and was raised some 15 feet above the surrounding swamplands and Weir Creek. In use from 1910 to 1959, when it was then obliterated by the Throgs Neck Expressway, sometimes it was called the Cinder Path.

CINELLI'S THEATRE. This was an old-time silent film house at Arthur Avenue and East 187th Street. To prevent children remaining all day long on Saturday, Mr. Cinelli collected tickets after each showing.

CITY ISLAND DIVISION. This Volunteer Station of the Life Saving Service of New York City was situated north of Fordham Street on the east shore of City Island. In 1917, one officer, two coxswains and fourteen surfmen were listed. Instead of the standard white painted frame bungalow, the unit was housed in Bracker's Casino on the shore.

CLAFLIN TERRACE. This is the former name of Reservoir Avenue. Horace B. Claflin was a wealthy landowner of this region.

CLANCY'S HOTEL. Patrick Clancy was listed as proprietor of this hotel in 1890 at Ogden Avenue and Orchard Street (now West 168th Street).

CLAREMONT. A village that was started in 1852 using the name of a local estate called "Claremont." See: Claremont Park.

"CLAREMONT" (1). This was an estate owned by John Devoe extending from Ogden Avenue down to Jerome Avenue in the vicinity of West 169th Street. It was surveyed by R. Henwood in 1852.

"CLAREMONT" (2). The estate owned by the Zborowski family, called "Claremont,"was later used to name the village of Claremont. See: Claremont Park, Claremont Parkway.

CLAREMONT, The. A silent movie house at East 174th Street, near Third Avenue, in business before 1910.

CLAREMONT AVENUE. See: Grant (Edward L.) Highway.

CLARK STREET. This short lane once ran from Valentine to Tiebout Avenues at East 186th Street according to an 1889 map.

CLARKE'S BREWERY. This is listed in the 1869 Directory of Westchester County as being located on Carr Street (now St. Ann's Avenue) Robert Clarke, brewer, is listed as living on Grove Hill, the elevation at East 156th Street and Cauldwell Avenue.

CLASON AVENUE. This is the former name of Beach Avenue from East Tremont Avenue to Westchester Avenue, and was once part of Old Clason Point Road. It was named for Isaac Clason, who purchased the east side of Cornell's Neck in 1804.

CLASON POINT AMUSEMENT PARK. Approximately a 50-acre triangular tract at the end of Clason Point, it was a focal point for entertainment from the early 1890s to the 1940s. Launches from Mott Haven and Port Morris, and excursion boats from Manhattan brought people, and the Park was further expanded when streetcars were introduced in 1910. Ferries from College Point, Queens furthered the popularity of the Park, but after World War II it fell into disuse. It was replaced by the Shorehaven Beach Club.

CLASON POINT DIVISION. This Volunteer Station of the Life Saving Service of New York City was situated at Clason Point in the standard frame bungalow equipped with lifeboats on launching rails. It was manned by four officers, medical attendants and surfmen, as listed in 1914, 1916 and 1917 reports of the Life Saving Service.

CLASON POINT MILITARY ACADEMY. This Catholic boys' school was located on the former Dominick Lynch estate, and its official name was the Sacred Heart Academy for Boys. The school was noted for its scholastic excellence, marching bands and simulated war exercises. From 1916 to 1927, a submarine called the Fenian Ram was set on the campus. It is now in Westside Park, Paterson, N.J., where its builder, John P. Holland, once lived. See: Fenian Ram. In 1927, the Academy removed to Oakdale, L.I., and took the name of LaSalle Military Academy.

CLASON'S CREEK. An 1889 map of Clason Point gives this as an alternate name for Pugsley Creek. See: Cromwell's Creek (1693), Wilkins' Creek (1770s), West Creek (1823).

CLAY AVENUE. This is the former name of East 187th Street from Cambreleng Avenue to Southern Boulevard. See: Jacob Street, Sanford Street.

CLEMENTINE STREET. According to a 1909 deed, Mrs. Clementine Brodbeck owned 3.2 acres of land along the White Plains Road (now Provost Avenue) north of Boston Road. In 1984 the street sign was superseded by a new one, reading Naclerio Plaza.

CLERMONT, The. This old-time silent movie house at East 180th Street and Mohegan Avenue was in operation prior to World War I.

CLEVELAND AVENUE. In the old-time settlement of Washingtonville, this was the name of East 243rd Street from White Plains Road to Barnes Avenue. See: Huguenot Street.

CLIFF STREET. Mapped in 1871 from Third Avenue to Prospect Avenue, we know it now as East 161st Street. See: William Street, Cedar Street (2).

CLIFFORD STREET. Until 1894, the name of East 234th Street from Woodlawn Cemetery to Webster Avenue.

CLIFTON AVENUE. This was the earlier name of Brook Avenue from East 134th Street to East 160th Street.

CLINTON AVENUE. This is the former name of Weeks Avenue in Tremont.

CLOVER STREET. In 1888, this was East 181st Street from Vyse Avenue to Boston Road. See: Ann Street, Irene Place, Irving Street, John Street, Ponus Street, 5th Street.

COBB AVENUE. This was the earlier name of White Plains Road from Westchester Avenue to Lafayette Avenue (where it once ran across the Cobb farm). Marcus Cobb bought part of Cow Neck farm from heirs of Oakley Pugsley in 1854, and sold it to the Catholic Protectory in 1880. See: Cow Neck, Larkin Avenue, Washington Street, Third Street (Wakefield).

COBWEB, The. An 1899 advertisement of this restaurant in East Morrisania stressed its superb wine cellar. Charles Walter was the owner.

CODLING ISLAND. This former island is now part of the left bank of the Hutchinson River, directly north of the New England Thruway bridge. It is partially covered by Huguenot Street. Originally, it was part of a picnic grounds belonging to G. F. Codling (1868 map) on the right bank of the Hutchinson River, but the Army Engineering Corps cut a deeper channel across the grounds detaching a tip of land which was called Codling Island. Gradually, the more shallow channel to the east silted up, and when Huguenot Street was laid across it, Codling Island disappeared under landfill.

COLD BLOW. This was a local nickname for the area around Schley Avenue and East Tremont Avenue in use from the 1890s to about 1910. This was an exposed section open to the Wintry winds of Long Island Sound, and the expression might have been in use even earlier, but it is now impossible to ascertain.

COLE STREET. This was the original name for the stretch of East 194th Street from Marion Avenue to Webster Avenue. Jacob Cole was Postmaster of Fordham and his property extended back from Fordham Road to East 194th Street.

COLES' BOSTON ROAD. This was an 1795 name for the present-day Boston Road which was laid out by John B. Coles at the instigation of Lewis Morris of Morrisania. This was so called to distinguish it from the original, older Boston Road that crossed the northern Bronx from Kingsbridge to Eastchester in pre-Revolutionary times. Coles' Boston Road follows Third Avenue from the Harlem River to East 164th Street, and then the Boston Road as we know it today, up to the city line.

COLES' ROAD. This was an alternate name for Coles' Boston Road.

COLFORD OVAL. This small racetrack and training course in use from the 1870s to the early 1900s was situated at the northern end of the John A. Morris estate on Throggs Neck. Its present-day location would be East Tremont Avenue from the Cross-Bronx Expressway to Schley Avenue. Matt Colford was a well-known horse trainer, and all the wealthy squires of the Neck brought their trotting horses and carriage horses to the Oval. Later, the grounds became a baseball field, and then was covered by a supermarket and parking lot.

COLLEGE AVENUE. The earlier name of Rider Avenue from East 138th Street to East 142nd Street at Morris Avenue, named for the Catholic College of the Sacred Heart.

COLLEGE STREET. See: East 191st Street.

COLLINS LANE. The alternate name of Split Rock Road harks back to a 19th-century landowner. This historic road across Split Rock golf course figured in the running battle of 1776 between Colonel Glover's men and the advancing British and Hessian troops. It was mapped at various times as Pelham Road, Prospect Hill Road, and Pell's Lane.

COLONEL'S. A 1754 deed describes "a place called Colonel's" in the vicinity of the west shoreline of the Hutchinson River now covered by Co-Op City. It may owe its name to Cornelius Jones who owned the land. Latter-day spelling was "the Kernels."

COLONIAL, The. This old time movie house on Willis Avenue and East 147th Street was owned by the Archer family. According to an Atlas dated 1912, it was a brick, one-story building on the west side.

COLONIAL INN, The. This was a famous restaurant of the early 1900s on the road to City Island. It was originally built by Elisha King in 1829 with stones quarried on High Island, It was razed in 1937. See: Marshall's Corner, High Island.

COLUMBIA AVENUE. This is the former name of Van Nest Avenue, from Adams Street to White Plains Road. See: Columbus Avenue.

COLUMBIA HOSE COMPANY. An old photograph of this volunteer fire department has a caption locating it on the grounds of the Catholic Protectory, now occupied by Parkchester. No other particulars were given.

COLUMBIA STREET. An 1851 map shows this as the former name of East 183rd Street, from Quarry Road to Southern Boulevard. See: Columbine Avenue.

COLUMBINE AVENUE. An earlier name of East 183rd Street in the village of Tremont.

COLUMBINE STREET. See: Columbia Street, Columbine Avenue.

COLUMBUS AVENUE. Although Van Nest Avenue was once called Columbia Avenue from Adams Street to White Plains Road, the eight-block continuation to Bronxdale Avenue was called Columbus Avenue. One old-timer hazarded the guess the first section appeased the Americans, and the extension appealed to the Italians who made up a sizeable minority.

COLUMBUS, The. Giuseppe Fusco operated the Columbus Vaudeville Theatre from 1905 to 1912 on East 151st Street between Park and Morris Avenues. Sixty years later the faint sign on the building on the north side of the street was still discernable. Sicilian puppet shows were presented between movies, which could loosely be termed "vaudeville." On the second floor was a prizefighting ring that was used for local boxing exhibitions.

COMET, The. This silent movie house, listed in 1912, was at 1015 Boston Road opposite the East 165th Street stairs down to Third Avenue.

COMFORT AVENUE. This is the earliest name of Baychester Avenue with possibly some connection with Old Point Comfort Park. See: Breinlinger's.

COMMERCE AVENUE. This former name of Cedar Avenue in West Fordham was noted in 1893 and in use up to 1938. See: Riverview Terrace.

COMMERCE STREET (1). On Clason Point, an early name of Compton Avenue alongside Pugsley Creek.

COMMERCE STREET (2). In Unionport, this was the former name of Bruckner Boulevard as it led to the commercial docks on Westchester Creek. Other names, in other sections, for this important street were Eastern Boulevard, Ludlow Avenue, Sixth Street and Willow Lane.

COMMERCIAL ROAD. This road began just north of Fordham Hospital and ran in a northwesterly direction to East 200th Street and Webster Avenue. It may once have been a boundary road between the Lorillard property (Botanical Garden) and the Cambreleng farm. See: Hospital Road.

CONCORD STREET (1). This is the former name of East 142nd Street.

CONCORD STREET (2). In Wakefield, this was an older name of Furman Avenue from East 236th Street to Nereid Avenue.

CONCOURSE PLAZA HOTEL. This hotel at the northeast corner of East 161st Street and the Grand Concourse was opened by Governor Alfred E. Smith in a ceremony in October, 1923. In its early days, it housed visiting baseball teams which played at nearby Yankee Stadium. Its ballrooms became the site of important social functions, such as political dinners, weddings and business meetings. Presidents Harry S Truman and John F. Kennedy made political speeches there. It was largely a residential hotel, and in the late 1960s, it housed families on welfare. Neighborhood opposition caused the hotel to be shut, and it was renovated and reopened as housing for senior citizens.

CONCOURSE PLAZA. See: Kilmer (Joyce) Park.

CONEY ISLAND. This was a small island, surrounded by swamps, touching Westchester Creek adjacent to Ferry Point Park. It is crossed by present-day Yznaga Place and landfill has made it a part of the mainland. It was noted in an 1814 deed and no doubt referred to the conies (rabbits) to be found there. In 1846, it was mentioned in a deed between Valentine Seaman and Charlton Ferris, mapped in 1862 as being 41/2 acres.

CONEY ISLAND CREEK. This creek once reached in from Morris Cove in a northwesterly direction to Westchester Creek in the vicinity of Yznaga Place. It is now covered by Ferry Point Park, Hutchinson River Parkway and Yznaga Place, but its mouth is in existence on the west side of the peninsula. See: Old Barn Creek.

CONGRESS, The. This small silent movie house was at 558 Southern Boulevard near East 149th Street. In the 1920s, it was enlarged to seat 1,800 and later became the Ace.

CONNERSVILLE. This was a settlement between the Hutchinson River and Stony Brook (the upper reach of Westchester Creek, near Pelham Parkway) as noted on an 1871 map, entitled "NEW YORK, From a Balloon." It is more precisely located on the Dripps' map of 1895 as being a cluster of houses around Eastchester Road just north of Pelham Parkway.

CONRAD'S HOTEL. An 1898 map shows this hotel overlooking the Harlem River north of Macomb's Dam Bridge, and being flanked by boat clubs.

CONSTABLE'S POINT. This is a name in use during the 1700s referring to Spuyten Duyvil. Constable Adrian van der Donck was the Dutch owner of the land in the preceding century when it was known as Konstabelsche Hoek, a name the English found too difficult to pronounce. See: Konstabelsche Hoek.

COOGAN'S ALLEY. This was a narrow lane in Irishtown, a settlement at West 258th Street in the vicinity of Riverdale Avenue. See: Irishtown (2).

COOLIDGE AVENUE. According to the postal authorities, Coolidge Avenue was completely wiped out during the building of the Throgs Neck Bridge in 1959-1960. Regulations prescribed that an address had to be kept in the Postal Directory for at least two years, and so the name still appeared in 1962.

COOPERSTOWN. This was a local name in Riverdale for a small settlement on a tract laid out in 1853 as Hudson Park. The real estate development was laid out along the Hudson River from West 234th Street to West 237th Street. See: Hudson Park.

COOSA AVENUE. An early Edenwald map carries this name on Eastchester Road, from Adee Avenue to Givan Avenue and is evidently a misspelling of Corsa. See: Coria Lane.

COPPEE STREET. This street was laid out, but never cut through, across the former Wilkins estate on Castle Hill Point. Paul Emil Koppe had a small truck farm off Zerega Avenue in the 1890s at present-day Castle Hill Houses.

CORDOVA PLACE. It is now absorbed by the Grand Concourse, but once was located in the vicinity of Ryer and Anthony Avenues. Also, it was noted on the east side of the Concourse at East 205th Street.

CORLAER AVENUE. See: Corlear Avenue.

CORNELL AVENUE (1). The former name of Wood Avenue, from the Cross-Bronx Expressway to Parkchester.

CORNELL AVENUE (2). This was the earlier name of Wellman Avenue in Middletown when it was first cut through the Cornell farm there.

CORNELL AVENUE (3). This was Tillotson Avenue's original title from Eastchester Road to Bruner Avenue overlaying Black Dog Brook.

CORNELL'S NECK. This was the name of Clason Point in 1646, when the land was granted to Thomas Cornell by the Dutch Governor, Kieft. See: Sheriff Willett's Point, Snakapins, Willetts Point, Woolet's Point.

CORRINGE AVENUE. This former name of McAlpin Avenue is thought to honor Commander H. H. Corringe who brought the Egyptian obelisk to America in 1877. McAlpin Avenue, in turn, was eliminated in 1958 by the State Mental Hospital, alongside the Hutchinson River Parkway.

CORSA AVENUE. This is the former name of Bedford Park Boulevard from the Concourse into the northern end of Fordham University (which once had been the Corsa farm). The Corsas were landowners ever since Dutch days, and traced their ancestry back to Hendrik Christianson Cortinsenze. The Corsa burial plot, unmarked, is directly east of St. John's Hall on the University grounds.

CORSA LANE. This was a former road that ran from Eastchester Road and East 221st Street diagonally northwest to Barnes Avenue and East 226th Street. It was noted on British Headquarters maps of 1778, on an 1850 map of Eastchester, and was still passable in 1960. Today, the entire route has been built upon. See: Coosa, Cusser.

CORTLANDT AVENUE. Te earlier name of West 237th Street, west of Henry Hudson Parkway, recalling Cortlandt Ridge or Upper Cortlandt, as it was called around 1770.

COSTER ESTATE, The. This is a former farm and orchard extending from Bruckner Boulevard to Lafayette Avenue, along the former Throggs Neck Road (now East Tremont Avenue). The eastern boundary was Weir Creek, now overlaid by the Throgs Neck Expressway. Henry A. Coster, vestryman (1875-1887) and warden (1888-1905) of St. Peter's P.E. church at Westchester Square, had a fine mansion which was demolished to make way for St. Benedict's R.C. school on Edison Avenue. His father, Daniel J. Coster, had also been a vestryman from 1843-1849, and his mother was Julia DeLancey (1806-1890), a daughter of Oliver DeLancey. The previous owner of the property was a Mr.

Brandon, who had purchased the 32-acre tract from Elbert Anderson around 1830. On maps of those days, the farm was known as "The Homestead."

COSTER STREET. This is the former name of Boyd Avenue, from East 233rd Street to Bussing Avenue.

COTTAGE GROVE STREET. This is the former name of Underhill Avenue, from Westchester Avenue to the Cross-Bronx Expressway. See: Tompkins Street.

COTTAGE ROW. This was a short dead-end street (now non-existent) that ran from West 230th Street north almost to Albany Crescent west of Bailey Avenue.

COTTAGE STREET. This was the name of East 146th Street, in 1871, . See: Grove Street.

COUNTRY CLUB LAND ASSOCIATION. It was mapped in 1892 and was bounded by present-day Bruckner Boulevard, Rawlins Avenue, Long Island Sound and Country Club Road. It was purchased in part from James M. Waterbury in 1888.

COUNTRY CLUB ROAD. Former name of Jarvis Avenue opposite the Indian Museum, as it led to the Country Club Land Association on Long Island Sound. See: Williams Avenue.

COURTLANDT AVENUE. In the 1870s, this was the name of Post Road off Broadway, and named after the Van Cortlandt [spelled this way] family. See: Newton Street.

COW NECK ROAD. Formerly McGraw Avenue, it was sometimes mapped as merely Old Road, and led to the Cow Neck farmlands.

COW NECK. This was the former farmlands of Stephen Leggett who owned 200 acres in 1794. It passed into the possession of the Pugsley family until 1854. Francis Larkin and Marcus Cobb farmed the land, but then sold out to the Catholic Protectory in 1880. It became Parkchester in 1938.

COWANGONG. This was the Siwanoy Indian name for present-day Gun Hill Road at the Bronx River. "The Boundary Beyond" is its translation, for the river did mark the limits of the Siwanoy territory.

COWBOY TREE. This was an ancient sycamore that was used as a gallows for "Cowboys" during the Revolutionary War. It stood on the Cortlandt Ridge, or Upper Cortlandt, which was sold in 1836 to James R. Whiting. In 1892, it became the property of the Sisters of Charity who ran the Mother Seton hospital.

COWSLIP ISLAND. This was a local name for a small island in the Bronx River north of East 205th Street off Rosewood Street. The official name was DuBois Island, and it is now part of the west bank of the river. See: DuBois Island, French Charley's.

COWSNECK. This is an early reference to a section of Clason Point in the vicinity of Pugsley Creek. Cornell's Neck was the early name of Clason Point, and in the records of the Society of Friends (Quakers) we read: "Phebe Cornel wife of Calib Cornell of Cowsneck Deceased in ffushing ye 2 Month ye - Day 1750."

CRAB ISLAND. A small marshy island near the mouth of Saproughah Creek in the Harlem River at High Bridge, used as a boundary marker between Daniel Turneur's land and the

Morris grant. It was used as a staging area for British troops in 1777. It is now joined to The Bronx mainland and covered by the Major Deegan Expressway.

CRAB LANE. City Island's first silent movie theatre was located on the west side of Main Street (now City Island Avenue) between Ditmars and Tier Streets. A 1908 Sandborn map shows a crooked alley next to it which was called Crab Lane.

CRABBE ISLAND. This is the earliest spelling of Crab Island on the 1675 map of the Lewis Morris grant.

"CRAGDON." This was the estate of the Seton family that is now incorporated into Seton Falls Park.

CRAIG LEA ISLAND. It is mentioned in a 1901 ordinance relating to the discharging of firearms as adopted by the Board of Aldermen. The island is described briefly as being in Pelham Bay, and the property of the Craig Lea Rod & Gun Club.

CRAIGHILL STREET. The former name of Stratford Avenue from Watson Avenue to Bronx River Houses.

CRAMES SQUARE. It was named in March 1923 to honor Private Charles Crames, who died in World War I. The Square was once crossed by an Indian trail that led to a village out on Oak Point Avenue. Crames is honored in a nearby park when this square was rededicated to msgr. Raul del Valle Square.

CRANE STREET. This was the earlier name of East 145th Street from St. Mary's Park to Southern Boulevard in East Morrisania. It was named after a landowner named John J. Crane, merchant, who lived there from 1866 to 1887. The name was changed in 1897. See: Elm Street.

CRAVEN STREET. See: East 156th Street.

CRESCENT FIELD. A baseball field used by Williamsbridge teams up to the 1920s, it was bounded by Paulding and Laconia Avenues, from East 213th Street to East 215th Street before those streets and avenues were cut through.

CRESCENT, The. A large silent movie house that was listed in 1922 at 1165 Boston Road and Home Street, it was in operation from 1914 to 1935, with a seating capacity of 1,639.

CRESTLINE STREET. This was a topographical name of Clay Avenue behind Claremont Park from East 169th Street to East 171st Street. See: Elliott Avenue.

CROMWELL'S CREEK (1). This was an arm of the Harlem River that ran inland at approximately Yankee Stadium and up River Avenue and Jerome Avenue to approximately West 178th Street. It was named after James Cromwell who had a mill on the stream and worked for Lewis Morris. He was born in 1752 and claimed descent from Oliver Cromwell.

CROMWELL'S CREEK (2). According to Westchester County Wills, there were "14 acres sold to Israell Honneywell in 1693, salt meadow on Castle Hill Neck bounded by John Hunt, John Butler, west by Cromwell's Creek and east by upland." The owner of this land had been John Cromwell, and the creek had other names. See: Clason's Creek, Pugsley Creek, Wilkin's Creek and West Creek.

CROMWELL'S FARM. This was one of the earliest farms to be established in the Morrisania Manorlands. James Cromwell's acreage was bounded by East 157th Street, West 168th Street, Jerome Avenue and Walton Avenue, including Cromwell's Creek (1).

CROMWELL'S NECK. This earliest name of Castle Hill Point was noted in 1685 when John Cromwell and Elizabeth, his wife, exchanged six acres of meadow with Thomas Hunt for eight acres of upland. See: Castle Neck, Screvin's Point.

CROSS STREET (1). This is the former name of West 162nd Street from Summit Avenue to Anderson Avenue. See: High Street.

CROSS STREET (2). An 1878 name for Rodman Place which crossed from Boston Road to West Farms Road. The original Grace Episcopal Church of West Farms was located at Cross Street, facing the Bronx River.

CROSS STREET (3). This was the earlier name of Bathgate Avenue from Fordham Road to the Fordham campus. See: Elizabeth Avenue, Madison Avenue.

CROTON AVENUE. This was the former name of West 190th Street as it led from the Croton Aqueduct to Jerome Avenue. In 1884, it was renamed St. James Street. See: Pipe Street, St. James Street.

CROTON PARK. This was the earlier name of Crotona Park after it had been acquired by the Department of Parks in 1889 from the Bathgate family. Originally planned as Bathgate Park, it was later decided to call it Croton Park, but because of confusion with Croton Aqueduct on the far side of The Bronx, it was finally named Crotona Park. See: Bathgate Park.

CROTON STREET. See: Croton Avenue.

CROTONA CASINO. This was a prominent dance hall of the early 1900s on Boston Road and East 169th Street. During a cold winter's night, it caught fire and burned down.

CROTONA, The. A silent movie house at 435 East Tremont Avenue near Third Avenue, it was opened in 1910 and closed in 1960. It had the impressive seating capacity of 2,210.

CROTTY'S WOODS. This was a former picnic grounds in Tremont, along Third Avenue from East 177th Street to East 178th Street run by Patrick Crotty, who also had the Railroad Hotel on the premises. Its span of operation was from 1885 to 1902.

CROW HILL. This was the 19th-century name of the hill now bisected by Mayflower Avenue and Middletown Road. It is occupied by the Tremont Terrace Moravian church, which replaced the mansion of Captain Watson Ferris. The steep slope of this hill facing East Tremont Avenue was sometimes called Bridge Hill, as it faced the bridge over Westchester Creek. This may also be the Red(d) Hill mentioned in 1686 lying near the Great Creek, which was an early name of Westchester Creek.

CULLINDEN AVENUE. This is a name once officially given to Wissmann's Lane around 1882, but never used. See: Morgan Avenue.

CURTISS AVENUE. This avenue was laid out to extend from Baychester Avenue to Tillotson Avenue, alongside an airport planned to be called Curtiss Airport. The avenue has since become absorbed by the New England Thruway, and the airport was never built. Instead, "Freedomland" was constructed there in 1959-1960, and now is the land upon which Co-Op City stands.

CUSICK'S SLIP. This is the narrow slip at the extreme eastern end of The Bronx at the tip of Fort Schuyler. The name was in use during the tenure of Captain Charles Ferriera, keeper of the Fort Schuyler lighthouse (1900-1932) and may also have been known to his father, Alexander, before him. Captain Ferriera thought the name originated with an Irish stone mason from Schuylerville who worked on that portion of the sea wall in the 1840s. At that point, the waters of the East River become that of Long Island Sound. In the 1970s it was filled in.

CUSSER. This is a misspelling of Corsa on the British Headquarters map of 1778 by Skinner & Taylor.

CUSTER STREET. This street was laid out prior to 1898 in Port Morris, running from the East River to Bruckner Boulevard, parallel to East 149th Street. The street was never cut through, and is part of the railroad yards today.

CUTHBERT LANE. Around 1870, this was the name of West 261st Street from Independence Avenue to Riverdale Avenue, and was named for Thomas Cuthbert, whose estate was bounded on the north by this lane. He came from Glastonbury, England, in 1864 and settled in the wooded estate overlooking the Hudson River. See: Randolph Lane.

CYPRESS AVENUE. This was the original name of Trinity Avenue when it was first surveyed in 1857. See: Grove Street, Delmonico Street, Passage Avenue.

CZECH CORNER. This was a local nickname for a small promontory east of the Bartow mansion in Pelham Bay Park facing Hunter Island. Czech (Bohemian) campers patronized this section for many years.

D.S.C. BLOCK. This became a local nickname for East 152nd Street between Melrose and Courtlandt Avenues, because the Department of Street Cleaning maintained its stables, wagon lots, blacksmith shop and harness-repair loft on that block. The nickname apparently was in use from about 1910 to around 1938.

D.S.C. HILL. See: D.S.C. Block, Elton Street, Rose Street.

DAIRY CREEK. This tidal stream that flowed in from the Bronx River in the vicinity of Harding Park on Clason Point was noted on several 19th-century maps, and was still navigable in the 1920s, but is now filled in.

DALY, The. This silent movie house on Boston Road at East 180th Street was in operation during the 1920s.

DALY THEATRE. A silent movie house on East Tremont and Honeywell Avenues. See: Vogue Theatre.

DALY'S DOCK. This was a busy wharf on the Harlem River at Third Avenue mentioned in City ordinances of 1898, 1899 and 1900. Excursion boats left from this pier bound for the amusement centers of Clason Point, Belden's Point on City Island, and to Kyle's Park at Highbridgeville. The ordinances concerned public safety, police coverage and maritime safeguards.

DAMIS AVENUE. This was the original name of Colgate Avenue from the Bronx River to Bruckner Boulevard. Sherrood Damis was one of the early Westchester colonists of 1656 who voluntarily pledged his allegiance to the New Netherlands.

DANGERTOWN. This was a nickname applied to Highbridgeville in the 1890s. The nickname was frequently encountered in the New York *Eagle*, a Bronx-based newspaper that was published from 1884 to 1896. See: Dangerville.

DANGERVILLE. This was a facetious name that was given to Highbridgeville by residents of neighboring villages. One belief as to its origin was that it was considered dangerous to venture among the Irish laborers that made up the bulk of the population. Another tale, told to this writer around 1926 by an old resident, was that a wealthy resident ordered wrought-iron letters spelling out GARDENVILLE to be set on his lawn overlooking the Harlem River. Some jokers climbed the hillside and transposed the letters to read D-A-N-G-E-R-V-I-L-L-E, to the amusement of the commuters on the New York Central R.R. that passed the estate. An alternate nickname was Dangertown. See: Devoe's Point, Highbridgeville, Nuasin, Turneur's Land.

DANIEL ORCHARD CREEK. This waterway ran through the Taber and Ferris farmlands into Westchester Creek in the vicinity of present-day Bruckner Circle. It is mentioned in an 1830 deed, and mapped in 1868.

DANIEL ORCHARD'S LAND. In the ancient town records of Oostdorp (now Westchester Square) the Archer family was frequently referred to as the Orchard family in 1664. The lands referred to were alongside Westchester Creek, near present-day Bruckner Circle.

DARK STREET. This is the former name of Lustre Street in the Edenwald sector. It was displaced on February 26, 1926, by an ordinance of the Board of Aldermen.

DARKE STREET. This is an earlier name of Heath Avenue, at West 230th Street named after Charles Darke, a landowner in the 1860s. See: "Blood Alley," Drake Lane.

DARLING'S SAIL LOFT. The first sail loft was established on City Island by William Darling around 1865. He made sails for sloops of the oystermen's fleet, as well as for ships built on City Island. His loft was located on the north side of Rochelle Street.

DASH'S LANE. In the 19th century, this was the name of West 240th Street. Bowie Dash had his mansion on Waldo Avenue, opposite Van Cortlandt Park. See: Spuyten Duyvil Road.

DASHWOOD AVENUE. This is the former name of West Burnside Avenue, named in honor of a landowner named G. L. Dashwood. See: Transverse Road.

DASHWOOD PLACE. On Morris Heights, this was the early name of West Tremont Avenue from the Harlem River to Sedgwick Avenue. An 1890 map indicates the property was owned by G. L. Dashwood.

DATER AVENUE. In the village of East Morrisania, Philip Dater was a wealthy grocery merchant. With his brother, he brought the Fleetwood Trotting Course in 1870. When streets were cut through from St. Mary's Park to Southern Boulevard, Dater Avenue was named in his honor, but later was changed to East 147th Street. See: Lexington Avenue, Lexington Street, Trinity Avenue (2).

DATER ESTATE. In 1888, this was owned by Philip Dater in what was then the 23rd Ward. The approximate bounds of his property were Westchester Avenue, Prospect Avenue, Southern Boulevard and Jackson Avenue. It was surveyed and laid out by John G. Van Horne in 1894. See: Dater Avenue. The project known as Adams Houses is erected on part of this estate.

DAUGHERTY'S POINT. This was a nickname in vogue around 1900-1910 for the area around the foot of Miles Avenue on Long Island Sound. The Daugherty family were nearby residents who sold bait, clams and illicit liquor to canoeists and campers. At one time, they were hired as caretakers on the Adee estate (now Park of Edgewater). This family occupied the same house and grounds from 1840 to 1962. See: Dinny's Bay.

DAVID'S NECK. This was an 1797 reference to a small neck of land on the west side of the Hutchinson River below Co-Op City. It is likely to allude to David 1. Pell, who was a landowner there.

DAVIDS AVENUE. For many decades, on Clason Point maps, this avenue has been depicted between O'Brien and Gildersleeve Avenues, running from Soundview Avenue to Betts Avenue. It has yet to be a reality.

DAWSON STREET. This is the former name of East 155th Street, from Wales Avenue to Tinton Avenue, now covered by Adams Houses. When the Dater estate was subdivided in 1894, this street was given the name of a local historian, Henry B. Dawson, but a few years later it became East 155th Street. See: Mary Street.

DAYTON PLACE. This name appears on an 1888 mapping of Loring Place north of the Bronx Community College.

DE LANCEY'S BLOCKHOUSE. During the Revolutionary War, this formidable fortress dominated the Boston Road at East 179th Street and also controlled any activity on the Bronx River. It was erected by Colonel DeLancey who had sided with the King of England, and who was later exiled to Canada at war's end. See: Mapes' Temperance Hall.

DE LANCEY'S MILLS. This was the pre-Revolutionary name for West Farms, for the mill and the manor house were located on the east bank of the Bronx River above East 180th Street. The property was confiscated after the Revolution. See: "Hive Town," Twelve Farms.

DE LANCEY'S PINE. A towering pine once stood north of East 180th Street on the left (east) bank of the Bronx River, parallel to the site of the vanished DeLancey mansion.

A poem written in 1830 called it "DeLancey's ancient pine." During the Revolutionary War, Colonial sharpshooters climbed the pine to give them a view of DeLancey's blockhouse, and perhaps a chance to shoot the Tory colonel himself.

DE MILT AVENUE. This is the earliest name of East 242nd Street, on the Penfield property east of the Bronx River. Anthony DeMilt was one of the first burghers of Nieuw Amsterdam, being listed in 1673 as a baker, although for one year he served as a Sheriff. Sarah and Elizabeth DeMilt were the last occupants, and they were related to the Penfields. See: Elizabeth Avenue (2).

DEADHEAD HILL. A wellknown nickname a century ago, it referred to a wooded ridge along Morris Avenue, north of Kingsbridge Road. It overlooked the Jerome Park Racetrack (1865 - 1894) so that spectators could watch the steeplechases and races without paying admission. "Deadheads" was the term applied to those who did not pay fares, or the price of admission to an event.

DEANE PLACE. In Morris Park, the early name of Bogart Avenue, from Sackett Avenue to Pierce Avenue.

DEBATABLE LANDS. This was a marshy tract of land between the Morris holdings in the West and the Hunt and Richardson grant in the East that was in constant dispute for centuries owing to a confusion in boundary creeks. See: Bound Brook, Bungay Creek, du Pont Street, Leggett's Creek, Sacrahong, Wigwam Brook.

DECATUR STREET. See: Decatur Avenue.

DECKERVILLE. See: Woodstock.

DEFENDER HOSE COMPANY. A photograph of this Eastchester firehouse shows a frame building with a sharply gabled roof. Before 1890, it stood on Boston Road where the Dyre Avenue railway line crosses it.

DEIDERMAN'S BREWERY. This was one of Melrose's earliest breweries, owned and operated by Charles Deiderman in the 1850s. It was located on the east side of Third Avenue, a short distance north of its junction with Westchester Avenue. In 1864, it was sold to Peter Kirchof. See: Kirchof's Brewery.

DELAFIELD LANE. The original name of West 246th Street from Riverdale Avenue down to Broadway. Delafield Lane continues as a private road from Riverdale Avenue to approximately 50 yards past Arlington Avenue in "Dodgewood, Inc."

DELAFIELD POND. This was an almost circular pond formerly located at Livingston Avenue and Waldo Avenue at West 248th Street. See: Duck Pond, Indian Pond.

DELANCEY STREET. This is the former name of Washington Avenue in Belmont from Third Avenue to Fordham Road. It is not named for Colonel DeLancey, the Tory, but for a former landowner named Delancey Powell whose "Cedar Hill farm" was situated there in the 1860s. He was related to an earlier owner, Reverend William Powell, who had married the widow Bayard. An earlier name of the property (1837) was the Bayard farm. The story is that the town fathers, who did not like the Tory connotation, so furthered Washington Avenue up from Tremont.

DELMONICO STREET. This former name of Trinity Avenue, from East 161st Street to East 166th Street, was mapped in 1868 in the settlement of Forest Grove. Prior to that time, it had been the farmland of H. P. DeGraaf. The name was superseded in 1875 by the numbered street sign. See: Avenue C (1), Grove Avenue.

DELMOUR'S POINT. This is a point of land on the northwestern shoreline of City Island facing Eastchester Bay. The property was once that of a prominent politician, "Whispering Larry" Delmour.

DELUXE. Originally this silent movie house, located at East Tremont Avenue and Belmont Avenue, was listed in 1922 as The Belmont, having a seating capacity of 1,458, and was in operation from 1920 to 1933, when it became the Deluxe, to around the 1970s.

DENISON'S LANE. The alternate spelling of Dennison's Lane. See: Dennison's Lane.

DENMAN PLACE. According to an 1879 Atlas, this was the early name of East 159th Street from Forest Avenue to Union Avenue. It was named for Hampton B. Denman, a prominent real estate developer of the time.

DENMAN STREET. This is the former name of East 150th Street in the village of Melrose. A February 1859 deed mentions Hampton B. Denman, who purchased lots from D. H. Carpenter. Later that year, a deed records a conveyance of 79 lots from Hampton B. Denman to Robert Elton. On this country street, the German Redemptorist priests built their first wooden church, the Immaculate Conception R.C. church.

DENNISON'S LANE. Charles Dennison owned property near Hunts Point in the 19th century, and for a time it was named after him. See: Leggett's Lane, White's Lane.

DEPOT PLACE. It led from Ogden Avenue at Union Place to the railroad depot below High Bridge, but now is under Highbridge Houses and the roadbed of the Major Deegan Expressway.

DEPOT SQUARE NORTH. This is the former name of Botanical Square North.

DEPOT SQUARE SOUTH. This is the original name of Botanical Square South.

DEPOT STREET. This is the former name of West 230th Street off Broadway.

DESBROW PLACE. Murdock Avenue, on the city line at East 242nd Street, once carried this name. Disbrow [spelled this way] is a name often encountered in the Huguenot records of that area.

DEVIL'S BELT. This was the colonial name for Long Island Sound. Belt is a geographical term referring to a strait, and the early settlers regarded the Sound as a dangerous, devilish body of water.

DEVINE'S LANE. The turn-of-the-century name of Taylor Avenue from East Tremont Avenue to Westchester Avenue, across what was earlier the Mapes farm. Some old-timers say the lane was really a boardwalk to raise it above boggy ground and had been built for a tavern keeper on the Westchester Turnpike named Devine.

DEVOE STREET. Noted in the 1871 Directory of Highbridgeville was this forerunner of West 165th Street, from University Avenue to Anderson Avenue. It harks back to Frederick Devoe who settled there around 1700.

DEVOE'S LANE. This was the former name of Oneida Avenue, from Woodlawn Cemetery to Van Cortlandt Park East and into the park itself. This lane figured in the Revolutionary War (1778) when Chief Nimham and his Indians fought Colonel Simcoe and his British cavalry. See: Devow, Fourth Avenue (2).

DEVOE'S POINT. The land between the Harlem River and Cromwell's Creek (Jerome Avenue and Cromwell Avenue) was called this during the 1700s. It was the original "Nuasin" of Indian days that was purchased by Daniel Turneur in 1671. Frederick Devoe wed Turneur's daughter and inherited the land. See: Bickley Patent.

DEVOW. This misspelling of Devoe appears on British Headquarters maps of 1778 by Skinner & Taylor.

DICKERT'S OLD POINT COMFORT PARK. This old-time amusement park and picnic grounds was situated at 4018 Boston Road, above Dyre Avenue. It accommodated a maximum of 5,000 visitors, and on its 4 acres boasted of four bowling alleys, a rifle range, a dance pavilion and hotel. The hotel had originally been a roadside tavern of the 1700s. The Park was operated by Henry Dickert in the 1920s, and was sold to the Breinlinger family in 1926. See: Breinlinger's, Odell's Tavern.

DICKEY STREET. This is the former name of Lafayette Avenue from Hunts Point Road to the Bronx River. C. D. Dickey was listed as the owner of a tract of land spanned by this avenue in 1879. The Dickey mansion stood for many years after the estate was sold, and it was used as a synagogue.

DICKINSON'S POND. An 1865 map shows this pond in the center of present-day Mosholu Parkway off Gates Place, on John Dickinson's land. The runoff ran eastward to feed into Mill Brook (Webster Avenue). See: School Brook, Schuil Brook.

DIEHL'S BREWERY. It is listed in the 1871-1872 Directory of Morrisania and Tremont and was located on the Westchester Road near Eagle Avenue. John Diehl, Sr., and John Diehl, Jr., were listed as owners. The low brick building stood for almost a century, its last 35 years as a chicken market.

DIMANS ISLAND. It is mentioned in the 1686 Records of Westchester as being owned by a Richard Osbourne. Once surrounded by marshy land, it is now part of Pelham Bay Park at the western end of the Pelham bridge. It carried no less than six other names, the last one being the most familiar to many Bronxites: Tallapoosa Point. See: Dorman's Island, Dormer Island, Dormont's Island, Hunt's Neck, Taylor's Island and Tallapoosa Point.

DINNY'S BAY. This is a localism for a small inlet at the foot of Miles Avenue on Long Island Sound. "Dinny" Daugherty— fisherman, clam digger farmer— lived nearby on the edge of the Adee estate (now the Park of Edgewater) and he and his father once worked as caretakers there. See: Daugherty's Point.

DIVISION AVENUE. This avenue than ran from St. Ann's church to Southern Boulevard was noted on maps of 1871 and 1872. Beech Terrace is slightly to the south of it.

DIVISION STREET. This is the former name of Longfellow Avenue from East 176th Street to Boston Road. Evidently, it marked the division between Thomas Minford's property and that of August Woodruff in the 1850s. See: Boone Street.

DOCK STREET (1). A former name of Ferris Place, as it led to the dock on Westchester Creek when the village of Westchester was an important riverport. See: Bowne Street.

DOCK STREET (2). Possibly the oldest street in The Bronx was this landing place for the Dutch outpost of Oostdorp (Westchester Square). It was demapped in 1989 to accommodate the expansion of the Schildwachter Fuel Oil Corporation. See: adjoining Kirk Street.

DOCTORS' ROW. This is a turn-of-the-century nickname for Alexander Avenue when it was a fashionable tree-lined street. Judges, doctors, business men and politicians lived in the well maintained brick homes, giving rise to a more jocular nickname, "Irish Fifth Avenue."

DODGE LANE. The former name of West 247th Street when it led to "Greyston," the 1870 estate of William E. Dodge. Cleveland H. Dodge was the chairman of the Henry Hudson Monument Committee in 1909.

DODGE POND. See: Alder Brook.

DOG BEACH. This is the south shoreline of what is left of Weir Creek, forming a part of the Park of Edgewater, It received its name when the former bungalow colony was divided into four sections: A, B, C and D. This occurred around 1919, and in the nomenclature of World War I, the areas were nicknamed Able, Baker, Charley and Dog. So the beach in front of D-Section got its name.

DOGWOOD BROOK. See: Alder Brook.

DONGAN STREET. This is the former name of East 163rd Street from Westchester Avenue to Hunts Point. It was displaced in 1908. Colonel Thomas Dongan, Governor of New York (1683), approved the Charter dividing the Province into twelve counties, among them being Westchester County from which The Bronx was formed. Dongan had served as Governor of Tangiers, and had commanded the Regiment Irlandais in the army of Louis XIV of France.

DONNYBROOK HILL. It is mentioned in 1886 as being located north of Kingsbridge Road in the Claflin estate overlooking Jerome Park Reservoir. It can be located as East 196th Street west of the Concourse. It is reputed to have been the scene of an Irish laborers' battle (a donnybrook!) during the construction of the Aqueduct. Other versions ascribe the name to latter-day fights at the Jerome Park racetrack.

DONNYBROOK STREET. This is the former name of East 196th Street from Jerome Avenue to the Concourse, and took its name from Donnybrook Hill. See: Ridge Street, Sherwood Street, Wellesley Street.

DOON AVENUE. In earlier times, this was the name of Ely Avenue. This might be a survival of the Seton estate, which was known as "Cragdon." See: Bayard Street (2), South 21st Avenue.

DORIC FARM. This is the land on the east side of the Throggs Neck Road extending down to Long Island Sound mapped around 1840. The bounds were East Tremont Avenue, Philip Avenue to a line between Sampson and Dewey Avenues, to the Sound. The tract was originally owned by Philip Livingston in the 1790s. A sister, Margaretta Livingston Bayard who next owned it, bequeathed the land to a nephew, Stephen Bayard Hoffman, in 1822. A Dr. Wright Post was the next proprietor, who sold to John D. Wolfe in 1827. The Doric mansion was built in 1830 by Ellery Anderson, and it stood 50 yards north of Schley Avenue in the current center of Edison Avenue. It received its name from the Doric columns that supported the impressive facade. Daughter Catherine Lorillard Wolfe sold the land to Bruce Brown sometime before 1895. The mansion was demolished around 1929. See: Wolfe's Hill.

DORMAN'S ISLAND. This small island was mapped prior to 1704 near the mouth of the Hutchinson River. See: Dimans Island.

DORMER ISLAND. This is another version of Dorman's Island. There is a possible connection with Captain Thomas Dermer [spelled this way], the first Englishman to explore Long Island Sound in 1619.

DORMONDS ISLAND. It is mentioned in an ancient deed of 1684 when a Captain Osbourne became the owner.

DORMONTS ISLAND. An 1823 deed refers to this island, which was Hunt's Neck in 1777, Taylor's Island in 1851 and, in the 1890s, Tallapoosa Point.

DOUCINE STREET. This is the former name of Chestnut Street in Bronxwood Park, a real estate development around 1900.

DOUGHTY STREET (1). This is the former name of Ruppert Place, alongside Yankee Stadium, harking back to a 19th-century landowner who gave his name to Doughty's Brook. The name was changed in 1933.

DOUGHTY STREET (2). This is the earliest name of East 166th Street from Jerome Avenue to Gerard Avenue. This street subsequently was overlaid by Mullaly Park. Doughty's Brook flowed into Cromwell's Creek (Jerome Avenue) a few yards north of this point.

DOUGHTY STREET (3). Back in 1871, this street ran past the property of Colonel George Doughty. It was eliminated by the expansion of the Major Deegan Expressway, some time after the other end was renamed Ruppert Place.

DOUGHTY'S BROOK. This brook originated in the vicinity of Featherbed Lane and fed into Cromwell's Creek (West 167th Street).

DOWNING'S BROOK. It originated in a series of springs on the former Hitchcock estate in the vicinity of Boston Road, Matthews Avenue and Bronxwood Avenue. It flowed through swamplands, now Pelham Parkway, through the Morris Park sector and into the Bronx River at West Farms. On 1835 maps, W. B. Downing was a proprietor of considerable property along Bear Swamp Road before Van Nest was laid out. The brook ran through his estate and took his name.

DRAKE LANE. Actually, this is a misspelling of the original name of Darke Street. In the 1860s, Charles Darke was listed as owner of the property across which Heath Avenue is laid out. See: "Blood Alley."

DROVERS' INN. The Revolutionary homestead of the Jennings family alongside Boston Road at Jefferson Place. Walls showed bullet holes and other sears of warfare. When the village of Morrisania grew up around it, the old homestead became a popular tavern with sufficient pasturage for cattle being driven to market.

DU BOIS ISLAND. This small island was once in the Bronx River north of East 205th Street on a line with Rosewood Street. When the river was straightened from 1912 to 1920, the island was attached to the west bank. The owner, listed in the 1880s, was E. DuBois, one of many Frenchmen who lived in the area. The island, once it became part of the mainland, was included in the ballfield known as "French Charley's." See: Cowslip Island.

DUCK ISLAND. Around 1900, it was known to the boys of Springhurst, a small settlement near Hunts Point. It lay offshore in the East River at approximately Tiffany Street and Viele Avenue where Leggett Creek once flowed into the river. Landfill blotted out Duck Island around 1910. See: Long Rock.

DUCK POND. This was another name for Delafield Pond. See: Indian Pond.

"DUNSNAB." An 1868 map of South Yonkers (now The Bronx), shows this estate of W. A. Butler extending alongside Riverdale Avenue at West 242nd Street.

DURELL'S HOTEL. It is identified as the first hotel in Morrisania and opened for business in 1855. The address, "North side of Halsey St.," means it was located at East 162nd Street west of Melrose Park (near the depot).

DUTCH BROADWAY. This was Courtlandt Avenue from East 145th Street to East 161st Street. Due to its heavily Germanic population, this main thoroughfare of Melrose and Melrose South earned its nickname. The Haffen family, wealthy brewers, lived at East 152nd Street, and one son became the first Borough President. Beer gardens, gymnast clubs, German social halls and Lutheran churches lined Courtlandt Avenue, and the mile-long funeral cortege of General Franz Sigel in 1902 was of national note.

DUTCH BURYING GROUNDS. This small cemetery belonging to the Dutch Reformed church of Fordham was situated on the southeastern side of Fordham Road and Sedgwick Avenue. At one time, Edgar Allan Poe's wife was buried there. The burial grounds were established on what was John Archer's Manor of Fordham around 1680.

DUTCH FIVE, The. The popular nickname for the volunteer firemen of Hose Company 5 of Morrisania, due to its preponderance of "Dutchmen," i.e. Germans. The nickname was in use prior to 1900. See: Honey Bees.

DUTCH PRISON, the Old. This was an irreverent nickname for the Immaculate Conception R.C. school on East 151st Street in Melrose. The parish was predominately German, and nearby Courtlandt Avenue was referred to as Dutch Broadway.

DUTCH REFORMED CHURCH, The. It was build in 1706 at the junction of present-day Fordham Road and Sedgwick Avenue. The land was that of James Valentine's farm, later that of Moses Devoe. The site became the property of Webb's Shipbuilding Academy of the 1850s.

DYCKMAN'S BRIDGE. This was another name for the Farmers' Bridge over the Harlem River at West 225th Street, built in 1758 to circumvent the tolls upstream at the King's Bridge. See: Farmers' Bridge, Free Bridge, Hadley's Bridge, Queen's Bridge.

EADS STREET. This was a short street laid out on maps directly after World War I, and apparently honoring James Buchanan Eads, the Civil War builder of the Union's Mississippi flotilla - eight ironclads. In 1937, the building of the Bronx-Whitestone Bridge approach caused it to be demapped.

EAGLE GROVE. This rustic picnic grounds were on the west bank of the Hutchinson River just north of Co-Op City. In 1868, it was the property of George F. Codling. See: Codling Island.

EAGLE ROCK. This is a prominent bluff on the east bank of the Hutchinson River 200 yards north of the Hutchinson River Parkway bridge. It once was a popular camp of the Siwanoy Indians, judging from the many aboriginal artifacts found there. It was also a popular camping spot for canoeists prior to World War II, as there is a small cove at its base.

EAST 140TH STREET was called Bronx Street in Old Morrisania and named after the first settler of 1641. The Mill Brook crossed from 508 East 140th Street diagonally to 507. It was 2nd Street in Port Morris.

EAST 178TH STREET was formerly known as Cedar Street, and also Mechanic Street in the village of Belmont, but ran only to present-day Hughes Avenue.

EAST 194TH STREET off Pelham Parkway was cut through the Paul farm of earlier times. See: St. Paul Avenue.

EAST 195th Street off Pelham Parkway was cut through the Paul farm of earlier times. See: St. Paul Avenue.

EAST 196th Street off the Pelham Parkway was cut through the Paul farm of earlier times. See: St. Paul Avenue.

EAST 197th Street off Pelham Parkway was cut through the Paul farm of earlier times. See: St. Paul Avenue.

EAST 236th Street from the Bronx River to White Plains Avenue was formerly 22nd Avenue in Wakefield, after it had absorbed the smaller settlement of Jacksonville.

EAST BELMONT STREET. This is a former name of Mount Eden Avenue, from Townsend Avenue to Walton Avenue. See: Wolf Street, Walnut Street, Jane Street.

EAST MELROSE. It adjoined Melrose proper on the east, and was a long narrow strip extending from East 149th Street to East 159th Street, Third Avenue on the west, and the Mill Brook (now Brook Avenue) on the east. See: Melrose, Melrose South.

EAST MORRISANIA. Originally, this was part of the Morris family holdings until sold by Gouverneur Morris II from 1853 to 1857. The land was surveyed by I. C. Buckout and divided into city lots. Principal purchasers were the Dater brothers, John J. Crane, George St. John, Captain Kelly and J. Arnold.

EAST ROAD. See: Stadium Avenue.

EASTCHESTER AVENUE. See: Eastchester Street.

EASTCHESTER BAY DIVISION. This Volunteer Station of the Life Saving Service of New York City was located at the foot of Barkley Avenue on Eastchester Bay. In 1914, five officers and seven surfmen were listed.

EASTCHESTER CREEK. See: Hutchinson River (Current).

EASTCHESTER LANE. This is the former name of Conner Street which runs down to the Hutchinson River which was also known as Eastchester Creek. See: Hutchinson River (Current).

EASTCHESTER STREET. This was the former name of East 233rd Street from Van Cortlandt Park to Webster Avenue. It was superseded in 1894. See: Eastchester Avenue.

EASTDORPE. This half-anglicized version of Oostdorp-now Westchester Square-was used on September 21st, 1663, by Governor Peter Stuyvesant: "Eastdorpe, by the English called Westchester." See: Oostdorp.

EASTERN AVENUE. This is the former name of Prospect Avenue from Westchester Avenue to Southern Boulevard. This was a fashionable thoroughfare cutting across the former Dater estate in East Morrisania, hence its name. See: Taylor Avenue.

EASTERN BOULEVARD. This is the former name of Bruckner Boulevard from Hunts Point to Pelham Bay Park. The name was changed in 1942 to honor a former Borough President, Henry Bruckner. See: Commerce Street, Ludlow Avenue, Sixth Street, Willow Lane.

EASTWOOD PLACE. This short lane at the northwest corner of Gun Hill Road and Webster Avenue was mapped in 1884, but incorporated into Woodlawn Cemetery. See: Sesgo Place.

EBLING'S BREWERY. Aurora Park was a spacious picnic grounds, advertised in the N.Y. *Eagle* in 1891 and 1892, located on St. Ann's Avenue at East 156th Street, and extending uphill to Cauldwell Avenue. Owner Anna Beck sold some property in 1868, to brothers Philip and William Ebling, who then established their brewery on St. Ann's Avenue from East 157th Street to East 158th Street. The hillside up to Eagle Avenue was hollowed out for the caverns for storage of their beer. They later built a hall, adjoining their brewery. The brewery eventually was converted into small shops, ice plants and trade rooms. See: Ebling's Casino.

EBLING'S CASINO. This became one of the landmarks of The Bronx at the corner of East 156th Street from St. Ann's Avenue to Eagle Avenue. From the 1890s on, the casino was famous for its folk festivals, operas, concerts, dances, carnivals and Oktoberfests. It ceased its activities during World War II.

EBLING'S PARK. See: Ebling's Casino.

ECHO PARK. The name was changed to Richman Park in August, 1973, to honor a civic leader, Julius J. Richman. See: Echo Place.

ECLIPSE STREET. This is the former name of East 207th Street from Reservoir Oval to Webster Avenue, mapped in 1888. There was a solar eclipse in Japan in June, 1887 and it was reported worldwide, but news took a bit of time in those days. Perhaps that phenomena was the reason for the name.

EDEN AVENUE. This was an earlier name of Selwyn Avenue, from Morris Avenue to Mount Eden Avenue. See: Third Avenue (3).

EDEN TERRACE. This formed the lower boundary of John H. Eden's estate, which he called "Edenwald," the name being noted on maps of 1903 and 1910. Rattlesnake Creek, mentioned in 1644 deeds, flowed alongside this terrace until the 1960s, and has now become a rain-fed culvert. The name was changed to Marolla Terrace in 1968 to honor a local resident prominent in Little League baseball activities.

EDENWOOD AVENUE. On a 1905 map, it appeared as the former name of Grand Avenue, from Fordham Road to Kingsbridge Road. See: Sixth Avenue (1).

EDGEHILL INN. This was a late 19th-century resort overlooking the Spuyten Duyvil Creek. The slope was once known as Tippett's Hill, and the hotel was built on the site of Revolutionary Fort No. 3. See: Fort No. 3, Sage House.

EDGEWATER CAMP DIVISION. This Volunteer Station of the Life Saving Service of New York City was on what had been called Adee's Point, which then became Edgewater Camp, a bungalow colony on Throggs Neck. Thirty-nine surfmen were listed in 1923, commanded by four officers, while three lifeboats constituted their equipment.

"EDNAM." Robert and Agnes (Thompson) Givan came to American shores from Kelso, Scotland, in 1795, and named their property in Westchester County after a county in Scotland. This tract was located east of present-day Givan Square overlooking the Hutchinson River valley, the approximate bounds being Givan's Creek at the foot of Mace Avenue, Morgan and Arnow Avenues, and Grace Avenue where it overlays Black Dog Brook.

EDSALL STREET. This was the original name of East 135th Street. Samuel Edsall bought the land from Harman Smeeman in 1664, and sold it to Lewis Morris in 1668. He was proficient in the Indian language, and acted as interpreter for the Dutch authorities. See: Orange Street, Seventh Street (1).

EEL HOLE, The. This was a deep section of Baxter Creek mentioned in deeds of 1794 and 1796. From earliest times, farmers on Ferris Point (now Ferry Point Park) staked out dead horses and cows in the creek to attract eels and crabs. Landfill blotted out Baxter Creek, Coney Island Creek and Morris Cove in the 1940s, and the Eel Hole is now under some 10 feet of soil in the new addition to St. Raymond's Cemetery.

EELPOT CHURCH, The. This is a longstanding nickname for the Westchester Methodist church on East Tremont Avenue near Silver Street, the oldest church of this denomination that was chartered in 1803 (some annals list 1808). Two previous churches on the spot had been destroyed by fire, so the parishioners decided a brick church would be safer. From 1913 to 1948, the congregation worshipped in the roofed-over basement that was 2/3rds below ground, just as an eelpot is 2/3rds submerged. In 1948, the church was finally constructed over the basement. See: Basement Church, The.

EGBERT STREET. This street in the vicinity of East 205th Street is now incorporated into the Botanical Garden.

EICHLER'S BREWERY. John Eichler was born in Bavaria in 1829 and teamed the brewers' trade before he came to New York in 1853. He worked in the Turtle Bay (Ruppert) Brewery until he had enough capital to buy the Kolb Brewery on Third Avenue and East 169th Street. By 1888, his brewery was judged the best equipped in America. Mr. Eichler died while on a visit to Germany, but was buried in Woodlawn Cemetery. The brewery was razed in 1965, but the Eichler mansion still stands on Fulton Avenue and East 169th Street.

EIGHTEENTH AVENUE. In Wakefield, this was the earlier name of East 232nd Street from Bronx Park East to White Plains Road.

EIGHTH AVENUE. This is the former name of East 222nd Street from the Bronx River to Bronxwood Avenue.

EIGHTH STREET (1). In the village of Morrisania, this was the first name of East 170th Street from Webster Avenue to Third Avenue in the 1860s.

EIGHTH STREET (2). When Unionport was laid out in Civil War days, the streets were given numbers, and the avenues given letters of the alphabet. This is Blackrock Avenue.

EIGHTH STREET (3). This is the former name of East 134th Street in Port Morris from Willow Avenue to the East River. See: Mott Street.

EISENHOWER PARK. This was an unofficial name used by some residents for the controversial park at Wilkinson Avenue and Hutchinson River Parkway. It was formerly a rocky stretch that was finally made into a playground and bocci courts in 1969 after neighborhood groups were active in its completion. The official name is Florence Colucci Playground honoring a civic leader. See: Quoglia Playground.

ELBERON AVENUE. This sharp incline, sloping to the east, is believed to have been the heights utilized by the Colonial troops to halt a British advance over upper Westchester Creek (Stoney Brook) in October 1776. The creek is now filled in and covered by the grounds of the State Mental Hospital. Originally Elberon Avenue was to have stretched from Pelham Parkway to Eastchester Road when a real estate development called Westchester Heights was planned. However, the section north of Morris Park Avenue was never cut through, and is now part of the Van Etten and Jacobi hospital grounds. Elberon Avenue was opened in 1924. Every street in the development was given the name of a prominent vacation resort, such as Newport, Narragansett, Saratoga, etc. Elberon, N.J. was, at that time, a gathering place for socialites. In the end, the section of the street which had been opened was absorbed into the State hospital site.

EL DORADO. This small silent movie house on Southern Boulevard, Freeman Street and Wilkins Avenue also had an open-air theatre leading out towards Intervale Avenue. It was owned and operated by the Whitman family, and was listed in 1912. It lasted until around the 1920s.

ELEVENTH AVENUE. This is the former name of East 225th Street in Wakefield.

ELEVENTH STREET (1). This is the former name of East 173rd Street from Webster Avenue to Third Avenue according to the "Map of Central Morrisania, being part of the Bathgate Farm, situated in the Manor of Westchester."

ELEVENTH STREET (2). In Unionport, this was the earliest name of Powell Avenue.

ELIZABETH AVENUE (1). This former name of Bathgate Avenue was possibly named for Elizabeth Bassford, wife of the landowner in this area. See: Cross Street, Madison Avenue.

ELIZABETH AVENUE (2). Elizabeth DeMilt was the daughter of the property owner in Wakefield, and this was the first name given to the street that eventually became East 237th Street. In 1897, Father Francis Moore celebrated the first Mass of St. Francis of Rome in a tent on this street. See: First Avenue (6), Oakley Street, Oakly Street.

ELIZABETH PLACE. An 1878 map shows this to be present-day Cameron Place.

ELIZABETH STREET (1). This is the former name of Rochelle Street on City Island. It is most likely named for Elizabeth Horton whose name appears on deeds of the 1840s. She was the wife of George Washington Horton, who once owned this property.

ELIZABETH STREET (2). Up to 1950, this was the name of Kilroe Street on City Island. Elizabeth King was the former property owner of a few lots to give her husband, Elisha King, landing rights from High Island. See: First Avenue (6), Oakley Street, Oakly Street.

ELIZABETH STREET (3). See: Elizabeth Avenue (2).

ELIZABETH STREET (4). On an 1889 map this was the former name of Cameron Place from Jerome Avenue to Morris Avenue. See: Elizabeth Place.

ELLA STREET. This was the earlier name of East 164th Street from the Concourse to Park Avenue, noted in 1871. Ella Morris was the wife of the owner of this tract, William H. Morris. See: Second Place.

ELLIOTT AVENUE (1). This is the former name of Baisley Avenue. David Blizzard Elliott was a former resident of Schuylerville, and his family dated back to the early 1800s.

ELLIOTT AVENUE (2). The earliest name of Olinville Avenue from Burke Avenue to Gun Hill Road.

ELLIOTT STREET. William Elliott Zborowski was the resident of "Claremont," an estate of the 1870s. When part of it was subdivided into city streets, one was named Elliott Street, but soon after became Clay Avenue. See: Anthony Street, Crestline Street.

ELM AVENUE. This is the former name of East 182nd Street from Crotona Avenue to Southern Boulevard. See: Fourth Street (4).

ELM STREET (1). This is the early name of East 145th Street from St. Mary's Park to Southern Boulevard. See: Crane Street.

ELM STREET (2). In the 1890s, this was Daly Avenue from East 176th Street to East 177th Street. See: Catherine Avenue.

ELM STREET (3). Third Avenue to Southern Boulevard. See: East 179th Street.

ELM STREET (4). A short street that ran from Belmont Avenue to Southern Boulevard between East 182nd Street and Garden Street on an 1874 map of West Farms.

"ELMHURST." This 17 acres of velvet lawns, summer houses, and a handsome villa belonged to Giovanni P. Morosini in the 1870s. The estate was bounded by the Hudson River, West 254th Street, West 256th Street and Riverdale Avenue. Morosini came to America practically penniless and died as one of the richest men in the country. His impressive mausoleum stands in Woodlawn Cemetery.

"ELMS, The"(1). Surveyed by Henry Lett in 1889, this rustic estate was south of Fordham Road at its junction with Kingsbridge Road. The name survives in Elm Place.

"ELMS, The"(2). This estate of Thomas Ash was recorded in 1831, and is found on an 1853 map of Throggs Neck. His 60 acres were bounded by the East River, Lawton, Hollywood and Swinton Avenues. The junction of East Tremont and Lawton Avenues was called Ash's corner in the 19th century.

"ELMSRING." The Waldo Hutchins estate (1868), west of Broadway near West 244th Street.

"ELMWOOD"(1). This was the 19th-century Paul Spofford estate on Hunts Point.

"ELMWOOD" (2). The Dashwood family of the West Bronx owned this estate in 1870 in the vicinity of University and West Burnside Avenues.

ELMWOOD PLACE. This is the former name of East 178th Street from Mapes to Marmion Avenues. See: Cedar Street, Mechanic Street.

ELSMERE, The. This silent movie house at 1926 Crotona Parkway was in operation from 1914 to around 1950. It had a seating capacity of 1,552, later enlarged to 1,721 when it converted to talking pictures. It was converted into a community center.

ELTON STREET. This was the earliest name of East 152nd Street from Park Avenue to Third Avenue in the village of Melrose South. The name was noted in 1870, and refers to Robert Elton, who was listed as early as 1848 when he purchased land from Gouverneur Morris II. See: D.S.C. Hill, Kelly Street, Rose Street, Willow Street.

ELTONA. This village was bounded on the west by Morrisania and on the south and east by Woodstock. It was formed from Robert Elton's small estate, and the boundaries were from Third Avenue to Prospect Avenue, from East 163rd Street to East 165th Street.

"ELTONA." An 1851 map shows the property of Robert Elton extending from Forest Avenue to Prospect Avenue around East 165th Street.

ELWOOD PLACE. It was laid out (on paper) for many years, from 1900 onward, across what had been the Kidd Estate on Eastchester Road. Finally, around 1967, it was cut through, but then was speedily renamed Cahill Place to honor a pastor of a nearby Catholic church. Finally, in 1973, Cahill Place became the name officially adopted.

EMBRIE PLACE. The former name of Lydig Avenue from Bronx Park East to White Plains Road, noted in 1913 when a realtor named Embrie Hill formed a corporation there.

EMERALD FIELD. A ballfield, the headquarters of the Emerald A.C., was located off Westchester Avenue at the foot of Herschell Street. A tributary of Westchester Creek limited the outfield. This diamond was laid out on the pre-Revolutionary farm belonging to the Seabury family. It was razed in the 1930s.

EMILY STREET. This is the early name of Roberts Avenue from Hutchinson River Parkway to Westchester Avenue. It carried this name from the late 1890s to the early 1900s when the section was called Tremont Terrace. It was most likely named for Emily Gainsborg, wife of the man who financed the realty venture. See: Tremont Road.

"EMMAUS." This was the land grant of Jonas Bronck whose house was then built on the site of the Indian camp of Ranaqua at East 132nd Street. The settler was undoubtedly religious and chose this name from the Bible.

EMMERICK PLACE. On University Heights, there was much Revolutionary War activity, and this was the especial territory of Lt. Colonel Andreas Emmerich [correct spelling] and his German Jaegers. Evidently, the people did not object to a street being named for a man who had fought "on the other side." Emmerick Place was incorporated into a stretch of Sedgwick Avenue at the edge of Devoe Park in 1905. See: Cedar Avenue.

EMMET STREET. See: Fordham Square [Current].

EMPIRE ENGINE COMPANY. This Washingtonville firehouse was located at Fulton Street and Westchester Avenue, the present-day East 240th Street and Richardson Avenue. The unit was disbanded in February, 1896.

EMPIRE ENGINE COMPANY, NO. 1. This Volunteer Hose Company was listed in the 1869 Directory of Westchester County. The firehouse was located alongside Westchester Creek at Westchester Square on a triangle of land now occupied by Dock Street. Later, the organization was known as the Westchester Exempt Firemen's Hose Company.

EMPIRE, The. The Empire was a silent movie house as early as 1894, and was listed in 1922 as a theatre with a seating capacity of 1,660. It was located on Westchester Avenue near East 161st Street on the west side. It was one of the first Bronx motion picture pictures to run Spanish films, as advertised in 1940.

EMPRESS, The. This silent movie house, once in operation on East Tremont Avenue, was listed in 1922.

ENDROW PLACE (1). *In Robinson's Atlas* of 1879, the former name of East 166th Street from Cromwell's Creek (Jerome Avenue) to Gerard Avenue. The street subsequently was overlaid by John Mullaly Park.

ENDROW PLACE (2). This was the earliest name of McClellan Street, as mapped in Bromley's Atlas of 1897. See: Maillard Place.

"ENGELHEIM." This 150-acre estate belonged to John Albert Morris in 1865. It ran along the East River shoreline of Throggs Neck from present-day East Tremont Avenue to Ferry Point Park, and was part of his father's lands. It later included subsequent purchases of lowlying swamplands bordering Baxter's Creek, now occupied by Throggs Neck Houses.

ERNESCLIFFE PLACE. A former name of East 205th Street, from Bainbridge Avenue to Webster Avenue.

ESCORIAZA FARM. The property of Virgil and Mercedes de Escoriaza along the Throggs Neck Road was mapped in 1868. The approximate bounds were East Tremont Avenue to Quincy Avenue, from Randall to Philip Avenues. It was purchased from Teofilo Gimbernat in 1865. Mrs. deescoriaza was a sister to Teofilo. Her daughters were named Ludovina and Corrine. See: Gimbernat Farm, Rhinelander Lot.

EUREKA BEACH. Advertised as the largest, most palatial and sanitary bathing spot in Greater New York, with first class canoe accommodations, and rowboats for hire, R. Kirchhein was the proprietor. It was situated on Eastchester Bay at the foot of Layton Avenue according to the 1920 ad.

EUREKA HOUSE. This was a small frame hotel on the shore of Eastchester Bay at the foot of Layton Avenue. A coach was sent to Schuylerville to meet the trolley cars and transport the guests to the hotel that specialized in seafood. It was in business as early as 1910, and later became a beach resort.

EVADNA STREET. This was the earlier name of St. Raymond's Avenue from Williamsbridge Road to Blondell Avenue, across the former Arnow property. Evadna Arnow married Daniel Mapes in the 1880s, and this realtor and auctioneer saw to it that his wife's name was given to the street when it was first cut through.

EVANS AVENUE. The former name of Randall Avenue from East Tremont Avenue to Long Island Sound.

EVELYN PLACE. The 19th-century name of Wilkinson Avenue, from the Hutchinson River Parkway to Westchester Avenue, across the former property of Robert and Ellen Wilkinson. Before them, it was the Revolutionary farm of the Paul family. Evelyn Stinard (related to the Pauls) was remembered for a few decades.

EXPOSITION PARK. This was the original name of Starlight Park on the Bronx River off West Farms Square. It was laid out in 1917-1918 as the exposition grounds of the New York International Exposition of Science, Industry and Arts. Later, it became an amusement park. See: Starlight Park.

EXTRA STREET. This was the former name of Irvine Street on Hunts Point, as shown on a 1912 land map.

EYVERST STREET. This was a former name of Lyvere Street, or - most likely - a mistake in spelling.

"FAIRLAWN" (1). Hugh N. Camp's home on the West Bronx slopes, as mapped in 1865.

"FAIRLAWN" (2). This was a small estate at Sackett Avenue south of Williamsbridge Road, overlooking the railroad tracks, belonging to the Baisley family. The land totaled but 3.6 acres.

"FAIRMOUNT" (1). The property of Robert Cochran in 1854 gave its name to the later settlement of Fairmount, and to present-day Fairmount Place. This tract belonged, later on, to a Colonel Dunham and Thomas B. Minford.

"FAIRMOUNT" (2). This was an estate in the possession of Colonel Dunham in the vicinity of present-day Crotona Park in the 1860s which still used the name, Fairmount, given by Robert Cochran. Part of this estate was sold to Thomas B. Minford, who called his portion "Minford Place." FAIRMOUNT ATHLETIC CLUB. Sometimes spelled Fairmont, it was once a prominent athletic center noted for its professional boxing matches of the early 1900s. Jack Dempsey fought there in an early preliminary bout, and Benny Leonard was another fighter who made his mark there. A local idol was Frankie Jerome, who fought anyone, anytime, at any weight. This fight club was housed in what once was a part of the Mott Iron Works, at 251 East 137th Street. Its location in the South Bronx attracted fight fans from Harlem, New Jersey (via the 125th Street crosstown trolleys), and even soldiers from Fort Schuyler who came downstream by launch. Boat service from Clason Point and Hunts Point supplemented the Third Avenue Elevated, and the newly opened New York, Westchester & Boston Railway. Billy Gibson ran the clubhouse. It was razed in the early 1960s.

FAIRMOUNT AVENUE (1). The former name of Crotona Park North appeared on an 1874 map of West Farms.

FAIRMOUNT AVENUE (2). This was an earlier name of East 175th Street from Crotona Avenue to Southern Boulevard. See: Fairmount Street.

FAIRMOUNT HALL. This social hall was located on the northwest corner of Prospect Avenue and Waverly Place (now Fairmount Place) in 1874.

FAIRMOUNT STREET. The former name of East 175th Street, from Marmion Avenue to Southern Boulevard was named after the village of Fairmount, which, in turn, harked back to a 19th-century estate called "Fairmount." See: Fitch Street, Gray Street, Oxford Street.

FAIRMOUNT, The. A 1912 Atlas lists this silent movie house at East 180th Street as a one-storied frame building.

"FAIRMOUNT-ON-SOUND." This real estate venture of 1874 was initiated by Patrick L. and Sarah Rogers, whose "Sunnyside" estate occupied the mainland behind Hunter Island. Avenues were planned, a pier was designed and Shore Road was to be the principal thoroughfare. It was surveyed by Rudolph Rosa, but the project was never completed. See: Rogers' Islands.

FAIRYLAND. A small amusement park near the end of Soundview Avenue and Stephens Avenue, opposite the far larger Clason Point Amusement Park. It operated in the summers of the 1920s and 1930s.

FALCONER STREET. Along with other Hunts Point avenues, this street honored a poet, William Falconer, born in Scotland in 1732. Falconer Street was part of the original 1663 "Corne Field Grant on Ye Planting Neck" of Jessup and Richardson, which passed to the Hunt family in Revolutionary times. An ancient Indian trail crossed Falconer Street at its junction with East Bay Avenue. In the 19th century the street was on the Fox lands, which later became the grounds of the Caswell Academy. Sometime around 1971, the street was covered over by the Hunt's Point Market.

FARADAY AVENUE. This is now Iselin Avenue, from West 246th Street in Riverdale to Grosvenor Avenue.

FARMERS' BRIDGE. A wooden bridge was built across the Harlem River at the foot of Kingsbridge Road to by-pass the King's Bridge toll. It was built in 1758 at West 225th Street at Exterior Street. See: Free Bridge.

FARRINGTON AVENUE. A 1910 map shows this short avenue plotted south Lafayette Avenue across the marshes between Zerega and Havemeyer Avenues. A 1947 Directory listed this avenue as well. It was never cut through.

FAY AVENUE. This is a former name of St. Raymond's Avenue from Westchester Creek (now filled in by the State Mental Hospital grounds) to Buhre Avenue. See: Evadna Street.

FEIGEL'S PARK. This German picnic grove, beer garden and dance pavilion was on East 169th Street and Park Avenue in the 1880s.

FENFELL AVENUE. This was the earlier name of Evergreen Avenue from the Bronx River to Bruckner Boulevard. William Fenfell was one of the Westchester colonists who pledged allegiance to the New Netherlands in 1656. Lt. Wheeler was their leader. The name was changed in 1911. See: Damis Avenue, Genner Avenue.

FENIAN RAM, The. This was a landmark for many years on the grounds of the Clason Point Military Academy. John P. Holland, Irish-born inventor of this 31-foot submarine, launched it in 1881, and it made many successful runs in New York harbor. In 1916, it was on exhibit in the original Madison Square Garden at the Irish Bazaar, and then was transported to the military school grounds. In 1927, the academy removed to Long Island and the submarine was acquired by the city of Patterson, N.J., where Holland had lived. It can be seen in Westside Park. See: Clason Point Military Academy, Holland Number Six.

FENWAY, The. Originally this silent movie house was located on Bathgate Avenue and Claremont Parkway from 1910 to around 1920. Becoming the Benenson, listed in 1922, located at 1546 Washington Avenue, near Claremont Parkway, where 1,312 patrons could be accommodated. The theatre was in operation from 1921 to 1932. Benjamin Benenson was a real estate developer of the early 1900s.

FERN BROOK. A small brook ran from Fern Brook Lake (now Mount St. Vincent) northeastward to the Yonkers line approximately in the bed of West 263rd Street.

FERN BROOK LAKE. A lake was formerly on the grounds of Mount St. Vincent, opposite West 263rd Street according to an 1868 map.

FERNCLIFF PLACE. A 1900 map shows this street near present-day Cromwell Avenue and East 151st Street, after it lost an earlier name of Waldorf Place. The land once was owned by John Jacob Astor, whose German ancestors came from a village called Waldorf. In 1955, Ferncliff Place was blotted out by the Bronx County Jail. See: Waldorf Place.

FERRIS AVENUE (1). This is the former name of Mace Avenue from Baychester Road to the Hutchinson River.

FERRIS AVENUE (2). Originally, it was called Ferris Lane. The name harks back to the original patentees of the Town of Westchester in 1667 when John Ferris' name was recorded. The name is said to be of Huguenot root: "Feriers." The Ferris family were 18th-century settlers on Ferris Neck or Ferris Point, and include landowners, patriots, sea captains, slaveholders, farmers and storekeepers. On December 7, 1938, Ferris Avenue had its name changed to Hutchinson River Parkway.

FERRIS DOCK. A 19th-century steamboat landing on Westchester Creek at present-day Brush Avenue north of Wenner Place. The road leading to this pier was mapped as Osseo Place, which was named after the principal steamboat that plied these waters. The dock was on the lands of the earliest Ferris homestead. See: Ferris House (1).

FERRIS HOUSE (1). The John Ferris house was a pre-Revolutionary homestead located on a slope between Westchester Creek and the present-day Hutchinson River Parkway. It was built by Thomas Hunt as a wedding gift to his daughter, Marianna, who married John Ferris (1733-1814). The house was demolished in 1941, and the tract around it - north of Wenner Place stood idle until 1969 when a complex of warehouses was built upon it.

FERRIS HOUSE (2). The Charlton Ferris house was an impressive mansion with shorefront property on the East River on the east side of Ferris Neck, now part of Ferry Point Park. The property had wrought-iron gates and a sweeping driveway. The mansion was sometimes called the Tabor House as Mrs. Tabor was a great-granddaughter of John Ferris and the last occupant of the house. Later, the estate became an amusement park and picnic grounds known as Pleasant Bay Park. Data was supplied by May A. Doherty, descendant of the Ferris family.

FERRIS HOUSE (3). The James Ferris house was located at the end of Rawlins Avenue, overlooking Eastchester Bay. Born in 1734, James, and his wife Charity, were Revolutionary War patriots who once were forced to give up this home to Admiral Lord Richard Howe who established his headquarters there in October, 1776. James Ferris was kept in the notorious prison ships maintained by the English, and died in 1780 as a result of the hardships he endured. In the late 1800s, the Ferris house was occupied over 30 years by Mrs. A.V.H. Ellis. It was demolished - practically overnight - in 1962.

FERRIS HOUSE (4). The Watson Ferris house was an imposing mansion off Middletown Road on Mayflower Avenue. Captain Ferris was a son of James and Charity Ferris and, ending his career as a sea captain, died in Panama in 1852. His body was brought home two years later, perfectly preserved in a hogshead of rum. The mansion was, for a short time, an inn called the "Marionette" - then became the Tremont Terrace Moravian Church It was razed in 1964.

FERRIS HOUSE (5). The Claiborne Ferris house was an exceptionally imposing home - the exact duplicate of Captain Watson Ferris' mansion off Middletown Road. This landmark was located on the corner of Eastchester Road and Pelham Parkway, but has been demolished - its place being taken by a stretch of lawn belonging to the Jacobi Hospital grounds.

FERRIS HOUSE (6). The David Ferris house was a large, barn-like structure on Middletown Road, off Edison Avenue. It was later occupied by farmer Cornell Ferris, a second cousin of Captain Cornell Ferris. Around 1891, it was owned by William Koch, a truck farmer, who had it torn down in order to build a more modern house.

FERRIS HOUSE (7). The Benjamin Ferris house was situated in Westchester Square near Thomas Street. This house descended first to his son, Benjamin, who was Supervisor for the Town of Westchester, 1802-1816 and 1819-1828. His son, Captain Cornell Ferris, inherited the property, and it was his widow, Mary Coleman Arthur Ferris, who retained it until 1894. The straightening of Westchester Avenue shaved off a portion of the land around 1898, and the Ferris house itself was replaced by a row of buildings.

FERRIS LANE. This very ancient lane led down to Ferry Point. It was flanked by Ferris farms for over a century, and later was designated Ferris Avenue. In 1938, it was taken over completely by a four-lane Hutchinson River Parkway that led to the Bronx-Whitestone Bridge.

FERRIS NECK. This was the Revolutionary name for Ferry Point. Members of the Ferris family lived there over several generations. See: Burial Point, Ferris Point, Lang's Point (1833), Lorillard Point, Old Ferry Point, Prime's Point (1851), Spicer's Neck, State's Point (1853) and Zeregors Point (1860).

FERRIS POINT. An earlier name, from the Ferris Family, for Ferry Point. See: Ferry Point Park.

FERRIS' WOODS. This was a mid-19th-century name for the wooded sector in the vicinity of present-day Plymouth and Roberts Avenues and Middletown Road, the former lands of Captain Watson Ferris. The tract was 5 1/2 acres in extent (1893 map) and later was subdivided into a real estate venture called Tremont Terrace. See: Tremont Terrace.

FERRY POINT. See: Ferry Point Park.

FIELD STREET. This is the former name of Delafield Avenue in Riverdale. See: Von Humboldt Avenue.

"FIELDSTON." These 250 acres were bought by Major Joseph Delafield in 1829 on what had been Upper Cortlandt. "Fieldston" was the name of his family seat in England. The estate was sold for city lots in 1893. See: Van Cortlandt's Ridge, Upper Cortlandt.

FIELDSTONE BRIDGE. This was the official name of a stone bridge over the Bronx River south of Allerton Avenue. The span was locally known as "the rubblestone bridge," and was replaced in the 1930s. See: "Rubblestone bridge."

FIFTEENTH AVENUE. This was a former name of East 229th Street from Bronx Park East to White Plains Road.

FIFTH AVENUE (1). This was the earlier name of Davidson Avenue, from West Fordham Road to West Kingsbridge Road.

FIFTH AVENUE (2). In the village of Wakefield, this was the 19th-century number of East 219th Street from the Bronx River to Bronxwood Avenue.

FIFTH AVENUE (3). This is the former name of East 241st Street from Van Cortlandt Park to the city line. See: Hyatt Street (1).

FIFTH STREET (1). In Port Morris, this was the former name of East 137th Street from Willow Avenue to the East River. See: Hendricks Street.

FIFTH STREET (2). When the village of Morrisania was laid out, this was East 167th Street from Webster Avenue to Third Avenue. See: James Street, Lyons Street, Overlook Avenue.

FIFTH STREET (3). This is the former name of East 181st Street from Jerome Avenue up to the Concourse. See: Ann Street.

FIFTH STREET (4). When the village of Unionport was laid out, this was Houghton Avenue.

FIFTH STREET (5). This is the former name of Bronxwood Avenue, from Gun Hill Road to Bussing Avenue.

FIFTH STREET (6). It once ran west from the Shore Road to the railroad depot of Bartow - now incorporated in Pelham Bay Park - between present-day Bartow Circle and the golf course parking lot.

FINDLAY PLACE. This is the former name of Marcy Place and obviously named after the former owner and one-time surveyor, Andrew Findlay, who lived on a small estate between present-day Jerome Avenue and the Concourse. The name was listed in the 1872 Directory of Morrisania and Tremont.

FINDLAY STREET. Andrew Findlay surveyed the settlement of Melrose in 1848, and one of the streets was named in his honor. Today, we know it as East 160th Street.

FINNEGANVILLE. This small settlement was on the eastern slope of West 246th Street from Riverdale Avenue down to Broadway. The localism was in use before 1900 by the scant population of Mosholu and Yonkers.

FIR STREET. This short street, now incorporated in the Botanical Garden, was west of the Bronx River near the Snuff Mill.

FIREMEN'S FIELD. This vacant lot was behind Fire Headquarters at 1129 East 180th Street near Devoe Street, next to the Bronx River. It was tended by and improved upon by local Little League teams, with assistance from members of the Fire Department.

FIREMEN'S HALL. Listed in the 1871 Directory of Morrisania and Tremont, it was located on Morris Street, between Railroad Avenue and Washington Avenue (East Tremont Avenue, between Park and Washington Avenues). Miss Sarah Tarbox, daughter of Hiram Tarbox the postmaster of Tremont, conducted a Temperance Society in the hall.

FIRST AVENUE (1). This is the former name of Willow Avenue in Port Morris.

FIRST AVENUE (2). This was an earlier name of Eastburn Avenue from the Concourse to Claremont Park.

FIRST AVENUE (3). This street, mapped in 1888, was renumbered East 192nd Street.

FIRST AVENUE (4). The former name of Briggs Avenue, from Bedford Park Boulevard to Mosholu Parkway.

FIRST AVENUE (5). In Wakefield, this avenue was changed to East 215th Street.

FIRST AVENUE (6). This is the former name of East 237th Street, from Kepler Avenue to Vireo Avenue. See: Elizabeth Avenue (2), Oakley Street, Oakly Street.

FIRST PLACE. The village of Morrisania was laid out in the 1850s, and this short street was the southern boundary. The Teasdale family lived there for decades, so sometime around 1875, when Morrisania was annexed to New York City, this became Teasdale Place.

FIRST STREET (1). In Port Morris, this was the original name of East 141st Street. See: Lowell Street.

FIRST STREET (2). This was the original name of East 163rd Street, mapped in 1860. See: Dongan Street, Helen Street, Strong Avenue.

FIRST STREET (3). In Unionport, this was the first street laid out from the boundary line (approximately Lafayette Avenue), but is now Turnbull Avenue.

FIRST STREET (4). The first street from the Westchester Turnpike at Westchester Square is what we call Tratman Avenue today.

FIRST STREET (5). This is the former name of Waterbury Avenue from Mayflower Avenue to Crosby Avenue.

FIRST STREET (6). This street was in the vanished settlement of Bartow that is now part of Pelham Bay Park. It ran from the railroad tracks to the Shore Road below present-day Bartow Circle.

FIRST STREET (7). This was the former name of Bullard Avenue as it was the first street from the railroad in the settlement of Washingtonville.

FIRST STREET (8). The first street in the village of Woodlawn Heights from the cemetery up to the city line later became Vireo Avenue.

FISHER'S LANDING ROAD. This was a very old name for modern East 233rd Street where it led down to the Hutchinson River. From the town records of Eastchester, we quote: "November ye 11th 1714 - One Highway laid out where William Fisher liveth." Well over a century later, Hannah Fisher kept a tavern near the landing for the travellers on the Boston Post Road as well as for the schoonermen of the Hutchinson River.

FITCH STREET. This is the former name of East 175th Street from Webster Avenue to Third Avenue and harks back to James T. Fitch, Tax Assessor in the village of Tremont in 1856. See: Gray Street, Oxford Street.

FIVE CORNERS, The. This was the unofficial name of the intersection of streets near St. Dominic's R.C. church. Victor Street, Van Nest Street and White Plains Road touch a small park called Van Nest Memorial Park. As a tribute to Msgr. Fiorentino, a move was made by Councilman Mario Merola in 1964 to name it "St. Dominic Square." See: Van Nest Memorial Park.

FLEETWOOD AVENUE. This former name of Teller Avenue recalls the Fleetwood Trotting course that was in operation from 1870 to 1898. The one-mile racetrack crossed the avenue at East 165th Street and East 167th Street.

FLEETWOOD ENGINE COMPANY NO. 3. This Exempt Firemen's headquarters was listed only as being on Morrisania Avenue in 1884. An 1889 booklet listed Jonas Volz and John Cordes as members who marched in a Grand Annual Parade and Picnic at Zeltner's Park.

FLEETWOOD HOTEL. Listed in an 1869 Directory as being located at the entrance to Fleetwood Park (present-day Morris Avenue at East 164th Street) and the owner was identified as G. P. Arcularius. This might be a misspelling, for a Charles P. Arcularius was also listed as owner of the Jerome Park hotel of West Morrisania.

FLEETWOOD PARK. In 1869, Louis A. Risse, surveyor and engineer, laid out this racetrack west of the Mill Brook at the northern limit of Melrose. The land had belonged to the Morris family for generations, a racetrack being mentioned as far back as 1750. The Dater brothers leased the property and engaged Risse to enlarge and modernize their undertaking, which they called the Fleetwood Trotting Course. However, the popular name was Fleetwood Park.

FLEETWOOD, The. This was a silent movie house for just a short time as it was built in 1925 with a seating capacity of 1,650. Talkies came into vogue in 1927, and within a year most theatres converted from silent to sound. It was located on Morris Avenue and East 165th Street on the east side of the avenue. It was converted to a supermarket in the 1950s.

FLEETWOOD TROTTING COURSE. See: Fleetwood Park.

FLETCHER STREET. This was the former name of East 182nd Street from Park to Washington Avenues when the Fletcher family lived on this street in 1892.

FLOWER STREET. In Olinville, this street was cut through in 1892 and named after Governor Roswell Pettibone Flower. By the turn of the century, after annexation, it was numbered East 213th Street. See: Arthur Street, Randall Avenue.

FLYING LADY, The. Nickname of the infamous monorail car. See: Pelham Park & City Island Railroad.

FOLEY'S HOTEL. This wayside hotel was on Southern Boulevard and Home Street on the former Fox estate. Mention was made of it in newspapers dated 1894.

"FONT HILL." This was the former estate of Edwin Forrest, the actor, overlooking the Hudson River. It was dominated by a castle, built in 1847, which was copied from an older English castle located near Salisbury. The original castle was called Fonthill-Gifford and had been erected in 1796 by an eccentric traveler and writer named William Beckford. Forrest evidently saw the castle, and had it copied for his own "Font Hill." Shortly before the Civil War, Forrest sold his estate to Archbishop Hughes, and it is today Mount St. Vincent.

FOOTE AVENUE. This short street overlaid the Indian burial grounds of the Siwanoys, which was later known as Burial Point. It was on Ferris family property from Revolutionary times to the mid-1860s when Valentine Seaman married a Ferris daughter. In the 1920s, it was a picnic resort called Pleasant Bay Park, and in 1938-1939 was a short-lived trailer camp for the New York World's Fair. Charles Berry Foote was a banker and an associate of two nearby squires, Morris and Huntington, and it may well be the street was named in his honor. Unfortunately for him, the street has been absorbed by Ferry Point Park where it touches the new St. Raymond's Cemetery.

FORD, The. See: Wading Place, The.

FORDHAM. This 17th-century grant to John Archer evolved into a recognized village. It had a separate political identity until it was annexed to New York City on January 1, 1874. See: 24th Ward (1).

FORDHAM, The. According to a 1912 listing, this was a frame building at 2508 Webster Avenue, between East 189th Street and Fordham Road, used as a silent movie house. It might have opened as early as 1910.

FORDHAM AVENUE. This was the former name of Third Avenue as it ran through Morrisania from East 163rd Street to East 179th Street, and so noted on a map of 1848. As early as 1871, the thoroughfare was being called Third Avenue, but the name was only officially changed in 1891.

FORDHAM HEIGHTS BURIAL GROUNDS. Once located at the intersection of Sedgwick Avenue and Fordham Road, tombstones were still standing in 1881 honoring the Berrians, Valentines, Devoes, Cromwells, and Laurences. It was also called the Dutch Burial Grounds.

FORDHAM HILL. This is an old name for the steep slope of Kingsbridge Road down to the Harlem River where the *original* village of Fordham was located. See: Break Neck Hill.

FORDHAM HOTEL. A small inn listed in the 1870 directory whose location was described as "Kingsbridge Rd n Depot, Fordham. Phil. Duffy, prop." Present-day location would be Fordham Road at Third Avenue.

FORDHAM LANDING An old-time name for the point of land on the Harlem River 100 yards north of the present-day University Heights Bridge. See: Berrian's Landing.

FORDHAM LANDING DIVISION. This Volunteer Station of the Life Saving Service of New York, with four officers and fifteen surfmen, was situated on the Harlem River at Fordham Road. The station was a two-storied frame building with a pitched roof, a pier and a launching slip. It was organized in 1897.

FORDHAM PLACE. On a 1906 map, this Place ran off West Tremont Avenue between Cedar and Sedgwick Avenues. In 1961, a sole reminder was a row of curbstones leading northward into an empty field.

FORDHAM ROAD. An 1851 survey shows this to be an early name for St. Ann's Avenue from Westchester Avenue to East 160th Street. In parenthesis, the same stretch was described as Benson Avenue, but Benjamin Benson did not purchase the land until 1853. Evidently the surveyor anticipated the sale. See: Benson Avenue, St. Ann's Avenue.

FORDHAM SQUARE. See: Rose Hill Park. (current).

FOREST GROVE. This was a former locality in the Township of Morrisania, and recognized as a postal address in 1866. It embraced the area of Cauldwell, Trinity and Forest Avenues, below East 163rd Street. It is partially covered by Forest Houses project. See: Grove Street, Woodstock.

FOREST STREET (1). This was the earlier name of Paulding Avenue from East Tremont Avenue to the railroad tracks. See: Chauncy Street, Sixth Street.

FOREST STREET (2). This was the original name of Liebig Avenue in Riverdale.

FORREST BROOK. This name was only encountered once, and that was prior to 1880. The brook originates in a meadow of Mount St. Vincent, formerly the property of Edwin Forrest, flows alongside West 261st Street westward and downward to the Hudson River. See: Randolph Brook.

FORREST'S CASTLE. See: "Font Hill."

FORT INDEPENDENCE. A Revolutionary fort once was located on present-day Giles Place. The fort was built upon the 75-acre farm of Captain Richard Montgomery, who later became Major-General during the Revolutionary War. The Continental Congress had resolved to build a fort to dominate the strategic road that led to Albany, and construction was started after General Washington had personally inspected the terrain. Colonel Rufus Putnam supervised the work, which was done by New York and Pennsylvania militia. Despite its importance and armament, the Americans abandoned the fort upon the approach of the Hessians, who greatly outnumbered and outgunned them, and the British held Fort No. 4 (as they numbered it) until 1779.

FORT NUMBER 1. This was an American fort near the crest of Tippett's Hill north of Henry Hudson Park at West 231st Street. On a British map, the location is spelled Spike & Devil Hill (Spuyten Duyvil). There, a Colonel Swartwout and his forces built a small, square fort surrounded by felled trees pointed outward. They abandoned it to a stronger British force, who then turned it over to a Hessian garrison.

FORT NUMBER 2. This was another small fort that was built by Colonel Abraham Swartwout (or Swarthout) on Spuyten Duyvil Hill in 1776. Later, it was overrun and then occupied by Hessians. The Americans had called it Fort Swartwout, but the British merely identified it by number, and set the Hessians to work on a fortified trench that eventually connected it to Fort No. 3. Reginald Pelham Bolton, noted historian, conducted excavations in 1911 and found many relics of the military occupation. See: Fort Swartwout.

FORT NUMBER 3. To cover the main road to Yonkers, Colonel Swartwout built this Revolutionary fort that was located on Tippett's Hill (Spuyten Duyvil Hill), overlooking Johnson Avenue at West 230th Street. It also had a view of Spuyten Duyvil Creek. It was a square earthwork fort, lined with an abatis - a thick wall of felled trees with their branches pointed outward - but it, too, was captured by Hessian Chasseurs. When the British withdrew to Manhattan Island in 1779, the fort was razed. The Americans re-occupied it, nonetheless, in its ruined state until 1781.

FORT NUMBER 4. See: Fort Independence.

FORT NUMBER 5. According to British military maps of the Revolution, this fort was perched off the uphill curve of Kingsbridge Road near West 195th Street. It was built by "Provincials," Americans who remained loyal to the British Crown, and for a time it was commanded by Colonel DeLancey, a wealthy West Farms landowner. Fort No. 5 had no heavy cannon, nor did it figure in any action. The little fort was destroyed and abandoned by the British in 1779. Three historians, Bolton, Hall and Calver, did important excavation work in 1910 before apartment buildings were built on the site, and their findings were donated to the New-York Historical Society.

FORT NUMBER 6. This fort was located near Reservoir Avenue and Claflin Avenue and was an earthwork affair according to Von Krafft's (translated) war diary. The garrison consisted of twenty-five Hessian privates, a sergeant and a drummer. These men were later replaced by soldiers of the Prince of Wales regiment and Bayard's Orange Rangers - both units comprised of Americans who sided with the British. See: King's Redoubt.

FORT NUMBER 7. A Revolutionary fort that was located at Morris Heights due north of the Hall of Fame, it overlooked the Harlem River. Military journals describe it as square, with the usual wall of felled trees. Inside was a raised mound for 20 field pieces and howitzers to dominate the waterway below. Von Kraft, a German officer, mentions in his war diary log cabins occupied by Hessian woodchoppers and foragers.

FORT NUMBER 8. This fort of Revolutionary times was the southernmost and largest fort on Fordham Heights on what was a bluff over the Harlem River. The land had been the ancestral home of the Archer family, that (in the 19th century), was the Schwab estate, the grounds of New York University, and is now the campus of Bronx Community College. Fort No. 8 was built in 1777 in the shape of a four-pointed star and was surrounded by the usual abatis. Throughout its existence it was British-held, being harassed over a long period of time by American raiders, but never captured by them. In October, 1782, the post was abandoned in orderly fashion. In the 1850s, a mansion was built by Justus Schwab and many relics of the British occupation were turned up. In 1960, when New York University constructed a building nearby, Dr. Theodore

Kazimiroff, official Bronx Historian, conducted an exhaustive search of the excavations and found still more military artifacts. See: "Archer Manor."

FORT ROAD, The. This was the earlier name of Pennyfield Avenue, dating from the 1850s when Fort Schuyler was being constructed.

FORT STREET. This is the former name of Wayne Avenue from Gun Hill Road to Woodlawn Cemetery. Revolutionary warfare did take place in this immediate area, but there was no fort built there.

FORT SWARTWOUT. This was a small fort that was built in 1776 on Tippett's Hill (Spuyten Duyvil Hill) due east of Henry Hudson Parkway at West 230th Street. It was named in honor of Colonel Abraham Swartwout, although on some records, the name is rendered Swarthout. See: Fort No. 2.

FORUM, The. See: Puerto Rico, The.

FOUNTAIN PARK. These picnic grounds were located at present-day East 147th Street and East 148th Street extending east from Willis Avenue (Robinson's Map of 1878). See: Germania Park, Karl's Park, Loeffler's Park, Phoenix Park and the Twenty-Third Ward Park.

FOUR CORNERS. This was the junction of Williamsbridge Road and Boston Road as noted on "Map of Boston Road, 1868 - Wm. Livingston, Engineer." See: Spencer's Corners (1906) and Spencer Square (1910).

FOURTEENTH AVENUE. This was the earlier name of East 228th Street in Wakefield.

FOURTEENTH STREET. In the village of Unionport, this was Newbold Avenue.

FOURTH AVENUE (1). This is the former name of East 218th Street, from the Bronx River to Bronxwood Avenue in Wakefield.

FOURTH AVENUE (2). In Woodlawn Heights, this was the earlier name of Oneida Avenue from the cemetery to the city line. See: Devoe's Lane.

FOURTH AVENUE (3). A 19th-century designation for East 240th Street, from Van Cortlandt Park to Vireo Avenue. Then it was changed to Holly Street, and in 1894 became East 240th Street. See: Holly Street.

FOURTH AVENUE (4). An 1877 Morrisania firehouse report indicated an un-mapped street in the vicinity of Park Avenue and East 174th Street.

FOURTH STREET (1). This was the first name of East 138th Street, in Port Morris.

FOURTH STREET (2). This is the former name of East 166th Street from Webster Avenue to Third Avenue in the village of Morrisania. See: George Street.

FOURTH STREET (3). When Unionport was laid out in 1851, this was Quimby Avenue.

FOURTH STREET (4). In Fordham, this was the former name of East 182nd Street from Jerome Avenue to the Concourse. See: Elm Street.

FOURTH STREET (5). When Wakefield was surveyed in 1853, this was the first name of Barnes Avenue from Gun Hill Road up to East 236th Street. See: Cedar Street, Madison Street.

FOURTH STREET (6). This was an East-West street in the former community of Bartow that is now inside Pelham Bay Park west of Bartow Circle.

FOWLER PLACE. Now out of existence, it once ran from Bantam Place to Allerton Avenue.

FOX AVENUE. This is the former name of Gunther Avenue, from East 233rd Street to Bussing Avenue, according to a 1905 map. See: Walnut Street.

FOX CEMETERY, The. This is another name for the Quaker Cemetery adjoining St. Peter's church on Westchester Avenue. The name honors George S. Fox, founder of the Society of Friends. There is a tradition that he preached there in 1672, and that the first Friends' Meeting in America was held at Westchester on the banks of a stream known as Indian Brook (now covered by Zerega Avenue).

FOX SQUARE. This is the former name of Benjamin Gladstone Square. The Fox mansion was situated nearby in the 19th century, but a long-time belief by the local residents was that fox hunts were assembled at this point during Revolutionary times.

FOX STREET (1). This was the earlier name of East 150th Street from St. Mary's Park to Southern Boulevard until 1894. See: Ungas Street.

FOX STREET (2). This was the former name of Barretto Street, for when the street was cut through, it was on Fox property.

FOX WOODS. This was an area bounded by the Bronx River, Farragut Street and East Bay Avenue - triangular in shape - and thickly wooded in the 18th century. George S. Fox was the owner in the 1840s, and he is buried in the Quaker Cemetery near Westchester Square.

"FOXHURST." A large estate near Hunts Point was owned by H. D. Tiffany, whose wife was a daughter of William Fox. Beers' *Atlas* of 1868 notes it was also known as Fox Woods. See: Simpson's Point.

FOX'S CORNERS. The intersection of Southern Boulevard and Westchester Avenue in Revolutionary times was already known by this name. William Fox, wealthy Quaker merchant, was a local landowner and he married into the Leggett family. His daughter married a Tiffany.

FOX'S LANE. On an 1891 map, this is an alternate name of Watson's Lane.

FRANCIS HOUSE, The. Once a landmark of Mott Haven, it was the later home of Olin Stephens, wealthy merchant, around 1900. It was situated at East 146th Street and Gerard Avenue, facing the Harlem River and was considered a gem of Greek Revival architecture. The U.S. Department of the Interior considered it a good example of 19th-century construction, but nevertheless it was demolished in 1941 or 1942.

FRANKLIN AVENUE (1). This is the former name of Crotona Avenue from Crotona Park north to East 177th Street. See: Broad Street, Broadway, Lafayette Street, Old Broadway.

FRANKLIN AVENUE (2). A tract of land behind Westchester Square was bought by the Franklin Society for Home Building & Saving, and so the avenue that traversed it received this name. Later, in the early 1900s, the Franklin Athletic & Social Club had its clubhouse on this avenue near Eastchester Road. The name was changed to Blondell Avenue in 1914.

FRANKLIN STREET. This was the earlier name of Marine Street, from City Island Avenue, east to the Sound.

FRANKLIN, The. This was a silent movie house located at Prospect Avenue and East 161st Street with a seating capacity of 2,855. It was in operation from 1920 to the 1950s, and then was converted into a supermarket.

FREDERICK STREET. In 1887, this was the name of Hughes Avenue in the village of Belmont above East 187th Street. See: Avenue St. John.

FREE BRIDGE. This was another name for the Farmer's Bridge that was built in 1758-1759 over the Harlem River as instituted by Benjamin Palmer. When the Ship Canal was dug past West 225th Street, the bridge was stranded and covered over with landfill in 1911.

FREEDLAND. This name was noted on a 1656 map of New Netherlands by A. Vanderdonck. It meant "Land of Peace" and was a variant of the archaic Dutch Vriedelandt, which was later allocated to John Throgmorton and his English colonists. The English called the territory Throgmorton's Neck, which became our Throggs Neck.

FREEDOMLAND. This $65,000,000 concept that was built upon 25 acres of reclaimed swampland and creeks along the Hutchinson River was part of the original Pell grant of the 17th century that became known as Pinckney's Meadows. Early plans called for a race track to be built there, later an airport, but neither materialized. The geographic outline of Freedomland was a map of America itself, and the amusement park was well patronized in Summertime only. Groundbreaking ceremonies took place in 1959 with Mayor Wagner officiating, but the extravaganza went bankrupt in 1965. The site is now occupied by Co-Op City.

FREEMAN, The. This was a silent movie house on Southern Boulevard, and Freeman Street, built in 1921 with a seating capacity of 1,605. It went out of business in 1970.

FREMONT, The. This was an old-time silent movie house on Southern Boulevard, but otherwise not further pinpointed. Its years of operation are uncertain.

FRENCH CHARLEY'S. A well-known ballfield and picnic grounds that were bounded by Webster Avenue, the Bronx River, East 203rd Street and East 203rd Street was the former grounds of a French restaurant, of which "French Charley" Mangin was the proprietor in the 1890s. His daughter married a Philip Bianchi, stonemason, and lived her entire life not far from where her father's restaurant had stood. At the upper end of this field was DuBois Island in the Bronx River, owned by another "French Charlie," DuBois, whose father was listed in 1888 as owner of the tiny island. See: Cowslip Island, DuBois Island.

FROCKES NECKE. This is a very early variant of Throggs Neck, as noted on a bill of sale from John Archer to Timothy Winter on March 27, 1668. See: Frog's Neck, Hog Neck, Throgsneck, Throggs Neck, Throgmorton's Neck, Vriedelandt.

FROG HOLLOW (1). This is the area of Morris Avenue around East 148th Street and East 150th Street, and was in use from Civil War times to the early 1900s. Its name derives from the swampy terrain that was bisected by a small brook variously called Ice Pond Brook or Morris Brook, and which undoubtedly abounded in frogs. In the 1870s, it was the name of a shantytown principally inhabited by Irish draymen and laborers and their families. An unsavory group of idlers called itself the Frog Hollow Gang, but confined itself to petty crime. The most publicized misdemeanor (and the one that caused its downfall) was an incident wherein the gang forced its way into Haffen's brewery one Sunday morning and demanded a keg of beer. Haffen happened to be there and ordered them out, but someone stabbed the brewer in the back and the gang fled. Haffen recovered, but the public indignation and subsequent demand for law and order caused the Melrose police to disperse the gang.

FROG HOLLOW (2). This was a nickname for the area of Woodlawn Heights extending east from Kepler Avenue down to the Bronx River. The name was current from about the 1880s to the early 1900s.

FROG POINT. This is a former name of Fort Schuyler. See: New Found Passage (1667), Frogge Point (1777), and Hammond's Point (1826).

FROGGE POINT. On a Revolutionary map, the tip of Throggs Neck was identified by this name. (Fort Schuyler was not yet built). A British officer by the name of Blaskowitz rendered the drawing.

FULTON STREET (1). This was the earlier name of Cambreleng Street. See: Monroe Street, Pine Street.

FULTON STREET (2). In Washingtonville, the Empire Engine Company of volunteer firemen had their firehouse on this street prior to 1896. It has since been renamed Richardson Avenue.

FURMAN'S LANE. This was the 1870s name of Middletown Road from Bruckner Boulevard to Eastchester Bay, when the Waterbury family owned the land. Kate Waterbury Furman inherited the estate from her father, and the country road was known as Waterbury Lane as well. On a map of 1910, it was still called Waterbury Lane, but shortly thereafter, Middletown Road was extended past Bruckner Boulevard to its terminus on Eastchester Bay.

GAINSBORG AVENUE. This is the former name of Parkview Avenue in Middletown. S. H. Gainsborg was a prominent realtor of the 1890s and lived in a fine mansion on this avenue. He came from Lima, Peru, and had eight children, all of whom were musically inclined. Mr. Gainsborg was president of the Bankers' Realty & Security Company, and purchased a tract of land that had been the former Ferris estate. An 1895 map shows the real estate located along Bruckner Boulevard. The name of the avenue was changed in April, 1924.

GALLAGHER AVENUE. This was a name proposed for Bailey Avenue at West 231st Street during a 1922 real estate development. See: Barry Street, Martin Terrace.

GALWAY'S SEA WALL. On a U.S. Army survey of 1874, this stone embankment was located on the right bank of the Hutchinson River at the point where the Boston Post Road crosses over. An 1863 map shows Thomas Galwey [spelled this way] owning 11 acres there. An 1868 spelling of the name has it Galloway.

GAMBRILL STREET. This was an earlier name of East 201st Street, from the Concourse to Webster Avenue.

GARDEN HALL. A Melrose social hall with outdoor garden owned by John Kircherer in the 1890s, it was located on East 149th Street near Westchester Avenue.

GARDEN PLACE. It is now incorporated into Furman Avenue at East 235th Street.

GARDEN STREET. This is the former name of East 143rd Street in North New York.

GARDEN, The. This silent movie house at 2755 Webster Avenue near Fordham Road was listed in 1914.

GARFIELD STREET. From the Concourse to Bainbridge Avenue, this was the earlier name of East 199th Street.

GARRISON HEIGHTS. The hill across which Spofford Avenue now cuts once carried this name in the 1880s. The Garrison family were the landowners there before Paul Spofford bought up the property for his estate.

GARRISON SQUARE. This is the former name of Cpl. Fufidio Square. Commodore John H. Garrison purchased part of the Fox estate in the 1820s, and his family lived there for three generations. The intersection of Longwood Avenue and Tiffany Street was named Garrison Square, but then was renamed Truxton Square, a name that was in use until April, 1953, when it was again renamed to honor a local man who had been killed in World War II.

GENNER AVENUE. John S. Genner was one of the early Westchester colonists who pledged allegiance to the New Netherlands in 1656. Other avenues harked back to fellow-colonists, Close, Ward, Damis and Fenfell, so Genner was also remembered. It was displaced by Boynton Avenue from Westchester Avenue to Lafayette Avenue.

GEORGE STREET (1). The former name of East 166th Street, east of Third Avenue.

GEORGE STREET (2). East 166th Street from Jackson Avenue to Tinton Avenue in the old village of Woodstock was once known as George Street. It is now the northern border of Forest Houses.

GEORGE'S POINT. In Van Cortlandt Park, this was a bend in Tibbett's Brook - 25 acres of upland with 1 acre of meadow, which is now classified as the Parade Grounds opposite West 250th Street, It was named for George Tippet who conveyed the land in 1691 to his brother-in-law, Joseph Hadley.

GERARD AVENUE. This is now Clarke Place.

GERARD STREET. This short street ran from Bergen Avenue diagonally to East 149th Street, and was named after Gerard Morris, 19th-century landowner. The building of the subway obliterated it save for a small indentation of the curbing.

GERMAN PLACE. It was called Balcom Avenue on an 1858 map of St. Ann's Avenue, off Westchester Avenue, but was never cut through until 1890. It was then named German Place, but due to the anti-German feeling of the first World War, in 1916 it was given the name of the first Bronx soldier killed in action: Arthur V. Hegney.

GERMAN STADIUM. This was the earlier name of Throggs Neck Stadium, on the former estate of Cora Morris at Schurz and Harding Avenues. The change occurred during World War I.

GERMANIA PARK. This picnic grove, amusement park and rifle range may have been the earliest establishment of its kind as it was mentioned prior to the Civil War. It was located between Willis and Brook Avenues at East 147th to East 149th Streets. An early print shows a casino built with turrets, and the entire park was fenced in except for the side facing the Mill Brook. See: Fountain Park, Karl's Park, Loeffler's Park (1), Phoenix Park, Twenty-Third Ward Park.

GESELLEN de. This Dutch word for Brethren alludes to North Brother and South Brother Islands, which have become separated in the long march of history. See: North Brother Island, South Brother Island.

GIEFFEN'S CREEK. This is a misspelling found on some maps. See: Givan's Creek.

GILES STREET. See Giles Place.

GILROY PLACE. This street was eliminated at East 208th Street by the building of the Jerome Avenue subway line. It was named after Mayor Thomas F. Gilroy who served in 1893-1894, and who is buried in Woodlawn Cemetery.

GIMBERNAT FARM. Florentino and Justa Gimbernat had five sons - Teofilo, Heraclius, Joseph, Charles and Jules - and two daughters named Mercedes and Julia. The family migrated from Sewaren, N.J., (near Perth Amboy) in 1860 and purchased land on Throggs Neck. See: Escoriaza Farm, Rhinelander Lot.

GINKO SQUARE. See: Muller (Maurice) Park.

GIVAN'S BASIN. This was the alternate name of Black Dog Creek where it widened into the Hutchinson River, now covered by Co-Op City. Robert Given of Scotland purchased the land from the Bartow family in 1794, and used the tidal run to power his mill. It subsequently was called Givan's Creek. He died in 1830, leaving the considerable property to his descendants - LeRoys, Palmers and Morgans.

GIVAN'S CREEK. This was a former name of the Hutchinson River that was used for water-power by Robert Givan, a 19th-century miller. It was earlier considered the northern boundary of Vriedelandt, which was Dutch territory in the 1600s. The creek is now covered by Co-Op City.

GIVAN'S DRIVE. This semi-circular driveway was plotted to traverse the Arden Estate, but did not survive the street-cutting of Mickle, Kingsland and Tieman Avenues. A short portion survived as a footpath from Givan Square on Eastchester Road, alongside the Catholic church.

GLEN STREET. This is the former name of East 168th Street from Boston Road to Prospect Avenue. See: Charles Place, Sixth Street.

GLENCOE STREET. This was an earlier name of Perry Avenue from Gun Hill Road to the cemetery.

"GLENN DALE." The Hatfield estate of the 1860s which took in the area between Bronxdale Avenue, Morris Park Avenue, Williamsbridge Road and the railroad - approximately the area of later Morris Park Racetrack.

GLOVER STREET. In the village of West Farms, this was East 179th Street.

GO-WA-HA-SU-A-SING This is the Algonquin name for the marshy peninsula projecting into Spuyten Duyvil Creek. When the Ship Canal was dredged through in 1895, the marsh was obliterated. Indian experts Tooker and Zeisberger agree the word means "Place of Hedges."

GOBLE'S POND. George S. Goebel [spelled this way] was listed in 1872 as an ice-dealer living on the east side of Anderson Avenue. The ice pond and ice house were situated at Jerome Avenue and East 173rd Street.

GODWIN ISLAND. See: Wading Place, The.

GODWIN'S CURVE. This former curve was in front of the Godwin mansion when Spuyten Duyvil Creek divided Manhattan from The Bronx. Its present-day location is West 230th Street and Broadway. Squire Godwin's mansion was built on the site of Verveelen's tavern.

GOLDEN RULE, The. This silent movie house also had an open-air annex for Summertime movies. It was located on the west side of Third Avenue between St. Paul's Place and East 171st Street in the 1920s. The owner's name was Mr. Golden.

GOODRIDGE POND. See: Alder Brook.

GOOSE ISLAND DIVISION. See: Goose Island.

GOUVERNEUR PLACE. This short street ran from the Bronx Kill to East 132nd Street, east of St. Ann's Avenue, but it is now covered by a railroad yard. It formerly led to the mansion of Gouverneur Morris.

GOUVERNEUR STREET. This was the first name of East 151st Street in Melrose South.

GRACE STREET. This is the former name of Glover Street from Castle Hill Avenue to Westchester Avenue.

GRAHAM SQUARE. It has been covered by Highbridge Houses since 1953.

GRAHAM STREET. The former name of Muliner Avenue, north of Morris Park Avenue.

GRAHAM'S POINT. An 1816 map shows this neck of land due south of Bungay Creek, which would be East 149th Street and the East River. James Graham was the father-in-law of Richard Morris.

GRAND AVENUE. The earlier name of East 233rd Street alongside Woodlawn Cemetery.

GRAND, The. At 5 West Fordham Road, this silent movie house had an impressive seating capacity of 2,430. As it opened around 1925, it speedily converted to "talking pictures."

GRAND VIEW HOTEL (1). E. Johnson was listed as proprietor when this hotel was in operation from 1885 to 1895 at East 170th Street and Jerome Avenue.

GRAND VIEW HOTEL (2). An inn which once stood on Tallapoosa Point at the Pelham Bridge from about 1870 to 1900. "Nanny" Blizzard and her brother Dennis Mahoney were the proprietors, while David Blizzard rented boats and fishing tackle. The hotel was torn down around 1908 when the new Pelham Bridge was built. See: Blizzard Island.

GRANITE PLACE. On a 1902 map, this Place was shown between Tiebout and Webster Avenues running south from East 184th Street. There was no sign of it in 1940.

GRANITE STREET. This is the former name of Rider Avenue, from the Harlem River to East 136th Street, according to an 1870 map.

GRANT AVENUE (1). This is the former name of Mohegan Avenue.

GRANT AVENUE (2). This was another earlier name of Hollywood Avenue, from Waterbury Avenue to Middletown Road.

GRANT STREET. Van Nest residents used this name instead of Mead Street in the 1880s.

GRANT, The. In 1920, this was a silent movie house at Tremont and Webster Avenues, with an astounding seating capacity of 3,500. The Theatre Historical Society could find no further record of this edifice, and concludes it was never built.

GRAPEVINE BROOK. This appears as a 1793 mention in a deed bounding the land of Anthony Lispenard Underhill, and possibly is a very early name for Barrett's Brook. See: Barrett's Creek.

GRAY AVENUE. This was the earlier name of Virginia Avenue, as shown on a 1910 map. The name was changed in 1924.

GRAY STREET. The former name of East 175th Street from the Concourse down to Webster Avenue. There is a possibility this was the edge of Francis Gray's property (1836) before he sold out and moved to Boston.

GREAT CREEK. This was the Colonial name of Westchester Creek. "In 1686, sold by Nathaniel Underhill to Thomas Mollinex a piece of land on Frogg's Neck by the Redd Hill, bounded on the west by the Great Creek." A later reference reads, "1794 - the Great Creek known by the name Westchester Creek. "

GREAT ISLAND, Ye. This name occurs in a conveyance (deed) of July 24, 1729, in which Thomas Pell, Sr., conveyed present-day City Island to his son, Thomas Pell, Jr.

See: Great Minneford, Minneford's Island, Menefors, Minnewits, Minnefers, Missives Island, Mullberry Island.

GREAT KILL. See: Hudson River [Current].

GREAT MINNEFORD. This was an early reference to City Island in 1685.

GREAT NORTH RIVER. See: Hudson River [Current].

GREAT RIVER. See: Hudson River [Current].

GREEN AVENUE. This was the former name of Roebling Avenue in 1896.

GREEN LANE. This is an earlier name of Zerega Avenue from Lyvere Street to Westchester Avenue. See: Avenue A.

"GREENBANK." This is the former estate on Hunts Point owned by C.D. Dickey.

GREENE AVENUE. The 1872 name of Miles Avenue from the Throgs Neck Expressway to Long Island Sound, named after a surveyor, George S. Greene, who also owned property along this avenue. Along with Luke Owens and Samuel Gelston, he sold the lots to George T. Adee in 1872. See: Ocean Avenue, Pennyfield Avenue.

GRENADA PLACE. It was once the name of East 206th Street at the Concourse, and may have a classical connection with present-day Lisbon Place and Cadiz Place that was eliminated by Mosholu Parkway South - all historically linked with St. George.

GREY MARE, The. A boulder is located on the northwestern shore of Hunter Island surrounded by salt marsh. It bears a resemblance to a horse if viewed from the mainland.

"GREYSTON." This is an estate of William E. Dodge, whose mansion was built in 1863. See: Dodge Lane.

"GREYSTONE." An 1868 map of South Yonkers (now Riverdale) shows this estate of William Dodge, Jr., midway between the Hudson River and Riverdale Avenue in the vicinity of West 242nd Street.

"GREYSTONES." This estate, owned by William H. DeLancey, is now incorporated into the Pelham Bay Park golf course. His daughter married John Hunter. The north border of this property was formed by Roosevelt Brook.

GRIDLEY AVENUE. This is the former name of Lafayette Avenue from East Tremont Avenue to Long Island Sound. During the Spanish-American War, at the battle of Manila Bay, Commodore George Dewey said: "You may fire when ready, Gridley! "

GRIMES' HILL. This is mentioned by Bronx old-timers as being inside St. Mary's Park at East 148th Street. In the 1880s, this was Janes' Hill.

GROOTE KILL. See: Hudson River [Current].

GROTIS PARK. A park was mapped in 1868 alongside the Bronx River from East 239th Street to East 242nd Street, which evidently was a picnic grove and amusement center.

GROVE AVENUE. This is the former name of Trinity Avenue from East 161st Street to East 163rd Street. See: Avenue C, Cypress Avenue, Delmonico Street.

"GROVE FARM." An 1831 deed mentions "Grove Farm" on Ferris Neck. It was owned by Thomas Hunt in 1686, whose grandson Josiah called it "Grove' Siah."

GROVE HILL. This locality name extended from Third Avenue to Cauldwell Avenue around East 161st Street, according to an 1866 map. See: Hupfel's Hill.

"GROVE HILL." This is the summit of Eagle Avenue and East 162nd Street, which was the estate of the DeGraaf family in the 1870s and 1880s. The immediate vicinity was also known as Grove Hill.

GROVE HILL PLACE. This lane was on the property of Clara Decker in 1876. See: Deckerville.

"GROVE 'SIAH." This is the name of a homestead belonging to Josiah Hunt in 1694 along Westchester Creek, the site being occupied by a Catholic high school at Brush Avenue.

GROVE STREET (1). This is the former name of East 144th Street from St. Mary's Park to Southern Boulevard, as mapped in 1871. See: St. Joseph's Street.

GROVE STREET (2). This earlier name of East 146th Street in Mott Haven was listed in 1871 as being in the Buena Ridge sector.

GROVE STREET (3). This was the original name of East 153rd Street from Third Avenue to Brook Avenue.

GROVE STREET (4). This was the former name of East 162nd Street, from the Concourse to Sherman Avenue, according to an 1871 Directory. See: Halsey Street, Union Lane, Union Street.

GROVE STREET (5). This was an early name of East 178th Street from Creston Avenue to the Concourse.

GROVE STREET (6). This was the former name of Bantam Place in 1911.

GUION PLACE. This was the earlier name of Guion Street.

GUTTENBERG STREET. When Colonel Hoe's estate, "Brightside," was partitioned in 1897, streets were given names connected with the printing industry: Hoe (who invented the rotary press), Aldus, who was a medieval printer, and Guttenberg. However, in 1902, the street name was changed to Bancroft Street, and then, in 1911, it became East 165th Street.

HADLEY AVENUE. William Hadley owned 257 acres of land extending from the Hudson River to the Albany Post Road (Broadway), 1761-1786. This tract was sold by Hadley's executors in 1829 to the Delafield family. Bromley's Map of 1891 calls this street Yonkers Avenue as it pointed north to that city. The upper end was briefly known as Half Moon Place, evidently recalling Henry Hudson's ship that once sailed in the area. It has since been eliminated as a city street and housing is on the site.

HADLEY'S BRIDGE. This was another name for the Farmers' Free Bridge over the Harlem River at West 225th Street. It was formally opened in 1759 with a barbecue on The Bronx side. Around 1911, it was demolished and covered over by landfill. See: Free Bridge.

HAFFEN'S BREWERY. This brewery was founded in 1856 by Mathias Haffen at the foot of Elton Street (now East 152nd Street) before Melrose Avenue was cut through. The storage vaults, with walls almost 4 feet thick, extended 230 feet back to Gouverneur Street (now East 151st Street). The business was inherited by two sons who ran it successfully until the coming of Prohibition. The brewery was razed in 1917.

HAIGHT ESTATE. This was a tract of 473 lots, surveyed by Charles A. Mapes in 1893 in the town and county of Westchester. The bounds were present-day Hutchinson River Parkway, Pelham Parkway, Bruckner Boulevard, Wellman Avenue, back to point of origin.

HAKE'S CORNER. It was listed in the Morrisania Directory of 1871 as being on the northwestern corner of Milton Street and Boston Road (now, East 158th Street and Third Avenue). Casper Hake was a Trustee of the German Savings Bank located in the building on that corner.

HAKE'S HALL. A social hall on present-day Third Avenue and East 158th Street was owned by Casper Hake. An 1875 notice advertised a meeting of a Temperance Society, Crystal Spring Division No. 89.

HALE AVENUE. This avenue was laid out on New York City maps of 1922, 1923 and 1924, but was eliminated by the new St. Raymond's Cemetery.

HALF MOON PLACE. This was the upper end of Yonkers Avenue in 1891, and obviously commemorated Henry Hudson's ship, *Halve Maen*.

HALF WAY BROOK. It is mentioned in a 1718 deed as being on the land of Thomas Baxter. This early pioneer on Throggs Neck was credited with 365 acres, but there is no clue to where this brook was.

HALL AVENUE. This is the former name of Wallace Avenue, from Allerton Avenue to South Oak Drive. See: Hickory Avenue.

HALSEY STREET. This was the earlier name of East 162nd Street, from Sherman Avenue to the Park Avenue railroad tracks, according to the Morrisania Directory of 1872. In the same book, Isaac Halsey is identified as a builder. See: Grove Street, Union Lane, Union Street.

HAMMER HOTEL. For an inn that formerly stood by the Post Road in the early 1860s, Edward C. Hammer was listed as the hotel keeper. The location was at East 161st Street and Third Avenue, on the site of the Magistrates' Court. See: the Stone Jug.

HAMMOND CREEK. This is an alternate of Hammond's Cove between Locust Point and Fort Schuyler. See: Scuttle Duck Harbor.

HAMMOND'S FLATS. A local name for the marshlands surrounding Hammond's Cove was in usage from the 1850s up to the 1900s. They were drained in 1923-1924, and now are partially overlaid by Longstreet Avenue south of Harding Avenue. Colonel Abijah Hammond was born in Boston in 1757 and served in the American Revolution before coming to New York. He made his fortune in Greenwich Village real estate, and bought land on Throggs Neck (present-day Silver Beach Gardens and a portion of Fort Schuyler). His first wife, Catherine (Ogden) died in 1814, and on his 59th birthday, he married Margaret Aspinwall. The Colonel died at his country estate in December, 1832.

HAMMOND'S POINT. According to records in the Bronx County Building, "Charles Henry Hammond and William Bayard sold 52 acres to the U.S. Government in November 14, 1826 that certain parcel called Throgs Point, otherwise Hammond's Point." This is present-day Fort Schuyler. See: Frogge Point (1777).

HAMPDEN AVENUE. This is the former name of West 183rd Street from Loring Place to University Avenue. G. W. Hampden was a realtor of the 1890s.

HANCOCK STREET. In Van Nest, this was the original name of Melville Street.

HANNOCK'S HILL. This was a rocky promontory jutting out from the mainland in the area of present-day Orchard Beach, and from it, a clear view was had out into Long Island Sound to observe the British fleet during the Revolutionary War. A City Island historian, Alfred Fordham, did yeoman work in tracing obscure locations that were mentioned in military or naval annals. This is one example.

HANSEN'S SHIPYARD. Adam Hansen bought a tract of land from the City Island Athletic Club on the east side of the island in 1902. This yard later became the Nevins' Shipyard.

HARGOUS CRESCENT. Peter A. Hargous owned property on the east side of the Concourse above East 196th Street which he purchased from Leonard Jerome in 1867. The Crescent was mapped in 1889 as extending in a semicircle eastward from the Concourse at that point.

HARLEM RIVER STATE PARK. This is the former name of Roberto Clemente State Park on the western rim of The Bronx paralleling the Harlem River. The Park was opened in 1970 and given this geographical name, but it was superseded in September, 1974, to honor the baseball player who lost his life on a mercy mission.

HARALD AVENUE. See: Dickinson Avenue.

HARRIET PLACE. See: Calhoun Avenue, Revere Avenue.

HARRIS, The. (Theatre) See: Bronx Opera House.

HARRISON AVENUE. The earlier name of Thieriot Avenue from East Tremont Avenue to Westchester Avenue.

HARRISON STREET. This was the original name of Taylor Avenue from Westchester Avenue to the Cross-Bronx Expressway.

HARTUNG'S PARK. This popular picnic spot of the late 1870s and 1880s was at East 147th Street and Southern Boulevard. See: Union Park.

HARVEY'S HILL. This early 1900 reference to the area at the foot of Willis Avenue seemed to linger until 1917. The land had previously been Brommer's Park of the

1890s, Christ's Park of the 1870s, the manorlands of the Morris family and the original farmland of Jonas Bronck. Pulaski Park occupies the spot at the present time, although much reduced in size and height due to the overpasses. It was also known locally as "Scratch Park" due to the derelicts that would lounge there.

"HASLEWOOD." This was an 1860 estate in Fordham owned by a Mrs. Lees.

HASSOCK MEADOW. This is mentioned in the first Indian deed to Jessup and Richardson in 1663 as a landmark. It is ascertained to have been a marshy stretch now overlaid by Southern Boulevard, south of East Tremont Avenue. In the 1800s, it was called Walker's Swamp, and around Civil War times was part of the Thomas Minford estate. See: Walker's Swamp.

HASSUCK MEADOW. This variant of Hassock Meadow is found many times in early deeds of West Farms.

HAUNTED CEDAR KNOLL. This wooded hill was alongside the Shore Road just short of the city line that was part of the Roosevelt estate of the early 1800s. It overlooked a portion of Long Island Sound and had the eerie reputation of being haunted by a band of Siwanoy Indians who had been beheaded by bloodthirsty Matinicock Indians from Long Island.

HAUNTED OAK, The. This was another name for the Spy Oak. See: Spy Oak.

HAUNTED WALL, The. This stone wall once stood on the west side of East Tremont Avenue from Schley Avenue to Roosevelt Avenue. It protected an orchard belonging to the Morris family. Many years ago "in the horse-and-buggy days" (date indeterminate), a murdered man was found tied in a sack and saturated with acid. Tracks in the mud indicated he had been tossed from a wagon. The wall, behind which the body was found, acquired a ghostly reputation for many people swore they saw the wraith of the murder victim walking atop the stone wall. The gardener actively fostered the ghost story to discourage boyish raids on Mr. Morris' orchard. See: Lamport Place.

HAVEMEYER'S POINT. This was a 19th-century designation for what had been, during the 1700s, Stephenson's Point. It was near the intersection of Hollywood and Schurz Avenues on the East River - and an ancient road still leads down to a former ferry slip. The Havemeyer family owned the property now called Silver Beach Gardens. See: Stephenson's Point, Stevenson's Point.

HAWKINS SHIPYARD. Once known as Hillman & Hubbee's Shipyard, it became the property of John P. Hawkins in 1887. It was situated at the foot of East Fordham Street, and it became famous in yachting circles.

HAWKINS STREET PARK. This is an unofficial name for the small plot at Hawkins Street and City Island Avenue on City Island. It was named in 1934 to honor members of the Leonard Hawkins Post who had served in World War I. In earlier times, the police station was situated on the spot. Park Department records list the plot as Memorial Park.

"HAWKSWOOD." This 19th-century estate on the City Island road at Rodman's Neck belonged to L. R. Marshall. It was originally part of the Pells' lands, and preserved the superstition that it was a lucky omen to build a home where hawks nested. See: Colonial Inn, Marshall's Corner.

HAWTHORN FARM. This was the 1814 name for the farmlands opposite St. Raymond's Cemetery on the Throggs Neck Road (now East Tremont Avenue). This farm was later absorbed into the Willow Lane farm that ran from LaSalle Avenue to Bruckner Boulevard in 1829. See: Walsh farm.

HAWTHORNE AVENUE. This is the former name of West 256th Street from Sylvan Avenue to Broadway. See: St. Vincent Avenue.

HAYNES AVENUE. It was laid out on city maps from 1924 to 1937, but then was eliminated by the Bronx-Whitestone bridge tollbooth area.

HAZEL, STREET. This was an earlier name of Ely Avenue from East 222nd Street to the Hutchinson River. See: Bayard Street, Doon Avenue, South 21st Avenue.

HEART ISLAND. This was a 1775 designation of present-day Hart Island in a book of Naval Hydrographic maps, called the *Atlantic Neptune*. Drawn by naval cartographers, it was presented to King George III. See: Little Minneford, Spectacle Island.

HEINE PARK. (Unofficial) See: Kilmer (Joyce) Park.

HEINTZ PARK. (Unofficial) See: Kilmer (Joyce) Park.

HELEN STREET. This was the former name of East 163rd Street from the Concourse to Park Avenue in 1897. See: Dongan Street, First Street (2), Strong Avenue.

HENDERSON'S CORNER. The Hendersons were prominent landowners of the village of Schuylerville, according to an 1809 map. At one time, they owned the triangle of land on which the Heye Foundation Indian Museum is now situated. The Henderson mansion and lumber mill was at the junction of present-day Bruckner Boulevard and the Throgs Neck Expressway; and in the days when Bruckner Boulevard was only a country lane known as Willow Lane, this was Henderson's Corner.

HENDERSON'S ISLAND. This was an early name of Hunter Island when it was owned by a Dr. William Alexander Henderson, a surgeon in the British navy. His ownership ran from 1794 to 1804, and in his will he bequeathed to a woman named Betsy two hundred and fifty dollars. She had resided with him in Bengal in the East Indies, and (rumor had it) had sent him their half-caste son to live with him on Henderson's Island. See: Appleby's Island.

HENDRICKS STREET. This is the former name of East 137th Street. An early owner named Hendricks bought the land from a J. Van Stoll in 1662 and sold it to H. Smeeman in the same year. See: Fifth Street (1).

HENNESSY PLACE. See: Fischer (Corporal) Place.

HENRY STREET (1). The former name of East 148th Street to Brook Avenue from Willis Avenue in the 1870s was probably named for Henry L. Morris, who owned one-third of Old Morrisania. See: Mott Street.

HENRY STREET (2). Nonexistent now, but once it was laid out across the Gerard Morris estate from Cromwell Avenue to the Concourse. Around 1880, the expanding freight yards along the Harlem River wiped it out.

HENRY'S BAY VIEW INN. This was a well-known restaurant at Layton Avenue on Eastchester Bay, on land that once was part of Lohbauer's Park, and reputed to have been constructed from parts of the original Lohbauer's Casino. It was gutted by flames in mid-April, 1958, and never rebuilt.

HERMITAGE, The. This 19th-century inn alongside the Bronx River in the vicinity of White Plains Road and Pelham Parkway was a noted resort of politicians, sportsmen and wealthy Bronxites. Its popular name was "Sormani's," for Joe Sormani was the genial proprietor. See: L'Hermitage.

HEROLD PARK. A small plot bounded by East Tremont Avenue, Lyvere Street and Castle Hill Avenue was surveyed in January, 1916. It had been the site of a tavern, Cannon's Hotel, with an outdoor garden. Later, it was an athletic training camp (Del Hawkins' Gym) and a ballfield. The tract is now occupied by an apartment house, and the rectory of St. Raymond's R.C. church.

HICKORY AVENUE. The former name of Wallace Avenue from North Oak Drive to Gun Hill Road. The name was changed in April, 1922. See: Hall Avenue, Rosewood Street.

HIGGS' BEACH. This is the former name of Harding Park on Clason Point. It was once the property of the Higgs family from approximately 1900 to 1916. William Higgs rented out campsites for Summertime residents, the forerunner of Harding Park. He was on the reception committee at the celebration in honor of the opening of the Municipal Ferry system, Clason Point-College Point, in August, 1921. See: Killian's Grove.

"HIGH COTTAGE." This 1855 estate in Fordham was owned by Mrs. Montgomery.

HIGH STREET. This was the earliest name of West 162nd Street from Summit Avenue to Anderson Avenue. See: Cross Street.

HIGHBRIDGE AVENUE. This was the original name of Ogden Avenue, from West 163rd Street to the High Bridge, according to an 1868 map.

HIGHBRIDGE ROAD. This was the former name of West Fordham Road from the Harlem River to its junction with Kingsbridge Road, according to an 1868 map. See: Cammann Street, Pelham Avenue, Union Avenue.

HIGHBRIDGE STREET. See: Fischer (Corporal) Place.

HIGHBRIDGEVILLE. This predominantly Irish settlement came into existence around 1848 when the Croton Aqueduct was finished on The Bronx side. According to Beers' Map of 1868, it was bounded by the Harlem River, High Bridge, West 170th Street and West 162nd Street (Cromwell's Creek). See: Devoe's Point, Dangerville, Nuasin, Turneur's Land.

HIGHWOOD AVENUE. This was the former name of Macomb's Dam Road (which has since been shortened to Macombs Road), according to Robinson's *Atlas*, 1888.

HILDA AVENUE. This was the former name of Brinsmade Avenue from Lafayette Avenue to Bruckner Boulevard, according to an 1888 map.

HILDER AVENUE. It was plotted in 1926 to divert traffic from Pelham Parkway to Gun Hill Road, but was eliminated by the Hutchinson River Parkway cloverleaf.

HILL STREET. This is the former name of Tyndall Avenue in Riverdale.

HILLHOUSE AVENUE. Noted on a 1911 map as the early name of Saxon Avenue.

HILLMAN AVENUE. See: Saxon Avenue.

HILLMAN & HUBBEE SHIPYARD. The second commercial shipyard to be established on City Island was this firm, located at the foot of East Fordham Street. Gustavus Hillman purchased land in 1877, and took Hubbee in as a partner. The business was sold to John P. Hawkins in 1887.

HILLSDALE AVENUE. An 1890 map shows this name on what is now Faile Street at Spofford Avenue. The "hill" and "dale" evidently referred to the hilly nature of the terrain. See: Garrison Heights.

HILLSIDE COTTAGE. An 1874 map of West Farms shows this house and property, belonging to R. Knighton, facing north to Fairmount Avenue, now Crotona Park North. The property is now inside Crotona Park.

HILTON AVENUE. The former name of Sackett Avenue from Bronxdale Avenue to Williamsbridge Road.

HINCHY'S BEACH. It was located at East 132nd Street and the East River, and in the 1920s, was inhabited by a houseboat colony of squatters. In the Prohibition Era (1918-1933), sheds there were used to berth rumrunners, speedy motorboats designed to outrun the Coast Guard cutters. Jack Hinchy ran a boat yard there, and in the 1930s, removed his business to the end of Castle Hill Point.

HIPPODROME, The. This silent movie house was located at 1313 Prospect Avenue near East 167th Street around 1910.

"HIVE TOWN." This was a Colonial nickname for the Bronx River settlement of West Farms. The inhabitants were "bustling like bees in a hive" and the low stone houses resembled the hive. See: DeLancey's Mills, Twelve Farms.

HIVE TOWN ROAD. To the early colonists, West Farms Road was known by this name. The pre-Revolutionary name actually was the Queen's Road, and was so marked on a 1723 conveyance. It was also called the Back Road and the Lower Road.

HOBART COURT. It formerly ran off Hobart Avenue north of Baisley Avenue.

HOBART STREET. See: Digney Avenue.

HOFFMANN'S CASINO. Martin Hoffmann operated a casino and picnic grounds at Avenue B (Havemeyer and Haviland Avenues) in the village of Unionport around 1900-1914. The building itself was listed as the Unionport Hotel. See Huber's Casino, Hoffmann's Park.

HOFFMANN'S PARK. It was located at East Tremont Avenue and Bruckner Boulevard, and was operated by the family of Adam Hoffmann from about 1913 to 1919. It was a noted picnic grounds of 40 acres, used for turtle feasts, clambakes, beer parties and barbecues. The Park was patronized by German singing societies, but also by Irish soccer players and Czech gymnasts. Part of the grounds are now occupied by the 45th Police Precinct. During the Revolutionary War, the site was used as a bivouac area by British troops, as there was a well near Bruckner Boulevard and Revere Avenue. In the 1860s, the estate belonged to the Stillwell family, who sold it to the Crosby family. They occupied the mansion and estate until the late 1890s. After Hoffmann's tenure, the Teutonic Realty Company subdivided the land into building lots. The mansion became a restaurant ("Capalbo's"), and finally was razed in 1971.

HOG NECK. This is so noted on a French map, dated 1777. The map is entitled "Province de New-York, 1777, par Montresor a Paris." It depicts Throggs Neck.

HOGARTH STREET. An 1884 map shows this lane running from Gun Hill Road and Gates Place to the Croton Aqueduct. It was absorbed by Van Cortlandt Park.

HOGUET AVENUE. A 1947 New York Directory lists this avenue running from Olmstead Avenue and Unionport Road to 2320 East Tremont Avenue, but that space is traversed by Odell Street and Purdy Street.

HOLGER ROCK ISLAND. This was a rocky ledge south of the Pelham Bridge off the western shoreline of Pelham Bay Park, according to a 1912 map. When this was the estate of John Hunter III in the 1880s, there was a footbridge connecting this island to the mainland. Then, it was known as Pine Island. In 1962 and 1963, the channel was filled in, and the island was joined to the shoreline. See: Buck Rock.

HOLLAND NUMBER SIX. This was a landmark for many years in Starlight Amusement Park in The Bronx. It was the first submarine commissioned by the U.S. Navy, and was built in 1888 by John P. Holland. The Navy retained it at Norfolk from 1900 to 1914 for training purposes, then sold it to a group of men who exhibited it in Philadelphia for two years. Next, it was on display in Atlantic City, finally to be on exhibit near West Farms Square in Starlight Park from 1918 to around 1933. Upon demolition of the amusement park, the Holland Number 6 was cut up for scrap metal. Frank Wuttge of The Bronx County Historical Society was an authority on this early submersible. See: Fenian Ram.

HOLLAND'S FERRY. During the Revolutionary War, this ferry system had its Bronx pier at approximately West 179th Street on the Harlem River.

HOLLERS ESTATE. Until July, 1961, this was a Bronx post office address that harked back to an earlier time when Hollers' Pond and Hollers' ice house were Eastchester landmarks. It is now covered by the Boston-Secor Housing Project.

HOLLERS' POND. This freshwater pond of some 15 acres was formed by damming Rattlesnake Creek as it flowed from Seton Falls Park. Wakefield and Eastchester children used it for ice-skating in the Wintertime. It was filled in around 1951, and is now covered by Hollers, Harper and Tillotson Avenues. In 1944, James W. Holier [spelled this way] was listed as owner of the property and he submitted a report to the Department of Health attesting to the fact the pond water was not used during the Summer months. In the Summertime, well water was used. The well, drilled to a depth of 600 feet, yielded water at 200 gallons per minute. Wrought iron pipes were used.

HOLLY PARK. This is the former name of an area within Van Cortlandt Park bounded by the golf course, Jerome Avenue, East 233rd Street and the Major Deegan Expressway. It received its name from the profusion of holly bushes there. The name was changed in 1967 to Allen Shandler Recreation Area.

HOLLY STREET. This was the earlier name of East 240th Street, from Van Cortlandt Park to Webster Avenue. The name was changed in 1894. See Fourth Avenue (3).

HOME STREET. When this was the settlement of Wilton of Civil War times, Cypress Avenue from East 134th Street to East 138th Street was called Home Street.

"HOMELAND." This was an estate along the Hunts Point Road at approximately Lafayette Avenue, according to an 1879 Map of the Annexed Districts.

"HOMESTEAD, The"(1). This was a 32-acre orchard on the Throggs Neck Road whose present bounds would be East Tremont Avenue, Bruckner Boulevard, Lafayette Avenue and the Throgs Neck Expressway. Elbert Anderson was its proprietor in the 1820s, and he sold it to a Mr. Brandon who, in turn, sold it to the Coster family. See: Coster Estate.

"HOMESTEAD, The" (2). This is the 133-acre property on Throggs Neck that once belonged to H.O. Havemeyer. The land and mansion was sold to Collis P. Huntington around 1893 for $250,000. The property was later subdivided, and there is a Huntington Avenue and a Collis Place. The mansion remained, and is now a Catholic high school for girls.

"HONEYBEES The." The nickname of the Volunteer Fire Company No. 5 of Melrose was used from the 1850s onward to 1898. It was organized in September, 1852, and had close ties with the Manhattan volunteers of Company No. 5. See: Dutch Five.

HOOKER AVENUE. In keeping with the neighborhood theme of using Civil War names on the avenues, this street was mapped in 1923 across Locust Point in honor of General "Joe" Hooker. It has since been renamed Longstreet Avenue.

HOPE HOSE COMPANY NO. 2. The 1864 Annual Report lists this volunteer hose company on Washington Avenue near 5th Street (East 167th Street and Washington Avenue). Oliver Tilden, first Bronx casualty of the Civil War, was a member.

"HOPE MILLS." An 1838 deed mentions John Benson, miller, of Westchester Village, and his place of business as "Hope Mills."

"HOPEDALE." This was an estate owned by E. Haight in 1868, bounded by Zerega, Glebe and Westchester Avenues. See: Lyon Court.

HOPKINS MORRISANIA FIRE DEPARTMENT. This was a volunteer firemen's company with quarters on the northwestern corner of East 149th Street and Robbins (now Jackson) Avenue. It was disbanded in 1874.

HORSE NECK. This was a very ancient name of Locust Point. In 1667, Colonel Nicolls granted Roger Townsend a small neck containing 15 acres called Horseneck at the southeast end of Throgmorton's Neck. See: Horse Point, Locust Island, Wright's Island.

HORSE POINT. This was the southeastern tip of present-day Locust Point, now under the Throgs Neck bridge. The name was in use from about 1840 to Civil War times, when Captain George Wright revived the ancient name of Horse Neck. Wright's Island was later linked to the mainland. See: Locust Island, Wright's Comers, Wright's Island.

HORSESHOE PARK. This was a local name for the semi-circular park at Rogers Place and East 165th Street, which had been part of the Keller estate of the 1880s. It was acquired in March, 1897, by the City, and placed under the jurisdiction of the Department of Parks.

HORSESHOE, The (1). This tavern was located at Third Avenue and East 138th Street in the late 1890s, with Edward Williams the proprietor.

HORSESHOE, The (2). The local name for the area at Hall Place, Rogers Place around East 165th Street was in use *circa* 1900. A small park there acquired the name of "Horseshoe Park," although its official name was Albert Dorey Park.

HORTON'S LANE. Shown on an 1888 map leading to the mansion of George Washington Horton on the southern tip of City Island, the property had already been sold to William Belden in 1885. The lane is now the driveway to the Morris Yacht Club.

HORTON'S POINT. This was an earlier name of the southern end of City Island, mapped in 1819 when it was the property of George Washington Horton. It is now known as Belden's Point.

HOSPITAL ROAD. As late as the 1950s this road, leading past Fordham Hospital, was noted on Bronx maps. On other maps it was labeled Commercial Road, but since 1981 both names have been superseded by Dr. Theodore Kazimiroff Boulevard.

HOTEL CLAUSEN. This hotel on Morris Avenue and East 146th Street was owned by Henry Clausen in the 1890s.

HOTEL GORBETS. This was a 19th-century inn at Gun Hill Road, overlooking the Bronx River. Mrs. Caroline Sage was the owner, and she later married John B. Lazzeri, who owned the adjoining property. Mr. Lazzeri was responsible for putting up the statue of a Civil War soldier on a pedestal in the Bronx River, which became a local landmark. The hotel was later converted into a small factory called the Gobelin Tapestry Works, and managed by William Baumgarten, who imported weavers from France.

HOUSE OF FUN, The. This silent movie house on Prospect Avenue near Westchester Avenue was listed in 1912.

HOWE'S CASTLE. This was a massive stone mansard house with an adjoining square tower that was built, *circa* 1870, on top of a rocky ledge, bounded by Westchester and Cauldwell Avenues, East 149th Street and Trinity Avenue, overlooking St. Mary's Park. Mrs. Edward Howe was the authoress of a cookbook in 1882. Their home was demolished in 1955 to make way for the St. Mary's Park Housing Project.

HOYT PARK. An unofficial park was used in the 1920s by the children of Highbridge. The plot, between West 166th Street and West 167th Street, was owned by John A. Hoyt who installed swings, see-saws and benches.

HUB, The. (1) This localism stood for the intersection of East 149th Street, Melrose, Willis and Third Avenues, as a hub of traffic. It is believed to have been coined around 1896 when Melrose Avenue was cut through to join Willis Avenue, but some old-timers insist the term was current in the 1880s.

HUB, The. (2) Listed in 1922, this was a silent movie theatre with a capacity of 550. It was located on the uptown side of Westchester Avenue, east of Bergen Avenue and was converted to "sound" and "talking pictures." It closed down around 1948.

HUBER'S CASINO. This was the 1930 name for what had been Hoffmann's Casino of an earlier era. See: Hoffmann's Casino.

HUBER'S HOTEL. George H. Huber was the proprietor of this three-storied, highly ornate inn at Jerome Avenue and East 162nd Street in the 1890s. Racetrack fans, on their way to the Jerome Park Race Track, frequented it in large numbers.

HUCKLEBERRY ROAD. The first street railway extending up Third Avenue from the Harlem River bridge acquired this nickname in the 1860s. So frequent were mishaps and delays that a writer of the 1864 New York *Herald* told how passengers would pick huckleberries whenever the streetcar was derailed.

HUDSON PARK (1). This was a small settlement on the Hudson River from West 234th Street to West 237th Street. It was once part of the Betts and Tippett lands, and was locally referred to as Cooperstown, despite its renaming in 1853. See: Cooperstown.

HUDSON PARK (2). A Baychester real estate venture of 1902 when Hudson P. Rose hired George Hollerith to survey an area in the vicinity of the city line and Wright Avenue.

HUGUENOT STREET. This is the former name of East 243rd Street to the city line. See: Cleveland Avenue, Huguenot Avenue.

HULL STREET. This was an earlier name of Marion Avenue from Bedford Park Boulevard to Mosholu Parkway. See: Virginia Avenue.

HUMMOCK ISLAND. Sometimes called Humock Island, this name was current in the 1600s for the island of Paparinemin from which Kingsbridge was formed. See: Paparinemin, Paparinemo.

HUNT ESTATE, The. The original Hunt farmlands were recorded in Revolutionary War days, and the property was held intact for more than a century later. In 1892, most of the tract was sold to a real estate developer named Ephraim B. Levy who incorporated it into "Van Nest Park." The Hunt estate was located between Morris Park Avenue and the railroad tracks from Unionport Road east to Matthews Avenue.

HUNT'S COVE. It is mentioned in a 1792 deed. Hunt's Cove is now filled in at the end of Hunts Point from the waterfront to East Bay Avenue.

HUNT'S ISLAND. See: Jesse Hunt's Island.

HUNT'S MOVIE HOUSE. This silent movie theatre was once located on Morris Park Avenue and Fillmore Street around 1908. John Hunt, the owner and operator, lived on nearby Garfield Street. See: Morris Park Movie Palace.

HUNT'S NECK. This point of land is now covered by Pelham bridge, but noted on the 1777 Blaskowitz map. The last owner of record was Peter Lorillard at the turn of the century, before the land was acquired by the Parks Department. See: Dimans Island.

HUNT'S POINT. This was a separate village until annexed to New York City on January 1, 1874. Its history goes back to the Indian grants of the 1600s.

HUNT'S POINT DIVISION. Around 1906, this was a volunteer lifeguard station on the East River facing Clason Point. The Consolidated Edison Company established a power plant at that point, and the unit (in 1914) was relocated at Edgewater Camp on Throggs Neck as an all-year station.

HUNT'S POINT PALACE. Once the social center of the East Bronx, this establishment flourished from the early 1900s to sometime in 1940. It was located on Southern Boulevard and East 167th Street. The ballroom was famous as countless civic events, weddings, bar mitzvahs, dance contests and amateur shows were held there.

HUNT'S POINT ROAD. This narrow tree-lined lane ran from Southern Boulevard to the East River. From Southern Boulevard to Lafayette Avenue, it was replaced by Hunts Point Avenue, but then the older road followed Coster Street to Oak Point Avenue, then east to Sacrahong Street and south to the Point itself. See: Sacrahong Street.

HUNTER AVENUE. This Baychester street was originally called Lorillard Street in the 1880s, but then was named after John Hunter III. It was eliminated in 1969 by Co-Op City east of the Hutchinson River Parkway.

HUNTER STREET. This is the former name of Bryant Avenue. See: Oostdorp Street, Walker Street.

HUNTLEY'S POND. It was mapped on the Dater estate in 1880, and its present location would be East 152nd Street, Kelly Street and Tinton Avenue.

HUPFEL'S BREWERY. This important brewery was located on St. Ann's Avenue and extended from East 158th Street to East 160th Street, the vaults extending back into the hillside of Eagle Avenue. A few old-timers swore the caverns tunneled under St. Ann's Avenue as well. The original brewery was owned by an Xavier Grant, who sold it to a Mr. Schilling. In 1863, Anton Huepfel [spelled this way] bought the brewery and it passed to his two stepsons who had taken his name. They were Adolph and John. The partnership dissolved in 1883, and Adolph became the sole owner. He had two sons and two daughters, the elder son studying brewery and bacteriology in Berlin and Copenhagen. This was Adolph G. Hupfel (he Americanized the name) who later adapted the brewery to a mushroom plantation during Prohibition. The family plot of this brewing family is adjoining those of the Eichler and Kolb families in Woodlawn Cemetery.

HUPFEL'S HILL. This was a local nickname dating from the 1870s to about 1920 for East 161st Street from St. Ann's Avenue up to Trinity Avenue. It was named after Hupfel's Brewery, which was alongside the steep hill. See: Grove Hill.

HUT HILL. This is the former name of Albany Crescent, and derives its name from the Revolutionary War times when Continental soldiers built barracks on the hillside. Later, the British and Hessian troops occupied the huts. See: Boston Avenue.

HUTCHINSON, ANNE. Anne Hutchinson was the first settler in the area around the river that bears her name. It is recorded she and her family reached that wilderness in 1643, but the Indians massacred her and most of the family and wiped the farm out of existence. She is remembered in many places. See: Hutchinson (all listings) Exact location of farm is not clear, but it covered a wide area.

HUTCHINSON AVENUE. The avenue was laid out across the marshes on the west bank of the Hutchinson River, but only on maps. It was filled in, in 1962, obliterating Barrow Creek. The lower end of Co-Op City was then erected on the reclaimed land, and Einstein Loop covers Hutchinson Avenue. See: Barrow Creek, Einstein Loop.

HUTCHINSON'S BAY. This was a 1666 reference to Eastchester Bay. It is mentioned in Pell's title to the land issued by Governor Nicolls.

HUTCHINSON'S BROOK. "Hutchinson's Brook, later referred to as Black Dog Brook," is noted in the 1664 Holland Society Records. It formed the ancient boundary between the Townships of Eastchester and Westchester. This small, but important, brook is not to be confused with present-day Hutchinson River, which is to the east. See: Black Dog Brook.

HYATT FARM, The. This large Revolutionary farm was bounded by the Bronx River on the east and Katonah Avenue on the west from Vireo Avenue to approximately East 241st Street. It was surveyed by H. H. Spiedler or Spindler in 1889 as an annex to Woodlawn Heights. General Washington stored guns in the Hyatt homestead at one time, according to historical accounts.

HYATT STREET (1). This is the former name of East 241st Street from Van Cortlandt Park to the city line. The name was changed in 1894. See: Fifth Avenue.

HYATT STREET (2). This short street ran eastward from Broadway at West 227th Street until it was eliminated by Marble Hill Houses.

"HYLAN'S FOLLY." Mayor Hylan (known as "Red Mike") was in office when the Bronx Terminal Market was instituted in the 1920s. Its location on the Harlem River was thought to be too remote, and Bronxites took to calling it "Hylan's Folly."

ICE POND BROOK. This brook originated in the locality of East 162nd Street and Morris Avenue, ran parallel to Park Avenue beneath the Concourse Village development, was dammed at West 158th Street to form Morris Pond (also called Melrose Pond), and finally emptied into the Harlem River below Canal Place. The Mott Haven Canal was formed of its mouth. The course of this brook was noted on an 1881 map of the Morrisania District.

IDEAL, The. This early silent movie house at the junction of Dawson Street and Intervale Avenue was opened in 1910, and stayed in business to sometime in the 1930s. It was later known as the Victor Theatre.

INCOG STREET. This is the former name of Old Albany Post Road in Riverdale for its final five blocks before entering Yonkers.

INDIAN BROOK. This was the former name of Seabury Creek. It originated in the Bear Swamp, flowed southeasterly around St. Raymond's R.C. church, across Lyons and Zerega Avenues and into Westchester Creek. See: Bain's Creek.

INDIAN CAVE. Several gorges lead from the steep hillside of the former Seton Hospital grounds atop Spuyten Duyvil. In one of the gorges, some overhanging rocks formed a natural cave known locally as "Indian Cave." Tradition has it that two of Chief Nimham's band of Stockbridge Indians hid there after their defeat near Woodlawn Heights in August, 1778.

INDIAN POND. This is an alternate name for Delafield Pond, or Duck Pond.

INDIAN ROAD. This is the former name of Livingston Avenue, south of West 250th Street. It was named after the picturesque Indian Pond it passed.

INDIAN ROCK (1). A prominent elevation on the west side of the Bronx River on the land owned by the Lorillard family, but now is the Botanical Garden. Its approximate location is on a line with East 200th Street. Since World War II, Lincoln Rock has become its local name due to a carved likeness on its crown. See: Lincoln Rock.

INDIAN ROCK (2). On the salt marshes that made up a great part of Clason Point were some rocky hummocks. One particular one was called "Indian Rock" which, so far as historians can find out, has no particular significance. It was still visible at the end of Blackrock Avenue near Bruckner Boulevard and Soundview Avenue up to 1965.

INDIAN ROCK (3). This is a large boulder, roughly 6 x 8, with a height of about 10 feet on a rocky ledge overlooking Indian Lake in Crotona Park. It is located southwest from the lake, and has four deeply-cut steps on its north face. These steps are evenly spaced, and resemble stirrups.

INDIAN STEPS. This huge rock slope at 5455 Fieldston Road was likened to giant steps.

INLET, The. This was the mouth of Cromwell's Creek as indicated on an 1852 map, which would be directly north of Yankee Stadium.

INTIMATE, The. This silent movie house at 2133 Boston Road near Bronx Park was listed in 1920, with a seating capacity of 2,840.

INWOOD. This was an older name of West Morrisania in the vicinity of Jerome Avenue and East 167th Street on maps of 1858 and 1867. It was recognized as a postal address in 1868. Records indicate it was an estate belonging to Julia (Morris) Stebbins.

IRENE PLACE. This is the former name of West 181st Street from University Avenue to Jerome Avenue. The name was changed in 1886. See: Ann Street.

"IRISH FIFTH AVENUE." It was a nickname of Alexander Avenue when it was a fashionable street, lined with trees and well-kept homes. A goodly number of judges, politicians, doctors and professional men lived there in the early 1900s. The street lost its genteel atmosphere in the 1930s when ethnic changes took place. See: "Doctors' Row."

IRISH TOWN (1). It was a nickname for the area of Woodlawn Heights from Van Cortlandt Park to Oneida Avenue. Irish laborers had worked on the nearby Croton Aqueduct, and they always referred to it as "The Pipe-Line. The nickname was used from about 1880 to the early 1900s.

IRISHTOWN (2). This settlement of the 19th century was at West 258th Street and Riverdale Avenue consisting chiefly of Irish families. See: Coogan's Alley.

IRVIN HOUSE. This was a small hotel with a lecture hall, ballroom and attached stables, as advertised in the 1854 Westchester County *Journal*. It was located on Railroad Avenue and Fifth Street in the village of Morrisania, which would be - today - Park Avenue and East 167th Street.

IRVINE STREET. The former name of Seneca Avenue. The name was changed in 1907.

IRVING PLACE. It ran, according to an 1872 Directory, from Morris Avenue to Park Avenue in the vicinity of East 165th Street.

IRVING STREET. The former name of East 181st Street from Park Avenue to Third Avenue. See: Ann Street.

ISAAC STREET. The earliest name of East 197th Street from Decatur Avenue down to Webster Avenue. Isaac Varian was the 1850 developer of this area, a work that was carried on by Isaac Varian, Jr., and Isaac Varian, III. See: William Street, Rosa Street.

"ISLAND FARM." In 1789, Alexander Macomb purchased the whole island of Kingsbridge, and gave it this name. His grandson, Robert Macomb, later inherited it.. The island was formed by two arms of Tibbett's Brook feeding into Spuyten Duyvil Creek, and was called by the colonists of the 1700s Hummock Island. See: Paparinemo.

"ISLAND HALL." This estate of Augustus Zerega diZerega, his sons and grandsons was situated on what is today Ferry Point Park. It consisted of 114 acres, with 2 1/2 miles of waterfront on the East River, Westchester Creek and Morris Cove. The manor house was built in 1853. It burned down on January 31, 1895, and a smaller mansion was built on the foundations the same year. The land was acquired by the City around 1937, and the mansion was demolished soon after.

ITTNER'S PARK. John Ittner was the proprietor of a Tremont hotel on Webster Avenue and East 177th Street. Later, a grove was utilized for picnics, lawn parties and bowling matches. The hotel faced Berrian Road, as Webster Avenue was called in the 1870s, and the grove sloped down to the Mill Brook (now Park Avenue).

IVES' LANE. This was a former lane running from present-day Aldus Street and Southern Boulevard to a point at Garrison Avenue and the Bronx River. A good-sized pond on the property yielded two crops of ice each winter. In 1890, a Mr. Ives was listed as the landowner.

J. L. MOTT HOOK & LADDER COMPANY, NO. 2. This Volunteer Fire Company was listed in the 1869 Directory of Morrisania. The location of the firehouse was given as East 138th Street west of Third Avenue.

JABURG AVENUE. In 1956, the City Council accommodatingly changed the name of a two-block stretch of Leggett Avenue to Jaburg Avenue. Hugo Jaburg, president of the Williams grocery chain, requested the change as it was embarrassing to have an address advertising a rival grocery firm, Francis H. Leggett & Co. Councilman Stanley Isaacs objected to this tampering with a historic Bronx name - and the street signs were never put up.

JACK'S ROCK DIVISION. This Volunteer Station of the Life Saving Service of New York was manned by four officers and eight surfmen, in the period 1896 to 1911. It was situated in Pelham Bay Park on the mainland south of Hunter Island, and consisted of the standard white frame bungalow, launching slip and lifeboats. Later, the unit was moved to Throggs Neck and became the White Cross Station.

JACKSON ENGINE COMPANY NO. 4. An 1862 Annual Report lists this volunteer fire company as being in a two-storied brick building on Boston Road, below Mott Street (Third Avenue below East 149th Street). Three of the firemen later served in the Civil War.

JACKSON STREET (1). This is the former name of Beaumont Avenue from Grote Street to Crescent Avenue.

JACKSON STREET (2). This was the earlier name of Odell Street.

JACKSON STREET (3). In Van Nest, this was the original name of Baker Street.

JACKSON, The. This small silent movie house at 745 Westchester Avenue opened in 1919. It was later called The Coleman, the Louis and finally Borinquen until it was razed around 1947.

JACKSON'S DOCK. A stone pier on Weir Creek was used by the Jackson family from the late 1870s to the early 1900s. Frederick Jackson married Sarah Laura Havemeyer, and he was a member of the Westchester County Historical Society, when Throggs Neck was part of Westchester County. The Jacksons are buried in Woodlawn Cemetery, and the pier is now under Bicentennial Veterans' Memorial Park.

JACKSON'S FARM. This 140-acre tract on Throggs Neck was purchased by Samuel D. and Julia Ann Jackson in 1835 from widow Hannah Brown. The bounds were roughly Miles Avenue to Schley Avenue, along East Tremont Avenue and eastward to Weir Creek. This farm was purchased by a nephew, Frederick Wendell Jackson (see Jackson's Dock) and he lived on a crest of land (Migel Place) which was the highest elevation on Throggs Neck. See: Jackson's Pond.

JACKSON'S LANE. This pre-Revolutionary lane was begun by the Baxter family in the 1650s leading from East Tremont Avenue at Migel Place down to Weir Creek. A latterday owner of the tract was Frederick and Sarah Jackson. In 1958, the lane was blotted out by the Throgs Neck Expressway and Bicentennial Veterans' Memorial Park.

JACKSON'S POND. An ice-pond on the Havemeyer estate of the 1860s was inherited by his son-in-law, Frederick W. Jackson. It was a spring-fed pond, located southwest of Miles Avenue and Throggs Neck Boulevard, and the overflow ran into Weir Creek. The pond was used for ice-cutting, and also contained carp. See: Soldier Pond.

JACKSONVILLE. This small and sparsely-populated settlement was on the east bank of the Bronx River from 233rd Street to East 238th Street, and eastward to Bussing Avenue. It was mapped in the 1880s as a separate entity, but shortly after 1900 it was absorbed by Wakefield. See: Washingtonville.

JACOB STREET. The former name of East 187th Street from Third Avenue to Hughes Avenue in the village of Belmont. This was once part of the Jacob Lorillard estate, "Belmont." See: Clay Avenue, Sanford Street.

JACOB'S SHIPYARD. At the foot of Pilot Street on City Island, this yard was begun a century ago when David Carll started in business. From 1886 to 1900, it was known as Piepgras' Yard, until Robert Jacob assumed ownership. The American Cup challengers were outfitted there, and many fine yachts were built in the yard.

JAEGER'S BREWERY. This small brewery was in operation from 1855 to 1860, when Wilhelm Jaeger sold it to a fellow German brewer named Heinrich Zeltner. It was located on what was to become East 170th Street and Third Avenue, on the northeastern corner. See: Zeltner's Brewery.

JAMES STREET. This was the earlier name of East 167th Street from Morris Avenue to the Concourse in 1871. See: Fifth Street (2), Lyons Street, Overlook Avenue.

JAMES' LANE. This is the former name of West 253rd Street, dating back to the time when J. B. James was listed in the 1880s as owner of the property. His mansion was located under present-day Henry Hudson Parkway 150 yards west of Broadway. See: Riverdale Lane.

JANE STREET (1). This was Mt. Eden Avenue from Weeks Avenue to Topping Avenue, according to an 1897 Atlas. See: Belmont Street (2), Walnut Street (1).

JANE STREET (2). This was the earlier name of Mt. Eden Avenue, from Walton Avenue to the Concourse. See: East Belmont Street, Wolf Street, Walnut Street.

JANES' HILL. The former name of the northern end of St. Mary's Park - roughly from East 145th Street to East 149th Street. Around 1858, Adrian and Adeline Janes purchased the tract from Gouverneur Morris II and lived in a fine mansion, which was noted on an 1872 map. He was a parntner of The Janes & Kirtland's Foundry.

JAXON POINT DIVISION. This Volunteer Station of the Life Saving Service of New York was built at the foot of Griswold and Outlook Avenues, fronting on Eastchester Bay. One officer and four surfmen manned the station, as listed in 1935 - with a Major Honore Jaxon on the rolls as an Honorary Member. The promontory on which the frame building stood led to Palmer Cove, so the name was not geographic. The station was destroyed by fire in 1937, but was rebuilt. In 1938, it was blown from its foundations by a windstorm, and never replaced.

JAY PLACE. This short Place was laid out across historic Ferris Point that was first settled by Brockett and Spicer, two associates of John Throgmorton. It passed through the Hunt, Ferris, Seaman and Waring families until acquired by the City in the 1930s. The Place and its name was carried for many years on maps, while extensive changes took place due to the Bronx-Whitestone Bridge's construction in 1938-1939. An Engineer in the Bronx County office hazarded a guess that the name is the initial "J," and there seems no reason to name it after John Jay, first Chief Justice of the Supreme Court in 1795. Augustus Jay was an associate of the Warings, Havemeyers and Huntingtons, all of whom owned estates in the vicinity. The question is academic, as Jay Place was eliminated from maps in 1970.

JEAFFERD'S NECK. During the Revolutionary War, Oak Point bore this name. It was also mentioned in the Graham grants of 1806. See: Arnold's Point, Leggett's Point, Oak Point.

JEFFERSON AVENUE (1). This is a former name of Hughes Avenue from Crotona Park up to East 187th Street. See: Frederick Street.

JEFFERSON AVENUE (2). This is an earlier name for Edenwald Avenue.

JEFFERSON STREET (1). This was the 1880 name of Jefferson Place off Boston Road. The original wooden church of St. Augustine was located on this street, until destroyed by a fire. After that, the brick and stone church was built at East 167th Street and Franklin Avenue.

JEFFERSON STREET (2). In Van Nest, this is the former name of Wallace Avenue from the railroad yards to Morris Park Avenue. See: Hall Avenue, Hickory Avenue.

JEROME PARK. This former racetrack and club grounds reached from Kingsbridge Road north to Van Cortlandt Park South, and from Sedgwick Avenue to Jerome Avenue. It was built in 1865, financed by the New York Jockey Club headed by Leonard W. Jerome, famous sportsman and financier, and flourished until 1894. See: Bathgate Farm (2).

JEROME PARK HOTEL. It was listed in the 1872 Directory of Morrisania and Tremont as under the directorship of C. P. Arcularius. An 1898 map locates this hotel at the junction of East 169th Street and Jerome Avenue. See: Arcularius Place, Charles Place.

JEROME SKATING LAKE. This fair-sized pond, east of Webster Avenue at East 198th Street, was mapped in 1888, but is now covered by the Penn Central railroad tracks. It was fed by a brook from Peter Briggs farm (past Decatur Avenue) and also by School Brook. See: School Brook.

JEROME, The. This silent movie house at 1 West Tremont Avenue opened in 1926 with 1,660 seats.

JESSE HUNT'S ISLAND. This was a former name of Hunter Island on a 1779 map. Jesse Hunt was a Sheriff of Westchester County in 1777, and in 1795, his son, Thomas, sold the island to John Blagge; however, four years later, Hunt's name was still carried on the charts. See: Appleby's Island.

JOHN STREET (1). This is the former name of East 159th Street from Brook Avenue to Eagle Avenue. The name was changed in 1888. See: St. John's Hill.

JOHN STREET (2). In the village of Belmont, this was the earlier name of East 181st Street from Third Avenue to Southern Boulevard. See: Ann Street.

JOHN STREET (3). This is the former name of Zulette Avenue in Middletown.

JOHN STREET (4). Sir John Pell sold City Island to John Smith in 1685, and a subsequent owner in 1755 was John Jones. All these squires qualify, but my choice is John Hunter III who, in the 1870s, dabbled in City Island real estate west of City Island Avenue with an idea of starting a racetrack there. Hunter Avenue is the present name.

JOHNSON AVENUE. This is the former name of Mapes Avenue, but sometimes it was called Johnson Street and probably harked back to a tavernkeeper of that name around 1860. See: Johnson Street (1).

JOHNSON STREET (1). This was an earlier name of Mapes Avenue from East Tremont Avenue to East 180th Street where a small tract belonged to the Johnson family. Johnson's Tavern at Boston Road and East 182nd Street was a well-known inn at the time of the Civil War, and its occupancy began before the Revolutionary War, when it was a stagecoach stop. See: Planters Inn.

JOHNSON STREET (2). This was the original name of Secor Avenue in Edenwald.

JOHNSON'S SEMINARY. In the 1870s, this select school for young ladies and children was located alongside the Boston Road on the exact spot where Morris High School stands.

JOHNSON'S TAVERN. See: Planters Inn.

JONES AVENUE. This was the former name of Grace Avenue. Robert Givan had owned the land since 1794, and his descendants lived on it until the late 1880s when the real estate development, Pelham Bayview Park, was inaugurated. The Jones family were fourth generation inhabitants, descending from a daughter, Margaret Givan Palmer. See: Myrtle Street.

JONES' LANE. All that is known is that this was an early name of West 250th Street from Riverdale Avenue to the Post Road, with no further information available.

JUDGE SMITH'S HILL. This was Shakespeare Avenue from West 168th Street down to Jerome Avenue. It was known by this name from about the Civil War days into the 1900s because of the roadhouse at the foot of the hill which carried that name. Each Winter, the first horse-drawn sled, called a cutter, to arrive at Judge Smith's received a magnum of champagne. See: Birch Street (2), Marcher Avenue.

JUDGE'S BEACH. A well-known and well patronized beach and picnic spot on Eastchester Bay at the foot of Layton Avenue flourished from 1910 to the 1930s. Prior to 1918, it was also headquarters of the Throggs Neck Volunteer Life Saving Division.

JULIA STREET. This is the former name of Crotona Park South, from Third Avenue to Fulton Avenue, recalling Julia Huerstel who owned the tract in 1889.

JULIANA STREET. This was Magenta Street's earlier name from the Bronx River up to White Plains Road, according to an 1884 map of Williamsbridge.

JULIET STREET. This was the original name of East 158th Street from the Harlem River to the Concourse in the 1870s, most of which is now covered by the Yankee Stadium. As this was Gerard Morris' lands, it is possible Juliet was (a) a daughter, (b) granddaughter, (c) maidservant.

JUMBO'S CORNER. This is now West 254th Street and Riverdale Avenue, and got this nickname from having been the location of the Riverdale Inn, known locally as "Jumbo's Hotel."

JUMBO'S HOTEL. The Riverdale Inn of the 1890s carried this nickname, as William Olms, the proprietor, was so huge and elephantine he was known far and wide as "Jumbo." See: Riverdale Inn.

KANE STREET. Former streets in this neighborhood carry names of naval men, and Elisha Kane was a physician, explorer and U.S. Naval Officer (1820-1857) who pioneered a route to the North Pole. The land through which this street runs was once called Fox Woods (early 1800s) and, later (1869), "Foxhurst" - the estate of H. D. Tiffany, whose wife's maiden name was Fox. It is now overlaid by the Hunts Point Market.

KANE'S PARK (1). This popular amusement park and picnic grounds, dance pavilion and swimming pool was at the foot of Soundview Avenue on Clason Point. Patrick Kane was the proprietor in the 1920s. The Park eventually went out of business, to be replaced by Shorehaven, a private Beach Club. See: Shorehaven.

KANE'S PARK AND CASINO (2). A picnic grounds, dance pavilion, bowling alleys was on the former Bayard farm that was acquired by John Hunter III in the 1880s. The resort was in operation from 1919 to about 1933 under the ownership of Martin J. Kane. The casino was a stone mansion, originally belonging to John Hunter III, and Mr. Kane kept it in excellent condition until the entire sector and mansion was acquired by the Parks Department and incorporated into Pelham Bay Park. A disastrous fire in September, 1939, gutted the mansion and so it had to be demolished. See: Pavilion Street.

KANE'S ROAD. This was the unofficial name of the entry road from Shore Road into Pelham Bay Park, and in use from the early 1920s to the 1960s. It once led to Martin J. Kane's casino, the former mansion of John Hunter III, which was demolished after a fire on September 28, 1939.

KARL'S HOTEL. It was an 1890 resort overlooking the Harlem River at High Bridge. It was advertised in the New York *Eagle* in 1891, 1892 and 1894. The hotel boasted of bicycle racks, a pasture for horses, and a boat slip for canoeists and rowers.

KARL'S PARK. This was an amusement park, with picnic grounds and ballroom, on Bergen Avenue from East 148th Street to East 149th Street, extending down to the Mill Brook (Brook Avenue). See: Fountain Park, Germania Park, Loeffler's Park (1), Phoenix Park, Twenty-third Ward Park.

KATONAH STREET. The former name of East 211th Street alongside Woodlawn Cemetery. See: Ruskin Street.

KELLY STREET. This was the earliest name of East 152nd Street at Wales Avenue, where it once crossed the Dater estate in the 23rd Ward. Captain Samuel Kelly had a farm there in the 19th century. The name was changed in 1894. See: Elton Street, Willow Street, Rose Street.

KEMBLE STREET. This is the former name of East 238th Street from Van Cortlandt Park to Vireo Avenue, and may be a remembrance of Nicholas Kemble, who was a landowner in 1800. See: Kimble Street.

KEMP PLACE. Noted in 1881 firehouse journals, the name was changed to Kemp Street.

KEMP STREET. In Highbridgeville, when the Anderson farm was cut through, this was the original name of West 164th Street from Summit Avenue to Anderson Avenue.

"KENILWORTH." This was the 17th-century name of the lands held by Nathaniel Underhill bordering on Eastchester Bay south of present-day Pelham Bay Park. The Underhills originated in Kenilworth, Warwickshire, England hence the name of his property. In 1922, when the Lorillard Spencer estate (the former Underhill land) was subdivided, one small street was named Kennelworth [spelled this way] Place.

KEOUGH STREET. This was a small street that was to run from Allerton Avenue north to Arnow Avenue through the Burke estate. However, it was closed out in December, 1926. See: Laverty Street.

KERNELS, The. On the west bank of the Hutchinson River, approximately the site of Co-Op City, a series of streets were surveyed in 1859 and 1868. Almost a century before that, it had been described in a deed as "a place called Colonel's." See: Colonel's.

KESKESKECH. This was a Weckguasegeeck Indian name for The Bronx, specifically the extreme western edge, flanking the Harlem River and the junction with the Bronx Kill. The name means "Stony Ground." See: North New York, North Side, Stony Island, Stony Point.

KETCHUM AVENUE. This was the former name of Fieldston Road when Alexander Phoenix Ketchum was a resident there in the 1890s. He was Commissioner of Education in 1895. See: Suydam Avenue.

KIAMIE ESTATE. Najeeb Kiamie purchased the northern part of the former John A. Morris property around 1921. He was a wealthy Lebanese silk manufacturer with three sons and one daughter. Mr. Kiamie was a pioneer in the silk business at Paterson, N.J., being the first to commercialize the use of silk into garments. A mansion on the property had a total of 40 rooms, some even soundproofed, which was a rarity in those days. When Sampson Avenue was cut through, the Kiamie mansion had to be moved, and the newspaper accounts pegged the cost at $150,000. The mansion remained for another decade, facing Revere Avenue, and was finally demolished around 1946.

KIDD BROOK. This brook flowed in a southeasterly direction through the estate of Harvey and Lucretia Kidd between Boston Road and Eastchester Road, according to an 1868 Map of Boston Road, by William Livingston, Chief Engineer. It coursed through the present-day Bronx Municipal Hospital Center (Van Etten and Jacobi Hospitals) and emptied into Westchester Creek (now the State Mental Hospital Grounds). See: Abbott's Brook.

KILLIAN'S GROVE. This former pavilion and beach resort belonging to Philip Kilian (one "l") was in operation in the early 1900s along the Clason Point Road. Mr. Kilian was on the reception committee when the ferry system was officially begun between

Clason Point and College Point on August 6, 1921. Killian's Grove was later incorporated into a bungalow colony called Harding Park. See: Higgs' Beach.

KILLIAN'S LANE. A hard-surfaced road once ran from Soundview Avenue west to the vicinity of O'Brien Avenue and through a summertime bungalow colony, described above. It terminated at the shoreline community.

KILLY BRIDGE. A wooden bridge over Weir Creek at Lafayette Avenue was in use from the late 1890s to sometime around 1920. The creek is now covered by the Throgs Neck Expressway. Killy Bridge was a favorite spot for the children who fished for killyfish there.

KILLY ROCK. This flattish boulder on the shoreline of Eastchester Bay at the Park of Edgewater was known to generations of children. See: Adee Point.

KIMBLE STREET. This is the former name of East 238th Street in Woodlawn Heights. The name was changed on March 12, 1890, by order of the Common Council. See: Kemble Street.

KINDERMANN PLACE. Back in the 1870s, this short street was called Anna Place after Anna Bathgate, who married Martin Zborowski of the adjoining estate. It was a typical working-class street in Morrisania off Webster Avenue, composed of two-storied frame houses. In 1913, the name was changed to Kindermann Place as the Kindermann Warehouse was built on its corner, and was a Bronx landmark with its tower that could be seen from Fordham. In 1960, the Borgia Butler Houses were built, and the Place was closed by the contractors. In 1962, it was physically closed and the firebox (last official remnant) was removed in January, 1963.

KING STREET. This was the former name of Adee Avenue from Bronx Park to Cruger Avenue, a distance of 4 blocks, according to a 1905 map. See: LeRoy Avenue.

KING, The. On Third Avenue between East 174th Street and East 175th Street, a small silent movie house opened in 1916. It was then called The McKinley. In 1930, it introduced "talkies" and took the name of The King.

KING'S BRIDGE, The. This former bridge at West 230th Street and Kingsbridge Avenue was the main passage from the island of Manhattan to The Bronx mainland. It was built and owned by Vredryck Flypsen (Frederick Philipse) in 1693, who made it a tollbridge onto his lands. During the Revolutionary War, it was the main military artery for both sides, and was under continuous attack. The bridge fulfilled its duties until 1916 when the Spuyten Duyvil Creek was filled in. During excavations in the 1960s, parts of the wooden span were brought to the surface by workmen, and are now in the possession of the Kingsbridge Historical Society.

"KING'S REDOUBT." A Revolutionary fort named by the British was located at West 190th Street overlooking the Harlem River. It was the forward portion of what the Americans called Fort Number 6. The Veterans' Hospital now covers the site. See: Fort Number 6.

KINGSBRIDGE BURIAL GROUNDS. This plot of several acres bordered Van Cortlandt Lake, marked by field stones only, with no inscriptions, or with only the initials of the earliest settlers. It is thought to be the final resting place of William Betts and George Tippett. In the 1870s, gravestones of the Berrians, Bashfords, Warners and Ackermanns still stood. The New York City & Northern Railroad cut through one end of the plot around 1890 after it consolidated with the West Side & Yonkers Railway.

KINGSBRIDGE CREEK. This name was sometimes applied to Spuyten Duyvil Creek because of the bridge that spanned it and the locality it traversed. It is now covered by West 230th Street.

KINGSBRIDGE ROAD (1). This most ancient highway was one over which our predecessors travelled across The Bronx from West to East. It is more or less under the roadbed of present-day Kingsbridge Road from West 225th Street to its junction with Fordham Road. The Old Kingsbridge Road continued to Third Avenue, ran down to East 182nd Street and through Belmont to Boston Road, thence down to East 180th Street at the Bronx Falls.

KINGSBRIDGE ROAD (2). This was the older name for Bussing Avenue. The Kingsbridge Road of Revolutionary times branched off White Plains Road, ran up Barnes Avenue and Bussing Avenue and on up into Westchester County to Boston. In Mount Vernon, it is still called Kingsbridge Road.

KINGSBRIDGE, The. According to the Journal of Theatre Historical Society, this was a silent movie house at 15 East Kingsbridge Road with a seating capacity of 1,125, which opened in 1920. It closed early in the 1950s, and was converted to a supermarket.

KINGSBRIDGE. This was a separate village on the east bank of the Harlem River whose roots go back to the Dutch days. It took its English name from the King's Bridge that spanned the Spuyten Duyvil Creek in Revolutionary times. It (the village) was annexed to New York City in 1874. See: "Island Farm," Paparinemo, Twenty-fourth Ward (1).

KINGSBRIDGEVILLE. This title was noted in 1870 in the official Post Office records, but the name was changed to Kingsbridge in 1872.

KINGSTON AVENUE. This is the name of Givan Avenue from Seymour Avenue to Kingsland Avenue, a matter of six blocks.

KIPP'S PARKWAY HOTEL. A fashionable restaurant was located on the southwestern corner of Southern Boulevard and Fordham Road, opposite the Bronx Zoo. Frank W. Kipp was the proprietor in 1913.

KIRCHOF'S BREWERY. In pre-Civil War days, this brewery was located on Third Avenue, east side, some 25 yards north of Westchester Avenue. In 1866, Peter Kirchof built Central Hall next to it on the corner of Third and Westchester Avenues. See: Central Hall.

KIRK STREET (1). This is the former name of Field Place. The name was changed in 1897.

KIRK STREET (2). This was a very ancient path leading from Westchester Creek, dating back to the 1600s. It was demapped in 1989 due to the expansion of the Schildwachter Fuel Oil Corporation. See: Dock Street.

KIRKSIDE AVENUE. This was the earlier name of Morris Avenue from Fordham Road to St. James Park, a stretch of two blocks. It was a descriptive name as it ran alongside the kirk of St. James. See: Avenue A, Second Avenue (2).

KLIPPEL'S BEACH. This small beach and picnic grounds at the foot of Layton Avenue on Eastchester Bay in the 1920s was also the headquarters of the Echo Canoe Club. Charles Klippel was Commander of the Throggs Neck Life Saving Station in 1914, the frame building being located on the beach itself.

KNICKERBOCKER INN. A hostelry of the late 1870s to the 1890s on the southeastern corner of Eastchester Road and Pelham Parkway had been the residence of the Ferris family (1830s) and James C. Cooley (1860s) and the Timpsons shortly before it became the roadside tavern. In the early 1920s after Prohibition - it was called the Pelham Health Inn, a roadhouse that lasted until 1947. In that year, it was destroyed by fire. The site is now occupied by Mother Butler R.C. High School.

KNICKERBOCKER, The. This was an early movie house at 879 Prospect Avenue between East 161st Street and East 163rd Street. It was opened in 1910, and in 1912 was listed as a brick building with adequate fireproofing.

KNOX PLACE. This is the former name of West 193rd Street at Bailey Avenue for one block. See: Vietor Place.

KNOX STREET. Until 1894, this was the name of East 239th Street from Van Cortlandt Park to Vireo Avenue.

KOLB'S BREWERY. One of the earliest breweries in the village of Melrose was in operation before the Civil War, and was located on Third Avenue - east side - at East 169th Street. The lager caverns ran into the steep hillside of Franklin Avenue. The brewery was purchased in 1865 by John Eichler. On White Oak Avenue in Woodlawn Cemetery ("Brewers' Row") are the family plots of Kolb, Hupfel and Eichler.

KONSTABELSCHE HOEK. This was a very early Dutch reference to Spuyten Duyvil. It was anglicized to Constable Point. Adrian van der Donck, owner of the lands in Dutch times, had been a Constable. See: Constable Point.

KOSSUTH AVENUE. This is the former name of East 239th Street in Wakefield from Bronx Park East to White Plains Road.

KOSSUTH PLACE. This was the earliest designation of Kossuth Avenue off Gun Hill Road. The story is that this street was intended to bear the name of Kosciusko Street as adjoining avenues honored foreign-born soldiers that aided the American Revolution. The town fathers voted against it on the grounds the name was too hard to pronounce and spell, so Kossuth Place was substituted, even though the Hungarian had no place in our country's history, except for a fund-raising visit in 1849.

KULLMANN'S PARK. This was a small picnic grounds once located at the northeast corner of East 167th Street and Park Avenue and very popular with the German population of Morrisania. It was in operation from 1890 to around 1902. See: Sylvan Park.

KUNTZ'S BREWERY. Founded by the Kuntz family in 1867 in the village of Morrisania at what is now East 168th Street and Third Avenue. Major Louis F. Kuntz was the proprietor, but if he won that title in the Civil War, or in the Prussian army is not known. In 1902 Michael Kuntz sold the business to the North Side Brewing Company.

KYLE'S PARK. This popular amusement center on the Harlem River just north of High Bridge was in operation from the 1880s to around 1914. Steamboats brought excursionists up from the Third Avenue bridge, and there was a pier capable of berthing three boats at a time. The railroad brought additional picnickers, and canoeists and scullers could stop by for a cold beer. Mr. Kyle once hired an aerial slack rope walker to cross from the Manhattan slopes to the Bronx side. His name was Leslie, and newspaper accounts tell of this feat whereby the acrobat carried a stove to the middle of the slack rope, cooked pancakes and threw them down to the boaters.

L'HERMITAGE. A famous French restaurant on the Bronx River at Gun Hill Road was once patronized by poets, sculptors, artists, writers and kindred spirits. It flourished in the late 1800s and may have had more than one owner, for there is mention of a Constant Aresene Baudouin, and also a Baudoin Laguerre.

LAAP-HA-WACH-KING. The Siwanoy Indian name for Hunter Island, which was their main source of shells from which they manufactured wampum. The name meant "Place of string beads." See: Appleby's Island.

LACONIA, The. In 1926, this silent movie house was built on White Plains Road and East 224th Street. It seated 1,160 people.

LADY WASHINGTON HOSE COMPANY NO. 1. This was the important volunteer fire company of Morrisania located on Fulton Avenue between East 167th Street and East 168th Street. The firehouse boasted of rugs, armchairs and "all the comforts of home." When the Civil War began, manpower was seriously depleted and many wives came out to fight the fires while their husbands were away. Frank Royal and Charles Hoffmeyer were killed at Antietam. David Hammill was wounded at the battle of Harrison's Farm, was brought home and died. See: White Ghost, Fulton Avenue.

LAFAYETTE AVENUE (1). This was the former name of Topping Avenue from Claremont Park to East 176th Street, according to an 1893 map.

LAFAYETTE AVENUE (2). Maps of the 1890 era show this to be the name of Castle Hill Avenue from Zerega Avenue to Westchester Avenue alongside the Catholic Protectory grounds. See: Avenue C in Unionport.

LAFAYETTE AVENUE (3). On City Island, this is the former name of King Street alongside the Pelham Cemetery, sometimes called the Fordham Cemetery.

LAFAYETTE LANE. This is the forerunner of Lafayette Avenue from Hunts Point Avenue to the Bronx River, noted on a 1900 map. General Lafayette traversed the lane in August, 1824, when on a triumphal tour of the East, and the townspeople gave it that name in his memory.

LAFAYETTE PARK. This was the original name of Soundview Park, named by Commissioner Robert Moses in 1937, but local residents asked to have it be given a better known name, locally. At that time, the newly acquired park was hardly half the acreage it has since acquired through landfill.

LAFAYETTE PLACE (1). This is the former name of Lafontaine Avenue from East Tremont Avenue to 180th Street, a distance of three blocks.

LAFAYETTE PLACE (2). Before 1912, this was the name of East 188th Street from Park Avenue to Third Avenue.

LAFAYETTE STREET. This was the earlier name of Crotona Avenue from Boston Road to Crotona Park. It is about the ancient boundary line between the Manorlands of Morrisania and the West Farms Patent - the actual dividing line being a brook that flowed south from present-day Crotona Park. See: Bungay Brook.

LANDING ROAD. This was the 19th-century name of Oak Point Avenue from Manida Street to Truxton Street as it led to a landing on Sacrahong Creek (on some maps) or on Leggett's Creek (as other maps called it).

LANE AVENUE. This is the former name of Longwood Avenue.

LANG AVENUE. This street was laid out, but eliminated by the State Mental Hospital grounds adjoining Hutchinson River Parkway.

LANG'S POINT. This was a name for present-day Ferry Point Park that was in use in 1816 when the peninsula was owned by Hugh Lang. It was also noted on a U.S. Army map of 1833. See: Ferris Neck.

LARKIN AVENUE. This is the former name of Watson Avenue from Commonwealth Avenue to White Plains Road. Francis Larkin bought Cow Neck Farm from Oakley Pugsley's heirs in 1854 and sold it to the Catholic Protectory in 1880. See: Cobb Avenue, Cow Neck, Ninth Street (2), Walter Street.

LAUREL STREET. This street was mapped in 1888, but is now incorporated into the Bronx Botanical Garden west of the Bronx River.

LAURIE AVENUE. This was an avenue that was laid out but never opened. The land is now occupied by the State Mental Hospital grounds behind Westchester Square.

LAVERTY STREET. This was a small street that was to run from Allerton Avenue north to Arnow Avenue across the Burke estate. It was surveyed, but closed out in December, 1926. See: Keough Street.

LAWRENCE AVENUE. An 1851 and an 1864 map of Highbridgeville shows the land owned by William Beach Lawrence. Lawrence Avenue formerly ran from University Avenue below West 166th Street to West 167th Street, but is now covered by Highbridge Houses. (Beach Avenue was the former name of West 167th Street).

LE ROY AVENUE. It is the former name of Adee Avenue from Eastchester Road to approximately Ely Avenue (where Black Dog' Brook once flowed). This was once the Givan lands, and Mrs. Elizabeth Palmer LeRoy was a great-granddaughter of Robert Givan who owned the land in 1795. The real estate development was called "Pelham Bayview Park." See: King Street.

LE ROY BAY. This shallow bay, between Rodman's Neck and Hunter Island, has been filled in to form the parking field for Orchard Beach. See: LeRoy Point.

LE ROY POINT. In the 1830s and 1840s, this point of land belonged to Jacob LeRoy, a wealthy New York merchant. His sister married Daniel Webster. His spacious grounds, facing City Island, is now incorporated into Pelham Bay Park, and the Point - once an Indian settlement - is now at the southwestern end of the Orchard Beach parking field.

LEBANON HOSPITAL. This imposing hospital was incorporated in 1890 by a group of Jewish citizens, and began functioning in 1892. It stood on a rocky bluff overlooking Westchester Avenue and Cauldwell Avenue. The hospital removed to the Grand Concourse in 1943. See: Bronx Garment Center, Ursuline Convent.

LEBANON STREET. See: East 179th Street.

LEE'S HOTEL. A hotel near the Fordham depot was in operation from 1900 to about 1910. Its address was East 189th Street and Third Avenue.

LEGGETT PLACE. It formerly ran from Westchester Avenue to McGraw Avenue, but was obliterated when the Cross-Bronx Expressway was built. Gabriel Leggett was an early landowner there, and some of his farmlands are now part of Parkchester. See: Grinnell Place.

LEGGETT POINT. This was an 18th-century name of Oak Point, when the Leggetts owned most of the land below Hunts Point. See: Arnold's Point, Jeffeard's Point, Oak Point.

LEGGETT'S BROOK. This creek wound through the Debatable Lands. The Morris family insisted this was Bound Brook, which would have shifted their boundary eastward - and resulted in more territory. See: Wigwam Brook.

LEGGETT'S LANE. This very ancient path, in use before the American Revolution, is now overlaid by Leggett Avenue. See: Dennison's Lane, White's Lane.

LENOX PLACE. It was laid out in 1905 as part of a real estate development called Westchester Heights along Pelham Parkway. It is now covered by the grounds of Jacobi Hospital in the vicinity of Seminole Avenue.

LENT'S ISLAND. See: Wading Place, The.

LESSER MINNEFORD. A 1713 deed, using this name, designated High Island on record of Register, Court of Westchester. This is not to be confused with Little Minneford, now Hart Island.

LEWIS STREET. It formerly ran from East 163rd Street to East 164th Street, between Sherman Avenue and Sheridan Avenue, as listed in 1871. There is a strong possibility it honored Lewis Morris, the original title holder to the land.

LEXINGTON AVENUE. In Mott Haven, this avenue once ran from East 138th Street to East 144th Street west of the Mott Haven Canal. It was mapped in 1871.

LEXINGTON PLACE. It is found "On a certain map, of Lexington Place at Williamsbridge Depot, Town of West Farms. Dec. 22, 1852." The Place was taken by the City of New York for the opening and widening of Gun Hill Road.

LEXINGTON STREET. This is the former name of East 147th Street from St. Mary's Park to Southern Boulevard, a distance of five blocks. It was listed in 1871. See: Dater Avenue, Trinity Avenue (2).

LIBBY STREET. This is the former name of Buhre Avenue in 1905, from the Hutchinson River Parkway to Westchester Avenue. Note that neither the Parkway nor Westchester Avenue existed at that date. See: Madison Avenue (3).

LIBERTY STREET. This was the earlier name of St. Theresa Avenue, a five-block stretch from the Hutchinson River Parkway to Westchester Avenue. When this property was auctioned off in 1892, the auctioning firm was H. C. Mapes of 59 Liberty Street. In 1910, the street was called Alice Avenue, and from 1920 to 1968, it was an extension of Morris Park Avenue. See: Alice Avenue, Morris Park Avenue.

LILLIAN PLACE. Formerly a carriage path connecting Boston Road and East Tremont Avenue due west of West Farms Square, in front of the botanist, Professor Wood's home, it was paved sometime around 1910 and named in honor of the daughter, Lillian. It was blocked off at Boston Road in 1956, and finally closed entirely in 1970. It has become a nameless concrete cul-de-sac next to a market.

LINCOLN ROCK. This is the high point in the Botanical Garden west of the Bronx River. Neighborhood children know the rocky peak by this name since a likeness of the President was cut into the rock sometime in the 1940s. See: Indian Rock (1).

LINCOLN STREET (1). This is the former name of Holland Avenue in Van Nest from Morris Park Avenue to the railroad. See: Maple Avenue, Pine Street, Post Avenue.

LINCOLN STREET (2). The one-block stretch of Hone Avenue from East Tremont Avenue to the railroad was once called Lincoln Street, according to a 1906 map.

LINCOLN, The. A 1912 Atlas lists a silent movie house at East 171st Street between Third and Washington Avenues.

LIND AVENUE. This is the former name of University Avenue at High Bridge. It was in use as far back as 1860, and it was only changed in 1917. The original name might honor Jennie Lind, the Swedish Nightingale, who came to New York in 1850 when the High Bridge was being completed. There was a "Jennie Lind Craze" wherein parasols, songs, candies, etc., were named for the singer. See: Aqueduct Avenue.

LINNAEUS STREET. A street to honor the Swedish naturalist was planned to run from Broadway at West 253rd Street east to the head of Van Cortlandt Lake, according to an 1890 map. It is now overlaid by the Henry Hudson Parkway.

LITTLE CREEK, Ye. This was the earliest name of Bain's Creek or Indian Brook. In the Westchester Records, a citation reads, "1727 . . . then to Ebenezer Haviland's land, then to John Forgason to ye Little Creek to ye Great Creek." This Little Creek was an arm of the Great Creek (Westchester Creek) that wound past Zerega Avenue. See: Bain's Creek.

LITTLE MINNEFORD ISLAND. An 18th-century name of Hart Island. See: Heart Island, Spectacle Island.

LITTLE PLAYHOUSE, The. This silent movie house was on East 180th Street, between Prospect Avenue and Southern Boulevard. No dates given as to its operation, but an educated guess would be before 1920.

LIZARD POND. This was a local nickname for a pond surrounded by swamps at what is now East 165th Street, Findlay Avenue and College Avenue, and the term seems to have been current from 1890 to 1910. The spring later supplied water to Sheffield Farms Milk Company on Webster Avenue.

LOCUST AVENUE (1). This is a former name of East Tremont Avenue from Third Avenue to Mapes Avenue. See: Main Street, Walker Avenue, Waverly Street (2).

LOCUST AVENUE (2). In the village of Belmont, this was once the name of East 177th Street from Third Avenue to Arthur Avenue.

LOCUST AVENUE (3). Formerly Rosewood Street, in Williamsbridge,. See: Post Street.

LOCUST GROVE (1). A real estate development, bounded by Morris Park Avenue, White Plains Road and East Tremont Avenue was auctioned off by John S. Mapes around 1900. The name never took root.

"LOCUST GROVE" (2). This Hudson River estate was owned by Assemblyman Russell Smith west of Riverdale Avenue at West 258th Street. The Hudson River steamer, *Henry Clay*, caught fire and was beached at this shoreline. See: Allen Place.

LOCUST ISLAND. This was an earlier name of Locust Point, as noted on the British war map of 1777, drawn by Colonel Blaskowitz. In those days, it was actually an island surrounded by salt marches and Hammond's Cove on the west and south, and by Long Island Sound on the north and east. See: Horse Neck, Wright's Island.

LOCUST POINT DIVISION. In the 1920s, this Station of volunteer life-guardsmen was built on Locust Point at the end of East 177th Street. It was manned by three officers and three surfmen, but in the 1930s, it was expanded to triple that strength in surfmen. It was disbanded around World War II, but the frame building stood until late in 1970.

LOEFFLER'S PARK (1). This amusement park and picnic grounds was on Bergen Avenue from East 147th Street to East 149th Street in the 1870s, when Anton Loeffler catered to a predominantly German clientele. In those days, Bergen Avenue was known as Retreat Avenue, and the park was a wooded retreat that sloped down to the Brook (now Brook Avenue). See: Fountain Park.

LOEFFLER'S PARK (2). This amusement park was on the west side of Westchester Avenue at Olmstead Avenue. It flourished in the early part of the 1900s, and went out of business sometime around 1935. Part of the tavern was used as a temporary Catholic church when Parkchester was built in 1939. The park is now incorporated into the schoolyard of St. Helena's School.

LOGAN STREET. This is the former name of East 212th Street from White Plains Road to Holland Avenue for one block.

LOHBAUER'S CASINO. A pleasure resort at approximately Wilcox Avenue and Fairmount Avenue on what was once the Layton estate of the 1860s. Frederick Lohbauer was born in Germany in 1853 and came to America as a young man. By the time he was 30, he was operating a small hotel on Throggs Neck, but then opened his famous resort in 1896, calling it Lohbauer's Grove. It was a favorite for clambakes, beefsteak parties and outings for political clubs. The casino specialized in shore dinners, for the waters of Eastchester Bay teemed with all kinds of fish and shellfish. Mr. Lohbauer died in 1904, and his widow and son-in-law, Augustus C. Miller, ran the business competently until the late 1920s, when the land was sold for building lots. He was a remarkable man who was instrumental in organizing the life-saving station of the Throggs Neck Division in 1903, was Commodore of the Oak Point Yacht Club, President of the Chester Taxpayers' Alliance in 1914, a school Commissioner, Chairman of the Sound View Taxpayers' Association in 1921, the Golf Supervisor of Bronx Parks in 1924, and a gifted musician and artist. See: Century Golf Club.

LONDON BRIDGE. The last will and testament of Lewis Morris, on November 6, 1794, gives Lewis Morris, Jr. the house, barns and following land in Morrisania: "beginning at old Dock in front of Dwelling House and running Westerly and Northerly along little Neek by the River and up Harlem River to opposite London Bridge, thence Easterly to Mill Brook." As no other deeds or indentures ever refer to this London Bridge again, no clues are forthcoming. The word "opposite" might mean that boundary marker was on the Manhattan side of the Harlem River.

LONG BRANCH. This is the former nickname of Tibbett Avenue, which followed the course of Tibbett's Brook that is now filled in. It referred to the New Jersey seaside resort because, now and then, the brook rose and flooded the front lawns and cellars of the Kingsbridge residents. See: River Street.

LONG NECK FARM. This 313-acre farm in Old Morrisania belonging to Colonel Lewis Morris was mapped in 1847. See: Moriseny, Old Morrisania.

LONG POND. This was once fed by Long Pond Creek (outlet of Rattlesnake Brook) and was located on the Eastchester meadows between the present-day New England Thruway and the Hutchinson River. It was covered over by Freedomland in 1960, and further buried under Co-Op City.

LONG POND CREEK. This was the northeastern arm of Rattlesnake Creek that extended almost to Connor Street. It was still in existence in 1962, as it was above Freedomland, and this writer navigated it that same year in a kayak. It was covered by Co-Op City in 1966.

LONG REACH. This was vacant land situated on the west side of Rattlesnake Brook in the northwest corner of the town of Eastchester, mentioned in 1690 and again in 1766. Its probable site was Bussing Avenue and DeReimer Avenue.

LONG ROCK. This oval rock was located offshore in the East River at the foot of Tiffany Street and Viele Avenue. In 1856, it was used as a property marker of the Benjamin Whitlock estate, as it lay in the mouth of Leggett's Creek. It is now incorporated in the bulkhead. See: Duck Island.

LONGWOOD PARK. This was a 19th-century estate owned by S. B. White, running along the Westchester Turnpike from Prospect Avenue to Intervale Avenue, Longwood Avenue to Southern Boulevard back to Intervale Avenue. The name was popular in Napoleonic times, as that was the name of the estate where the Emperor was exiled on St. Helena.

LONGWOOD, The. This silent movie house was listed in 1912 on Longwood Avenue between Prospect Avenue and Hewitt Place. It was a wooden, one-story building with no capacity given.

LORAL PLAZA. In 1945, the Bronx Board of Trade requested Borough President Lyons to name the area occupied by the Loral Electronics Corporation. The plant was bounded by the Bronx River, East River, Colgate Avenue and Story Avenue. The matter was hot for a few years, but the president of the corporation, Leon Alpert, never had the pleasure of seeing a street sign put up proclaiming it Loral Plaza. The land had once been part of the Ludlow "Black Rock" farm, and the electronic plant overlaid Ludlow Island that had once been in the Bronx River.

LORELEI POINT. This was a local name for the southeastern point of Hunter Island around 1900, now incorporated into Orchard Beach. Evidently, some German was the Summer vacationer who painted that name on a huge boulder which in Indian days was reputed to have been the center of religious rites. See: Mishow.

LORET PLACE. Although it was mapped in 1922 at the foot of Cabot Street and Barry Street, adjoining the railroad yards, this Place was never opened.

LORILLARD AVENUE. This was the former name of Huntington Avenue from Bruckner Boulevard to the Cross-Bronx Expressway, a matter of three blocks, in 1900.

LORILLARD POINT. This was a former name of Ferry Point Park in 1868. A mansion and an estate, "All Breeze," was owned by Colonel Jacob Lorillard, along with his private wharf on Westchester Creek. See: Ferris Neck.

LORILLARD TERRACE. This was a street on Pierre Lorillard's land, now incorporated in the Bronx Botanical Garden west of the Bronx River.

LOUISE STREET. This is the former name of Hunt Avenue in Vast Nest, noted in 1900.

LOVERS' LANE. This was the unofficial name of Valentine Avenue, during the 1890s and 1900s, north of East 194th Street. It led to Valentine Spring, an idyllic little spot noted by James Reuel Smith who wrote *Springs and Wells of Manhattan and the Bronx*.

LOWELL STREET (1). This was the earlier name of East 141st Street, until it was changed in 1901.

LOWELL STREET (2). This name was changed to Fielding Street off Eastchester Road in March 1950 to avoid confusion with another Lowell Street near Whitlock Avenue.

LOWER CORTLANDT. The original grant of "Colen Donck" (to Van der Donck) was passed to Frederick Philipse, and to the Van Cortlandt family before the American Revolution. It is now the parade grounds of Van Cortlandt Park. See: Upper Cortlandt.

LOWER PLACE. This was the earlier name of East 202nd Street from Webster Avenue to the railroad.

LOWER ROAD, The. This was an early Colonial name for West Farms road, and was the name frequently used in ancient deeds before the official Queen's Road was used. See: Back Road, Hive Town Road.

LOWER YONCKERS. The name of Van der Donck's holdings in 1650, which is now Van Cortlandt Park.

LOWERRE'S LANE. This was a private road extending from present-day Westchester Avenue in the 1850s to the farm of Samuel Lowerre. It is now overlaid by Havemeyer Avenue and the farmhouse site is occupied by the Odd Fellows' Home. See: Avenue B.

LOWMEDE STREET. This street is no longer in existence since it was covered by the Bronx River Parkway south of Gun Hill Road.

LUCY PLACE. One short street mapped in 1900, but covered by Bronx Park East just north of Pelham Parkway.

LUDLOW AVENUE. This was the former name of Bruckner Boulevard from Hunts Point to Soundview Avenue in the early 1900s, honoring a Clason Point family who lived on the "Black Rock" farm in the 1850s. Later, Ludlow Avenue was used from Castle Hill Avenue to Zerega Avenue superseding Sixth Street and Commerce Street in Unionport. Eventually, the entire stretch was called Eastern Boulevard until 1942, when it became Bruckner Boulevard.

LUDLOW CREEK. Barrett's Creek was its name in the 1700s. It was mapped in the 1850s as flowing through the Ludlow farm and entering the Bronx River near Seward Avenue, now covered by Soundview Park.

LUDLOW ISLAND. It was once part of the township of West Farms, and was situated close to the left bank of the Bronx River, south of present-day Bruckner Boulevard. It is now part of the mainland near Soundview Park. See: Loral Plaza.

LUDLOW STREET. See: Echo Place.

LYCEUM, The. A tavern and meeting hall at the lower end of Crotona Park was advertised in 1919. It was Niblo's Garden in 1912, and a part of Zeltner's Park of the 1880s to 1900.

LYDIG FALLS. This was an alternate name of DeLancey's Falls, a natural water fall nearly 12 feet in height at the lower end of the Bronx Lake. It was used by the DeLancey family in Revolutionary times to power their mills, and by the Lydig family after them. The falls can be seen from East 180th Street and the Bronx River.

LYDIG'S WOODS. This was the 19th-century name of Bronx Park. David Lydig purchased the lands in the 1830s, and the property was inherited by his son, Philip.

"LYON COURT." This was once a small estate owned by a Darius Lyon in the 1890s, which was only part of a much larger estate in Civil War days, and known then as "Hopedale." Lyon Avenue is a survival of the name, and perhaps Doris is too.

LYON STREET. This was the 19th-century name of East 167th Street from Westchester Avenue to Southern Boulevard. The Lyon estate was next to the William W. Fox property. See: Fifth Street (2), James Street, Overlook Avenue.

LYONS NICOLODEON, The. On the west side of Westchester Avenue below Parker Street, this movie house was owned by Mr. Lyon, whose home stood at Lyon Avenue and Doris Street. See: Westchester Theatre.

LYONS STREET. According to Bromley's *Atlas* of 1897, this was the former Home Street. The James L. Lyons estate was between Home Street and Boston Road. He was a Seargent in the Civil War, and also a charter member of the Bronx Oldtimers.

LYRIC, The. This silent movie house also had an open-air garden for Summer nights. It was in operation during the 1920s on Third Avenue and East 180th Street.

LYVERE FARM, The. It was in the village of Westchester at the junction of Bear Swamp Road and Walker Avenue in 1881. The present-day names are Bronxdale Avenue and East Tremont Avenue, and Lyvere Street behind the junction is a reminder of the former residents.

MACEDONIAN HOTEL. This wooden hotel on the eastern shore of City Island was made of the timbers of an American frigate called the *Macedonian*. The ship was built in 1836 at Gosport, and broken up in 1874 at Cow Bay on the north shore of Long Island. Soon after, the timber was transported to City Island.

MACOMB STREET. This is the former name of West 231st Street in Kingsbridge. Alexander Macomb purchased, in 1789, the entire island of Paparinemin (Kingsbridge) and called it "Island Farm." The Macomb mansion stood at West 231st Street (Godwin Terrace) and remained in his possession until 1848. See: Morrison's Lane.

MACOMB'S DAM. In 1813, Robert Macomb erected a dam across the Harlem River at approximately 161st Street and, on it, a tollbridge. Only small boats could pass through a lock. In 1838, Lewis R. Morris claimed it was illegal to block off a navigable stream and tested the legality of Macomb's Dam. He sent his boatmen upstream in a flat-bottomed boat, called a periauger, that was too wide for the lock. The boat, named *Nonpareil*, could not pass, so the crew tore a hole in the dam and passed through. Later, Morris won his case in court.

MACOMB'S DAM ROAD. The former name of University Avenue from West 176th Street to Kingsbridge Road.

MACOMB'S ISLAND. See: Wading Place, The.

MADISON AVENUE (1). The earlier name of Bathgate Avenue from East 172nd Street to East 182nd Street in the village of Tremont. See: Cross Street, Elizabeth Avenue (1).

MADISON AVENUE (2). This is the former name of Benson Street from Westchester Square to East Tremont Avenue. See: Adee estate (1).

MADISON AVENUE (3). In Middletown, this was the name of Buhre Avenue from Westchester Avenue to Bruckner Boulevard, according to a 1905 map of "Tremont Terrace." See: Libby Street.

MADISON STREET. This is the former name of Barnes Avenue from Morris Park Avenue to the railroad. Van Nest specialized in Presidential names. See: Cedar Street, Fourth Street (5), Old White Plains Road.

MAENIPPIS KILL. This is a mixed Indian-Dutch name for a creek that ran inland from the Harlem River in the area of the Yankee Stadium, extending northward along Jerome Avenue to West 178th Street. Historian Bolton thinks it may be derived from the Algonquin "Manunne" (slow) "Nepis" (water). The Dutch used the Indian name, but added "Kill" (creek). See: Cromwell's Creek, Doughty's Brook, the Inlet, Mentipathe.

MAGENTA PLACE. This short street, south of Magenta Street, is now absorbed by Cruger Avenue. See: Timpson Avenue.

"MAGNOLIA COTTAGE." A small estate belonging to H. M. Lee on Fairmount Avenue (now Crotona Park North) facing the park, according to an 1874 map of West Farms.

MAHICAN. See: Hudson River [Current].

MAHKANITTUK. See: Hudson River [Current].

MAILLARD PLACE. See: McClellan Street.

MAIN AVENUE (1). In the village of Tremont, this was the former name of Sheridan Avenue from East 167th Street to Mount Eden Avenue.

MAIN AVENUE (2). This was a 19th-century name of Belmont Avenue from East 180th Street to Grote Street, a matter of two blocks.

MAIN STREET (1). This is the former name of East 144th Street.

MAIN STREET (2). In the 19th century, West Farms Road carried this name. It was lined with stores, warehouses, wharves, factories, coalyards and chandleries. See: the Back Road, Hive Town Road, the Lower Road.

MAIN STREET (3). This was the earlier name of City Island Avenue.

MAIN STREET (4). This was the 19th-century name of Williamsbridge Road from Westchester Village (now the Square) to approximately Morris Park Avenue.

MALI ESTATE. This estate of considerable size in the West Bronx extended from Sedgwick Avenue to University Avenue above West 180th Street. The 36 acres belonged to Henry W. T. Mali, who came from Verviers Belgium in 1826 as a wine importer. He became Belgian Consul in 1830, and the post remained in the family until J. T. Johnson Mali died in 1950. A Charles Pierre was Consul in 1879, and a John Mali in 1893. The Mali estate was sold to New York University for $600,000, a property that has devolved to the Bronx Community College.

MANINKETSUCK. This was the Siwanoy Indian name for a brook that rises in the northeastern corner of Pelham Bay Park and runs under the Shore Road into the lagoon behind Hunter Island. Professor Tooker, an expert on Indian tongues, renders it Ma-nun-ket-e-suck, "a strong flowing brook." See: Roosevelt Brook.

MAPES AVENUE. This was the former name of Mayflower Avenue from Maitland Avenue to Zulette Avenue over what had been a Mapes farm in the 1860s.

MAPES LANE. East 176th Street from Southern Boulevard to Boston Road has been obliterated by the Cross-Bronx Expressway, but when it ran to the Woodruff property in the 19th century, it had been called Mapes Lane. See: Mole Street, Orchard Street, Prospect Street, Woodruff Street.

MAPLE AVENUE. This is the former name of Holland Avenue from Gun Hill Road to East 214th Street, some four blocks. See: Lincoln Street, Pine Street, Post Avenue.

MAPLE PLACE. In 1940, this short Eastchester street was changed to Schorr Place.

MAPLE STREET. Now covered by the New England Thruway, this street formerly was laid out in the plan called "The Kernels."

MARBLE STREET. This was the former name of East 178th Street from Webster Avenue to Third Avenue, when it was once part of Quarry Road.

MARCHER AVENUE. This was the earlier name of Shakespeare Avenue from West 168th Street down to Jerome Avenue. An 1868 Atlas lists Rebecca Marcher as owner of "Rockycliff," an estate she had purchased from John Poole three years earlier. She was an ardent admirer of Shakespeare, having named her horses and dogs after characters in Shakespeare's plays. Statues on the lawn immortalized Puck, Romeo and Juliet, and Falstaff - according to a former stableboy. See: Judge Smith's Hill.

MARIAN STREET (1). This was a misspelling of Marrin Street. See: Marrin Street.

MARIAN STREET (2). This was the former name of Bronx Boulevard from Nereid Avenue to East 241st Street on a 1905 map. It was also spelled Marion Street.

MARION AVENUE. This is the former name of Marmion Avenue in West Farms. The name was changed slightly to avoid confusion with Marion Avenue in Fordham.

MARION STREET. This was the earlier name of Bronx Boulevard that was sometimes spelled Marian Street. See: Railroad Terrace.

MARKET SQUARE. This irregular "square" formerly lay between East 139th Street and East 140th Street next to the Mott Haven Canal basin. Today, it would be Rider Avenue to Canal Place.

MARRIN STREET. The original name of Waterbury Avenue from East Tremont Avenue down to Westchester Creek. This was the boundary of the Seton Homestead Land Company, and Joseph J. Marrin was one of its directors. As late as 1930, Emily Marrin was listed as a property owner on Balcom Avenue. See: First Street (5), Marian Street (1).

MARSHALL'S CORNER. This is the former junction of the City Island Road and the monorail. It is now incorporated into Pelham Bay Park, Orchard Beach division. Levin R. Marshall bought real estate from Elisha King, who had built a mansion there in 1829. This Georgian mansion later became the Colonial Inn that was razed in 1937. See: Colonial Inn, "Hawkswood," High Island, Pelham Park & City Island Railroad.

MARTIN TERRACE. This was a proposed name for West 231st Street from Albany Avenue to Bailey Avenue during a 1922 real estate development. Michael J. Martin was to be honored, but the bill was rejected. See: Barry Street, Gallagher Avenue.

MARTINDALE PLACE. An 1873 map shows this short Place running from "Claremont," the estate of Martin Zborowski, to Sherman Avenue, evidently after the ill-famed Black Swamp had been filled in. Its approximate location is East 172nd Street. James Martindale, a property owner is honored. See: the Black Swamp.

MARY STREET (1). This was the former name of East 155th Street across the village of Melrose in 1860. It is most likely named for Mary Nathalia Morris, widow of Henry Morris of Melrose South, as noted in an 1854 will. See: Dawson Street.

MARY STREET (2). This was the 19th-century name of Chesbrough Street, which is only one block long.

MARY'S PARK. This was the original name of St. Mary's Park, when it was plotted and mapped in 1868.

MAURITIUS RIVER. See: Hudson River [Current].

MAYER'S BREWERY. David Mayer's establishment was on the east side of Third Avenue, between East 168th Street and East 169th Street. It stood between Eichler's Brewery on the north and the American Brewing Company on the south.

MAYOR STREET. A 1910 Bronx map shows this short street running diagonally from Bruner and Astor Avenues to DeMeyer Street and Stillwell Avenue. It owed its name to the fact that all streets in this area carry names of former New York mayors.

McALPIN AVENUE. This land was once the swamps of upper Westchester Creek (Stoney Brook). It was platted as Corringe Avenue on city maps of 1891. Around 1900, it was decided to name the street after Edwin Augustus McAlpin, who entered the Civil War as a private and ended as Major General. He also served in the Spanish-American War. He was Vice President of the Hygeia Distilled Water Works near Westchester Square. The avenue was eliminated by the State Mental Hospital grounds in 1958. See: Corringe Avenue.

MAILLARD PLACE. This is the former name of McClellan Street. See: Endrow Place.

McDOWELL PLACE. Civil War generals are particularly honored in this part of Throggs Neck, so the Place was a reminder of General Irwin McDowell (1818-1885). It was part of the original landgrant of John Throgmorton in 1643, and later was property of the Baxter family of the 1700s and the Morris family of the 1800s. It was absorbed by the eastern edge of Ferry Point Park.

McGRAW'S HILL. Old maps of the Westchester Turnpike locate this incline as present-day Westchester Avenue from Prospect Avenue to East 163rd Street. The name was in use in the 1840s.

McKINLEY, The. This silent movie house was opened in 1916 with a seating capacity of 1,800. In 1930, the theatre became The King, but is now a church located at 1319 Boston Road.

McLAUGHLIN'S HOTEL. An 1899 ad located this hotel on Third Avenue and East 158th Street. It boasted of a wine room and a sporting ticker.

McLEAN'S BROOK. An 1878 Area Map shows a meandering brook in the George W. McLean estate (Yonkers) following Katonah Avenue into Woodlawn Cemetery. Intended as an ice pond, McLean's Lake was used for neighborhood fishing and swimming from the 1860s to about 1890. The brook entered Woodlawn Cemetery at its northeastern corner and then ran down into the Bronx River.

McLEAN'S THEATRE. In 1905, this tiny silent movie house opened at 732 Westchester Avenue near Tinton Avenue. A year later, Mr. McLean opened the Nicoland, with a seating capacity of 100.

McTEAGUE'S CORNERS. This was a Revolutionary War reference to the junction of the Old Boston Road, and the road to White Plains. This would be East 233rd Street, Barnes and Bussing Avenues.

MEADOW LANE. This is the former name of Haswell Street.

MECHANIC HALL. A fashionable social hall on the corner of Morris Avenue and Elton Street (East 152nd Street) in the 1860s and 1870s. William Reinhardt was the host.

MECHANIC STREET (1). This was the 19th-century name of East 178th Street from Third Avenue to Hughes Avenue. See: Cedar Street, Elmwood Place.

MECHANIC STREET (2). This is the former name of Bradley Street in the Penfield estate, almost surrounded by Mount Vernon.

MELROSE. Originally part of the Morris family manorlands, it was laid out on November 16, 1850, by surveyor Andrew Findlay from East 156th Street to East 163rd Street, from Park Avenue to Third Avenue. It was incorporated as a village in 1851, and owes its name to the popularity of Sir Walter Scott's novels wherein Melrose Abbey is frequently mentioned. Andrew Findlay was of Scottish descent, and two prominent realtors of the 1850s were Hampton Denman and Robert Elton of Scottish-English antecedents. See: Melrose South.

MELROSE GUARDS. This quasi-military organization under the leadership of Captain Louis Sauter protected the Immaculate Conception church during the anti-Catholic riots of 1851-53. The unit of riflemen also marched in the Corpus Christi processions which were held outdoors along Courtlandt Avenue and Melrose Avenue. The group was still in existence around 1906, and held annual rifle contests. See: Protection Hall.

MELROSE HOUSE. This hotel was in the vicinity of East 162nd Street and Park Avenue according to an 1854 ad in the Westchester County Journal: "Peter Duffy respectfully calls the attention of the public to his new hotel at the Melrose Depot. Good accommodations for boarders."

MELROSE POND. It was once located at East 158th Street, west of Morris Avenue, and now covered by Concourse Village. An earlier name was Morris Pond, as it was on the Gerard Morris lands known as "The Cedars." It was fed by Ice Pond Brook, noted on an 1881 map. See: Ice Pond Brook.

MELROSE SOUTH. The incorporated village (1851) ran from East 148th Street to East 156th Street, from Park Avenue to Third Avenue. See: East Melrose.

MELROSE STREET. See: East 156th Street.

MELROSE, The (1). Prior to World War I, this was a silent movie house on the southwestern corner of Melrose Avenue and East 154th Street. Later, the same owner opened a larger theatre with the same name on East 161st Street. See: Melrose, The (2).

MELROSE, The (2). It was listed in 1922, and was located on the north side of East 161st Street between Melrose and Elton Avenues. In the 1940s, it was converted to a banquet hall and ballroom.

MELROSE TURNVEREIN. This well-known community center of old-time Melrose was on the west side of Courtlandt Avenue between East 150th Street and East 151st Street. This frame building was popular with the German gymnasts from the late 1870s to around 1906. General Franz Sigel was a member, and when he died in 1902, the funeral was held from the *Turnhalle*, with gymnast classes taking part in the funeral procession. It later became the headquarters of the Swiss-American Turnverein before their own clubhouse was built in the lower Bronx. In the 1920s, the hall was turned into a Baptist church.

MELROSE VOLUNTEER HOSE COMPANY NO. 1. It is listed in the 1871 Directory of Morrisania, and was located at Elton Street near Boston Road, which would be East 152nd Street near Third Avenue.

MEMORIAL PARKWAY. In 1926, the Gold Star Mothers put on a determined drive to give this name to the Grand Boulevard and Concourse. The motion was not passed. See Woodrow Wilson Boulevard.

MENEFORS ISLAND. A 1729 conveyance from Thomas Pell, Sr., to his son of the same name describes City Island as the "said Island commonly called by the name Menefors Island or ye Great Island. . . ." See: Great Island, Ye.

MENIVES ISLAND. This was an early name of City Island, noted on a 1708 map, with the Royal Patent for Eastchester, commonly called the Long Reach Patent. See: Great Island, Ye.

MENTIPATHE. This was a Weckguasgeeck Indian name for a creek that ran inland from the Harlem River. See: Cromwell's Creek, Doughty's Brook, Maenippis Kill.

METRO, The. This silent movie house was located on Webster Avenue and East 183rd Street in the 1920s. It had a seating capacity of 1,500, and was razed in 1968.

METROPOLIS, The. William Seitz opened the Metropolis on East 142nd Street at Alexander Avenue in 1904 as a first-class theatre. The building also housed a Rathskeller in the basement, and a roof garden - and it was an instant success with the social set. It flourished for ten years, featuring Broadway headliners Francis X. Bushman, Leo Dietrichstein, Clara Kimball Young and Pat Rooney. In 1913, a rival theatre, The Bronx Opera House, was built and, being more centrally located in "the Hub," it spelled the doom of the Metropolis. It went downhill until, in 1926, it closed its doors. This legitimate theatre was partially razed in the 1940s, with one section remaining as a scenic studio.

MIANNA STREET. This is the former name of Rhinelander Avenue from Bronx Park East to Barnes Avenue. Lott G. Hunt was a landowner of this part of Van Nest, and his aunt was Myanna Hunt, who died in 1802.

MICHAEL'S HALL. Listed in 1862, this hall stood at East 155th Street and Courtlandt Avenue in the village of Melrose. A German Lutheran congregation met there prior to erection of the first St. Matthew's church in 1863. See: Rooster Church (2).

MIDDLEBROOK. In June, 1777, Colonel Glover's brigade was ordered by General Washington to Middlebrook, described as three miles from Fort Independence. This would be the area of the Negro Fort overlooking Middle Brook or, as it was also called, School Brook. See: Negro Fort.

MIDDLE BROOK (1). This small brook, originating in springs off Gates Place running eastward to Webster Avenue, according to Robinson's *Atlas*, 1888, was the main source of the Mill Brook. See: School Brook, Schuil Brook.

MIDDLE BROOK (2). This small brook originated near Middletown Road and Gillespie Avenue and flowed into Weir Creek, according to the 1777 Blaskowitz map. See: Turnbull Creek, Ware Creek, Wear Creek, Weir Creek, Wire Creek.

MIDDLEBROOK ROAD. This is the former name of Mosholu Parkway South.

MIGEL PLACE. The highest spot on Throggs Neck, some 29 feet above sea level, this was once part of the Baxter homestead of the 1600s. The slope was occupied by Hessians during October, 1776. In the 1800s, it was marked "Ludlam's Land" and the "Barker Homestead" and was the property of Frederick Jackson. After World War I, it was included in lots acquired by a syndicate that promoted Tremont Heights, whose associates had streets named for them. M. C. Migel was a philanthropist, especially in the field of work for the blind. In 1958, Migel Place was cut off by the Cross-Bronx Expressway and shortened to approximately 24 feet. In 1972, the Place disappeared entirely when a firehouse was built upon it. See: Jackson's Farm.

MILDRED PLACE. In 1979, the name was superseded by Santo Donato Place. Mildred (Paul) Laurie's ancestors were pre-Revolutionary settlers.

MILES SQUARE ROAD. This well-known Colonial highway was the forerunner of present-day Van Cortlandt Park East. See: Mount Vernon Avenue.

MILITARY HALL. An 1853 advertisement in the Westchester County Journal listed George Horn as owner of this hotel as Third Street near Washington Avenue (East 165th Street and Washington Avenue). It boasted of a commodious ballroom.

MILL BROOK. This was about the most important brook within the confines of The Bronx as it served as boundaries between various heirs of the Morrisania Manorlands, the Bathgate heirs, and other 19th-century squires. Its headquarters were in the north Bronx (Gates Place) and it ran roughly on a course down Webster Avenue and Brook Avenue to empty into the Bronx Kill. The early Dutch referred to a water course by its Indian name, "Armenperal," but it is not certain if they meant this stream. See: Mill Runn, Morrisena Creek, Morris Mill Brook, Roger Michelsen's Brook, Saw Mill Creek.

MILL CREEK. This is a former name of that stretch of Rattlesnake Creek that ran from Cole's Boston Post Road to Reid's Mill. See: Rattlesnake Creek.

MILL RUNN. A map of the John Archer Patent (Manor of Fordham) dated 1684 gives this name to the stream now covered by Webster Avenue and Brook Avenue. See: Mill Brook.

MILTON STREET. This is the early name of East 158th Street. See: Cedar Place.

MINER'S-IN-THE-BRONX (1). This theatre was originally located on Third Avenue and East 156th Street from 1910 to 1915. It was famous for its Sunday concerts, nightly burlesque and vaudeville acts, its Country Store Night, and the Friday night wrestling matches. In 1915, Miner's transferred to "the Hub," while this location was made available to Loew's Victory. See: Victory Theatre.

MINER'S-IN-THE-BRONX (2). This Bowery institution was transplanted into The Bronx in 1910, at East 156th Street and Third Avenue, but in 1915, it transferred to the west side of Melrose Avenue off East 149th Street. The theatre featured first-class burlesque, Broadway run dramas and revues and vaudeville. Stars that played there included Jimmy Savo, "Sliding Billy" Watson and Cecil Spooner. In the 1930s, it became a motion picture theatre, and finally went out of business around 1947.

"MINFORD PLACE." The former estate of Thomas Minford ran from Boston Road to West Farms Square. From 1892 to 1903, the eight-room mansion was occupied by the Bolton family who rented it at $900 a year. This is a rare instance whereby an estate name "Minford Place" was retained by the City as a street, Minford Place. See: The Bleach.

MINNEFORD'S ISLAND. This is a former name of City Island, and reputed to be named after an Indian chief. See: Great Island, Ye.

MINNEWITS. This is thought to be a corruption of Governor Minuit's name given to City Island. Peter Minuit was born in Wesel, Germany, a city that harbored many English, Dutch and French Protestant refugees. The Minuits were originally French, according to pastor-historian Sardemann of Wesel. See: Great Island, Ye.

MINNIEFORD PARK. A 1915 map shows a Minnieford [spelled this way] Park, east of City Island Avenue, from Cross Street to Minneford Avenue to Sutherland Street, but not along the waterfront. This was not a public park, but a name given to a real estate development. It was surveyed by George C. Hollerith.

MINNIFERS ISLAND. This is another variant of the early name of City Island.

MIRACLE, The. See: Boro, The.

MIRROR, The. This was reportedly an old-time silent movie house on Westchester Avenue with no further pinpointing as to place and years.

MISHOW. This is a prominent boulder, once on the southeastern tip of Hunter Island. According to legend, it figured in religious ceremonies of the Siwanoy Indians. Landfill operations almost covered it around 1936, but Borough Historian Dr. Kazimiroff persuaded the contractors to leave the crown exposed. Today it sets in the lawn near the eastern end of Orchard Beach promenade. See: Lorelei Point.

MITCHELL'S HILL. This is the former name of the lower end of St. Mary's Park.

MODEL, The. This small silent movie house was on Freeman Street, between Hoe and Vyse Avenues.

MOHAWK AVENUE. See: Garrison Avenue.

MOLE STREET. This was an earlier name of East 176th Street from Webster Avenue to Third Avenue. See: Mott Street, Orchard Street, Prospect Street, Woodruff Street.

MONAGHAN AVENUE. An 1886 name of DeReimer Avenue. See: St. Mary's Avenue.

MONITOR ENGINE COMPANY NO. 2. An 1862 Annual Report lists this volunteer Fire Company as being located on Westchester Railroad Street in East Morrisania. It was a two-storied frame building on East 149th Street near Jackson Avenue.

MONROE STREET (1). See: East 179th Street.

MONROE STREET (2). This was an earlier name of Cambreleng Street. See: Fulton Street, Pine Street.

MONTE CARLO. This was the official designation of a private park at the end of City Island as noted in ordinances of 1914. Marching clubs asked for a temporary suspension of a ban on fireworks when they held celebrations in this park. See: Chateau Laurier, Horton's Point, Monte Carlo Hotel.

MONTE CARLO HOTEL. This was originally the home of Stephen D. Horton on the southern end of City Island. When William Belden lived there, he called it the Chateau Laurier, and upon his death it became a famous hotel. In 1937, it became the Morris Yacht Club. See: Chateau Laurier.

"MONTEREY." This was the 1866 estate of the Ryer family. A story is that Samuel Ryer was an officer in the Mexican War, and so named his lands as a reminder of the battle he participated in. When the Ryer estate was cut up into city streets, the name of the estate was given to Monterey Avenue, and East 180th Street was first called Samuel Street.

MONTGOMERY AVENUE. According to an 1886 map, this was Heath Avenue. The original name harked back to Richard Montgomery who worked a farm in this vicinity, but was killed in Quebec.

MONTGOMERY PLACE. The former name of Cannon Place. See: Montgomery Avenue.

MORGAN AVENUE. This was an 1860 name of Wissman Avenue on Throggs Neck. Edwin D. Morgan, merchant, purchased a tract from Herman LeRoy Newbold in 1854. In 1856, he bought additional land to the south from John T. Wright. Later, the avenue was called Cullinden Avenue. See: Wright's Comers.

MORISENY. This is Morrisania as spelled on "A Draught of New York in 1749," a map by Mark Tiddeman.

MORRELL PARK. This was a real estate venture, listed in 1895 by Louis F. Haffen, Parks Commissioner, to be annexed. However, instead of it becoming a city park, two men named Morrell and Pinchot organized the Westchester & Van Nest Land Company and laid out a 70-acre tract for residences. This land, until 1920, was largely undeveloped, and was used as a picnic grounds by church groups and boy scouts. Most of the property is now occupied by Jacobi and Van Etten Hospitals. See: Westchester Heights.

MORRINECK RIVER. A 1698 mention is made of this river, presumably the Hutchinson River, in a petition concerning the Post Road going to Mamaroneck. The name does not occur again. It was also called the Pelham River and Eastchester Creek.

MORRIS AVENUE (1). An 1849 deed mentions the Post Road (Third Avenue) leading through the Morris lands from the Harlem River up to East 149th Street as Morris Avenue. This main thoroughfare was the later boundary between Mott Haven and North New York.

MORRIS AVENUE (2). This was the original name of Grandview Place, but was discarded when present-day Morris Avenue was extended to the west side of the Concourse.

MORRIS COVE. This former cove extended into the marshes from the East River but is now filled in to form Ferry Point Park. Francis Morris (no relation to the older Morris family of the South Bronx) owned the waterfront estate in the 1850s, which passed to his son, John A., and grandson, Alfred H. Morris, from the years 1850-1922. See: Ox Meadow Cove, Ox Pasture Cove, Turkey Bay.

MORRIS ESTATE. Some 80 acres that were purchased by Francis Morris around 1850 extending along the East River from Balcom Avenue into Morris Cove (now Ferry Point Park) to East Tremont Avenue to Lawton Avenue to Balcom Avenue. His son, John A., and daughter-in-law, Cora, inherited the section north of Lawton Avenue and built a 40-room mansion at Sampson Avenue and Revere Avenue. Later, they bought additional land in 1865 to total a 150acre estate which they called "Engelheim." John A. Morris died in 1895, but his widow lived on the estate until her death in 1922. A Najeeb Kiamie purchased the mansion and only two acres of land immediately around it. The southern end of the Morris lands was owned by a grandson of the original proprietor, Alfred Hennen Morris, who married Jennie Harding of Philadelphia. Their mansion was located on Schurz Avenue facing the East River, but was razed around 1940. Part of the property belongs to the Throggs Neck Stadium Association. See: Kiamie Estate, Colford Oval, German Stadium.

MORRIS LANDING This was a pier belonging to Lewis G Morris on the Harlem River at West 180th Street. It was noted on an 1868 map. See: Mount Fordham.

MORRIS LANE (1). This was a late 19th-century and early 20th-century name for the road leading to the Morris estates, specifically from East Tremont Avenue down to Balcom Avenue. The name was noted as late as 1921 on Bronx Title of Assessment. It is now Lawton Avenue.

MORRIS LANE (2). An 1875 firehouse report mentions this road, unmapped, near present day Sedgwick Avenue on Mom is Heights.

MORRIS MILL BROOK. This was noted in 1868 as the most prominent water course of the Manorlands. See: Mill Brook.

MORRIS PARK AVENUE. This is the former name of the five-block stretch from the Hutchinson River Parkway to Westchester Avenue that is now St. Theresa Avenue. The change occurred in December, 1968. See: Alice Avenue, Liberty Street.

MORRIS PARK MOVIE PALACE. This silent movie house occupied the upper floor of a two-story building erected in 1908 by Anton Landgrebe. It sat about 200 people, and provided only organ music to accompany the movies. It closed in 1923 to become the temporary home of St. Dominic's Church. As of August, 1974, the original screen of the old movie could still be seen painted on the rear inside (north) wall of 671 Morris Park Avenue between Victor and Amethyst Streets. See: Hunt's Movie House.

MORRIS PARK RACETRACK. This was the Hatfield estate of the 1860s, called "Glenn Dale," which passed to Eliza Macomb. She conveyed the property of 152 acres to John A. Morris who laid out a racetrack in 1888. The area covered by this track, grandstand, clubhouse and stables embraced Bronxdale Avenue, Rhinelander Avenue, Williamsbridge Road and the railroad. It closed in 1904, but then was used for automobile and motorcycle races. As late as 1908 the field was used for balloon ascensions, and the world's first airplane meet took place there in November, 1908. See: Aviation Field (1).

MORRIS PLACE. It formerly ran off Park Avenue to Washington Avenue between East 169th Street and East 170th Street, its eastern end being blocked by Washington Palace, a social hall. Morris Place was incorporated into Gouverneur Morris Houses in 1958.

MORRIS POINT. This was a former point of land belonging to Cora Morris in the 1880s in the vicinity of Harding Avenue and Emerson Avenue facing Morris Cove. It is now incorporated into Ferry Point Park. An earlier name was Ox Pasture Point as noted on an 1811 deed.

MORRIS POND. See: Melrose Pond.

MORRIS STREET (1). This was a former name of Mount Hope Place as it was the northern sector of General Staats Morris' holdings.

MORRIS STREET (2). This was the earlier name of Burke Avenue from Boston Road to Gun Hill Road, a matter of five blocks.

MORRIS STREET (3). This is the former name of East Tremont Avenue from Jerome Avenue to the Concourse.

MORRISANIA. The Manorlands of the Morris family was measured at 1,920 acres in the 18th century, from the Bronx Kill at East 132nd Street up to West 170th Street on the Harlem River, across the bottom of Crotona Park to the East River at The Debatable Lands. In the 19th century, numerous villages were laid out from this immense property: Forest Grove, Bensonia, Eltona, Melrose, Morrisania, Mott Haven, North New York, Highbridgeville and Tremont.

MORRISANIA AVENUE. This was an avenue that ran from East 161st Street to the Fleetwood Trotting Course at East 165th Street west of Sherman Avenue.

MORRISANIA CEMETERY. Reputed to have been an Indian burial ground, this 19th-century cemetery was once situated along St. Ann's Avenue. Its six footpaths were called St. James, St. John's, St. Luke's, St. Mark's, St. Peter's and St. Paul's. See: Bensonia Cemetery.

MORRISANIA FIRE DEPARTMENT. See: Hopkins Morrisania Fire Department.

MORRISANIA HALL. This social center was once located at the northeastern corner of East 165th Street and Washington Avenue. During the Civil War, Oliver Tilden, the first Morrisania casualty, was laid in state there before burial in Woodlawn Cemetery.

MORRISANIA HOOK & LADDER COMPANY, NO. 1. This volunteer Fire Company was listed in the Westchester County directory of 1869 at Fourth Street near Washington Avenue. A. Stockinger was Foreman. East 166th Street would be the present-day location.

MORRISANIA HOOK & LADDER COMPANY, NO. 4. In 1864, this volunteer Fire Company was located on Washington Avenue at Fifth Street (East 167th Street).

MORRISANIA PARK. This popular beer garden and picnic area of the 1880s was on the south side of East 170th Street between Third and Fulton Avenues. It was opposite Zeltner's Brewery. This gathering place is not to be confused with the Morrisania Schutzen Park off Boston Road and Jefferson Place. See: Zeltner's Park.

MORRISANIA SCHUTZEN PARK. This pleasure park was located on Boston Road above McKinley Square facing Jefferson Place. It was popular with German riflemen who held spirited contests there. The word "*Schützen*" connotes a shooting match. See: Sylvan Association.

MORRISANIA TURNER HALL. Listed in the 1869 Directory, it was on Fordham Avenue near Sixth Street (Third Avenue near East 168th Street). "*Turner*" is a term denoting gymnastics in German speaking circles.

MORRISENA CREEK. This was another name for the Mill Brook, noted on a 1778 map by "Thos. Kitchin, Senr. Hydrographer to his Majesty." See: Mill Brook.

MORRISON'S LANE. A 19th-century name of West 231st Street from Independence Avenue to Manhattan College Parkway on an 1889 map. See: Macomb's Street.

MORSE AVENUE. This was an early name of Third Avenue from East 168th Street to East 172nd Street.

MOSHOLU (1). This was a Weckguasgeeck Indian name for Tibbett's Brook. Professor Tooker thinks it might refer to "smooth stones" or "small stones." See: Uncas River .

MOSHOLU (2). This was an old hamlet and post office alongside the Albany Post Road on an 1887 map. It harbored a general store, a wagon shop and a smithy, schoolhouse, Methodist church and a cluster of homes. It was known in the earlier 1800s was Warner's, or Warnerville. The general site was Broadway at West 242nd Street.

MOSTYN STREET. This street in Melrose was listed in the New York Guide of 1947 running from 500 East 149th Street to 420 East 161st Street, or in the roadbed of the railroad yards.

MOTT AVENUE. The former name of the Grand Concourse from East 135th Street to East 149th Street in the former village of Mott Haven. The name was changed in 1927. Jordan L. Mott bought the land from the Morris family in 1849, some 100 acres between the Harlem River and Third Avenue from East 132nd Street to East 148th Street. Mott Avenue was called "The Ridge" in the vernacular of the 1890s due to its height.

MOTT HAVEN. This village was incorporated in 1848, and was annexed to New York City in January, 1874. Mottville was also a name for Mott Haven, as shown on an 1849 map of lower Westchester. Jordan L. Mott, who purchased the land, invented the base-burner stove (1833) which could use anthracite coal and founded the Mott Iron Works on the Harlem River. He died in 1915.

MOTT HAVEN CANAL. This was once an important canal extending inland from the Harlem River parallel to Rider Avenue and reached up to East 140th Street. It was an enlargement of Ice Pond Brook that meandered down from the former Fleetwood District. In 1903, the canal was filled in, leaving only a short basin south of the Major Deegan Expressway. This basin was filled in during 1964-1965. See: Ice Pond Brook, Canal Avenue.

MOTT HAVEN, The. This small silent movie house was once located on East 138th Street off Alexander Avenue, according to a 1912 Atlas. It is thought to have been the forerunner of the Peerless.

MOTT-HAVEN CORNER. This tavern on Third Avenue at East 136th Street advertised in the N.Y. *Eagle* of 1892, *"Das eleganteste Lokal ueber dem Harlem River ist."* Wilhelm Toebing was the host.

MOTT STREET (1). This is the former name of East 148th Street, which marked the northern limit of Jonas Bronck's farm, and also that of the village of Mott Haven. Mott Street in the 1870s ran from the Harlem River to Third Avenue. See: Henry Street (1).

MOTT STREET (2). This was the earliest name of East 176th Street from Webster Avenue to Third Avenue in the village of Tremont. Jordan L. Mott was one of several men who purchased land there in the 1850s. See: Mole Street.

MOTTVILLE. See: Mott Haven.

"MOUNT FORDHAM." This former estate of Lewis G. Morris, mapped in 1868, was incorporated into the grounds of New York University, now Bronx Community College. The land was inherited by Charles Popham, whose mother was Mary (Morris) Popham. See: Archer Manor, Morris Landing.

MOUNT MORRIS HOTEL. Mount Morris is in Harlem, but this hotel was to be found on Third Avenue at East 132nd Street in the 1890s, the property of the Mullen brothers.

MOUNT PLEASANT. It was a gradual hillside that was surveyed by S. E. Holmes in May of 1867. It appeared on a map of the Village of Tremont in 1871 in the vicinity of East Tremont Avenue and Anthony Avenue.

MOUNT SHARON. This is the hill whose summit is Fordham Road and the Grand Concourse as it appeared on an 1853 map. An earlier name was believed to be Union Hill.

MOUNT VERNON AVENUE. This is the former name of Van Cortlandt Park East, which was the Mile Square Road of Revolutionary fame. The name was changed in 1894.

MUD WEST. This was a derisive nickname for Van Nest in use around 1890 to about 1902, according to W. M. Dohmann. See: The Nest.

MUDDY SLOUGH. This was a drainage brook meandering through the Cobb and Larkin farms of the 1880s. It most likely seeped into the swamps at the head of Pugsley Creek in the vicinity of today's Pugsley and Watson Avenues.

MULBERRY ISLAND. This was a 1685 reference to City Island. See: Great Island, Ye.

MUNN AVENUE. A deed, dated April 1, 1852, mentions Sarah Munn whose house stood by the Westchester Turnpike near Seabury Creek. The avenue was named in 1916. Its western end was located at Glover Street and Westchester Avenue, but the whole length was eliminated in or around 1980 when the Zerega Industrial Park was planned. See: Wellington Avenue.

MUNROE AVENUE. This is a former name of Tomlinson Avenue from the railroad tracks at Sackett Avenue to Morris Park Avenue.

MURPHY PLACE. James R. Murphy. See: Fischer (Corporal) Place.

MUSCOOTA STREET. This is the former name of West 225th Street, alongside the Harlem River at the foot of Kingsbridge Road. See: Farmers' Bridge.

MUSCOOTA. This was the Indian name for the Harlem River, meaning "River among the green sedge."

MYRTLE AVENUE. In the village of Tremont, this was the original name of Park Avenue from East 175th Street to East 178th Street. See: Railroad Avenue, Vanderbilt Avenue.

MYRTLE STREET. This is the former name of Grace Avenue, from East 222nd Street to the Hutchinson River. See: Jones Avenue, South 20th Avenue.

N.Y., N.H.& H. HOTEL. John Hubert was the proprietor of this hotel at East 133rd Street and Third Avenue in the 1890s, when it catered to travelers on the New York, New Haven & Hartford Railroad.

NANNY GOAT HILL. West 258th Street, west of Riverdale Avenue, was known by this nickname as it was a favorite spot for the neighborhood goats. The name was in effect before 1915.

NANNY PIPER SPRING BROOK. This is a rivulet running through the Delafield land - West 246th Street - to the Hudson River. It was fed by a spring called Nanny Piper, which served the first Delafield at the end of Delafield Lane. In a dry summer it dries up, but with a hard rain it gains considerable volume.

NARRAGANSETT DRIVE. This is the former name of Seymour Avenue, now incorporated in the Bronx Municipal Hospital Center. It is not to be confused with Narragansett Avenue.

NASSAU RIVER. See: Hudson River [Current].

NATALIE AVENUE. This is a former name of Kingsbridge Terrace from Kingsbridge Road to Albany Crescent. The wife of Lewis Gouverneur Morris - a landowner in the neighborhood - was Natalie (Lorillard).

NAVY AVENUE. This was the original name of Paul Avenue. See: Army Avenue, President Avenue.

NEGRO FORT. A small redoubt was built atop a hill overlooking what is today the Grand Concourse off East 206th Street during the Revolutionary War. It guarded the strategically important Boston Road (Van Cortlandt Avenue East), and water was available at the foot of the hill in Middle Brook. Military dispatches mention the Negro Fort in January, 1777, and it was also noted on a map of 1782. All that is known is that the Black soldiers hailed from Virginia and were presumably "free men of color."

NEGRO HILL. This was the Revolutionary name for the elevation off the Grand Concourse called St. George Crescent. See: Negro Fort.

NELSON AVENUE. In Edenwald, this was the former name of Grenada Place.

NEPTUNE ENGINE COMPANY NO. 3. A volunteer Hose Company listed in 1869, on Riverdale Avenue.

NEREID VOLUNTEER HOSE COMPANY. This volunteer Hose Company once served the village of Washingtonville. The name suggests a water nymph, as was the custom in those days. Other engines rejoiced in the names of Neptune, Oceanus, Mermaid, etc. The firehouse was located off East 238th Street and White Plains Road.

NEST, The. This was a local nickname for Van Nest. See: Mud West.

NEUMANN-GOLDMAN PLAZA. This is the former name of Neumann-Goldman Memorial Plaza. The word "Memorial" was added on December 22nd, 1962. See: Neumann-Goldman Memorial Plaza.

NEW CITY ISLAND. This is a former name of City Island. See: Great Island, Ye.

NEW FOUND PASSAGE. According to Frederick Wendel Jackson, a Westchester historian, it was a channel which existed in Colonial days separating present-day Fort Schuyler from the mainland. In 1667, there was record "that Colonel Nicholls granted Roger Townsend a certain parcel of land at ye southeast end of Throgmorton's Neck, containing 15 acres commonly called New Found Passage."

NEWPORT PLACE. It ran northwestward from Morris Park Avenue, but now is inside the Bronx Municipal Hospital Center. It is not to be confused with Newport Avenue.

NEW STREET (1). It is listed in the 1871 Directory as a street running from Concord Avenue to Union Avenue. There is no trace of it today.

NEW STREET (2). This is a street that is now incorporated into the Botanical Garden west of the Bronx River.

NEW STREET (3). This was a short driveway that ran from Stable Street near the beach in Pelham Bay Park. See: Kane's Park and Casino (2).

NEW VILLAGE. Two hundred acres of cleared land were purchased in 1847 by Jordan L. Mott & Associates from Gouverneur Morris II, their bounds being East 161st Street to East 169th Street, from the Mill Brook to Boston Road. The name was short-lived, however, and was superseded by the Village of Morrisania.

NEW YORK ATHLETIC BALL GROUNDS. This field in Mott Haven was mapped in 1879 extending from East 150th Street to East 152nd Street along the shoreline of the Harlem River and inland to present-day Walton Avenue. Geographically speaking, the limits of Mott Haven were defined at East 148th Street.

NEWBOLD'S CORNER. This is the former junction of Hollywood Avenue and Lawton Avenue where a road led down to the estate of Herman LeRoy Newbold, in the 1870s. The estate was known as "Pennyfield."

NEWELL STREET. An important thoroughfare of the early 1900s when it led south from Gun Hill Road east of the Bronx River to such famous resorts as "French Charley's," "Voelker's Schutzen Park" and L'Hermitage, a French restaurant. This tree-lined lane, and the establishments mentioned, were wiped out by the Bronx River Parkway. Later, a short ramp leading to the Parkway was given the name of Newell Street.

NEWTON STREET. A 1905 map shows this to be the name of Post Road in Riverdale. See: Courtlandt Avenue.

NIAGARA AVENUE. See: Neill Avenue.

NIBLO'S GARDEN. This hall and picnic gardens attracted thousands of visitors to Third Avenue and East 170th Street after it had been purchased from the Zeltner family. The high point of its existence occurred in 1910 when Prince Henry of Prussia and his entourage stopped there for a reception. Louis Haffen was a member of the reception committee. See: The Lyceum, Zeltner's Park.

NICKELAND, The. This silent movie house was located on East 162nd Street and Third Avenue. It was in operation from sometime in 1917 to the late 1920s. This theatre was an enlargement of an earlier movie house called The Nickelodium.

NICKELODIUM, The. This was one of the earliest silent movie houses in The Bronx. It was located on the east side of Third Avenue and East 162nd Street and started operations in 1908. Later, it became The Nickeland.

NICOLAND, The. This very early silent movie house was built around 1906 on Westchester Avenue near Tinton Avenue. The owner, a man named McLean, had a tiny movie house next door one year earlier. See: McLean's.

NIGGER WOODS. This wooded area of the former Gerard W. Morris estate, called "Cedar Grove," extended along the Concourse from East 158th Street to East 161st Street. On the south, it bordered Cedar Park, which was renamed Franz Sigel Park in 1902. In the 1920s, the woods were cleared and athletic clubhouses occupied most of the land before the Bronx County Court House was built. The origin of the name is obscure, but old-timers repeated vague stories that the woods were a hideout of runaway slaves. Use of this localism seemed to be confined to the years 1890 to 1916.

NILES STREET. On an 1868 map, this was the former name of Rochambeau Avenue at its junction with Bainbridge Avenue in front of the church of the Holy Nativity. Nearby lived William W. Niles, vice president of the Bronx River Parkway Commission. See: Bronx River Parkway.

NINDHAM PLACE. It was laid out on December 18, 1895, but was never used. The tiny Place, running from Kingsbridge Road up to Kingsbridge Terrace, honored the chief of the Stockbridge Indians, who were allies of the Continental troops. His name is usually spelled Nimham. The Place was de-mapped on January 4, 1967, by the City Planning Commission and sold to realtors.

NINETEENTH AVENUE. This was the 19th-century name of East 233rd Street from Bronx Park East to White Plains Road. See: Fisher's Landing Road.

NINTH AVENUE. In Wakefield, this was East 223rd Street from Bronx Park East to White Plains Road.

NINTH STREET (1). This was the original name of East 133rd Street in Port Morris.

NINTH STREET (2). In the village of Tremont, this was the early name of East 171st Street.

NINTH STREET (3). The former name of Watson Avenue in Unionport. See: Larkin Avenue, Walter Street.

NIPINICHSEN. This was the Weckguasgeeck Indian village that was located on Spuyten Duyvil. It is Algonquin for "small pond."

NIPNICHSEN. This is a variant of the Indian village described above.

NOLAN'S HOTEL. A popular hotel at Fordham Road and Decatur Avenue in the decade before World War I.

NOLL'S SCHUTZEN PARK. See: Morrisania Schutzen Park.

NOORDEN KILL. See: Hudson River [Current].

NORMAN AVENUE. See: Hillman Avenue, Harald Avenue.

NORTH BEDFORD PARK. This is an area roughly bounded by Jerome Avenue, Gun Hill Road, Webster Avenue and Mosholu Parkway. It was renamed Brendan Hill in 1910 by the Board of Aldermen. See: Norwood Heights.

NORTH CHESTNUT DRIVE. The former name of Chestnut Street. See: Doucine Street.

NORTH MELROSE. On an 1850 survey by Andrew Findlay, this was the strip of land from East 161st Street to East 163rd Street, Park Avenue to Brook Avenue. See: Baby Park.

NORTH NEW YORK. This was another name for Old Morrisania, and was recognized as a postal address in 1862. It ran from East 134th Street to East 147th Street, east of Third Avenue, and was divided into city lots during the 1860s. It was annexed to New York City in 1874.

NORTH RIDING. This was a pre-Revolutionary name for the Town of Westchester. Shortly after Nieuw Amsterdam was subjugated by the English in 1664, the towns of Westchester, Hempstead and Oyster Bay were formed into the North Riding. Riding is a corruption of the Scandinavian "Thriding" (third part of a Shire). This was the official name of The Bronx region until 1683, when the County of Westchester was created. See: Westchester.

NORTH RIVER. See: Hudson River [Current].

NORTH ROAD. This is the former name of Country Club Road.

NORTH SIDE. The territory north of the Harlem River was The Bronx. This expression was much in vogue upon annexation to New York City from 1874 to around 1902, and was used by the North Side Savings Bank, The North Side Board of Trade and the *North Side News*. At times, The Bronx was referred to as The Great North Side.

NORTH SIDE BREWING COMPANY. Originally, this was the Kuntz Brewery on the west side of Third Avenue at East 168th Street. This company was formed in 1902 under George Gminder, president. See: Kuntz Brewery.

NORTH SIDE HOTEL. This German-oriented hotel was located on Brook Avenue at East 144th Street in the 1880s and 1890s. In 1892, the New York *Eagle* (a Bronx newspaper) noted that a social club met there calling themselves the Konstantinopoloschedudelsackpfeifers (Bagpipers from Constantinople).

NORTH STREET. This is the former name of Henwood Place in West Tremont, and may have been so-called as it was at the northern end of that district. The name was changed in May, 1916.

NORTHERN TERRACE. In Riverdale, this was the earlier name of West 239th Street from Henry Hudson Parkway to Independence Avenue.

NORWOOD. This real estate development was surveyed in 1889 by Josiah Briggs between Mosholu Parkway and Woodlawn Cemetery. Carlisle Norwood, a friend of Leonard Jerome and a scion of Old New York, is believed to be thus honored, but some people believe the name is a contraction of North Woods. See: North Bedford Park.

NORWOOD AVENUE. See: Norwood, Decatur Avenue.

NORWOOD HEIGHTS. See: North Bedford Park, Norwood.

NUASIN. The Weckguasgeeck Indian name for the area of High Bridge. The word is interpreted as "middle place," "the land between" the Muscoota (Harlem River) and Mentipathe (Cromwell's Creek.) See: Devoe's Point.

"NUMBER EIGHT." This was the estate of Gustav Schwab on Fordham Heights overlooking the Harlem River and now the grounds of the Bronx Community College. See: Fort Number 8.

OAK AVENUE (1). Formerly it ran parallel to Willow and Locust Avenues in Port Morris, but now is covered by the railroad yard.

OAK AVENUE (2). This street in the settlement of Bartow was mapped in 1900. The settlement is now part of Pelham Bay Park due west of the Bartow Circle.

OAK POINT. This promontory on the East River was an amusement park, ballfield, picnic grove and bathing beach from Civil War times to around 1908. The site is now occupied by the railroad yards above Port Morris. See: Arnold's Point, Jeffeard's Neck, Leggett's Point.

OAK STREET. This street, four blocks long, running parallel to Reid's Mill Road to Rattlesnake Creek was mapped in 1882, but is now covered by the New England Thruway. It was part of "The Kernels."

OAKES AVENUE. This is the former name of Bruner Avenue. See: South 22nd Avenue, Willis Place.

OAKLEY GROVE. This farm at Third Avenue and East 178th Street belonged to Miles Oakley, who was a vestryman of St. Peter's church in 1702 and Mayor of Westchester in 1730.

OAKLEY STREET (1). Up to 1894, the name of East 237th Street from Van Cortlandt Park to Vireo Avenue.

OAKLEY STREET (2). This is the former name of Amethyst Street in Van Nest.

OAKLY STREET. Spelled slightly differently, this was a continuation of Oakley Street (1) of Woodlawn Heights. Prior to 1896, it was the name of East 237th Street in Wakefield.

"THE OAKS." This wooded spot was used for May parties and June walks before 1915, near the Devoe house at 1423 Jesup Avenue.

"OAKSHADE." This estate of Richard L. Morris, grandson of the Signer, [Lewis Morris] was on the City Island Road in the 1870s. It was located midway between Glover's Rock and Bartow Circle. See: Boston Road.

OCEAN AVENUE. On an 1897 map of Pennyfield, this was the name of Miles Avenue. See: Greene Avenue, Pennyfield Avenue.

OCHILTREE AVENUE. This is the earlier name of Kings College Place, then south around Reservoir Oval to East 207th Street. This *might* be named after Tom Ochiltree, a famous racing thoroughbred (*circa* 1877) owned by George L. Lorillard, that raced in the nearby Jerome Park racetrack.

ODD FELLOWS' HALL. It was listed in the 1872 Directory of Morrisania as being located at Fifth Street, off Railroad Avenue (East 167th Street off Park Avenue).

ODELL AVENUE. This is the former name of Spencer Avenue in Riverdale.

ODELL'S TAVERN. This inn at Boston Road dating back to the early 1830s was probably on the site of an earlier tavern, judging by the immense trees planted at regular intervals on the grounds. It is shown on an 1868 map in the Eastchester region. See: Breinlinger's Old Point Comfort Park, Dickert's Old Point Comfort Park.

OGDEN, The. This fair-sized silent movie house was at 1431 Ogden Avenue with a seating capacity of 1,370. It was built in 1922.

OGDEN'S WOODS. This region of West 176th Street and University Avenue, including Ogden Avenue, was mentioned in an 1897 newspaper. The news item referred to the wooded estate of William B. Ogden.

OINK SQUARE. In the 1920s a large grassy triangle was caused by newly-opened East 177th Street where it crossed East Tremont Avenue. A street sign was installed indicating the intersection, but one night a prankster, in a highly professional manner, repainted the enameled sign to read "Oink Sq." The local residents never noticed the substitution until it was called to their attention by bewildered strangers. The Traffic Department was duly notified, but it took many weeks of red tape before the sign was removed. Among the younger set, the name was used long after the original story behind it was forgotten.

OLD ALBANY POST ROAD. The former name of Bailey Avenue on a map of 1886.

OLD BARN CREEK. This was an alternate name of Coney Island Creek, now filled in by Ferry Point Park. It is mentioned in deeds of 1814 and 1825, the 1854 deed of the diZerega purchase, and was mapped in 1874. See: Coney Island Creek.

OLD BOSTON ROAD. This was the earlier name of Third Avenue from the Harlem River to East 163rd Street.

OLD BROADWAY. This was a former name of Crotona Avenue from Boston Road to Crotona Park. See: Broad Street, Broadway, Franklin Avenue, Lafayette Street.

OLD CLASON POINT ROAD. This was Beach Avenue from East Tremont Avenue to the Cross-Bronx Expressway, a matter of three blocks.

OLD DUTCH PRISON, The. See: Dutch Prison.

OLD FERRY POINT. This is a former name of Ferry Point Park. See: Ferris Neck.

OLD HUNT'S POINT ROAD. This was the original name of Oak Point Avenue from Drake Cemetery to Sacrahong Street, a distance of seven blocks.

OLD MORRISANIA. According to Beers' 1868 *Atlas*, it is that portion of the Morris Manorlands below East 148th Street. It was later called North New York. The boundary coincided with the original Bronck grant.

OLD POINT COMFORT. It was located between the Boston Road, Hollers' Pond and Reid's Mill Lane, and may have had some connection with Comfort Avenue (now Baychester Avenue). As it was named at the time of the Civil War, it most likely honors Fortress Monroe on Old Point Comfort, Virginia, that remained in Union hands while the rest of the Commonwealth seceded. It was a noted resort of the early days. See: Breinlinger's, Dickert's.

OLD POKEY. See: Port Morris & Spuyten Duyvil Railroad.

OLD ROAD, The. This is the former name of McGraw Avenue. It ran from White Plains Road to Unionport Road between the Catholic Protectory and the McGraw farm. See: Cow Neck Road.

OLD STONE CHURCH, The. This was a nickname for the Mott Haven Dutch Reformed church, as it was a rarity in the days of wooden churches. It was organized in 1851 on land donated by Jordan L. Mott on the corner of East 146th Street and Third Avenue. In 1914, the church was moved to the center of the block because of the noise and vibrations of the elevated railroad.

OLD WESTCHESTER HOTEL. This was an inn near the Harlem River bridge in the lower Bronx, when it was still part of Westchester County. It was located at about East 136th Street in the 1860s.

OLIN AVENUE. This was the former name of Gun Hill Road from the Bronx River to White Plains Road, where it formed the boundary line between Olinville No. 1 and Olinville No. 2. The communities were named after Bishop Stephen Olin of the Methodist Church.

OLINVILLE AVENUE. This is the former name of Burke Avenue from the Bronx River to White Plains Road. See: Morris Street, Syracuse Avenue.

OLINVILLE NO. 1. Laid out in 1854, this community ran from Gun Hill Road north to East 215th Street, from the Bronx River to White Plains Road.

OLINVILLE NO. 2. This small community alongside the Bronx River was surveyed and laid out in 1854 from Burke Avenue to Gun Hill Road. The eastern boundary line was White Plains Road.

OLIVER AVENUE. The former name of Oliver Place when it was much longer, extending across Webster Avenue into what is today the Botanical Garden. That was in 1868. The Oliver family lived nearby, and one member was a prominent Bronx magistrate.

OLIVER STREET. In the 1880s, Oliver Avenue was shortened to Oliver Street. In 1893, it became Oliver Place.

OLOFF PARK. This was an 1869 real estate development of 100 acres, named after Oloff Stevenson Van Cortlandt. It is now a part of the Jerome Park Reservoir and Van Cortlandt Park. See: Orloff Avenue.

ONE-ACRE PARK. See Melrose Park.

ONE-TREE ISLAND. This was a local nickname for Catwillow Island, due east of East Twin Island, *circa* 1920. See: Two-Tree Island.

ONONDA PLACE. An 1888 survey of the Godwin estate showed a projected Place extending from West 230th Street to Verveelen Place, but it was absorbed by Broadway when it was widened at this point.

OOSTDORP. In the 1600s, this was the Dutch name for their outpost on Westchester Creek in the area of today's Westchester Square. It meant "the east village" as it was east of their Manhattan possessions. See: Eastdorpe.

OOSTDORP STREET. This was a former name of Bryant Avenue from East Tremont Avenue to Bronx Park, mapped in 1888. See: Hunter Street, Walker Street.

OPDYKE STREET. This was the original name of East 236th Street from Van Cortlandt Park to Webster Avenue until it was changed in 1894. George Opdyke laid out the village of Woodlawn Heights in 1873 from part of the Gilbert Valentine farm. He was a self-made man. Born in New Jersey of poor parentage, he was a millionaire before his 58th birthday. He made his fortune in textile importing, a chain of clothing factories and a munitions plant. He was also Mayor of New York from 1862 to 1864. See: Cadiz Place.

ORANGE STREET. This was the earlier name of East 135th Street in Mott Haven. Without doubt this name refers to the Dutch royal family, the House of Orange, for no oranges could ever grow in that climate. See: Edsall Street, Seventh Street (1).

ORCHARD AVENUE. This is the former name of Honeywell Avenue from East Tremont Avenue to Bronx Park.

ORCHARD POND. It formerly lay at East 147th Street and St. Ann's Avenue in St. Mary's Park. It was drained in 1892 by Park Commissioner Louis F. Haffen to make way for a bandstand.

ORCHARD STREET (1). This was the earlier name of East 176th Street from the Concourse to Anthony Avenue, thence from Webster Avenue to Third Avenue. See: Mole Street, Prospect Street, Woodruff Street.

ORCHARD STREET (2). This is the former name of West 168th Street from Ogden Avenue down to Jerome Avenue. See: Birch Street.

ORCHARD STREET (3). It formerly ran from Oliver Place to East 199th Street, one short block. It most likely referred to an orchard on the Briggs farm.

ORCHARD STREET (4). The original name of Hawkins Street on City Island. David Schofield gave it the name as it led to his orchard, and it was known by that name in 1858 when he made the conveyance to Robert Vail. The name was changed in 1950.

ORCHARD STREET (5). This short lane ran north from East 182nd Street to Garden Street, but is now incorporated into Belmont Avenue. It was shown on an 1874 map of West Farms.

OSCEOLA, The. Listed in 1922, this silent movie theatre was located on St. Ann's Avenue between East 138th Street and East 139th Street.

OSSEO HOTEL. This well-known tavern and inn of the 1880s was on the Westchester Turnpike south of Westchester Square, opposite St. Peter's churchyard. Old-timers claim it took its name from the "Osseo," a steamboat that plied between Westchester village and Peck Slip.

OSSEO PLACE. It was laid out across the last Ferris farm on Ferris Neck, facing Westchester Creek. On old maps, it was a lane leading down to a landing where the steamer "Osseo" docked on its daily run. The Place received its name from that fact. "There was a beautiful but small sidewheel steamboat called the Osseo. She was painted white, and had an Indian chief with all his feathers on him, painted on the paddlewheel boxes,"ran an old-time description by Matthew Husson. See: Ferris Dock.

OVERING CREEK. This creek that flowed into the marshes behind Morris Cove on the East River. Henry C. Overing's estate was at Lafayette Avenue, from East Tremont Avenue to the marshlands.

OVERLOOK AVENUE. This is the former name of East 167th Street from Jerome Avenue to the Concourse. See: Fifth Street (2), James Street, Lyons Street.

OX MEADOW COVE. This is mentioned as bounding the land of Thomas Baxter, Jr., in a 1718 deed, and is now filled in by Ferry Point Park. See: Morris Cove, Ox Pasture Cove, Turkey Bay.

OX PASTURE COVE. It is referred to in an 1811 Livingston deed, and alludes to the body of water at the western end of his land. This became the Morris Cove of the 1880s, and is now covered by the landfill of Ferry Point Park. See: Morris Cove, Ox Meadow Cove, Turkey Bay.

OX PASTURE POINT. There is an 1811 reference to this promontory when the East River met Schurz Avenue and Emerson Avenue. The point is now part of Ferry Point Park. See: Morris Point.

OXFIELD STREET. This was the earlier name of Putnam Place from Gun Hill Road to Woodlawn Cemetery.

OXFORD PLACE. Maps of New York University in the 1890s show this short street leading north from Hall of Fame Terrace, a bit west of Loring Place.

OXFORD STREET. This is a former name of East 175th Street from Jerome Avenue up to the Concourse. See: Gray Street, Fairmount Street, Fitch Street.

OYSTER STREET. This was the original name of the western half of Ditmars Street on City Island running from the main street to Eastchester Bay.

OZARK STREET. This is the former name of East 209th Street from Perry Avenue to Webster Avenue. It is believed to be named after the Federal gunboat of the Mississippi Squadron. The *Ozark* was very much in the news during the Civil War when this street was laid out.

OZONE PARK. This small picnic park was once located at East 147th Street and Southern Boulevard. See: Hartung's Park.

PALISADE AVENUE. The 19th-century name of Independence Avenue in Riverdale. See: Yonkers Avenue.

PALMER BOULEVARD. When the extensive Givan acreage between Boston Road, the Hutchinson River, Eastchester Road and Mace Avenue was subdivided, this main boulevard was laid out running eastward from present-day Givan Square. It was named for the last owners, fourth generation descendants of Robert Givan. It was never cut through however.

PALMER BROOK. This was a small spring, mentioned in an 1856 deed, on lands of Lorillard Spencer fronting on Eastchester Bay. The brook fed into Palmer Cove, situated at Radio Drive of today.

PALMER HOUSE. This hotel was listed in the 1890s New York *Eagle* as being located at Third Avenue and East 133rd Street, F.C. Palmer, proprietor.

PALMER PARK. A 1915 map shows a Palmer Park east of City Island Avenue bounded by Carroll Street, the Sound and Hawkins Street, but it was not an amusement park, but a real estate development. Benjamin Palmer owned the island during the American Revolution.

PALMER STREET. This was the former name of Buckley Street on City Island.

PAPARINEMIN. This is a variant of Paparinemo, the aboriginal name of Kingsbridge.

PAPARINEMO. This was a Weckguasgeeck Indian name for present-day Kingsbridge, which was then an island formed by twin branches of the Mosholu River (Tibbett's Brook) and Spuyten Duyvil Creek. The name meant "Place of False Starts," alluding to the double tide caused by the Hudson and Harlem Rivers. See: Hummock Island, "Island Farm," Paparinemin.

PARADISE, The. This small, silent movie house at East 167th Street and Findlay Avenue was in operation from about 1919 to the mid-1920s.

PARK AVENUE (1). This was the earlier name of Parker Street from Castle Hill Avenue to Westchester Avenue, a distance of five blocks.

PARK AVENUE (2). This is the former name of Willett Avenue from Gun Hill Road to East 219th Street.

PARK AVENUE (3). This was an earlier name of Blackstone Avenue in Riverdale.

PARK STREET. This was the former name of Cauldwell Avenue from East 149th Street to Westchester Avenue, referring to St. Mary's Park. See: Avenue B.

PARK, The (1). A silent movie house on East 169th Street between Park and Washington Avenues. See: Blenheim, The.

PARK, The (2). The 1912 Atlas lists this small frame silent movie house on East 180th Street between Mapes Avenue and Southern Boulevard.

PARK VERSAILLES. The Archer-Mapes farm of the 1860s was auctioned off by John S. Mapes and given this name. The bounds were Westchester Avenue, White Plains Road, East Tremont Avenue and Croes Avenue.

PARKVIEW AVENUE (1). When Pelham Bay Park was invested (taken over) by the City in October, 1888, this was a well-chosen name. Later, it was renamed in honor of Aaron Burr, a former landowner.

PARKVIEW AVENUE (2). This is the former name of Valles Avenue in Riverdale. The name was changed in May, 1916.

PARKVIEW PLACE. This was the original name of Devoe Terrace.

PARKWAY RESTAURANT. See: Kipp's Parkway Restaurant.

PARKWAY, The. This large silent movie house once operated on Third Avenue between East 172nd Street and East 173rd Street. It seated 2,000, and was built in 1926. In the 1960s, it was turned into a factory.

PAROLE PLACE. This was the former name of East 187th Street from the Concourse to Tiebout Avenue on an 1884 map. See: Clay Avenue, Jacob Street, Sanford Street.

PARSONAGE ROAD. This was an early name of Unionport Road where it passed through what is today Parkchester. An 1844 map of the Leggett farm (now Parkchester) shows this road leading to the Parsonage at present-day Castle Hill Avenue and Westchester Avenue, where the parsonage of St. Peter's P.E. church stood. The site of this building is now occupied by the Y.M.C.A. See: Sheep Pasture Road.

PARSONS AVENUE. This avenue was laid out over former Ferris farmlands, but has since been absorbed by the new St. Raymond's Cemetery. An Edward Parsons married Mary Ferris sometime prior to 1885.

PARSONS' WINDMILL. An 1893 map shows E. W. Parsons & Co. nursery and greenhouses located alongside Westchester Creek north of the Bruckner Circle. The windmill could be seen from a considerable distance downstream and mariners noted it on their charts.

PARTHEON, The. Some old-timers locate this silent movie house at Burnside and University Avenues, but records show a theatre of that name a short distance south at Harrison and West Tremont Avenues.

PASCAL PLACE. It once ran from West 246th Street to West 247th Street northwestward from Waldo Avenue in Riverdale.

PASS ROCKS. This rocky ledge north of Hunter Island is described in *The New Parks Beyond the Harlem* by John Mullaly, 1887. This narrow ledge is in the passage (or pass) between Glen Island and Hunter Island, now called Hog Island.

PASSAGE AVENUE. This roadway ran diagonally through St. Mary's Park to become Trinity Avenue, according to the "Map of Wilton, Port Morris & East Morrisania, July 12, 1857." See: Avenue C, Cypress Avenue, Delmonico Street, Grove Street.

PASTIME ATHLETIC CLUB. This athletic and social club was once located on the site of the Bronx County Court House from 1910 to around 1931. The members staged distance races along the Concourse almost every Sunday.

PASTURE HILL. A 1702 reference to this hill, and a Pasture Hill Cemetery, in the Ferris family annals identifies both to be near the foot of Commerce Street, behind Westchester Square.

PATTRY'S HOOK. In October 1673 Nelis Mattysen and Christine Laurens offered to purchase this neck of land (Hoek) "between Lewis Morris' Lane and the Two Brothers." Their petition referred to what is today Port Morris and North Brother and South Brother islands. Vingboom's Map of Manatus (Manhattan) in 1639 depicts Pieter Schorstinveger's Plantation at the location. Piet and Petrus are both versions of Pieter, so that Petrus Hock became anglicized to Pattry's Hook. Incidentally, Schorstinveger is the Dutch word for a chimneysweep.

PAUL ESTATE (1). This 23-acre tract, mapped in 1886, belonged to Philip and S. Paul, whose ancestors had a farm there before the American Revolution. The approximate bounds were Hutchinson River Parkway and Pelham Parkway to St. Paul Avenue, and then diagonally northwestward to Evelyn Place.

PAUL ESTATE (2). This was a 5-acre tract belonging to H. Paul north of Pelham Parkway from Lodovick Avenue to Astor Avenue to the railroad tracks.

PAUL'S BEACH. A 1930 ad located this beach at the foot of Philip Avenue on Eastchester Bay. Paul Burgdorf was the owner, and he continued in business into the 1940s.

PAVILION STREET. This was a driveway in Pelham Bay Park, behind the present-day Victory monument, that paralleled Shore Road to Tallapoosa Point at the Pelham bridge. Kane's Casino that closed in 1930 had a pavilion at the midway point of this street.

PAYNE STREET. This street did lay athwart the original West Farms purchase of 1663 by Edward Jessup and John Richardson. Elizabeth Jessup married Thomas Hunt, whose descendants gave the Point its name. Payne Street was once on the Jessup farm, mentioned in Revolutionary diaries, and, in the 1800s, was part of what was loosely called Fox Woods. These woodlands were owned at different times by the Tiffany and Fox families. Around 1900, it was on the lands of the Caswell Academy. Today, it lies under the Hunts Point Market. As many avenues in this area honored poets and writers, it is safe to say this avenue was a reminder of John Howard Payne, 1792-1852, who is chiefly remembered for his poem, "Home, Sweet Home."

PEARY GATE. This was a local name in use from the 1900s to around 1920 for the entrance to Bronx Park at Fordham Road and the Bronx River. It was so-called because Admiral Robert E. Peary of Arctic fame and some Eskimos used that gate when bringing polar specimens to the Zoo. See: West 175th Street.

PEERLESS, The. This silent movie house on East 138th Street, between Third and Alexander Avenue was in operation after 1912. It is believed the Mott Haven was its succeeding name, and in 1932 it was called The Haven.

PELHAM AVENUE. The former name of Fordham Road from Webster Avenue to the Bronx River. The name persisted until 1915. See: Highbridge Road, Union Avenue.

PELHAM BAY. This shallow body of water once lay south of Hunter Island, which was noted on 17th- and 18th-century maps. It is now covered by the Orchard Beach parking lot. See: LeRoy Bay.

PELHAM BAY DIVISION. This volunteer lifeguard station was listed in 1914 as being located at the foot of Layton Avenue on Eastchester Bay. This station was one of the oldest in the service, and was moved from place to place more times than any other unit.

PELHAM BAY NAVAL TRAINING CENTER. In 1917, the U.S. Navy built barracks and facilities for 5,000 men - in two weeks! It was on Rodman's Neck, facing Eastchester Bay. At the war's end, the camp had expanded to 16,000 men and more officers were trained at the Center during 1917-1918 than there had been in the entire Navy before World War I. The Naval base was closed down in 1919. See: Ann's Neck.

PELHAM BAYVIEW PARK. Around 1880, the Givan farm was sold and subdivided into a tract that was called Pelham Bayview Park. See: Givan's Basin.

PELHAM FIELD. See: Trojan Field.

PELHAM HEATH INN. This was a former roadside tavern and inn on the southeastern corner of Eastchester Road and Pelham Parkway. It was occupied in the late 1890s by the Knickerbocker Inn and, prior to that, was the residence of James C. Cooley, the Timpson family and the Ferris family of the early 1800s. The Reverend John Bartow owned the property in 1720. Pelham Heath Inn was destroyed by fire in 1947, and the site is now occupied by Mother Butler High School.

PELHAM PARK & CITY ISLAND RAILROAD. This was the official name of the monorail connection, a one-car shuttle, between the settlement of Bartow and City Island. It was developed by H. H. Tunis an engineer, and financed by August Belmont. The Flying Lady was the nickname of the monorail car that ran between Bartow station in Bartow village and Marshall's Corner at the end of the City Island bridge, from the Spring of 1910 to July 10th of the same year, when an accident occurred.

PELHAM POINT. This was an 18th-century name of Rodman's Neck. See: Ann's Neck.

PELHAM RIVER. See: Hutchinson River (Current).

PELHAM ROAD (1). This earlier road ran from Westchester Village through the settlements of Middletown and Stinardtown to Pelham. It began at the Westchester Creek bridge, ran along Ericson Place to Libby Place, thence to Buhre Avenue to Westchester Avenue, then to Pelham Bay. See: Pelham Road (2), Pelham Road (3), Pelham Road (4), Pelham Road (5).

PELHAM ROAD (2). The former name of Ericson Place to Libby Place.

PELHAM ROAD (3). The earlier name of Libby Place, a distance of two blocks.

PELHAM ROAD (4). This was Buhre Avenue's original name from Libby Place to Westchester Avenue, a stretch of three blocks.

PELHAM ROAD (5). This is the former name of Westchester Avenue from Buhre Avenue to Pelham Bay Park.

PELHAM ROAD (6). This is an 18th-century name of Split Rock Road inside the golf course. See: Collins Lane, Pell's Lane, Prospect Hill Road.

PELICAN ISLAND. A very ancient name of Hunter Island. "My Island called & known the name of Pelican Island and being over against the east side of planting neck in ye said Manur as also one small Island known by the name of Twins lying over against the east side of the sd Pelican Island," wrote Jno. Pell in 1713. See: Appleby's Island.

PELL STREET. This was the earlier name of Osgood Street at the Mount Vernon line.

PELL TREE INN. This roadhouse of the early 1900s was on the Shore Road of Pelham Bay Park. The name referred to the Treaty Oak on the grounds of the nearby Pell-Bartow mansion. See: Ramblers' Inn, Treaty Oak.

PELL'S ISLAND. This is the first European name of Hunter Island, noted in 1654 as part of the Royal Patent and Indian Grant.

PELL'S LANE. The former name of Split Rock Road. See: Collins Lane, Pelham Road (6).

PELL'S NECK. This was the 18th-century name of Rodman's Neck. See: Ann's Neck.

PENFOLD STREET. This is a former name of Crotona Park East.

"PENNYFIELD." This estate fronting on Long Island Sound was owned by a succession of squires, including John T. Wright, Herman LeRoy Newbold, Edwin D. Morgan and Frederick DeRuyter Wissmann. It was bounded by Miles Avenue, Harding Avenue, Meagher Avenue and the Sound. The name is found on deeds of 1793, and the origin goes even further back to Indian days when 17th-century colonists purchased the fields for a copper penny. The Indians coveted the metal.

PENNYFIELD AVENUE. This was the former name of Miles Avenue from East Tremont Avenue eastward to Long Island Sound, *circa* 1920s. It was so called as it led to the "Pennyfield Estate" south of Miles Avenue. It was mapped as Greene Avenue in 1866, and as Ocean Avenue in 1880.

PENNYFIELD LANE. This would be Meagher Avenue leading from Harding Avenue to the Park of Edgewater, although in 1874 when it was mapped, it was Pennyfield Lane leading from Fort Schuyler Road to the Adee estate. See: Adee's Lane.

PENNYFIELD ROAD. This was the former name of Meagher Avenue, as it once led to an estate called "Pennyfield," owned by Frederick DeRuyter Wissmann. See: Adee's Lane, Pennyfield Lane.

PERSIMMON POINT. This is the southern end of West Twin Island, so-called for its stand of persimmon trees. This point was largely destroyed when Orchard Beach was joined to the Twins and Hunter Island in 1935. See: West Twin Island.

PESKAR PLACE. This name was submitted to the City Council in November, 1945 to supplant Arnow Place. The motion was not carried.

PHOENIX PARK. An 1890 War Department map gives this name to the amusement park and picnic grove bounded by East 147th Street, Brook Avenue, East 148th Street to Willis Avenue. See: Fountain Park.

PICTORIUM, The. This very small silent movie house was on East 180th Street between Daly and Vyse Avenues in the early 1920s.

PIEPGRAS' SHIPYARD. Henry Piepgras operated a shipyard on City Island at the foot of Pilot Street, from 1886 to 1900. It had been the property of David Carll, who founded the business at the end of the Civil War. See: Carll's Shipyard, Jacob's Shipyard.

PIER AVENUE. This is the former name of Mulford Avenue. It either recalls a pier on Westchester Creek, or a former landowner, Emma Peere Miles, listed in 1899.

PIGEON HILL. In Riverdale, this is now West 260th Street and Liebig Avenue.

PIGEON ROCK. This was a prominent rock in the Bronx River well-known to generations of West Farms and Van Nest boys that stood downstream from the East Tremont Avenue bridge. It stood in water deep enough for diving. Around 1935, the river was diverted and Pigeon Rock was bulldozed to one side, eventually to be buried under tons of landfill.

PILGRIM, The. This motion picture theatre was built for talkies and had never been a silent movie house. It was located at Buhre and Westchester Avenues, on the west side, and closed in the 1950s.

PINCKNEY'S HUMACK. Philip Pinckney was one of the settlers who in 1662 agreed to "sett down at Hutchinsons." The hummock was a mass of rock out on the marshlands, 200 yards east of Baychester Avenue and was used for centuries as a reference point. The hummock is now surrounded by Co-Op City. See: Prickley Pear.

PINE ISLAND DIVISION. A Volunteer Station of the Life Saving Service of New York was manned by 11 surfmen, four officers and three coxswains. A 1914 listing locates this station on Pine Island near Tallapoosa Point, Pelham Bay Park. See: Pine Rock.

PINE ROCK CREEK. This tidal creek wound through Dominick Lynch's land (1803), and the Ludlow farm (1850), the Schieffelin estate (1870s) and the Sacred Heart Academy (1880s). The stream entered the Bronx River at Lacombe Avenue. Part of Soundview Park covers this ancient waterway.

PINE ROCK. A long rock off the shoreline of Pelham Bay Park, near Tallapoosa Point, was sometimes called Pine Island. It had no trees at all on it, but possibly there were some hardy pines on it in 1823, when it was mentioned in a deed describing William Bayard's farm. In 1912, an official map of the Parks Department called it Holger Rock Island. In December, 1963, tons of landfill were dumped into the narrow channel, and Pine Rock was attached to the mainland. By late February, 1964, the island was entirely swallowed up and covered over.

PINE STREET (1). This is the former name of Holland Avenue from Adee Avenue to South Oak Drive. This street was in the general neighborhood of Corsa and Oakley Streets, and both were named after Westchester Guides. John Pine, too, was a Westchester Guide and was attached to the French troops. See: Lincoln Street, Maple Avenue, Post Avenue.

PINE STREET (2). Although not on a map of the times, an 1882 firehouse report mentions a Pine Street near present day Jerome Avenue below Fordham Road.

PIONEER PARK. These picnic grounds were once located at Stebbins Avenue and Home Street in the 1890s, and earlier had been part of the William W. Fox estate.

PIPE STREET. See: East 190th Street.

PISER'S THEATRE. This silent movie house was in operation prior to World War I on East Tremont Avenue between Clinton and Prospect Avenues.

"PLAISANCE." This was an estate owned by L. Waterbury on Eastchester Bay in the 1880s, and is now incorporated into Pelham Bay Park and used by the Police Department. The bounds were Middletown Road, Eastchester Bay, Pelham Bay Park and Rice Stadium. Squire Waterbury had his favorite horses and dogs buried on this estate, and the tombstones can still be seen. See: "Pleasance," Waterbury Lane.

PLANT ROAD. The fiction about this old road in West Farms is that East Tremont Avenue in the neighborhood of Honeywell Avenue was a swamp. Brushwood was thrown on the bogs to form a passable road, but it sank several times. Every springtime, plants grew luxuriously on the boggy road, and this was the local nickname for what became East Tremont Avenue. The fact is that Henry W. Plant owned four acres in West Farms, according to an 1837 deed, and his small farm was bounded on the south by Plant Road.

PLANTATION OF PIETER SCHORSTINVEGER. Johan Vingboom's Map of Manatus (Manhattan) in 1639 depicts the lands of Pieter Andriessen, known as Pieter Schorstinveger (Peter the Chimneysweep). The tract was east of Jonas Bronck's land, approximately north of Port Morris. Andreissen ran a tavern in Manhattan and also did chimney-sweeping. His land, crops and livestock on the Bronx mainland were tended by Negro slaves. See: Port Morris, Stony Island, Stony Point.

PLANTERS INN. West Farms was the stopping place for stage coaches from Danbury and Mamaroneck, and the inn of Robert Hunt stood at East 182nd Street and Boston Road, where the travelers rested and the horses were rested. In 1800, it was called Planters Inn. Earlier, right after the Revolutionary War, the inn had been run by Levi Hunt. In May, 1792, John Ryer shot and killed Constable Isaac Smith during a dispute in the tavern, and subsequently was hanged in White Plains. See: Johnson's Tavern.

PLAZA, The. Located on Washington Avenue and East 180th Street, this was a silent movie house in operation from 1912 to the early 1920s. The silent films were accompanied by a piano and, later, an organ.

"PLEASANCE." This was an estate owned by Archer Milton Huntington that is now incorporated into Pelham Bay Park and used by the Police Department. The main house was three-storied, with a valuable library that was donated in 1904 to the Hispanic Museum of New York. About the same time, the estate was given to the Parks Department. An earlier owner, Mr. Waterbury, had called the estate "Plaisance."

PLEASANT AVENUE. This is the former name of Olinville Avenue from Gun Hill Road to East 219th Street.

PLEASANT BAY PARK. This picnic park was in use during the 1920s and 1930s on the eastern side of Ferris Point facing Morris Cove. At the turn of the century it had been a picnic grove owned by the Hupfel family of brewers, which was later sold to Adam Hoffmann, another brewer. The transaction took place at high tide and Mr. Hoffmann envisioned excursion steamers coming up the East River and docking in Morris Cove. At low tide, Morris Cove was a sea of mud and Mr. Hoffmann's plans bogged down. He did succeed in bringing electricity to his Pleasant Bay Park by putting up his own light poles from Eastern Boulevard (now Bruckner Boulevard). His grandnephew, Edward Wolf, collected memorabilia of this remarkable man and his numerous enterprises. See: Hoffmann's Park.

PLUMB TREE COVE. In a deed dated May 3rd, 1712, Sir John Pell refers to this cove without further clarification.

PLUMB TREE HAMMOCK. Sir John Pell refers to this mound in a deed dated May, 1712, but no one today knows where the landmark might have been.

POCAHONTAS CUT, The. This railroad cut belonging to the Port Morris & Spuyten Duyvil Railroad Line was located at Brook Avenue and Third Avenue. This gully was so called, as the line had been called the Pocahontas Line by its financier, Gouverneur Morris II.

POCAHONTAS LINE. See: Port Morris & Spuyten Duyvil Railroad.

POCAHONTAS RAILROAD. See: Port Morris & Spuyten Duyvil Railroad.

POILLON STREET. It was surveyed in 1890 southeast of Casanova Station, but was covered by the railroad tracks leading up from Port Morris. It was to have run from Avenue St. John to the East River.

POINT, The. This was a localism for Locust Point when the Summer bungalow colony came into existence around 1925.

POKEY HANNES. See: Port Morris & Spuyten Duyvil Railroad.

POND STREET. One hundred yards in length, this lane led west from City Island Avenue, opposite Reville Street, but is gone today.

PONTIAC STREET. This was the former name of East 151st Street from Jackson Avenue to Tinton Avenue, according to the 1872 Directory of Morrisania. It is thought to have been named after a political club of that era. See: Beck Street.

PONUS STREET. An earlier name of East 181st Street from Southern Boulevard to Vyse Avenue, a distance of four blocks. Ponus (or Ponns) was the Indian chief who sold land to the settlers in 1640. See: Ann Street.

POPE PLACE. This "paper street" was surveyed, and planned to extend from Shore Drive to Sampson Avenue, but is not yet in existence. It is most likely named for the Civil War general, John Pope, commander of the Army of Virginia. All avenues in this area carry names of Civil War generals. The Place is part of the pre-Revolutionary farmlands of Edward Stephenson, which went to one of his four sons. It was part of the 19th-century property of Herman LeRoy Newbold, which went to the Fox family. George T. Adee owned it up to the 1900s, and then Richard Shaw purchased the property and started a Summer tent colony in 1914, now the Park of Edgewater.

POPHAM STREET. This is the former name of Mount Hope Place from Jerome Avenue to the Concourse, marking the lower end of the Popham estate. The owner was a grandson of Richard Morris who had owned all of University Heights. See: Washington Place (2).

PORT MORRIS. This extreme southeastern corner of The Bronx was originally advanced by Gouverneur Morris as a sea port. The upper portion (East 141st Street to East 149th Street), then a swampy area bounded by Bungay Creek, was sold to the New York & Harlem Railroad in 1853. The lower tract (East 132nd Street to East 141st Street) was conveyed in 1868 to the Port Morris Land Improvement Company, and it was regarded as a separate village until annexed to New York City in January, 1874. See: Stony Island, Stony Point.

PORT MORRIS, The. This old-time silent movie house was located on East 138th Street between St. Ann's Avenue and Cypress Avenue.

PORT MORRIS & SPUYTEN DUYVIL RAILROAD. Laid out at the close of the Civil War (1865) it ran between Port Morris and Spuyten Duyvil. While it had a long official name its "official" nickname, Pocahontas Railroad, stemmed from the fact that Gouverneur Morris II wished to honor his mother, Ann Randolph, who was a fifth generation descendant of Pocahontas. Pokey Hannes was a German-English nickname for the Pocahontas Railroad Line where it ran through Melrose. It is said that Gouverneur Morris II, who named the Pocahontas Line, would fly into a rage if he heard the disrespectful nickname Old Pokey, [which also sounds slow] mentioned. See: Pocahontas Cut.

POST AVENUE. The former name of Holland Avenue from Adee Avenue to South Oak Drive. This might be named for Captain Martin Post, one of the Westchester Guides in the American Revolution, as the lower portion of this same street was once called Pine Street, and John Pine was also a Westchester Guide. See: Pine Street.

POST, Lands of Wright. There were two estates on Throggs Neck that belonged to Dr. Wright Post as mapped in 1822 and 1825. One was an 18-acre plot containing house, farm, barn and orchards that faced the East River and were bounded by Lawton Avenue, East Tremont Avenue and Calhoun Avenue. This tract was leased to his brother William, who was a butcher and who raised cattle there. The other lands of Wright Post were 47 acres facing Eastchester Bay and bounded by Sampson Avenue, East Tremont Avenue and Philip Avenue. In 1827 the doctor sold this property to John D. Wolfe, and this became "the Doric Farm." Dr. Post was Chief Surgeon of the New York Hospital for 35 years. He was born in 1765 and died on his Throggs Neck country seat in 1828. See: Doric Farm.

POTTER PLACE (1). This is the former name of East 204th Street from Jerome Avenue to the Concourse. In 1884, it formed part of the property owned by George Opdyke, Mayor of New York, who had purchased it from the Potter brothers in 1874. See: Cadiz Place, Scott Street.

POTTER PLACE (2). Its approximate location was at the foot of Waterbury Avenue and Eastchester Bay, now on the grounds of the Catholic institution called Providence Rest.

It was last mapped in 1913. Emily C. Potter had purchased the land from the Westchester Land Corporation in 1893.

POWELL PLACE (1). This was the former name of West 178th Street at the Harlem River, mapped in 1897, but now eliminated by the Major Deegan Expressway.

POWELL PLACE (2). This was the earlier name of East 189th Street from Third Avenue to Washington Avenue, harking back to Reverend William Powell's farm, 1830s-1850s. See: Welch Street, Webster Avenue.

PREBEL STREET. This street was laid out through the former Tiffany estate called "Foxhurst," which, before it was cultivated, was locally known as Fox Woods. A few streets here recalled commodores and admirals, and this street honors Rear Admiral Edward Prebel who blockaded the port of Tripoli in 1803. His 1804 diary is a valuable source of history. Today, the street is covered by the Hunts Point Market.

PRESIDENT AVENUE. This street was planned, but was eliminated when Hunter College was built. Since then, the grounds and buildings have been renamed Lehman College. See: Army Avenue, Navy Avenue.

PRESIDENT, The. This old-time silent movie house was listed in 1912 and located at East 163rd Street and Westchester Avenue.

PRICKLY PEAR, The. Out on the marshes in the bend of Hutchinson's River, about 200 yards east of Baychester Avenue, a mass of rock stands up, a last relic of Colonial times when this landmark was used by the colonists. It was called by the settlers of 1662 the Prickly Pear, due to its resemblance to the little cactus known by that name. After division of the Eastchester Planting Grounds, the marshes surrounding the rock became known as Pinckney's Meadow or Pinckney's Humack, which today is now covered by Co-Op City.

PRIME'S POINT. This former name of Ferry Point Park was used in 1851 maps after Edward Prime had purchased the peninsula from Oliver State in 1848. See: Ferris Neck.

PRIMROSE STREET. This is the former name of West 192nd Street from Aqueduct Avenue to Jerome Avenue. See: First Avenue (3).

"THE PRIORY." This estate was owned by Miss N. Bolton as mapped in 1868. It is now incorporated into the Split Rock golf course and bridle path of Pelham Bay Park.

PROSPECT AVENUE (1). This was the former name of Anthony Avenue on a 1905 map. See: Avenue C (2).

PROSPECT AVENUE (2). This became the earliest name of Decatur Avenue because of the prospect (view) it afforded of St. John's College, later renamed Fordham University. See: Decatur Street, Norwood Avenue.

PROSPECT AVENUE (3). This was the 19th-century name of Coddington Avenue, which had a fine view of Westchester Creek. See: Third Street (6).

PROSPECT AVENUE (4). This is the former name of Napier Avenue from Woodlawn cemetery to the City line.

PROSPECT HILL (1). The summit of this hill, which was at present-day East 182nd Street and the Concourse, was on a farm owned by Adam Cameron, in 1853.

PROSPECT HILL (2). This was a steep hill facing the marshes of the Hutchinson River north of the New England Thruway and part of Pelham Bay Park.

PROSPECT HILL ROAD. A former name of Split Rock Road. See: Collins Lane, Pell's Lane, Pelham Road (6).

PROSPECT STREET (1). The earlier name of East 157th Street in Melrose from Park Avenue to Third Avenue.

PROSPECT STREET (2). In 1878, the Grace Episcopal church was located on Prospect Street and West Farms Road, but now the street is known as East 176th Street. See: Mole Street.

PROSPECT STREET (3). This is a former name of Carroll Street on City Island.

PROSPECT STREET (4). According to a history book on City Island, a photograph of Ditmar Street is identified, "formerly called Prospect St." Allen Flood wrote it in 1949.

PROSPECT TERRACE. This is the former name of Lowerre Place at East 227th Street. This Terrace afforded a fine view of the Bronx River valley.

PROSPECT THEATRE, The. One of the finest theatres in The Bronx, it was located at Westchester Avenue and Prospect Avenue. In the 1920s, it featured stars of the Yiddish stage - and on opening nights, limousines, fur capes and evening dress were commonplace. The last drama to play there was "Broken Hearts," with Maurice Schwartz in the lead role, in 1926.

PROTECTION ENGINE COMPANY NO. 5. This volunteer Fire Company was organized in 1852 and was housed, at first, in a frame building "down in the hole" at Melrose depot (Park Avenue between East 161st Street and East 162nd Street). One of the founders was Matthias Haffen, father of the first Borough President. On September 1st of that year, other local volunteer companies as well as Harlem outfits and the New York Engine Company No. 5, met at Ward's Hotel in Mott Haven to formally welcome the new Fire Company. The New York company called itself "the Honey Bees" and the Melrose company, having the same number, adopted this nickname. However, because so many of the Melrose firemen were either German or of German descent, the Melrose residents used to call the company "the Dutch Five." An 1862 Annual Report listed this volunteer Fire Company as being housed in a two-storied brick building on the east side of Elton Avenue near East 156th Street.

PROTECTION HALL. This social hall and tavern formerly was on Courtlandt Avenue near East 152nd Street. Matthias Haffen, brewer and also owner of this hall, was a charter member of Protection Engine Company No. 5 and, in time, his four sons became members. Protection Hall became the hub of social life from the 1850s to the 1880s. The Melrose Guards, a quasi-military band, also met in the hall. In the 1880s, a man named Faulhaber was the proprietor and he, in turn, sold out to another tavernkeeper named Grundhoefer. See: Dutch Broadway, Melrose Guards.

PROTECTORY AVENUE. This was the southern and eastern bordering road of the Catholic Protectory, as mapped in 1910. This would be approximately present-day McGraw Avenue and Olmstead Avenue that bound Parkchester.

PUDDING ROCK. It once stood on the slope near Cauldwell Avenue and the Boston Road, rising "like a pudding in a bag," 25 feet high and topped by a group of cedar trees. It was frequently mentioned in 19th-century records and newspapers. When Boston Road was widened, it was demolished and the house that stands on the spot carries the address of 1074 Boston Road. See: Tramps' Rock.

PUDDLERS' ROW. The Johnson Iron Works on Spuyten Duyvil Creek employed foundrymen, called "puddlers," from 1852 to 1923. The men and their families lived along what is today Johnson Avenue and Kappock Street.

PUERTO RICO, The. Originally The Forum, this silent movie house was open around 1921 and later became modernized for "sound" and finally "talking pictures." It was located at 490 E. East 138th Street near Brown Place. It became The Puerto Rico in the 1940s.

PUGSLEY CREEK. This was the original Indian waterway leading in from the East River to within walking distance of their village of Snakapins. It was called Cromwell's Creek in 1693, Wilkins' Creek in the 1770s, West Creek in 1823, and Pugsley Creek from the 1850s onward. It once wound inland to Watson Avenue, but then was cut off at Lacombe Avenue. In 1972, it was filled in entirely. See: Clason's Creek.

PUNETT STREET. This is the former name of Walton Avenue from East 177th Street to Burnside Avenue, a one-block stretch. Punnet [spelled this way] & Chrystie were a landholding corporation around 1875, and Louis Risse - who later laid out the Concourse - surveyed their property in the 1880s. See: Butternut Street, Sylvan Avenue.

PUTNAM AVENUE. An 1888 Robinson map shows this as the former name of Swinton Avenue from Bruckner Boulevard to Lafayette Avenue.

PYNE STREET. This was an earlier name of Cambreleng Street in the village of Belmont. See: Fulton Street (1), Monroe Street (2).

QUAGLIA PLAYGROUND. This large playground at Wilkinson Avenue and Hutchinson River Parkway was named Florence Colucci Playground in 1969. In 1973, a drive was made to rename it for a World War II serviceman, a local resident named Cosimo Quaglia, who was killed in action in France, 1944. See: Eisenhower Park.

QUAIL AVENUE. This was the former name of Kepler Avenue on an 1890 map. See: Third Street (9).

QUAKER TRACT. This 10-acre tract was donated by Josiah Quinby to the Friends in 1808. It was bounded by Middletown Road, Merry, Coddington and Crosby Avenues. It was sold in 1922 for building lots.

QUARRY FARM. Sometimes referred to as Turner's farm, it was situated above High Bridge, overlooking the Harlem River. The property was originally purchased by Daniel

Turneur [spelled this way] from the Indians in 1671. An 1868 map lists both spellings. See: Nuasin, Turneur's Land.

QUARRY ROAD (1). In the early 1800s, a marble quarry was worked in the area now covered by St. Barnabas Hospital. A southern extension of this old road (now East 178th Street) ran westward to present-day Webster Avenue and was called Marble Street. In 1850, this road was the western boundary line of Jacob Lorillard's estate. The quarry then became the cellar of the Lorillard mansion, a building that in 1866 was transformed into the Home for Incurables.

QUARRY ROAD (2). This was a 19th-century name for Palisade Avenue. The granite quarry was located near Riverdale Avenue and West 247th Street, and many greystone mansions of Riverdale were built from the stone that were quarried there.

QUARRYTOP. This granite quarry was worked in the 1820s and 1830s to construct the mansions of Riverdale, including Fonthill Castle (now Mount St. Vincent). The quarry was located in the vicinity of Riverdale Avenue and West 247th Street.

QUEEN PLACE. A former name of Reville Street on City Island. See: Reville Street.

QUEEN'S BRIDGE. During the British occupation of New York, in the Revolutionary War, the name of the Farmers' Free Bridge at West 225th Street over the Harlem River was changed to honor the Queen of England. See: Farmers' Bridge.

QUEEN'S ROAD, The. This is a very early name of West Farms Road, with some reference to "ye Queen's Road leading to ye West Farms" being found in a conveyance dated 1723. Claude Sauthier's military map of October 1776 shows a "military road" alongside the Bronx River designated as Queen's Road; it undoubtedly was so named to honor Queen Anne. See: Back Road.

QUIMBY'S NECK. It is mentioned in various ancient deeds, specifically in the 1675 will of William Betts wherein he deeded to his son John 1 acre and 13 rods of meadows lying on the south end of Quimby's Neck. It is probably the same tract mentioned in the first patent of Westchester "beginning at the west part of the land called Bronx land, adjoining Harlem River and extending eastward to Annhooks neck."

QUINNAHUNG. This was a Siwanoy Indian name for Hunts Point, although ordinarily the Siwanoys did not cross to the west bank of the Bronx River. It meant "The Planting Neck."

QUINSHUNG. "...the land was under the control of the Dutch, and from them, Throckmorton took a grant. The land had been called Quinshung in the Indian language, but was destined from that time on to bear his name," notes Sarah Comstock in her book, *Old Roads from the Heart of New York*, describing Throggs Neck. See: Frockes Necke.

RADIO. Originally, this theatre was The Royal Photoplays, a silent movie house with 600 seats, at 1350 Southern Boulevard which ran from 1917 through the mid 1920s. When talking pictures displaced silent films, the theatre changed its name to The Bronx Playhouse. From around 1930 to 1935 it was the New Royal, and from 1935 to around 1940, the early heyday of radio, it was the Radio.

RAILROAD AVENUE (1). This street is now incorporated into the Botanical Garden west of the Bronx River and adjoining the railroad tracks.

RAILROAD AVENUE (2). This is the former name of Starling Avenue, but the whereabouts of a railway system has puzzled a generation of name searchers.

RAILROAD AVENUE EAST. This is the former name of Park Avenue from East 156th Street to approximately East 170th Street. It was in use until 1896.

RAILROAD AVENUE WEST. This is the earlier name of Park Avenue, on the west side, from East 156th Street to East 165th Street. The name was changed in 1896. See: Myrtle Avenue, Vanderbilt Avenue.

RAILROAD TERRACE. This was the original name of Bronx Terrace from East 234th Street to East 237th Street, and it evidently took its name from the location overlooking the railroad lines as seen on an 1898 map. See: Marian Street, Marion Street.

RAMBLERS' INN. This roadhouse of the 1920s was located on the Shore Road of Pelham Bay Park. The famous band leaders Tommy and Jimmy Dorsey were featured there in 1925 with their California Ramblers. See: Pell Tree Inn.

RANACHQUA. This was the Indian name of the tract of land that was acquired by Jonas Bronck. Experts in the Indian tongue cite "Wanachquiwi-Auke" meaning End Place, which described the peninsula where it met the waters of the East River and the Harlem River.

"RANAQUE." This former estate at the end of Oak Point was owned by B. G Arnold. "Ranaqua" was one of the spellings of the Indian name for this region, as well as "Ranachqua" and "Ranaque." It meant The End Place.

RANDALL AVENUE (1). This was the earlier name of East 213th Street from White Plains Road to Barnes Avenue. See: Arthur Street, Flower Street.

RANDALL AVENUE (2). In Edenwald, this was the former name of Strang Avenue. A Leslie F. Randall of Yonkers was the landowner in the 1890s.

RANDOLPH AVENUE. See: West 172nd Street.

RANDOLPH BROOK. Edmund D. Randolph owned property below West 261st Street in Riverdale, and a spring originated in a meadow of Mount St. Vincent's Academy a few rods north of West 261st Street (formerly known as Randolph Lane). A brook flowed alongside the road westward and downhill to the Hudson River. See: Forrest Brook.

RANDOLPH LANE. This is the former name of West 261st Street from the Hudson River up to Independence Avenue. See: Cuthbert Lane.

RAT ISLAND. Unofficial name for South Brother Island. See: South Brother Island.

RATTLESNAKE BROOK. This very ancient name was frequently mentioned in the 17th-century Deeds of Eastchester. It flowed through Seton Falls Park under Boston Road, was dammed to form Hollers' Pond and emptied into the Long Pond off the Hutchinson River. It was finally submerged by Freedomland, and now is diverted into storm sewers far beneath Co-Op City. See: Long Pond Creek, Sidney's Brook.

RED HILL. This hill was mentioned many times in land transfers of the late 1600s in the town of Westchester, and there are references to its proximity to Ye Great Creek (Westchester Creek). The sole sizeable hill is that now traversed by Dudley Avenue and Harrington Avenue. See: Bridge Hill, Crow Hill.

REDD HILL. As it was spelled in Colonial times in the Westchester Records in White Plains, it referred to the prominent hill southeast of Westchester village: "1686. Sold to Nathaniel Underhill Sr by John Winter and Hunny, his wife, 4 acres being the meadow beyond the Redd Hill...." See: Bridge Hill, Crow Hill.

REDOUBT AVENUE. It was laid out to run east and west across the southern end of Woodlawn Cemetery to Heath's Redoubt (a temporary fort). It is now inside the cemetery.

REED PLACE. Before 1916, this was the name of West 229th Street between Heath Avenue and Bailey Avenue.

REID'S MILL ROAD. This is the former name of Provost Avenue from Boston Road to the City line. It followed the general line of Steenwyck Avenue to Rattlesnake Creek on which the mill was located.

REID'S MILL. This was the first tidal mill to be erected on Eastchester Creek, or the Hutchinson River. The year was 1739. John Reid (sometimes spelled Reed) was the miller in 1790, and his son, Robert, continued on until the 1850s. After the Civil War, it was abandoned and stood forlornly on the salt meadows for decades, finally to be blown down in a storm in 1900. Its site would be roughly the center of Co-Op City.

REISS'S POND. In the early 1906's, this pond was situated on the grounds of the Reiss family. Their mansion stood near the Bronx River in the center of Pelham Parkway, which necessitated it being razed. The pond was filled in and made into a playground at Bronx Park East off Reiss Place.

RETREAT AVENUE. This was the former name of Bergen Avenue in 1871. The street was the western boundary of Karl's Germania Park, a wooded retreat for holiday crowds of those times.

RHINELANDER LOT, The. This squarish tract of 15 acres lay on the west side of East Tremont Avenue when it was known as Fort Schuyler Road in 1840. It was in the area of Philip Avenue and extended westward to the fringe of Morris Cove. Bernard Rhinelander was listed as the owner in 1836, and he sold the property to James D. Wolfe. This later became the Gimbernat farm of the 1860s, and finally the Escoriaza farm as mapped in 1888. See: Escoriaza farm, Gimbernat farm.

RICE PATH. This was a pedestrian/cyclist path that was laid out in 1933-1934 by Works Progress Administration (W.P.A.) labor. It began at Westchester Avenue where it terminated at Pelham Bay Park, and continued parallel to Bruckner Boulevard, around Rice Stadium and past the Victory monument and back to the subway station. Most of this path has been obliterated.

RICE STADIUM. Isaac Leopold Rice, LLB, was born on February 22nd, 1850 in Germany and was brought to America at the age of 6. He graduated from Law School of Columbia College in 1880, published essays on music, and became a successful lecturer and lawyer. He amassed wealth as a counsel to many railroad and mining companies, and was president of the Forum Publishing Company. He founded four different chess clubs and was the innovator of an opening in chess called Rice's Gambit. Rice was also president of the Electric Storage Battery Company, and helped John Holland in the construction of the first submarines. Mrs. Rice was the organizer of the Society for the Suppression of Unnecessary Noise, and also the originator of the Sane and Safe Fourth of July. When Isaac Rice died, he left a considerable fortune to the Parks Department which developed Rice Stadium and playing fields. As a result, the flanking streets have electrical connotations such as Ohm, Watt, Radio and so on.

RICHARD STREET. This former street is now incorporated into the Botanical Garden west of the Bronx River. It is alleged to have been named for a son-in-law of Daniel Allerton, who was named Ricardo. He changed it to Richard.

RICHFIELD STREET. This street was a part of a 70-acre development called "Westchester Heights" that was subdivided around 1892. Richfield Street is now part of Yates Avenue near Morris Park Avenue.

RICHMAN PLAZA. See: Thompson (Dennis A.) Plaza, Playground, Pavilion.

RIDER STREET. This is the former name of College Avenue in Mott Haven. Rider & Conklin owned 600 lots there in the 1870s.

"RIDGE LAWN." This 10.5-acre estate of Reverend R. W. Dickinson appeared on an 1893 map in the vicinity of University Avenue and West 188th Street.

RIDGE STREET (1). This is the former name of East 196th Street from the Concourse to Marion Avenue. See: Donnybrook Street, Sherwood Street, Wellesley Street.

RIDGE STREET (2). On an 1887 map, this is the name of Popham Avenue.

RIDGE STREET (3). Once extending into the Botanical Garden next to the Lorillard mansion, it is now gone.

RIDGE STREET (4). This was the former name of East 211th Street along the lower end of Woodlawn Cemetery, according to a map of 1884.

RIDGE, The. This was a localism for the high ground in the vicinity of the Grand Concourse and East 149th Street around 1900. A monied class of home owners, including Franz Sigel, lived there and employed Italian laborers in their gardens, and Irish maids in the kitchens. A nearby farmer delivered milk to their homes, and had a special cow set aside for their sole consumption. See: Buena Ridge.

RIO DE GUAMAS. See: Hudson River [Current].

RIO SAN ANTONIO. See: Hudson River [Current].

RITZ, The. See: Art. (movie theater).

RIVER AVENUE. This is the former name of West 254th Street from Independence Avenue up to Riverdale Avenue, a distance of seven blocks.

RIVER STREET. This is the earlier name of Tibbett Avenue from West 230th Street to West 231st Street. Up to 1890, it was alongside Tibbett's Brook, and served as a landing for boats that came up from Spuyten Duyvil Creek. See: Long Branch.

RIVERDALE. Financed in part by the Goodrich and Spaulding families, this was a real estate venture of 1856 when it was laid out and given the name "The Park, Riverdale."

RIVERDALE DIVISION. This volunteer life-saving unit with four officers and twelve surfmen officially was Station 2 of the Fordham Landing Division. The frame building was on the Hudson River about a mile above Spuyten Duyvil.

RIVERDALE INN. This hotel on Riverdale Avenue and West 154th Street was operated by William Olms, a very fat Alsatian. The first telephone in Riverdale was installed in this hotel in 1897. See: Jumbo's Hotel.

RIVERDALE LANE. This is the earlier name of West 253rd Street from Riverdale Avenue down to Broadway. See: James' Lane.

RIVERVIEW TERRACE. This is the former name of Cedar Avenue. An 1886 map shows this aptly-named street overlooking the Harlem River. See: Commerce Street.

RIVIERE VENDOME. See: Hudson River [Current].

RIVINIUS' BREWERY. Listed in the Westchester County Directory of 1869 was Charles Rivinius, brewer, Fordham Avenue near 7th Street (Third Avenue near East 169th Street), with his home on nearby Fulton Avenue near East 168th Street. He was listed three years later without change.

ROBBINS AVENUE. This is the former name of Westchester Avenue from St. Ann's Avenue to Jackson Avenue, but the name was changed prior to 1857.

ROBBINS STREET. According to an 1852 map, Rowlanda Robbins owned a 24-acre farm east of St. Mary's Park through which the street was cut. In 1905, the name was changed to Jackson Avenue.

"ROBERT PLACE." An 1874 map shows Christopher Rhinelander Robert's 9.14-acre estate of this name on the East River. It extended inland to Throggs Neck Boulevard and terminated at Harding Avenue. On the south is was bordered by the estate of William Whitehead, that is now Silver Beach Gardens. Mr. Robert later founded Robert College in Istanbul, Turkey.

ROBERTSON'S SHIPYARD. Archibald Robertson started a small shipyard on City Island in 1882 at the foot of Fordham Place.

ROBIN AVENUE. This is the former name of Mahan Avenue from Middletown Road to Buhre Avenue.

ROCK STREET. The 19th-century name of West 259th Street east of Riverdale Avenue.

ROCK, The. This was a localism for the huge rocky elevation that was bounded by East 180th Street, Lafontaine Avenue, Oak Tree Place and Quarry Road. It stood on the former Ryer farm of the 1860s. The plot was purchased by the Transit Company in

1899. In 1940, the company and its assets became part of New York City's transportation system.

ROCKFIELD STREET. This is a former name of East 203rd Street from the Concourse to Mosholu Parkway. See: Signal Place.

"ROCKLANDS." This 19th-century estate was owned by T. A. Vyse in the vicinity of Boston Road between Crotona Park and the Bronx River.

"ROCKWOOD." The estate of Samuel E. L yon in East Morrisania, mapped in 1886.

"ROCKYCLIFF." This is the estate of Mrs. Marcher that led to the naming of Shakespeare Avenue. Mrs. Rebecca Marcher purchased the property from John Poole in 1865, and had the grounds decorated with marble statues. These represented characters from Shakespearian plays, ranging from Othello to Falstaff, and even the dogs and carriage horses carried names like "Pyramus,""Thisbe"and "Queen Mab." When the estate was sold and cut up into lots and avenues. one of them was Marcher Avenue, later renamed Shakespeare Avenue.

RODMAN PARK. A 1915 map shows a Rodman Park west of City Island Avenue from Pell Street to Carroll Street, facing Eastchester Bay. It was a real estate development that was surveyed by George Hollerith that same year.

RODMAN'S NECK DIVISION. This volunteer lifeguard station was erected on Rodman's Neck under great difficulty, as it was fully one-half mile from any road. It was reached only by wading through swamps, or by boat from City Island. One officer and ten surfmen were listed in 1920, with a notation their station was in operation all year round. Their frame building was set up on piles, as shown on a 1914 photograph.

ROGER MICHELSEN'S BROOK. This name was noted in 1694 on a map of Fordham Manor, designating a stream that coursed down present-day Webster, Park and Brook Avenues. Roger Michelsen was the owner of a 102-acre tract purchased from John Archer. The name is also given as Ryer Michielsen on a 1736 deed, and descendants adopted Ryer as a surname. See: Mill Brook.

ROGERS' ISLANDS. Two small islands were so designated on an 1874 map when Patrick L. Rogers was the owner of an estate called "Sunnyside" on the Shore Road. The islands were behind Hunter Island, close to the mainland and were accessible at low tides. In 1964, the U.S. Olympic Committee designed the area as a rowing course, and the Parks Department widened the channel. The resultant lagoon blotted out the islands.

ROOSEVELT BROOK. This rivulet runs southeasterly into Pelham Bay near the city limits, after originating in a spring near the bridle path in Pelham Bay Park. It crosses under Shore Road opposite the northern end of Hunter Island. See: Maninketsuck.

ROOSEVELT COVE. This small inlet behind Hunter Island was once the waterfront property of the Roosevelt family in the early 1800s. Roosevelt Brook, originating in present-day Pelham Bay Park, empties into this cove. See: Shoal Harbor, Vagabond Bay.

ROOSEVELT ROCK. This granite ledge is on the shoreline of the former Roosevelt estate behind Hunter Island. Some 50 yards from the city limits, this rock faces Roosevelt Cove, and on its flank is engraved: "Isaac Roosevelt. 1833. "

ROOSTER CHURCH, The (1). St. John Evangelical Lutheran church bore this nickname, which was a common one due to the custom of using a metal rooster on the weather vane. The church was organized in 1860, and in 1864, it was erected on the west side of Fulton Avenue, north of East 169th Street as a small wooden edifice, seating 150. The tower with the rooster weathervane was added in 1876.

ROOSTER CHURCH, The (2). St. Matthew's Evangelical Lutheran church was built on Courtlandt Avenue between East 154th Street and East 155th Street in 1863. The weathervane in the form of a rooster gave it the nickname. See: Michael's Hall.

ROSA STREET. This is the former name of East 197th Street from Bainbridge Avenue to Decatur Avenue. The real estate development belonging to Isaac Varian was surveyed in the 1870s by Rudolph and William Rosa. The Dutch ancestor of the Rosa family was Albert Roosa, who came to America in 1666. See: Isaac Street, William Street (4).

"ROSE BANK." The estate of W. H. Leggett, on Leggett's Point, was noted on maps of 1835. It was next owned by Benjamin Whitlock in the 1850s, and sold to Innocencia Casanova at the time of the Civil War.

"ROSE HILL" (1). This was the name of John D. Poole's estate that stretched from University Avenue down to Grand Avenue in the vicinity of West 174th Street.

"ROSE HILL" (2). This 200-acre estate once was owned by the Corsa family. Later, it passed through the ownership of the Watts, Brevoort and Mowatt families until it was sold to the Catholic Diocese and became St. John's College. The manorhouse, built in 1838, served as the administrative building. We know the college today as Fordham University. See: Fordham Square. (Current).

ROSE ISLAND. In a deed of 1691, it is described as being bound by Hutchinson Creek. In the 19th century, it was further described when it was surveyed and mapped by Robert Findlay - no date given - for the Givan family as ". . . that certain piece or parcel of land and salt meadow commonly called Rose Island containing 14 acres more or less; being the same lot and premises conveyed by John and Theodosius Bartow to Robert Givan by deed dated 1st July 1802 . . . ." Jemima Watson was last owner of record, in 1905. Rose Island lies under the southern end of Co-Op City.

ROSE HILL PLACE. See: Fordham Square [Current].

ROSE PLACE. This is the former name of Steam Street from Castle Hill Avenue to Parker Street. Hudson P. Rose was the property owner prior to 1899.

ROSE STREET (1). This was the earlier name of East 152nd Street, east from Third Avenue, as noted in 1871. The name was changed in 1905 to the numbered street. See: Elton Street, Kelly Street, Willow Street.

ROSE STREET (2). This is the former name of Matthews Avenue, one block north of Morris Park Avenue. See: Adee Avenue.

ROSE, The. This small "hole-in-the-wall" movie house featured silent movies around 1910. Music was provided by a hand-cranked phonograph. The Rose was located on Prospect Avenue south of Home Street.

ROSEDALE LANE. This is the former name of Revere Avenue from Roosevelt Avenue to Lamport Place. This was part of a real estate development around 1905 called "Tremont Heights," and Benjamin Lamport was one of the backers.

"ROSEDALE" (1). This former estate was owned by Hudson P. Rose in the vicinity of St. Lawrence Avenue and Rosedale Avenue.

"ROSEDALE" (2). The estate, mapped in 1868, of P. Bruner ran from Eastchester Road down to Givan's Creek (now the junction of Gun Hill Road and Hutchinson River Parkway). Rose Island was part of the tract in earlier times.

"ROSELAND." This was the estate of Oswald Cammann, vestryman of St. James. P.E. church of Fordham. According to the maps of 1868, "Roseland" extended from the Harlem River to approximately University Avenue, and from West 183rd Street to West Fordham Road. See: Cammann Street (1), Cammann Street (2).

ROSEWOOD STREET. This was the original name of Wallace Avenue from Arnow Avenue to South Oak Drive, a distance of three blocks. See: Hall Street, Hickory Street, Jefferson Street (2).

"ROUND MEADOW." The name of a 19th-century farm belonging to John DeLancey Neill in the vicinity of Bronxdale Avenue and Morris Park Avenue. (A Delancy Place and Neill Avenue are on this land, today).

ROYAL MOVIES, The. This silent movie house is believed to have been opened in 1912 on Jennings Street and Southern Boulevard.

ROYAL PHOTOPLAYS, The. See: Bronx Playhouse, The.

RUBBLESTONE BRIDGE, The. This was a local nickname for a fieldstone bridge over the Bronx River south of Allerton Avenue. The localism persisted into the 1930s, although the official name was "Fieldstone Bridge" in Parks Department listings. The span was replaced around 1934. See: Fieldstone Bridge.

RUPPERT'S ISLAND. Unofficial name for South Brother Island. See: North Brother Island, South Brother Island.

RUSH, The. This was once a well-known skating pond near East Tremont Avenue and Clinton Avenue in the village of Fairmount. No explanation is forthcoming on this localism, but it may have some connection with rushes or swampgrass.

RUSKIN STREET. This is the former name of East 211th Street from White Plains Road to Holland Avenue. See: Katonah Street.

RUSSELL'S HOTEL. This hotel of lower Fordham was situated on Third Avenue at East 183rd Street around 1900. At the time of the cycling craze, Mr. Russell had bicycle-racks installed in front of his hotel to the annoyance of patrons who arrived in buggies and found no hitching posts.

RUSSIAN HILL. This was a nickname for the eastern slope of Hunter Island, where Russian-American groups congregated in the 1920s and 1930s.

RUSTIC AVENUE. This is the former name of Clinton Avenue from East 180th Street to Bronx Park. The name was self-explanatory.

RYAR. This is a misspelling of Ryer on British Headquarters maps of 1778 by Skinner & Taylor. See: Ryer Place.

RYER PLACE. This is the former name of Belmont Avenue from Crotona Park to East 181st Street. Samuel Ryer, descendant of pre-Revolutionary settlers, owned a large estate called "Monterey." It was north of Crotona Park.

SACKETT LANE. It once led from the Shore Road to the Sackett farm (*circa* 1870) which now is incorporated into Split Rock golf course. The path is used by hikers.

SACKETT'S FARM. This early 19th-century farm belonged to the Sackett family that was located in present-day Van Nest. Located above Bear Swamp Road (Bronxdale Avenue), it was marked "Wetherbee Farm" on an 1860 map. See: Wetherbee Farm.

SACRAHONG STREET. This street preserved the aboriginal name of the stream that divided West Farms from Morrisania. It was rendered as Sackwrahung, and also carried the alternate name of Bungay Brook. Some historians claim the entire shoreline of the Bronx River at Hunts Point was called Aquahong. Sacrahong Street partially overlaid an ancient waterway that led in from the Bronx River to an Indian settlement at East Bay Avenue. Today, the Hunts Point Market overlays it.

SAGE HOUSE. This fashionable hotel and resort was located on the site of Fort No. 3, overlooking Spuyten Duyvil Creek, before 1900. See: Edgehill Inn, Fort No. 3.

ST. AGNES AVENUE. See: Palmer Avenue.

ST. AUGUSTINE'S PARK. An oblong plot of 0.25 acres between Fulton and Franklin Avenues, below East 167th Street, this land was ceded to the City for park purposes on November 8, 1864. It received its name from the Roman Catholic church of that name which was, at that time, located on Jefferson Place. It was named Hines Park in 1929, and dedicated to Colonel Frank Hines, 1868-1929, "A Sterling Citizen Soldier."

ST. BONIFACE'S INN. A wayside hotel on the Westchester Turnpike (now the northwest corner of Rowland Street and Westchester Avenue) that was well known in the 1870s to the end of the century. A Mr. Withers was the host, and Randall Comfort, an early historian, noted a sign over the door that read "No really destitute person need pass this house hungry."

ST. JAMES STREET. One former name of both West 190th Street and East 190th Street from the Croton Aqueduct to Grand Avenue. See: East 190th Street, West 190th Street.

ST. JAMES' STABLES. On Jerome Avenue, near East 192nd Street, was a popular livery, run by Harry R. Haskin, nephew of John B. Haskin of early Bronx politics. It took the name from St. James' Park, which was opened in 1901.

ST. JOHN AVENUE. This was mapped in 1905, and was an alternate name of Avenue St. John. This street did not honor a saint, but referred to George and Catherine St. John who were property owners there from the 1850s to the early 1890s. See: Avenue St. John.

ST. JOHN'S COLLEGE. See: Fordham Square. (Current).

ST. JOHN'S HILL. This is the former name of steep East 159th Street from St. Ann's Avenue up to Eagle Avenue. See: Cedar Street, John Street.

ST. JOSEPH AVENUE. On an 1888 Robinson map, this avenue ran from Bruckner Boulevard to Lafayette Avenue across the former Charlton Ferris farm. The name stemmed from its proximity to St. Joseph's Deaf and Dumb Asylum, now known as St. Joseph's School for the Deaf. Today, St. Joseph Avenue is part of Balcom Avenue.

ST. JOSEPH'S STREET. This is the original name of East 144th Street from St. Mary's Park to Southern Boulevard. See: Grove Street.

ST. MARY'S AVENUE. This is the former name of DeReimer Avenue from Bartow Avenue to Mace Avenue, a distance of three blocks. See: Monaghan Avenue, South 17th Avenue.

ST. MARY'S PARK HOTEL. This was a late 19th-century hotel on the southwestern corner of East 148th Street and Willis Avenue that was run by Anton Loeffler's sons. It was the last remaining house (3 stories) on what had been a more extensive picnic grounds, known as Loeffler's Park. See: Loeffler's Park (1).

ST. PAUL'S PLACE. This was once the extreme southwestern corner of the Bathgate farm (1), and constituted the boundary of Lewis Morris' lands. It was named in honor of St. Paul's church that was organized in May, 1853. The Place was officially closed in 1960, and physically closed in 1962 from Webster Avenue to Brook Avenue, and is now part of Borgia Butler Houses.

ST. RAYMOND PARK. This was a real estate development in 1908 between present-day Castle Hill Avenue and Zerega Avenue in the vicinity of St. Raymond's R.C. church. Hudson P. Rose was the realtor.

ST. VINCENT AVENUE. This is the former name of West 256th Street from Fieldston Road to Sylvan Avenue. It referred to nearby Mount St. Vincent's Academy. See: Hawthorne Avenue.

ST. VINCENT'S POINT. This point of land is at West 264th Street and the Hudson River, below Mount St. Vincent's Academy, and practically on the Yonkers town line.

SAMUEL STREET. This was the former name of East 180th Street from the Concourse to Southern Boulevard on an 1891 map. This street formed the southern boundary of Samuel Ryer's farm, and the northern line of the L. S. Samuel property. At its eastern end, it flanked the lands of Samuel Purdy. In 1841, the West Farms Cemetery was laid out along this street west of Boston Road. See: Talmadge Street.

SAND STREET. An earlier name of Olmstead Avenue from Parkchester to Westchester Avenue. See: Avenue D.

SANDER'S LANDING. This was formerly the end of Reid's Mill Lane at Eastchester Creek (Hutchinson River), as noted in 1668. In 1739, a tidal mill was erected there by Shute and Stanton. This was run by John Bartow in 1766, and passed into the ownership of John Reid in 1790, and to his son Robert.

SANDS AVENUE. The former name of Sands Place as noted on maps of 1900 and 1905. The Sands were a very early Bronx family, intermarried with Schuylers, Hunters, Ferris and Bartows. An 1873 map shows the area as "land of William Bowne Sands."

SANFORD STREET. This was a 19th-century name of East 187th Street from the Concourse to Third Avenue. See: Clay Avenue, Jacob Street.

"SANS SOUCI." Joseph Husson purchased a 55-acre tract on the right bank of Pugsley Creek, part of the Stephens estate that had been purchased from Isaac Clason. The homestead was named "Sans Souci," and later the property was divided among seven children. A daughter, Susan Husson Rudd, inherited the mansion.

SAPROUGHAH. The Indian name for a creek that once flowed into the Harlem River slightly south of present-day High Bridge. It formed the lower boundary of Fordham Manor, and in the Algonquin tongue meant "Land Spread Out." See: Biggart's Canal.

SARATOGA AVENUE (1). This is the former name of Seminole Avenue where it runs alongside the Bronx Municipal Hospital Center. A 1917 real estate development gave names of resorts to the avenues: Newport, Narragansett and Saratoga.

SARATOGA AVENUE (2). This is the former name of Morris Park Avenue from Williamsbridge Road to Eastchester Road. See: Liberty Street, Seminole Avenue Watson's Lane.

SAVOY, The. This medium-sized silent movie theatre at 234 Hughes Avenue opened in 1923. It had a seating capacity of 1,141.

SAW MILL BROOK. This alternate name of Stony Brook, noted on an 1897 map, might well be a reminder of the Cornell mill that was located downstream in 1807. See: Abbott's Brook, Kidd Brook, Stony Brook.

SAW MILL CREEK. This was noted on an 1809 map by Charles Loess, City Surveyor, and again in 1853 by Andrew Findlay as an alternate name of the Mill Brook. See: Mill Brook.

SAXE STREET. This is a former name of Leland Avenue from Westchester Avenue to the Cross-Bronx Expressway. Simon P. Saxe purchased a section of the Mapes farm in 1889, and in 1898 Ira and Hulda Saxe bought land from W. and H. Mapes near present-day Wood Avenue.

SAXON STONE, The. This huge boulder lies along the eastern shoreline of Hunter Island, and the name seems to have been a localism of the Summer campers. Possibly there was an explanation for this name in a poem that was once engraved on the stone, but weather has eroded the lettering.

"SCABBY INDIAN." In the Will of John Bartow, 1725, we read "In consideration that my beloved wife bring up my children, I give unto her the sole use and benefit of my dwelling house and homestead, and all my land at Scabby Indian, bounded southeasterly by the land of John Williams, westerly by the country road, northerly by the road that goes to Thomas Haddon's saw mill, and by Daniel Turner's land." The approximate bounds of this tract would be Eastchester Road, Pelham Parkway, Baychester Avenue and the Hutchinson River.

SCENIC, The. The first silent movie house on City Island's Main Street in 1913.

SCHIEFFELIN'S LANE. This is the former name of Eden Terrace, for it led to the Schieffelin estate facing Boston Road.

SCHMIDT'S PARK. This was a small picnic grounds in Unionport around 1900, located on Westchester Avenue below Olmstead Avenue. It featured bowling alleys and an outdoor dance hall.

SCHOOL BROOK. This was one of the sources of the Mill Brook. It originated south of Gates Place and Mosholu Parkway, and flowed southeasterly to Webster Avenue and into the Jerome Skating Lake. On an 1860 map of Westchester, it was depicted running between the Dickinson lands and the Peter Bussing farm. There was no school nearby -the name being derived from an earlier Dutch name "Schuil" meaning hidden. See: Middlebrook (1), Schuil Brook.

SCHORSTINVEGER, Pieter. See: Plantation of Pieter Schorstinveger.

SCHOTT'S BREWERY. Gottlieb Schott is listed in the 1869 *Directory of Westchester County* as owner of a brewery on Carr Avenue, corner of Beck Street (St. Ann's Avenue, corner of East 156th Street). It was next to Ebling's Brewery that began business in 1868.

SCHUETZEN HALL. This social hall, indoor rifle range and bowling alleys was listed in 1871 on Third Avenue between East 165th Street and East 166th Street.

SCHUIL BROOK. "Schuil"has several connotations in Dutch, such as "hidden" "covered over" and "protected," which would describe a small brook almost hidden by shrubbery. It may have gotten its name from a Hollander named Schuyler who purchased land from John Archer in the 1690s in he neighborhood of today's Mosholu Parkway. See: Middle Brook (1) and School Brook.

SCHUYLER STREET (1). This is the former name of East 153rd Street in the village of Melrose South.

SCHUYLER STREET (2). This is an earlier name of Crosby Avenue from Baisley Avenue to Waterbury Avenue.

SCHUYLERVILLE. This former settlement at present-day East Tremont Avenue and Bruckner Boulevard came into existence around 1840. Composed principally of Irish, the men worked as stone masons and laborers on the neighboring estates, on Fort Schuyler, and even on Stepping Stones Lighthouse. Others were gardeners, coachmen and grooms, while some became mounted policemen. The women hired themselves out as maids, cooks and laundresses on the estates of Throggs Neck.

SCHUYLERVILLE PARK. This is the former name of the public square at the end of Meyers Street that is known officially as Throgs Neck Park. It was acquired as a public place in the year 1836, and in the 1870s and 1880s was known as Bolen Green and Bowling Green. See: Bolen Green.

SCOTCH HILL. A small 19th-century settlement of Scottish people at Old Post Road and West 246th Street. The area is now covered by the Horace Mann school.

SCOTT STREET. See: East 204th Street.

SCRATCH PARK. This derisive nickname for a rundown park at the foot of Willis Avenue in the early 1900s was still in use around 1924. A local name for it (1912-1917) was Harvey's Hill. It was situated on part of Brommer's Park, a noted resort of the 1890s - Christ's Park of a still earlier era - and the 18th-century manorlands of Lewis Morris, and the original farm of Jonas Bronck. Today, the official name is Pulaski Park.

SCREVIN'S POINT. The former name of Castle Hill Point when John Screvin (*circa* 1880) lived there. He was a son-in-law of Gouverneur Morris Wilkins who had owned the property before him. See: "Castle Hill."

SCUTTLE DUCK HARBOR. This is the earliest reference to Hammond's Cove, as it appeared on a 1696 deed.

SEA VIEW AVENUE. This is the former name of Boller Avenue in Baychester.

SEABREY'S CREEK. This was an 1850 spelling of Seabury Creek.

SEABURY CREEK. The Seabury family were already listed as residents of Westchester village in 1700, and one of their farms was behind St. Peter's P.E. Church in the vicinity of Zerega Avenue. A brook, originating in Bear Swamp, flowed southeasterly under Zerega Avenue and into Westchester Creek. As it flowed through the Seabury farm, fresh water and salt water mixed, and the brook became a creek. See: Bain's Creek.

SEAMAN AVENUE. The Seaman family was a prominent one in Colonial days, and were related to the Valentines, Sands and other old families. Anna Seaman married Charlton Ferris (see St. Joseph Avenue). Amelia Seaman inherited part of John Ferris' lands (see Burial Point). The avenue was laid out in the vicinity of the tollbooths of the Bronx-Whitestone Bridge in 1948, was never cut through and was finally eliminated in September, 1964.

SEAN-AUKE-PE-ING This was the aboriginal name of the Siwanoy village once located at Leland and Lacombe Avenues on Clason Point. It was excavated by Allanson Skinner and an Indian assistant, Amos One-Road. A loose translation would be "River, Land and Water Place" (Bronx River, Clason Point, and Pugsley Creek). Later, the name was shortened to Snakapins.

SEASIDE BOULEVARD. This was a former name of Shore Road from Hunter Island north to the city line, according to an 1873 map advertising Pelham Manor.

SEBERRY CREEK. This was a variant of Seabury's Creek as noted on a deed of 1844.

SECOND AVENUE (1). This is the former name of Walnut Avenue in Port Morris.

SECOND AVENUE (2). In the village of Tremont, this was the earlier name of Morris Avenue from Claremont Park to the Concourse. See: Avenue A, Kirkside Street.

SECOND AVENUE (3). The original name of Bainbridge Avenue from Bedford Park to Mosholu Parkway.

SECOND AVENUE (4). This is the former name of Bronx Boulevard from Gun Hill Road north to East 219th Street. See: Bronx Terrace, Marion Street.

SECOND AVENUE (5). In Wakefield, this was the earlier name of East 216th Street from the Bronx River to Bronxwood Avenue.

SECOND AVENUE (6). In 1889, this was the former name of Katonah Avenue from the cemetery up to the city line in Woodlawn Heights.

SECOND PLACE. When the village of Morrisania was laid out in the 1850s, this was a short street running from the Mill Brook to Third Avenue. See: Ella Street.

SECOND STREET (1). In Port Morris, this was the former name of East 140th Street from Bruckner Boulevard to the East River. See: Bronx Street.

SECOND STREET (2). This is an earlier name of East 164th Street in Morrisania.

SECOND STREET (3). Field Place's original name in Fordham. See: Kirk Street.

SECOND STREET (4). In Wakefield, this was the earlier name of Carpenter Avenue from Gun Hill Road to East 233rd Street. See: Catherine Street.

SECOND STREET (5). This is the former name of Hermany Avenue in Unionport.

SECOND STREET (6). The 19th-century name of Frisby Avenue from Zerega Avenue to East Tremont Avenue.

SECOND STREET (7). This was the former name of LaSalle Avenue from Edison Avenue to Crosby Avenue when it was cut through the Seton Homestead and was designated part of the village of Schuylerville.

SECOND STREET (8). In the 1890s, this street was mapped in the settlement of Bartow running from the railroad tracks to Shore Road, but it is now part of Pelham Bay Park.

SEDGE ISLAND. This was a lowlying island in the Bronx River near its left bank, adjoining the lands of William Watson in the 1870s. Its approximate location would be at Story Avenue, south of the Eastern Boulevard Bridge.

SEEBECK'S HOTEL. It was located at the lower end of an amusement park at East 148th Street and Brook Avenue around 1890. See Fountain Park.

SEIDE PLACE. Historians believe this place was part of Ane Hutchinson's holdings, which later was granted to the Ten Families by Pell in 1662 for farming purposes. Maps of the 1800s list this tract as part of the Provoost farm, and the Prevost farm. Today, it is an industrial site. The Place was named after 1. Lincoln Seide, a realtor, who owned the property around 1905.

SELECT, The. This old-time movie house that was off Westchester Square near East Tremont Avenue and Williamsbridge Road showed films accompanied by piano music only. The theatre operated from sometime around 1915 to 1925. See: Village, The.

SEMINOLE AVENUE. This is the former name of Livingston Avenue, but the name was changed in May, 1916, by request of local residents who wanted to honor an old Riverdale family. See: Indian Road.

SENECA PARK. This triangular real estate venture of 1901 was bounded by Eastchester Road, East 222nd Street and Boston Road. Chester Avenue was its main thoroughfare, but it has since been changed to Chester Street.

SESGO PLACE. This was a short lane at East 211th Street near Webster Avenue on an 1884 map. It has since been incorporated into the southeastern corner of Woodlawn Cemetery. See: Eastwood Place.

SETON STREET. This was the former name of Edison Avenue from Baisley Avenue to LaSalle Avenue, according to an 1868 Beers' *Atlas*. The Seton family were landowners from there to Westchester Creek.

SEVENTEENTH AVENUE. In Wakefield, this was formerly East 231st Street from Bronx Park East to White Plains Road.

SEVENTH AVENUE (1). This is the former name of Grenada Place in Edenwald.

SEVENTH AVENUE (2). In Wakefield, this used to be East 221st Street.

SEVENTH STREET (1). This was the earlier name of East 135th Street from Willow Avenue to the East River in Port Morris. See: Edsall Street, Orange Street.

SEVENTH STREET (2). This is the former name of East 169th Street from Webster Avenue to Third Avenue. See: Arcularius Place.

SEVENTH STREET (3). This is the former name of Chatterton Avenue in Unionport.

SEWANHACKY. This was the Siwanoy Indian name for the waters south of Hunter Island, now covered by the Orchard Beach parking lot. "Sewant" was the Indian name for wampum, and Sewanhacky meant "Great Bay of the Island of Shells." See: LeRoy Bay, Pelham Bay.

SEWARD PLACE. The original name of Sycamore Avenue in Riverdale, until it was replaced in May, 1916.

SHARON PLACE. This short street once connected Newport Place and Elberon Street off Morris Park Avenue. Now all three are inside the grounds of the Bronx Municipal Hospital Center, and are no longer in existence.

SHATEMUC. See: Hudson River [Current].

SHEEP PASTURE ROAD. This was the pre-Revolutionary name of Unionport Road, the stretch that passes through Parkchester heading to the common pasturage.

SHEEP PASTURE, The. In Revolutionary times, this was the name for the meadows on which Unionport was built. In the Charter of 1721, it is denoted as 400 acres owned by Westchester Village for free pasturage. In 1825, the town trustees sold these commons to Martin Wilkins. In 1851, his grandson established a building association which became the village of Unionport. See: Cromwell's Neck, Screvin's Point.

SHEEPSPEN ROCKS. This cluster of rocks off Tallapoosa Point in Pelham Bay Park was likened to a sheep's pen. They were buried under landfill in 1963.

SHEIL STREET. This is the former name of East 214th Street from Barnes Avenue to Bronxwood Avenue, a distance of two blocks. It is a misspelling of Shiel Avenue. See: Avenue A (4), Shiel Avenue.

SHERBORN FARM. This 40-acre tract was owned by Robert Watts in the Manor of Fordham, and was surveyed by Robert Findlay in May, 1811. The approximate bounds were Webster Avenue, Fordham Road, Third Avenue, East 178th Street. This became the Bassford farm in 1836. See: Union Hall.

SHERIDAN STREET. On a 1905 map, this was the earlier name of Britton Street.

SHERIFF WILLETT'S POINT. A Revolutionary War map shows this name at the tip of Clason Point. See: Cornell's Neck, Snakapins, Willetts Point, Woolet's Point.

SHERMAN STREET. Mapped before the Civil War as today's East 136th Street, it is evidently an error, for afl other maps list it as Smeeman Street.

SHERWOOD FOREST. North of Miles Avenue and Prentiss Avenue, the name was in verified use since 1925 and might have been used still earlier when it was part of an apple orchard. The trees were used by generations of Edgewater and Pennyfield boys for hideouts, forts and tree-houses. The name fell into disuse in 1961 when a block-square woods alongside Miles Avenue was cleared for homes.

SHERWOOD HALL. It was located in the heart of the old-time village of Morrisania, present-day Third Avenue and East 166th Street. The Oliver Tilden Post, honoring the first man of the village to be killed in the Civil War, held their first meeting in this hall. Robert R. Sherwood, proprietor, was a member of Lily (Masonic) Lodge, as was Oliver Tilden.

SHERWOOD STREET. See: East 196th Street.

SHIEL STREET. This is the former name of East 214th Street from Barnes Avenue to Bronxwood Avenue. Judge Peter Ashwin Shiel was active in Wakefield and Williamsbridge public improvements, notably the extension of the subway into the area. His name was in the news from 1890 to around 1910. See: Avenue A (4).

SHINGLE PLAIN. See: Woodstock.

SHIRMER STREET. This was the former name of North Oak Drive in 1894, when Charles and Lily Shimmer owned some property there.

SHOAL HARBOR. This bay, partly in Bronx County and partly in Westchester County, is north and west of Hunter Island. An 1811 mariners' chart gives the depth at less than a fathom. See: Roosevelt Cove, Vagabond Bay.

SHOLE HARBOUR. This name is mentioned in a deed of Sir John Pell, dated May 3, 1712. See: Shoal Harbor.

SHORAKAPKOCK (1). The Weckguasgeeck Indian name for the hillside and plateau of Spuyten Duyvil. Indian scholars interpret the name differently as "Sitting Down Place" or "Wet Ground." See: Berrian's Neck, Konstabelsche Hoek, Tippett's Neck.

SHORAKAPKOCK (2). This Indian name was also applied to Spuyten Duyvil Creek west of the Wading Place. See: Kingsbridge Creek.

SHORE DRIVE DIVISION. This station of the Life Saving Service of New York was situated at the foot of Randall Avenue on Eastchester Bay. It was organized in 1923 or 1924, and was manned by three officers and eight volunteer surfmen. Oddly enough, the Shore Drive was not yet in existence, but was on the drawing boards of the Planning Commission.

SHOREHAVEN. After the demise of Kane's Amusement Park in the early 1940s, the land at the foot of Soundview Avenue lay dormant. Dr. Joseph Goodstein and Mal Deitch interested a group of investors in building a beach club on the site, and from approximately 1947 to 1986 a clubhouse, swimming pool, handball courts, card rooms and entertainment attracted thousands of members. It was replaced in 1988 by the residential complex of the same name.

"THE SHRUBBERY." This estate was owned by G. Prevost in the 1870s. It extended from Split Rock Road to the Hutchinson River to Boston Road. It was practically blotted out by the New England Thruway-Hutchinson River Parkway cloverleaf.

SIDNEY STREET. The former name of West 227th Street from Henry Hudson Park to Netherland Avenue.

SIDNEY'S BROOK. An 1868 "Map of Boston Road" by Chief Engineer William Livingston listed this name on Rattlesnake Brook as it emerged from Seton Falls Park. See: Long Pond, Rattlesnake Brook.

SIGNAL PLACE. This is the former name of East 203rd Street from Webster Avenue to the railroad. The name might hark back to a time when a railroad signal was installed there, but at the time this street was laid out, during the Civil War, a Union gunboat, *U.S.S. Signal* was very much in the news. See: Ozark Street.

SILLECK'S LANDING. This town landing is mentioned in the annals of Eastchester, specifically in 1671, and is to the north of Fisher's Landing (East 233rd Street) on the Hutchinson River.

SILVER BEACH DIVISION. Six officers, two coxswains, ten surfmen and one mascot were listed on the volunteer rolls of this Life Saving Station at the northern end of Silver Beach, a Summer bungalow colony on the East River. Honorary members were

Commissioner John Flynn and William Peters, a co-owner of the property that was rented for Summer sites in 1920.

SILVER BEACH PLACE. The western slope of this little street was originally an Indian campsite. This area was also the scene of the 1776 landing of General Howe's British and Hessian troops. In 1983 it was renamed Monsignor John Halpin Place to honor the pastor of St. Frances de Chantal on his 60th year in the priesthood. Important artifacts were excavated there.

SILVER LAKE. This was an early name of the lake inside Bronx Park impounded by the Lydig dam, according to an 1850 map. In 1889, it carried the same name, but then was changed to Lake Agassiz.

SILVER LANE. This is the former name of Silver Street that was mapped by Ezra Cornell, Highway Commissioner in 1807. It was the southeastern boundary of a 5-acre homestead belonging to the Cornells that was later sold to Andrew Arnow. The origin of the name is lost in antiquity, but it has been suggested that the sale of land was consummated in silver currency in an era when barter was far more common.

SIMPSON'S POINT. An 1872 Police report noted this location on Bronx Creek near the East River, where an unidentified body floated in. The Simpson estate of Hunt's Point was on the right bank of the Bronx River. It was called "Foxhurst" and was a 120-acre stud farm for polo ponies and Shetlands that was operated by William Simpson and his two sons. Earlier, the land belonged to the Tiffany and Fox families. See: Bronx Oval.

SINGER'S BEACH. In the 1920s and 1930s, this was an Eastchester Bay beach resort at the foot of Philip Avenue.

SISTERS' POND. In the early 1900s, it was fed by Downing's Brook, and was located at approximately East 180th Street and Morris Park Avenue. Its overflow discharged into the Bronx River at a point now covered by the Fire Department signal station. The name is strictly a localism, as a nearby family named Rochi had many girls and their playmates called them all "Sister" and "Sis." The pond was on their property .

SIXTEENTH AVENUE. This is the former name of East 230th Street in Wakefield.

SIXTH AVENUE (1). In 1893, this was the name of Grand Avenue from West Fordham Road to West Kingsbridge Road. See: Edenwood Avenue.

SIXTH AVENUE (2). This was the earlier name of East 220th Street in Wakefield.

SIXTH STREET (1). (Port Morris) (Mott Haven) See: East 136th Street.

SIXTH STREET (2). This was the original name of East 168th Street from Webster Avenue to Third Avenue. See: Charles Place, Glen Street.

SIXTH STREET (3). In Unionport, this was the first name of Bruckner Boulevard. See: Commerce Street.

SIXTH STREET (4). This was another and earlier name of Paulding Avenue from Gun Hill Road north to Bussing Avenue. See: Forest Street (1).

SIXTH STREET (5). This East-West street once was mapped in Bartow between the present-day Shore Road and the railroad, below the golf course parking lot.

SIXTH STREET, WEST. See: West Sixth Street.

SKANEHTADE. See: Hudson River [Current].

SLAVE CEMETERY, The. This small plot was marked on several ancient maps, and is located south and west of the Joseph Rodman Drake Cemetery on Hunt's Point. It is now covered by a factory.

SLOCUM AVENUE. This is the former name of Anthony Avenue from the Cross-Bronx Expressway to East Tremont Avenue.

SMALL CREEK, The. This was an arm of Coney Island Creek, flowing eastward and joining the latter approximately at the toll booths of the Bronx-Whitestone bridge.

SMEEMAN STREET. See: East 136th Street.

SNAKAPINS. This is the accepted Indian name for Clason Point. Not wholly accepted is that it means "River, Land, Water Place" for some students think the name meant "Place of Ground Nuts." See: Sean-Auke-Pe-Ing.

SNAKE HILL (1). This was a local name for East 184th Street, as it winds S-fashion from Tiebout Avenue down to Webster Avenue.

SNAKE HILL (2). This name was in use in the late 1800s pertaining to West Tremont Avenue, winding down from University Avenue to Jerome Avenue.

SNAKE HILL (3). This rocky hill was on the southwestern side of Hugh Grant Circle at Virginia Avenue. This local name was in use in the 1930s by sleigh riders to describe the zigzag route. The hill was demolished in the Spring and Summer of 1962, and the flattened site is occupied by a drive-in bank. See: Sullivan's Hill.

SNUFFMILL LANE. This was the former name of Bronxdale Avenue as mapped in 1851. It led from the village of Westchester to Lorillard's Snuffmill (now in the Bronx Botanical Garden). See: Bear Swamp Road.

SOLDIER POND. This was a Civil War name for a pond on the Theodore Havemeyer estate on Throggs Neck that was located approximately southwest of the intersection of Miles Avenue and Throggs Neck Boulevard. Local lore has it that a soldier stationed at Fort Schuyler was waylaid, robbed and murdered for his money. His body was found in the pond. See: Jackson's Pond.

SOLDIERS' PLOT. This was an unofficial name for a burial ground on Hart Island that contained Civil War veterans who had died of an epidemic. In complete records indicate possible burials as early as 1861. A G.A.R. Post held simple ceremonies every Memorial Day at the well-kept plot, but in 1941 Soldiers' Plot was vacated and the remains brought to Cypress Hills National Cemetery.

SOLON PLACE. Solon L. Frank was a well-known and active realtor in and around Unionport and Clason Point in the years before 1910. As his own first name was that of an ancient Athenian law giver, it was Mr. Frank who gave Greek and Latin names to streets still surviving in the housing development which oddly enough has eliminated the street named for him.

SOMMER STREET. This small street ran east from White Plains Road between East 240th Street and East 241st Street, and was mapped in 1901.

SORMANI'S. This was a local name for the Hermitage Inn at the junction of White Plains Road and Pelham Parkway. Joseph Sormani was the wellknown proprietor of the roadhouse that catered to the carriage trade. See: Hermitage, The.

SOUND HEIGHTS. This was the name given by a real estate fin-n prior to 1909 to the former Havemeyer property on Throggs Neck. Due to its proximity to Long Island Sound and its elevation, the name was well chosen, but the early residents decided on Silver Beach Gardens a few years later.

SOUND, The. This was the earliest name of the East River along the Bronx shoreline, mentioned in Royal Charters for Cornell's Neck in 1667, Westchester in 1686 and "Grove Farm," 1686.

SOUND VIEW PARK. A 1915 map shows a Sound View Park east of City Island Avenue from Ditmars Street to Beach Street. This was a real estate venture.

SOUND VIEW PLACE. This is the former name of Wilder Avenue at the city line. Due to its elevation, Long Island Sound could be glimpsed. The Place was mapped in 1904. See: South 16th Avenue.

SOUP CHURCH, The. This nickname for the First Presbyterian Church of Throggs Neck originated during the Civil War. Soldiers, en route to their homes or back to duty, were sure of a bowl of soup if they came to the church. Young persons were sent to the Westchester depot to coax the soldiers past the numerous saloons. The first (wooden) church was built in 1855, and was destroyed by fire in 1875. It was replaced the same year by a brick edifice.

SOUTH BELMONT. On a map of West Farms, 1874, this district was bounded by East 182nd Street to East 184th Street (Kingsbridge Road) from Belmont Avenue to the Lydig estate (now Bronx Park).

SOUTH CHESTNUT DRIVE. This is the former name of South Oak Drive.

SOUTH EIGHTEENTH AVENUE. The former name of Baychester Avenue in Edenwald. See: Comfort Avenue.

SOUTH ELIZABETH STREET. On City Island, the earlier name of Rochelle Street.

SOUTH FIFTEENTH AVENUE. This was the original name of Murdock Avenue.

SOUTH FOURTEENTH AVENUE. In Edenwald, the former name of Hill Avenue.

SOUTH NINETEENTH AVENUE. Edson Avenue's earlier name. See: Bracken Avenue.

SOUTH RAILROAD AVENUE. This is the former name of Baldwin Street in the tiny enclave of Bronx territory surrounded by Mount Vernon.

SOUTH ROAD. This was the former name of Rawlins Avenue which, in the 1700s, was the driveway to the Ferris house, and to the Van Antwerp mansion of the 1800s.

SOUTH SEVENTEENTH AVENUE. This was the former name of DeReimer Avenue. See: Monaghan Avenue, St. Mary's Avenue.

SOUTH SIXTEENTH AVENUE. The earlier name of Wilder Avenue. This entire district was once the section known as South Mount Vernon. See: Sound View Place.

SOUTH STREET. This is the former name of West 252nd Street, west of Riverdale Avenue. It was the southern boundary of Thomas Cuthbert's estate in the 1870s.

SOUTH THIRTEENTH AVENUE. In Edenwald, Monticello Avenue's former name.

SOUTH TWELFTH AVENUE. Seton Avenue, when it was still part of South Mount Vernon, was known by this numbered avenue.

SOUTH TWENTIETH AVENUE. Grace Avenue's earliest name. See: Jones Avenue.

SOUTH TWENTY-FIRST AVENUE. This was the original name of Ely Avenue. See: Doon Avenue.

SOUTH TWENTY-SECOND AVENUE. Bruner Avenue's first designation. See: Oakes Avenue, Willis Place.

SPAKENT HEILL. This was the German spelling of Spuyten Duyvil as noted in Lt. Von Krafft's war diary of 1780. The origin of the name Spuyten Duyvil is veiled in history and mystery, and the story with the widest acceptance was written by Washington Irving. He wrote that when the British captured Nieuw Amsterdam in 1664, Governor Peter Stuyvesant dispatched his trumpeter, Anthony Van Corlaer, [spelled this way] to the (Bronx) mainland for reinforcements. The trumpeter could not find the ferryman to transport him across the stormy creek, so he impatiently forced his horse into the stream, shouting in Dutch "I will cross, en spijt den Duyvil (in spite of the Devil)." Halfway across, an enormous mossbunker was seen to rise to the surface and swallow Van Corlaer, trumpet and all. Reinforcements never came to Stuyvesant's aid, and so Nieuw Amsterdam became New York. No less than fourteen different spellings of this name have been noted on official documents! The reader is urged to see Speak Devil, Speight den Duyvil, Speit den Duyvil, Spike & Devil, Spiting Devil, Spilling Devil, Spiten deuval, Spitten Divil, Spittin Debell, Spitting Devil, Spitton Divil and Spouting Devil. Spuit den Duyvil gives a plausible explanation.

SPARROW AVENUE. This was the former name of Martha Avenue on an 1890 map of Woodlawn Heights. Adjoining streets were Quail and Vireo Avenues.

SPAULDING LANE. It was once located at West 249th Street and Riverdale Avenue. The lane refers to a wealthy family who lived in the area.

SPEAK DEVIL. This is Spuyten Duyvil as noted on a Hessian military map of 1776. Known as the Wiederholdt Diary, it provides an interesting account of the Revolutionary Bronx from a German viewpoint. See: Spakent Heill.

SPECTACLE ISLAND. This became the former name of Hart Island when, in the 1700s, its shape - rounded on both ends with a narrow waist - resembled a pair of eyeglasses. See: Heart Island, Little Minneford.

SPEIGHT DEN DUYVIL. Still another variant of Spuyten Duyvil, noted in the 1660s. See: Spakent Heill.

SPEIT DEN DUYVIL. A 1653 deed, in Dutch, read: "Papparinemin, byde onse speit den duyvil gesaght;" "Papparinemin, by us called speit den duyvil." See: Spakent Heill.

SPENCER ESTATE. The Lorillard Spencer tract of 129 acres were purchased from Robert Morris in 1856. It is partially in Pelham Bay Park of today, extending south to Country Club Road. In previous decades (1870s), it had extended to Rawlins Avenue. The other boundaries were Eastchester Bay and Bruckner Boulevard.

SPENCER PLACE. The former name of Anthony J. Griffin Place which is east of the Post Office on the Concourse and East 149th Street. Originally part of the 872-acre Manorlands of the Morris family, it remained in the hands of this family from 1746 to 1950. Lewis Spencer Morris, who died in 1944, was the last owner.

SPENCER SQUARE. A 1910 Bronx Atlas lists this at the junction of Boston Road and Williamsbridge Road. An 1868 map shows Lorillard Spencer owning a square tract east of Williamsbridge Road, extending from a point north of Mace Avenue to Adee Avenue. See: Four Comers, Spencer's Comers.

SPENCER'S CORNERS. See: Four Comers, Spencer Square.

SPICER AVENUE. This is the former name of Chesbrough Avenue.

SPICER'S NECK. This was a 1668 and 1680 reference to today's Ferry Point Park. Micah Spicer was granted 30 acres in 1669 by Governor Nicolls. Combined with Brockett's Neck, it became part of Thomas Hunt's "Grove Farm" of 1686. See: Brockett's Point.

SPIKE & DEVIL. Spuyten Duyvil on a 1778 British war map. See: Spakent Heill.

SPILING DEVIL. Another spelling of Spuyten Duyvil, noted on the Sauthier map(1776).

SPILLING DEVIL. A 1776 map uses this spelling of Spuyten Duyvil.

SPITEN DIVIL. Yet another reference to Spuyten Duyvil that was noted on the Sir Henry Clinton map of 1777.

SPITENDEUVAL. An 1829 deed has this version of Spuyten Duyvil.

SPITTEN DEBELL. John Archer's Patent of 1684 has this reference to Spuyten Duyvil in a claim of Fordham Manor.

SPITTING DEVIL. A 1683 variant of Spuyten Duyvil erroneously attributed to the Indians' description of Henry Hudson's *Half Moon*, in which they compared the cannonading to a devil spitting flames. It is not a likely story as the Indians spoke neither English nor Dutch.

SPITTON DIVIL. This was noted on the Nicolls Map of 1664 on which the Bronx mainland was identified as "Part of the Contenent of America."

SPOFFORD'S POINT. This was the 19th-century name of Hunt's Point, on which was situated Paul Spofford's estate, "Elmwood."

SPOON BEACH. This is the former name of a stretch of beach next to Locust Point and the tollbooths of the Throgs Neck bridge facing Long Island Sound. The beach was formed when John T. Wright extended a causeway out to Wright's Island (now Locust Point) which was the property of his father, Captain George T. Wright. This took place in the 1880s. The concave (spoon-shaped) aspect of the beach suggested the name, according to some old-timers. See: Spooners' Beach.

SPOONERS' BEACH. John T. Wright owned an estate called "Pennyfield" on Throggs Neck shortly after the Civil War, and in the 1880s he had a log-lined causeway extended to Wright's Island where his father lived. Wright's Island is now Locust Point. A sand barrier was built up over the years so that, at the turn of the century, a beach had formed which was a favorite spot for canoeists, campers and young couples who were "spooning." This nickname gained more currency than Spoon Beach. See: Spoon Beach, Wright's Corners.

SPOUTING DEVIL. On a 1777 British map, this was the English approximation of Spuyten Duyvil. See: Spakent Heill.

SPRING LOT, The. This was a 6-acre plot on the Throggs Neck Road (now East Tremont Avenue) owned by Elbert Anderson in the 1820s. It contained a spring that evidently served the Indians in centuries past, for in that area (now part of St. Raymond's Cemetery) many traces of a Siwanoy encampment have been found. Elbert Anderson owned a series of farms from LaSalle Avenue to Dewey Avenue. See: Centu farm, Doric farm, Hawthorn farm, Homestead, the, Willow Lane farm.

SPRING PLACE. The former name of East 166th Street at Franklin Avenue, from 1849. The spring was on the Allendorph estate, now occupied by the Armory, and it (the spring) was located by James Ruell Smith as being 20 feet in from Franklin Avenue on East 166th Street. Pumps in the Armory contend with the spring to this day.

SPRING STREET. Noted in 1885, this was the earlier name of East 174th Street, from the Concourse to Webster Avenue. This area was noted for its lush meadows and plentiful springs. Cows were pastured there until early in the 20th century. See: Beacon Avenue, Twelfth Street (1).

SPRINGFIELD STREET. The former name of East 154th Street in Melrose South.

"SPRINGHURST." This Hunt's Point estate was owned by George S. Fox, as mapped in 1874. Later, a small settlement of the same name came into existence. It consisted of two groceries, a barbershop and a dairy in the area of Baretto and Manida Streets. A farmer named Duffy and his sons maintained a herd of cows there before World War I.

SPUIT DEN DUYVIL. A 1647 reference to Spuyten Duyvil, and the word "Spuit" can be compared to the English "Spate," a freshet, or flow of water. Although not as colorful as the Trumpeter story (see Spakent Heill) this explanation is the most valid one on the meaning of Spuyten Duyvil, thanks to Dr. Ray Kelly who has delved into the 17th-century Dutch origin of the name. He is convinced the early settlers referred to a strong flow of water. The creek, subject to a double tide, was in almost constant flux which was incredible to Hollanders accustomed to the daily ebb and flow of the sea; this spate they ascribed to the Devil, which was a common European practice of their day.

SPUYTEN DUYVIL & PORT MORRIS RAILROAD. This railway line was laid out around 1865, connecting those divergent parts of The Bronx. Gouverneur Morris II was the principal stockholder. It had been surveyed 10 years earlier by Andrew Findlay See: Branch Railroad, Pocahontas Railroad.

SPUYTEN DUYVIL CREEK. This waterway formerly divided Marble Hill (Manhattan) from Kingsbridge (Bronx) and figured prominently in early Dutch history and the American Revolution. It is now filled in, and covered by Exterior Street, West 230th Street and John F. Kennedy high school. See: Kingsbridge Creek.

SPUYTEN DUYVIL DIVISION. This lifeguard station that was listed in 1911 was based at Fordham Landing on the Harlem River. Officers and surfmen patrolled the treacherous waters of the Hudson River where it joined the Harlem River which, at that point, was called Spuyten Duyvil Creek.

SPUYTEN DUYVIL PARKWAY. See: Manhattan College Parkway.

SPUYTEN DUYVIL ROAD (2). West 240th Street's former name. See: Dash's Lane.

SPUYTEN DUYVIL ROAD (1). This is a very ancient road in this area. The lower end was renamed Irwin Avenue on May 7, 1929, to honor a World War I soldier.

SPY OAK. This oak tree was a well-known landmark that was already full-grown during the Revolution. It stood on the west side of the Pelham Road (Westchester Avenue, north of Buhre Avenue) and was believed to be haunted by the ghost of a spy that had been hanged there. During the next century, many tales of the apparition were recorded so that they were collected in a book called *Reminiscences of an Old Westchester Homestead* by Charles Pryer. The tree was cut down in 1933. See: "Stony Lonesome."

SQUAREHEAD HILL. This was a jocular nickname for Sampson Avenue hill when it sloped down to the salt marshes of Morris Cove and when Swedes, Norwegians and some Finns made up the bulk of the population there. The name seemed to be in circulation from approximately 1920 to the year 1946.

STABLE STREET. This unpaved road once ran from the Shore Road to Eastchester Bay starting near the Victory monument in Pelham Bay Park. The stables referred to had belonged to Archer Milton Huntington, whose estate was incorporated into the park.

STAR, The. Opened in 1914, this was a silent movie house on Southern Boulevard and Hunt's Point Road. It seated 600.

STARLING ATHLETIC CLUB. See: Starling Avenue.

STARLIGHT PARK. This large amusement center, with rollercoasters and firework displays, games of skills and chance, and an impressive swimming pool, was an attraction for Bronxites for almost twenty years. A radio station was installed there, and one of the permanent exhibits was an early submarine. Among the transient attractions was an embalmed whale, and on two occasions young couples were married standing in the mouth of the mammal. This park was located east of the Bronx River, at the foot of Devoe Avenue off West Farms Square. It was originally laid out for the 1918 New York International Exposition of Science, Industry and Arts. Starlight Park was demolished in 1933 and 1934. See: Exposition Park.

STATE'S POINT. An 1853 map lists Oliver State, Jr., as the proprietor of what is today Ferry Point Park. See: Ferris Neck.

STATION PLACE. It is now out of existence, being covered by the Bronx River Parkway's southbound lane below Gun Hill Road.

STEBBINS PLACE. This was the former name of Elliot Place off the Concourse when the property was owned by a Mrs. Stebbins in 1868.

STEERS PLACE. This was a Baychester street, mapped in 1890, and no doubt honored George Steers who designed yachts, sailboats and clipper ships. Although he was prominent in City Island affairs, he did not live on the island.

STEPHENS' POINT. In the 1880s, Clinton Stephens owned the property at the foot of White Plains Road on the Bronx River. His name is perpetuated in Stephens Avenue next to White Plains Road. See: Burial Point.

STEPHENSON'S POINT. This point was mentioned in the Revolutionary records when Howe's officers ferried their equipment from that spot to the British warships in the East River. The landing of the Hessian and British troops had taken place in October, 1776, at present-day Schurz Avenue, and Stephenson's Point was most likely the steep bluff at the northern end of Silver Beach Gardens. Edward Stephenson was the owner of the lands, now Silver Beach Gardens, Locust Point and Fort Schuyler. See: Havemeyer's Point, Stevenson's Point.

STERLING OVAL. This baseball field of the Sterling B.B.C. was located at Teller Avenue and East 165th Street from the 1920s to the late 1930s. During that period "Pop" Auer, the manager, had two sons who were pitchers. One, Ken, pitched for the Fire Department, and the other, Don, did likewise for the Police Department.

STEVENS PLACE. This was the former name of Longstreet Avenue at Miles Avenue on an 1868 map. In 1872, John B. Stevens sold out to Theodore Havemeyer and to a real estate development called "Villa Sites." See: "Pennyfield" and "V illa Sites."

STEVENSON'S POINT. This is a misspelling found in the British war journals of 1776.See: Stephenson's Point.

STINARDTOWN. On an 1868 map, this was a small settlement above St. Theresa Avenue and west of Westchester Avenue along what used to be the Pelham Road of the 19th century. The principal landowners of long ago had been the Stennards or Stinards - the spelling varied - since 1787 when James Baxter sold a tract to Mary Stennard, Spinster. Some Stinard/Stennard property is now incorporated into Pelham Bay Park, north of the bridge. The usage of this name continued until around 1920.

STONE JUG, The. An old-time hotel and wayside tavern on the west side of Third Avenue at East 161st Street, dating back to Civil War times. Later, it became the Hammer Hotel, before it was razed to make way for the Magistrates' Court. See: Hammer Hotel.

"STONEHURST." This was an estate extending from Sycamore Avenue down to the Hudson River around West 250th Street. It was laid out in the 1840s by Frederick Law Olmstead (who designed Central Park), and the mansion and extensive grounds were occupied for decades by the Robert Colgate family. The next owners were the Perkins, to be followed by Spruille Braden, once the Ambassador to Argentina.

STONY BROOK. This was the northernmost extension of Westchester Creek that is now covered by the New England Thruway and Pelham Parkway. See: Abbott's Brook, Kidd Brook, Saw Mill Brook.

STONY ISLAND. This former island, once part of the Morris Manorlands, was located below East 130th Street and Willow Avenue in Port Morris. Lewis Morris, first Proprietor, may merely have translated the Weckguasgeeck Indian name for the same place, "Keskeskech," "the S tony Ground." In the 1850s, Gouverneur Morris II employed jobless workmen to fill in a causeway across the marshlands and join the island to the mainland. See: Stony Point.

"STONY LONESOME." Bordering the Pelham Road in the 1830s was situated "S tony Lonesome" belonging to the Drake family. On the edge of this farm stood the Spy Oak that figured in many supernatural occurrences. See: Spy Oak.

STONY POINT. This former promontory at East 130th Street and Willow Avenue was originally Stony Island before it was joined to the mainland in the 1850s. See: Keskeskech, Plantation of Pieter Schorstinveger, Port Morris and Stony Island.

STORROW STREET. This street formerly ran from Unionport Road to McGraw Avenue, but was eliminated in 1938 when Parkchester was built.

STRONG AVENUE. This name appears on an early survey of Eltona, a village of Morrisania whose boundaries were Third Avenue and Trinity Avenue, around East 163rd Street. It was first mapped in 1851. T. W. Strong was listed as a resident of Eltona, and his trade was given as lithographer. See: Dongan Street, First Street (2), Helen Street.

STUART'S LANE. This narrow country road has been swallowed up by West 263rd Street from Riverdale Avenue down to Broadway alongside the Yonkers city line. It serviced the Jonathan Odell farm of the early 1800s, which became the Lispenard Stewart [spelled this way] estate of the 1890s.

SUBURBAN STREET. This is the former name of East 201st Street from the Concourse to Webster Avenue. The name was changed on May 2nd 1919. See: Gambrill Street.

"SUGAR GROVE COTTAGE." This home and property was listed in 1874, on a map of West Farms on the northeastern corner of Third Avenue and East Tremont Avenue, belonging to Pat Crotty. See: Crotty's Woods.

SUICIDE HILL. This steep hill was located south of Morris Park Avenue facing Eastchester Road. Insofar as records show, but one small Revolutionary War skirmish took place there when Americans stopped a British party of scouts attempting to cross Westchester Creek. Because of its grade, Suicide Hill was a favorite sledding course in Wintertime and was known by that name to at least three generations of children. Today, the slope is occupied by the Albert Einstein College of Medicine.

SULLIVAN'S HILL. This rocky knob that was situated on the southwestern side of Hugh Grant Circle at Virginia Avenue. Mr. Sullivan's home was directly below it, and the story has it that the avenue honors his little daughter, Emily Virginia Sullivan. The hill was blasted down to street-level in the Spring of 1962. See: Snake Hill (3).

SULZER'S WESTCHESTER PARK. This was a well-patronized picnic grounds at East 180th Street near Morris Park Avenue and was widely advertised in the 1890s. It was the property of the Ebling Brewing Company.

SUMMIT STREET. This was the former name of East 202nd Street from the Concourse down to Webster Avenue. See: Lower Place.

"SUNNY BRAES." Before and during the 1920s, this was an estate owned by a wealthy family named Fairchild at West 230th Street and Sedgwick Avenue down to Kingsbridge Terrace.

"SUNNYSIDE." This was the name of the estate of Patrick L. and Sarah Rogers west of Shore Road, behind Hunter Island, in the 1870s. Plans were projected to subdivide the estate into sumptuous Summer homes, in a development called "Fairmount-on-Sound," but they never materialized. See: Rogers' Islands.

"SUNNYSLOPE." This estate of W. W. Gilbert in 1872, extended from the Hunt's Point Road to the Bronx River in the general vicinity of Bruckner Boulevard and Lafayette Avenue. Later, the land became the property of P. A. Hoe, who changed the name slightly to "Sunny Slope."

SUPERIOR, The. At 930 East 172nd Street, a small silent movie house built in 1910.

SUTTON PLACE. This short street formerly ran from Boston Road to Cauldwell Avenue between East 165th Street and East 166th Street. It was named for Thomas E. Sutton, who was a Commissioner in Morrisania in 1871.

SUYDAM AVENUE. This is the former name of Fieldston Road. See: Ketchum Avenue.

SWAINS' BRIDGE. This term was used in connection with the small wooden bridge that once spanned Westchester Creek outside Westchester Square. The nickname was in use in the 1870s and 1880s, when young men (swains) met village girls there for privacy. For another version, see Swaying Bridge.

SWAMP HOLLOW. This tract was bounded by West 260th Street, Huxley Avenue and Mosholu Avenue. The name appeared on no maps, but was strictly a local designation.

SWAN INN, The. A well-known hotel on the Westchester Turnpike at its junction with Unionport Road outside the old village of Centerville. It was in operation during the 1880s and 1890s, and local lore has it that the proprietor had formerly been a jockey.

SWAYING BRIDGE. This was an unverified nickname of the wooden bridge outside Westchester Square that spanned Westchester Creek. Some old-timers averred it swayed according to the ebb and flow of Westchester Creek. See: Swains' Bridge.

SWEDISH OLD PEOPLE'S HOME. A 23-room mansion located at Lafayette Avenue and Revere Avenue that had been the property of the Brinsmades, Overings and Tallmages in the 19th century. Going back in history, the site had first been occupied by the Baxter family in pre-Revolutionary times. The mansion (presumably built in the 1830s) and ground were sold in 1920 to the Eastern Missionary Society, a Scandinavian religious organization for $30,000. The Swedish pensioners published their own newspaper *Afton Timman* (The Evening Hour) and engaged in horticulture and a little farming on the grounds. In 1925, the City staked off new streets and extended existing ones, and one - Lafayette Avenue - would run through the mansion. The Home was turned and moved so that it faced Lafayette Avenue instead of East Tremont Avenue, but in 1964, the Society moved out entirely and the mansion was razed.

SWISS CHURCH, The. This was the nickname for a Lutheran church on the northeastern corner of Elton Avenue and East 156th Street, with a Swiss pastor and many Swiss parishioners. The term was in use from the 1890s to 1918.

SWITCHER'S ROCK. This rocky point, called "Switcher's Rock," was known in Colonial days, extending into the Hudson River at approximately West 256th Street below Mount St. Vincent. Some local residents believed the name applied to railroad workers who tended switches and used the point for midday fishing, but the railroad was built many years after the name was in use. There is a possibility there may have been an early settler named Switcher, Schweizer or Switzer.

SYCAMORE STREET. This was the former name of Edson Avenue from East 222nd Street to Bartow Avenue. See: Bracken Avenue, South 19th Avenue.

SYLVAN ATHLETIC ASSOCIATION. This pleasure park and shooting range was located opposite Jefferson Place and Boston Road. It was purchased in 1892, when it had been known as the Morrisania Schutzen Park.

SYLVAN AVENUE. This was an earlier, and more descriptive name, of Walton Avenue from East 167th Street to East 176th Street. See: Butternut Street, Punnet Street.

SYLVAN PARK. In the early 1880s, this was a picnic grove, *bierhalle* and rifle range located at Park Avenue and East 167th Street in the village of Morrisania. Outdoor gymnastic contests on the parallel bars, rings and horses were regularly held there among the various German athletic clubs. In 1890, the park changed hands and became known as Kullmann's Park.

SYRACUSE AVENUE. This was the earlier name of Burke Avenue from Eastchester Road to Black Dog Brook (Bruner Avenue). See: Morris Street, Olinville Avenue.

TABER AVENUE. This was the earlier name of Brinsmade Avenue from Bruckner Boulevard to the Cross-Bronx Expressway when it was part of the Charlton Ferris farm. Augustus Taber owned 56 acres, mapped in 1893, along Westchester Creek back to present-day Bruckner Boulevard. Lorinda Taber is buried in Woodlawn Cemetery with Ferrises and Parsons, all of whom owned land once in that area.

TACKAMACK PLACE. In 1888, the Godwin estate was mapped out for future city streets, and this Place was platted from West 230th Street to West 23 1st Street. The name honored a local political club that had taken the name of an Indian chief. It has since been renamed Godwin Terrace.

TALLAPOOSA CLUB. This was a political and social club of Woodstock that was organized around 1879, with a clubhouse on East 161st Street and Tinton Avenue. Charles Praetorius was listed as the manager in an advertisement in the 1892 New York *Eagle* wherein he called attention to the extensive wine cellars. In 1888, the Parks Department acquired title to the lands of Pierre Lorillard and the southern end of the Pelham bridge, and leased the Lorillard mansion to the Tallapoosa Club for a stated number of years. The grounds were then used for clambakes, barbecues and picnics for members of the club and their guests, some of whom arrived by yacht and naphtha launch. The name is said to have originated in the years immediately following the Civil War, as some of the charter members had seen action in Tallapoosa, Georgia. Its long-term fame gave rise to the point of land acquiring the name of Tallapoosa Point. See: Tallapoosa Point.

TALLAPOOSA POINT. This was the name of a point of land, next to the Pelham bridge, now part of Pelham Bay Park, and which received its name from a famous clubhouse that stood there in the 1890s. Originally it was an island, owned in the early part of the 1600s by a Captain Richard Osbourne who willed it to his grandson, also named Richard Osbourne, in 1683. It was known as Dimans Island some years later, and at the time of the Revolutionary War it was in the possession of Jesse Hunt. As the island was accessible at low tides, it was sometimes charted as a peninsula, and on charts of 1780, it was marked "Hunt's Neck." It was next part of the Bayard farm that was purchased by John Hunter III, who sold this portion to Pierre Lorillard. In 1888, the Parks Department acquired title to the lands of Pierre Lorillard and the southern end of the Pelham bridge, and leased the Lorillard mansion to the Tallapoosa Club for a stated number of years. The Tallapoosa Club took over the Lorillard mansion for its summer quarters, until its lease was up, whereupon the Parks Department used the mansion for storage. See: Tallapoosa Club, Taylor's Island.

TALMADGE STREET. This was the original name of East 180th Street from Webster Avenue to Third Avenue. See: Samuel Street.

TAN PLACE. Quite a few reasons are given for the naming of this short street, overlaying the former course of Westchester Creek opposite Lehman High School. One, is that a tannery was once located there. Two, a boss teamster in charge of filling in the creek wanted to have his name on the Place, but Timothy A. Naughton had to be satisfied with just his initials on the street sign. A Bronx County Engineer said it was the abbreviation of Tangent, a very apt name for the Place that angled off East Tremont Avenue. It is now little League Place, alongside a ballfield near Westchester Square. It received its new name in November 1984 and was duly dedicated in June of the following year.

"TANGLEWOOD." A tract owned by the Kidd family of the 19th century, roughly bounded by Laconia and Allerton Avenues, Fish Avenue and Pelham Parkway. The name persisted until the 1920s, long after the estate had been subdivided into building lots.

TAPPEN STREET. This is the former name of East 195th Street from Marion Avenue down to Webster Avenue.

TAVERN LOT. An 1854 deed refers to a sale of land from Edwin Morgan to Charles Reed, near the gateway to the Hammond estate (now Silver Beach Gardens); said land containing a small stone cottage or lodge. The tavern referred to a small inn that was maintained in the 1700s and into the early 1800s as a stop-over for travelers using the Whitestone ferry. The ferry road and ferry slip were still visible in the 1970s.

TAYLOR AVENUE. In the village of Belmont, this was the former name of Prospect Avenue from Grote Street to East 189th Street. See: Eastern Avenue.

TAYLOR'S BRIDGE. It is mentioned in deeds, dated 1847 and 1852, concerning a wooden span south of the Pelham bridge leading onto Taylor's Island.

TAYLOR'S ISLAND. This small island near the mouth of the Hutchinson River at the southern end of the Pelham bridge had long been known by many other names. An 1851 map lists T. Taylor as the owner. See: Dimans Island, Tallapoosa Point.

TEE TAW AVENUE (1). This is the former name of Webb Avenue. This street runs through the former farm of Dominie Jean Pierre Tetard of the late 1700s.

TEE TAW AVENUE (2). This was the earlier name of Devoe Terrace, and is a corruption of the name Tetard, a French-Swiss minister who farmed on that hillside in Revolutionary War times.

TELLER PLACE. On an 1888 map, this was a short thoroughfare extending from the Melrose depot to Courtlandt Avenue. It is now absorbed by Melrose Park.

TEMPLE ADATH ISRAEL. According to the American Jewish Yearbook of 1907-1908, Communal Register, this was the first temple in The Bronx, being organized in 1889. It was located on East 169th Street just east of Third Avenue.

TEMPLE HAND-IN-HAND. Believed to be the first Bronx synagogue, according to Randall Comfort, eminent historian. It was established first as a religious school around 1885 by Wilhelm Daub, superintendent of Lebanon hospital, and then organized as a Temple in 1895. The synagogue was built on East 145th Street between Brook Avenue and Willis Avenue, and Mr. Daub lived on the same street in a private house.

TEN FARMS, The. In 1665, ten settlers were granted land along the Hutchinson River by Thomas Pell. The land later became known as Eastchester, as it lay to the eastward of Westchester. Part of this grant is now under Co-Op City.

TENTH AVENUE. This is the former name of East 224th Street from the Bronx River to Bronxwood Avenue in Wakefield.

TENTH STREET (1). In Port Morris, this was the earlier name of East 132nd Street from Willow Avenue to the East River.

TENTH STREET (2). This is the former name of East 172nd Street from Webster Avenue (the Mill Brook) to Crotona Park (the Bathgate farm).

TENTH STREET (3). In Unionport, this was the original name of Haviland Avenue.

TERRACE PLACE (1). This street once led up and over a rocky ledge to a plateau overlooking St. Mary's Park. It was listed in the 1871 Directory as encircling "View Mount,"a wooded estate shared by several fine homes. This rocky elevation was blasted down to street-level in the 1950s to serve as a foundation for St. Mary's Park Houses.

TERRACE PLACE (2). This narrow alley once ran north from Pontiac Place at approximately 208 Pontiac Place. Around 1898, it lost its identity and was used as a backyard for the apartment house at 620 Trinity Avenue.

TERRACE PLACE (3). This is the former name of Park Avenue from East 148th Street to East 156th Street in Melrose South.

THIRD AVENUE (1). In Port Morris, this was the early name of Locust Avenue from East 132nd Street to East 141st Street.

THIRD AVENUE (2). This is the former name of Selwyn Avenue from Mt. Eden Avenue to the Concourse, a distance of four blocks.

THIRD AVENUE (3). In Bedford Park, the earlier name of Pond Place. See: Ursula Place.

THIRD AVENUE (4). In Wakefield, this was the original name of East 217th Street.

THIRD STREET (1). In Port Morris, this was the former name of East 139th Street from Bruckner Boulevard to the East River. See: Van Corlaer Street.

THIRD STREET (2). This street was laid out in the late 1840s in the village of Morrisania, but has since taken the number of East 165th Street. See: Bancroft Street, Guttenberg Street, Wall Street.

THIRD STREET (3). This was the earlier name of East 183rd Street from Jerome Avenue up to the Concourse. See: Columbine Avenue.

THIRD STREET (4). In Unionport, this is the former name of Story Avenue.

THIRD STREET (5). This was St. Raymond Avenue's earlier name from Zerega Avenue to East Tremont Avenue. See: Evadna Street.

THIRD STREET (6). This was the original name of Coddington Street from Edison Avenue to Crosby Avenue. See: Prospect Avenue.

THIRD STREET (7). This street once ran from Bartow station to the Shore Road, but is now incorporated into Pelham Bay Park and is part of Bartow Circle.

THIRD STREET (8). The former name of White Plains Road north of Gun Hill Road. See: Washington Street (2).

THIRD STREET (9). This was the original name of Kepler Avenue from Woodlawn Cemetery to the city line. See: Quail Avenue.

THIRTEENTH AVENUE. In Wakefield, this is the former name of East 227th Street.

THIRTEENTH STREET. This was the original name of Ellis Avenue in Unionport.

THOMAS STREET (1). As this was the former name of Kirk Street, it might be sheer coincidence that an early settler of "ye Burrow of Westchester in ye Collony of New York in 1699" was a Thomas Kyrke. The Thomas family were early residents of Westchester, and Charity Thomas married James Ferris before the Revolution.

THOMAS STREET (2). The earlier name of Webster Avenue from East Tremont Avenue north to Fordham Road. In the 1830s, this street was the western boundary of Thomas Bassford's 44-acre farm. See: Berrian Avenue, Bronx River Road, Worth Avenue.

THORN'S BASIN. From the early 1880s to the early 1900s, Thomas and William Thorn's fuel and feed yard was located on Spuyten Duyvil Creek west of present-day Kingsbridge Avenue and West 230th Street. Thorn's Basin was a large pond deep enough for tugs and barges, and was at the junction of the creek and Tibbett's Brook. The area is now filled in, and the Basin would be in the vicinity of Irwin Avenue.

THROCKMORTON'S NECK. John Throckmorton, Englishman, was permitted to settle his colony in what was Dutch territory in 1642. The Indians called the peninsula Quinshung. In the deed, the name is spelled Jan Trockmorten. Over the ensuing centuries, the name underwent changes, among them, Throgmorton's Neck, to Throggs Neck. and others.

THROGG'S NECK DIVISION. This volunteer lifeguard station was situated at the foot of Layton Avenue on Eastchester Bay. This had been the 19thcentury town dock of Schuylerville. In the 1920s, this station was manned by five officers, fourteen surfmen and two coxswains. Charles Klippel was Commander in 1914, and the unit was based on Klippel's Beach.

THROGMORTON'S NECK. See: Throckmorton's Neck.

THROGSNECK. This is another variation of Throggs Neck.

TIEMAN AVENUE. This is the former name of Kingsland Avenue from Adee Avenue to Givan Avenue. See: Cedar Street.

TILDEN AVENUE. In Williamsbridge, this was the earlier name of Laconia Avenue from Gun Hill Road to East 222nd Street. It is a reminder of the Blodgett-Tilden estate, mapped in 1902, owned by William Blodgett and Marmaduke Tilden.

TILLOTSON AVENUE. This short street is now incorporated into the Botanical Garden west of the Bronx River.

TILLY'S ROCK. A large boulder on the easternmost ledge off Hunter Island was known by this name for many years by the hardy campers. The origin of the name is obscure, and it might be named after someone named Matilda. A second theory is that it served as a navigational marker for old-time sailors to take tiller and change course. They called it Tiller Rock.

TIME THEATRE, The. It was listed in the 1912 Atlas as a silent movie house at 786 Courtlandt Avenue near East 157th Street. It was a brick building.

TIMPSON AVENUE. On a 1905 map, this was the former name of Cruger Avenue from Adee Avenue to North Oak Drive. Timpson, Adee & Co. was a well-known firm of auctioneers of the 1870s, and Edward and William Timpson were early landowners in the area. See: Brown Avenue, Magenta Place.

TINTON AVENUE. This is the original name of Wales Avenue from East 142nd Street to East 149th Street.

TIPPETT'S HILL. Prior to the Revolution, Spuyten Duyvil Hill was called Tippett's Hill. See: Berrian's Neck, Konstabelsche Hoek, Shorakapkock, Tippett's Neck.

TIPPETT'S NECK. This was the 17th-century name of Spuyten Duyvil Hill.

TOBIN'S UNIVERSITY. This large tract of land at East 138th Street east of St. Ann's Avenue was a well-known ballfield of the 1880s and 1890s. The name was a subterfuge to enable the local boys to play against college baseball nines. "Professor" Tobin's ballfield graduated a goodly number of big league ballplayers who received their early training there. One old-time story is that Mr. Robin allowed visiting collegiates to think nearby Lincoln hospital was the "University."

TOM PELL'S POINT. In October, 1776, General Sir Henry Clinton wrote in his journal that one of his frigates had been ordered to Tom Pell's Point. This is a variant of Pell's Point - the present-day Rodman's Neck. See: Ann's Hook.

TOMPKINS STREET. This is the former name of Underhill Avenue on Clason Point.

TONE'S LANE. On an 1870 map, Tone's Lane is shown running from the present-day eastern end of Gun Hill Road to the Shore Road down to Eastchester Bay, but it has long since been eliminated. Thomas Tone had some holdings west of the Shore Road.

TOWN DOCK ROAD. This was the 19th-century name of Layton Avenue.

TRACK RESTAURANT. A tavern that stood on the northwestern corner of Williamsbridge Road and Eastchester Road catering to the racing fans who attended the Morris Park racetrack in the 1890s. The spacious house had been the earlier home of Captain Charles Coleman, a sea captain of the 1850s. The Track Restaurant was demolished in 1959; this writer saved the outdoor sign for The Bronx County Historical Society.

TRAMPS' ROCK. In the 1880s, mention was made of this rock alongside Boston Road concerning the disappearance of George Leonidas Leslie, alias "Western George," bank robber. The law firm of Howe & Hummel defended him. Later on, his decomposed corpse was found on Tramps' Rock. It may well be the Pudding Rock of an earlier age, which was frequented by vagrants according to newspaper items. See: Pudding Rock.

TRANSVERSE ROAD. See: East Burnside Avenue.

TRASK ESTATE. This was an 80-acre tract belonging to Benjamin Trask in three separate parcels on Clason Point. They extended from the Bronx River to Clason Point Road, and then eastward from Westchester Avenue, according to an 1886 map.

TRAVERS STREET. This is the former name of East 198th Street from the Concourse to Webster Avenue. Maria L. Travers was the owner of this property in 1887, being related to the Jerome family who possessed much of the real estate in this region.

TREATY OAK. This was the tree under which the original proprietor, Thomas Pell, bought his lands from the Indian sachems. It stood in front of the Pell-Bartow mansion, but was destroyed in 1906 by fire. A circular iron fence remains to mark the spot, and a new tree has been flourishing in the plot.

TREMONT. Originally called Upper Morrisania, the new name was chosen by Postmaster Tarbox because three mounts were in the town area: Mount Hope, Mount Eden and Fairmount. This is the most widely accepted motive, but there is a story that John Metcalf, a resident of Upper Morrisania and a manufacturer of fire hose, was an ex-Bostonian and suggested the town be named after Boston's main street.

TREMONT HALL. This was a large, ornate social hall, with restaurant attached, on the southwestern corner of Park Avenue and Tremont Avenue, *circa* 1895.

TREMONT HEIGHTS. This was a modest real estate development, according to 1905 maps, along East Tremont Avenue on Throggs Neck. It was bounded by the aforementioned avenue, Roosevelt Avenue, Schley and Calhoun Avenues. Benjamin Lamport was one of the financial backers.

TREMONT ROAD. This is a former name of Roberts Avenue from Crosby Avenue to Bruckner Boulevard, a distance of five blocks. It was so-called because it traversed "Tremont Terrace," a real estate venture developed from the Mapes farm and Ferris property of the 1920s. See: Emily Street.

TREMONT, The. Around 1916, this silent movie house advertised in The Bronx *Home News*. It was located on Webster Avenue off East 178th Street.

TREMPER'S POND. This was the name of Van Cortlandt Lake from 1870 to 1900 when George R. Tremper had an ice business there. Along the eastern and southern banks he had storage houses, as well as his own mansion. Shortly after the Civil War, he organized the Mosholu Division of the Temperance Society, and contributed to a Temperance Hall at Broadway near West 240th Street. Later, it became Beck's Hotel. See: Beck's Hotel.

TRINITY AVENUE (1). In Port Morris, this is the former name of Cypress Avenue from the Bronx Kill to St. Mary's Park.

TRINITY AVENUE (2). This was the earlier name of East 147th Street from St. Mary's Park to Southern Boulevard. See: Dater Avenue, Lexington Street.

TRINITY HALL. This beer garden and saloon on Trinity Avenue and East 156th Street was owned by Adam Hoffmann around 1904 to 1910. It was popular with residents of Woodstock, Bensonia and East Morrisania who organized singing societies, rifle clubs and bowling teams. See: Hoffmann's Park.

TROJAN FIELD. Bounded by Unionport Road, Bronx River Parkway and Fordham Road, its name was derived from the Trojans ball club which set up the field in the early 1930s. It was also called Pelham Field. Later, the Parks Department and the W.P.A. turned the open field into a park for the city of New York.

TROWBRIDGE AVENUE. In 1874, the Rogers estate, "Sunnyside," was to be subdivided into suburban plots and called "Fairmount-on-Sound." One of the projected avenues was to be called Trowbridge Avenue and, according to the map, would have cut across Split Rock golf course and the Shore Road, to end at the shoreline behind Hunter Island. Before the plan could be implemented, the Parks Department acquired the estate and added it to Pelham Bay Park. See: Rogers' Islands.

TROY STREET. When the Johnson Iron Works on Spuyten Duyvil expanded in the 1890s, this short street was eliminated and rows of Workmen's Boardinghouses had to be razed. It owed its name to the Bank of Troy, N.Y. that financed some Johnson transactions.

TRYON ROW. This was the former name of Seddon Street in 1895.

TURKEY BAY. This was a local name for the bay once situated between Ferry Point and the Morris estate, but now covered over by Ferry Point Park. The official name at that time (*circa* 1890) was Morris Cove, but the other name was used by residents of Westchester village. See: Morris Cove, Ox Meadow Cove, Ox Pasture Cove.

TURNBULL CREEK. This was noted in 1850 as the former name of Weir Creek from Layton Avenue to Long Island Sound, where it flowed through the Robert Turnbull estate of 50 acres. Turnbull Creek persisted on the Beers' *Atlas* of 1868, and Bromley's map of 1897. See: Middle Brook (2).

TURNER'S FARM. Also designated as the Quarry Farm, it was shown on an 1868 map as extending from the Harlem River to approximately Aqueduct Avenue. Daniel Turneur purchased 80 acres from the Indians in 1671 above what is now High Bridge. His descendants retained the farm for over two centuries, but changed the spelling of his name. See: Turneur's Land.

TURNEUR'S LAND. The Tourneurs had fled their native France and settled in Holland in 1650. From there, the family migrated to the New World and settled in Nieuw Haarlem in 1658. Their original French name was slightly changed in the transition. Daniel Turneur purchased the land called Nuasin from the Indians, between the Harlem River and Maenippis Kill (Cromwell Avenue). See: Bickley Patent, Devoe's Neck, Highbridgeville, Nuasin, Quarry Farm, Turner's Farm.

TURTLE BROOK. This is evidently a very early name for Tibbett's Brook that flows through Van Cortlandt Park, for in the Wills and Testaments of early Westchester we find: "Frederick Van Cortlandt of the Little Yonckers, Infirm and weake, my body to be buried in a family vault which I intend to build on my Plantation on the little hill northeastward of Turtle Brook." The will is dated 1749.

TURTLE CREEK. This creek formerly wound down through Ludlow's "Black Rock farm" entering the Bronx River at Story Avenue. Its silt formed Sedge Island. See: Sedge Island.

TUXEDO STREET. Once part of the Denton Pearsall estate called "Woodmansten,"this was later the Ferris property on Eastchester Road of the 1870s. It was developed into a residential area around 1905 called "Westchester Heights" and the streets were given names of Summer resorts, such as Saratoga, Narragansett and Newport. Tuxedo Street was laid out, but never cut through, and is now covered by the grounds of Jacobi Hospital.

TWELFTH AVENUE. This is the former name of East 226th Street in Wakefield.

TWELFTH STREET (1). This was an early name of East 174th Street from Webster Avenue to Third Avenue. See: Beacon Avenue, Spring Street.

TWELFTH STREET (2). This is the former name of Gleason Avenue to Unionport.

TWELVE FARMS, The. This was the original patent of West Farms of March, 1663, that was subdivided into twelve farms. As the patent lay west of Westchester, it was commonly called the West Farms.

TWENTIETH AVENUE. In Wakefield, the 19th-century name of East 234th Street.

TWENTY-FIRST AVENUE. This is the former name of East 235th Street in Wakefield.

TWENTY-FOURTH WARD (1). This political unit was created in 1874 out of the area west of the Bronx River above 170th Street, and included the villages of Fordham, Kingsbridge, Tremont, Spuyten Duyvil and Woodlawn Heights. See: Annexed District.

TWENTY-FOURTH WARD (2). This was the political designation of that part of the Bronx which was annexed to New York City in 1895. It became a part of the Twenty-fourth Ward annexed in 1874, and took in the territory east of the Bronx River, and the villages included City Island, Schuylerville, Stinardtown, Unionport, Wakefield, Westchester and Williamsbridge.

TWENTY-SECOND AVENUE. This is the former name of East 236th Street from the Bronx River to White Plains Road.

TWENTY-THIRD WARD FIRE COMPANY. This was a volunteer Engine Company that was strategically located on Third Avenue and East 147th Street so that it could service Melrose, Mott Haven and North New York. It was disbanded in 1874 when the district was annexed to New York City.

TWENTY-THIRD WARD. This was the political designation of that part of The Bronx annexed to New York City in January, 1874. The bounds were the Harlem River to the Bronx River, south of 170th Street. The villages affected were Highbridgeville, Hunt's Point, Melrose, Mott Haven, Morrisania, Port Morris, West Farms and Wilton. See: Annexed District, North Side.

TWENTY-THIRD WARD PARK. This picnic glade, dance pavilion and amusement park was under the ownership of Jacob Cohen in the 1880s and was located at the extreme northern end of Old Morrisania (East 148th Street and Willis Avenue). Each year, Mr. Cohen gave three days' receipts to charity. When his daughter was married, the newspapers carried accounts of tropical blooms lavishly embellishing the dance pavilion, and the charming aspect of an Oriental innovation - Chinese lanterns. See: Fountain Park.

TWIGGS PLACE. This short Place was surveyed and planned to extend from Shore Drive to Longstreet Avenue, but it is not yet in existence. Major David Emanuel Twiggs (1790-1862) of the Mexican War may be honored some day.

TWO BROTHERS. A map of Morrisania in 1874 gives this name to North Brother and South Brother Islands. See: North Brother Island, South Brother Island.

TWO-TREE ISLAND. This was a local nickname for Catwillow Island, due east of East Twin Island, *circa* 1915. See: One-Tree Island.

TYNBURY AVENUE. This was the former name of Huntington Avenue from Bruckner Boulevard to Lafayette Avenue, according to an 1888 map.

U. THANT PARK. The State and City of New York joined forces and purchased the acreage of the former Douglas-U. Thant estate at Palisade Avenue north of 232nd Street. Located on the slope overlooking the Hudson River, the estate has reverted to a mix of meadow and woodland. It is named for the former head of the United Nations. Later renamed Raoul Wallenberg Park.

UNCAS RIVER. This is a former name of Tibbett's Brook, and is thought to be derived from "Yonkers." It is mentioned in John Adams' diary. See: Mosholu, Yonkers River.

UNCAS STREET. Spelled "Ungas St." on some maps, this short street of East Morrisania was named in honor of the Indian "Last of the Mohicans." It was the former name of East 150th Street from St. Mary's Park to Southern Boulevard. See: Fox Street (1), Ungas Street.

UNDERCLIFF HOTEL. This fashionable hotel with an impressive carriage driveway, south of High Bridge, was in operation from the 1870s to the 1890s. Its present-day location would be Undercliff Avenue and West 170th Street.

UNDERCLIFF PLACE. This Place was surveyed by George Hollerith in 1896, although records indicate it was already named in 1889 while still in the planning stage. It was a topographical name. The Place was eliminated by the massive cloverleaf connecting the Cross-Bronx Expressway with the Major Deegan Expressway and by the approach to the Alexander Hamilton Bridge.

UNGAS STREET. This street was cut through the former Dater estate of East Morrisania and given the name of the Last of the Mohicans.

UNION AVENUE (1). This was the former name of Fordham Road from Third Avenue to the Bronx River, according to Beers' Atlas of 1868. See: Cammann Street (2), Highbridge Road, Pelham Avenue.

UNION AVENUE (2). This was the 19th-century name of St. Peter's Avenue from Westchester Avenue to East Tremont Avenue. At one time (1890s), it formed the western boundary of William T. Adee's estate.

UNION BASEBALL GROUNDS. This athletic field of the 1880s was near the corner of East Tremont Avenue and Third Avenue in Crotona Park.

UNION HALL. An 1867 reference to this social center locates it on Courtlandt Avenue and Mary Street (East 155th Street) when Hans Kasschau was the proprietor.

UNION HILL. It is mentioned in 1822 and 1826 deeds as being located on the Cox farm. Surveyed still earlier in May, 1811, by Robert Findlay, it was defined as west of the Sherborn farm (Webster Avenue to Third Avenue, East 178th Street up to Fordham Road). This would then place Union Hill at Fordham Road and the Concourse - known as Mount Sharon on later maps.

UNION LANE. This is the former name of East 162nd Street from Brook Avenue to Third Avenue, a distance of one block. See: Grove Street, Halsey Street, Union Street (1).

UNION PARK. New York *Eagle* advertisements of 1889 and 1892 place this picnic park on Southern Boulevard and East 147th Street. See: Brommer's Union Park, Hartung's Park.

UNION PLACE. It was once part of Union Street, so named because it connected the center of the village of Highbridgeville with land to the east. It was eliminated when Highbridge House was built.

UNION STREET (1). This is the former name of East 162nd Street from Park Avenue to Brook Avenue. From there to Third Avenue, it became Union Lane.

UNION STREET (2). This was the earlier name of West 167th Street from Ogden Avenue east to Jerome Avenue. See: Beach Street, Wolf Street, Woolf Street.

UNION STREET (3). This is the former name of Arnow Avenue.

UNIONPORT. This was a separate village until it was incorporated into New York City in 1895. When it was laid out in 1851, all streets were numbered, and all avenues lettered from A to E. Upon annexation, confusion with already existing streets and avenues in New York City compelled the town fathers of Unionport to give names of early settlers to their thoroughfares.

UNIONPORT HOTEL. This inn was owned and operated by Martin Hoffmann at Avenue B (Havemeyer and Haviland Avenues), and was part of the property known as Hoffmann's Casino. Every year he placed an ad in the German-language Odd Fellows' Journal. See: Huber's Casino.

UNITED STATES, The. This large silent movie house was built in 1916 on Webster Avenue and East 195th Street. For many years, it was billed as the longest theatre in The Bronx, and had a seating capacity of 1,627. The name was changed to The Decatur in 1946, and closed some ten years later. The building became a warehouse.

UNIVERSITY AVENUE. See: Hall of Fame Terrace.

UNIVERSITY HEIGHTS EAST. This thoroughfare once ran across the campus of New York University on a line with present-day Andrews Avenue. No date is given on the New York University map, but it is a fair guess the year would be around 1895. This road is now incorporated inside the grounds of Bronx Community College, but remnants of paving could be seen as late as 1970.

UNIVERSITY, The. This old-time silent movie house was built in 1920 at 33 West Fordham Road. It had a seating capacity of 1,252.

"UNTER DEN LINDEN." Despite its German name, "Under the Lindens," this saloon and picnic grove catered to all nationalities. It was located on the corner of Westchester Avenue and Bruckner Boulevard at the end of the subway line, and was a popular spot for crowds returning from City Island or Orchard Beach. The picnic grounds were built upon around 1950, and the saloon was demolished in 1959.

UPPER CORTLANDT. The area in the vicinity of West 234th Street overlooking the Hudson River was so called to distinguish it from the Van Cortlandt holdings lower down in what is now the park. The name occurred frequently in Revolutionary journals. See: Fieldston, Lower Cortlandt, Van Cortlandt Ridge.

UPPER MORRISANIA. This land was sold by Gouverneur Morris II and Charles Bathgate to home owners in 1848. Its name was changed to Tremont upon the 1874 annexation to New York City.

UPPER VILLAGE HOTEL. In 1854, R. H. Thompson advertised that his hostelry had accommodations for permanent or transient guests; horses were boarded and carriages and wagons were to let. His inn was at the West Farms depot.

URBACH'S HOTEL. In the 1890s, this hotel was on Third Avenue at East 170th Street.

URSULA PLACE. This is the former name of Pond Place and was so mapped in 1888 when the Ursuline Academy was built nearby. See: Third Avenue (3).

URSULINE CONVENT. The Ursuline Sisters purchased some rocky acreage (Westchester Avenue and Cauldwell Avenue) from Henry Purdy around 1850. A convent was built atop the high bluff and the Sisters taught there until 1888, when the staff and students moved to the present location in Bedford Park. See: Bronx Garment Center, Lebanon Hospital.

VAGABOND BAY. The name of this inlet, between Hunter Island and the mainland, was in use during the early 1900s, but is not found on any official maps. It was probably named by the beachcombers who camped on Hunter Island before it was joined to the mainland. See: Roosevelt Cove, Shoal Harbor, Vagabond Island.

VAGABOND ISLAND. This was a small island, accessible at low tide, that was directly east of the Bartow mansion, facing the Orchard Beach parking lot (formerly LeRoy Bay). Although the name does not occur in any records, it was in use prior to World War I, and is strictly a localism used by the beachcombers and canoeists who used to camp on the little island. When the Olympic tryouts for racing shells were held in 1964, the lagoon behind Hunter Island was widened, and most of Vagabond Island was dredged away. The western part remaining is now part of the mainland.

VALE STREET. This was a street that formerly ran from River Avenue to Gerard Avenue north of East 162nd Street, but was eliminated by Mullaly Park.

VALENTINE AVENUE (1). This was the former name of Lodovick Avenue from Pelham Parkway up to Gun Hill Road. The Valentine lands were located on Eastchester Road and extended to Givan's Creek (Demeyer Street) on an 1891 map.

VALENTINE AVENUE (2). Originally this was the name of Throgmorton Avenue, until it was superseded in 1916. The Valentine family was an influential one in Westchester village, and many family tombstones can be seen in St. Peter's Episcopal churchyard.

VALENTINE STREET. See: Roberts Avenue.

VALENTINE'S BROOK. This stream originated in a spring on the Valentine farm, which was north of Fordham Road and west of St. James Park. The northern rim of Devoe

Park shows the curve of this ancient brook that flowed southwestward into the Harlem River near Landing Road. See: Valentine's Spring, West 190th Street.

VALENTINE'S SPRING. Mentioned in early deeds, this spring was located on the Valentine farm near Grand Avenue and East 190th Street. In his 1893 book on *Springs and Wells of New York and the Bronx*, James Reuel Smith mentioned it flowed copiously, and was sheltered by a spring house. See: Valentine's Brook.

VAN CORLEAR STREET. The former name of East 139th Street, reminding residents of Arent Van Corlear who had bought the land from Jonas Bronck's widow in 1644. He then sold it to Jacobus Van Stohl, or Van Stoll, in 1651. See: Third Street (1).

VAN CORTLANDT PARK HOTEL. John Mueller was the proprietor of this hotel on Jerome Avenue near Woodlawn Cemetery before 1900.

VAN CORTLANDT RIDGE. The name in use during the Revolutionary War for the area of West 232nd Street along Riverdale Avenue. The land was sold by the Van Cortlandt family to J. R. Whiting in 1836, and in 1892, it was bought by the Sisters of Charity who founded Seton Hospital. See: Cowboy Tree, Upper Cortlandt.

VAN CORTLANDT STREET. This was the 1907 name of Bantam Place in Eastchester. No amount of research has ever unearthed any connection with the Van Cortlandt family whose holdings were in the West Bronx. See: Grove Street.

VAN NEST HOSE COMPANY NO. 1. This was a volunteer Fire Company on the corner of Van Nest Avenue and Unionport Road. It was disbanded in 1910 with a colorful parade that stretched for over a mile. All civic, sport and social organizations turned out in force, and the affair was well-covered by reporters from the *North Side News*. See: Van Nest Hose Company No. 2.

VAN NEST HOSE COMPANY NO. 2. This was a volunteer Fire Company that was housed at East 177th Street and Rosedale Avenue. It was disbanded in 1910. See: Van Nest Hose Company No. 1.

VAN STOHL STREET. This was the former name of East 138th Street, and it honored Jacobus Van Stohl (or Van Stoll) who owned the land from 1651 to 1662 when he sold out to a man named Hendricks.

VAN WYCK AVENUE EAST. According to a 1926 City map, this avenue was to run from West 198th Street to Bedford Park Boulevard, but was covered instead by the Jerome Avenue subway yards. The avenue was to honor Robert A. Van Wyck, then mayor, who turned the first spadeful of earth inaugurating construction of New York City's first subway on March 24, 1900.

VAN WYCK AVENUE WEST. Mapped in 1926, this avenue was to run from West 198th Street to Bedford Park Boulevard, but instead was covered by the Jerome Avenue subway yards. See: Van Wyck Avenue East.

VAN WYCK'S NECK. A sale by Peter Remsen to Joseph Leggett in January 1815 names it "on the south by the road to Van Wyck's Neck and the sheep pasture." The sheep pasture became Unionport, and most of the Leggett farm became today's Parkchester

VANDERBILT AVENUE. On maps, this was the 19th-century name of Park Avenue from East 161st Street to East 167th Street at the time of the building of the railroad. Frederick William Vanderbilt was Director of the New York Central System of Railroads. Below East 161st Street, the street was known as Railroad Avenue, and the name was good enough for the citizens of Melrose, but it was said the people of Morrisania objected to Railroad Avenue, and wanted a more dignified name.

VARIAN AVENUE. In 1792, Isaac Varian purchased 260 acres in The Bronx from the Dutch Reformed Church, north of Fordham Road and extending from Jerome Avenue to the Bronx River. His son, Isaac, became a Mayor of New York in 1839-1841. Another son, Michael, became a prominent realtor of the mid-1800s, and is believed to have named Martha Avenue after his wife. A baseball field and trees are now in the vicinity of this once-mapped street, but an outline of its old route may still be seen from the air.

VARIAN STREET. This was the former name of West 234th Street. Dr. William Varian was an old-time Kingsbridge physician who lived on the corner of West 234th Street and Broadway. He started his practice in 1850, and retired around 1896.

VASA HALL. This large and commodious social center was particularly patronized by Scandinavians from the early 1900s to the late 1950s. Nordic festivals such as St. Lucia's Day, May parties and Christmastime "Gloggfests" were held there for many years. The Hall was situated on East 149th Street, west of the Concourse, and had smaller rooms for wedding receptions, Masonic meetings, banquets and private parties. Like all Bronx institutions, it was not confined to one ethnic group, for German masquerade balls and Greek festivals were held there, too. The name Vasa (or Vaasa) is prominent in Swedish chronicles.

VIA SANCTORUM ANGELORUM. During a critical need for water, a spring was found on the grounds of Mount St. Vincent on October 2nd, 1867. This date happened to be the Feast of the Guardian Angels, so the discovery was named the Angels' Spring, by the Sisters of Charity. As it flowed down West 261st Street to the Hudson River, the roadway was called the Avenue of the Holy Angels. In their bulletins, the Sisters used the Latin term. See: Randolph's Lane.

VICKERY LANE. This was a lane that once ran from City Island Avenue at Bay Street to the southern end of the cemetery. It is now eliminated. William Vickery was noted on records of 1866.

VICTOR, The. A movie house. See: Ideal, The.

VICTORY PARK. The northwestern corner of Crotona Park received this name during World War I days, and it lasted until World War II. The unofficial bounds were East Tremont Avenue, Arthur Avenue, East 175th Street to Third Avenue.

VIETOR PLACE. At one time, West 193rd Street carried this name. George F. Vietor was an importer who lived on Bailey Avenue and West 193rd Street around 1900.

"VIEW MOUNT." Listed in 1871, it was a shared property bounded by Westchester, Eagle and Trinity Avenues and St. Mary's Park. It was a rocky plateau occupied by several fine homes. In the 1950s, the mansions were razed and the rocks blasted down to street-level to make way for St. Mary's Park Houses.

VIGILANT HOSE COMPANY NUMBER 1. The 1864 Annual Report lists this volunteer hose company as being located on Elton Street near Boston Road (East 152nd Street and Third Avenue) in a two-storied frame house.

"VILLA BOSCOBEL." The former estate of William B. Ogden that was mapped in 1860, but is now occupied by Sedgwick Houses. In 1884-1885, Louis A. Risse surveyed and laid out the land for city streets. Mr. Risse later designed the Grand Concourse.

VILLA PLACE. This was the former name of East 145th Street from College Avenue to Third Avenue in Mott Haven. It was named after the villa of Charles Van Doren, legal counsel of Jordan L. Mott, and was situated on the corner of this Place and Third Avenue. He was an Abolitionist, and the villa was one of the "stations" of the Underground Railroad for escaping slaves.

"VILLA ROSA." An 1882 map shows property so called at present-day West 256th Street and Independence Avenue.

"VILLA SITES." This was a small development that was begun in 1872 by Messrs. Greene, Owens and Gelston on a tract of land purchased from John B. Stevens. It was bounded by Wissman Avenue, Meagher Avenue, Longstreet Avenue, and Miles Avenue. See: Greene Avenue, Stevens Place.

VILLAGE, The (1). This was a localism for present-day Westchester Square dating back to the days when it was the village of Westchester.

VILLAGE, The (2). This was the everyday name for present-day West 167th Street, Ogden Avenue and Union Place, dating back to 1850-1890 when it was the center of the village of Highbridgeville.

VILLAGE, The (3). This was an old-time silent movie house that was located on Williamsbridge Road and East Tremont Avenue in the years before 1915. See: The Select.

VINCENT AVENUE. In the village of Middletown, the former name of Merry Avenue.

VINEYARD BATHING BEACH. This was a small resort on the Harlem River at East 151st Street that was popular in the 1880s and 1890s. It featured a raft in midstream and, ashore, there were children's playgrounds, picnic tables, clothes lockers and an oyster bar. An open pavilion was provided for dancing, and the clientele consisted mainly of Melrose and Mott Haven residents. See: Ferncliff Place, Waldorf Place.

VIRGINIA AVENUE. This was the former name of Marion Avenue from West 184th Street to Fordham Road. See: Bainbridge Street, Hull Street.

VOELKER'S SCHUETZENPARK. Few people recall that this well-known German picnic grounds was located on the Bronx River below Gun Hill Road. It lay between L'Hermitage and "French Charley's" in the early 1900s, and was served by two roads, Post Avenue and Newell Street. "Schuetzenpark" is a German word denoting a park used for shooting matches. See: Williamsbridge Schuetzenpark.

VOGUE THEATRE. Originally the Daly Theatre, a silent movie house on East Tremont and Honeywell Avenues which was in operation during the 1920s.

VOLUNTEER ENGINE COMPANY NUMBER 1. See: Lady Washington Hose Company.

VON HUMBOLDT AVENUE. The earlier name of Delafield Avenue in Riverdale. As the name-change took place in 1916, it was most likely due to strong anti-German feeling during World War I. See: Field Street.

VREEDLANDT STREET. This street was plotted in 1897 across the former Seton estate alongside Westchester Creek and downstream from Westchester Square. This area was originally Dutch territory called "Vriedelandt" (Land of Peace) by the Dutch in the 1600s. By a happy coincidence, Enoch Vreeland was one of the trustees of the real estate firm that subdivided the Seton estate, and it took but a slight change to Vreeland Avenue from Vreedlandt Street to please him, and still appease the historically-minded people who knew of Vriedelandt. See: Vriedelandt.

VRIEDELANDT. "The Land of Peace" was so designated by the Dutch authorities in the 1620s, and it was bounded by Westchester Creek, the East River, Long Island Sound, Hutchinson River and Givan's Creek (which is now covered by Co-Op City's lower portion). In 1642, this vast peninsula was taken over by John Throgmorton, to be known henceforth as Throggs Neck.

WADDINGTON POINT. Joshua Waddington was listed in 1800 as a New York merchant. An 1809 map lists this old name for Barretto Point at Truxton Street and the East River.

WADING PLACE, The. This was an ancient Indian passage from Manna-hatta to the island of Paparinemin (Kingsbridge) which were separated by Spuyten Duyvil Creek. The approximate location of this wading place was West 230th Street, slightly west of Broadway at the foot of Godwin Terrace. The English settlers called it the Ford, and from it was derived "Fordham." Gradually, a sandbar built up, and in the ensuing centuries, various landowners cultivated it, planted shrubbery and trees and it became a verdant island. It was known as Lent's Island in 1785. Abraham Lent, Jr, was one of three men who bought the confiscated Phillipsburgh Manorlands in 1785. He settled in the village of Kingsbridge, and gave his name to this narrow island in the center of Spuyten Duyvil Creek. He sold his holdings, which included the island, to Alexander Macomb in 1795. When Alexander Macomb purchased what was the entire area of Kingsbridge, it too, included in the sale of the narrow island. which was connected to both Manhattan and (what became) The Bronx by a wooden bridge. When Spuyten Duyvil Creek was filled in at West 230th Street, Lent-McComb-Godwin Island disappeared under landfill, and is now under Marble Hill Houses.

WAGLER'S BATHING BEACH. This was a small resort with lockers, refreshment stands and a picnic area that was located on Eastchester Bay in the 1930s and 1940s. It was situated at the foot of Philip Avenue.

WAGNER'S SCHUETZENPARK. Located on the southeastern corner of Castle Hill Avenue and Westchester Avenue from 1902 to about 1916, this German-American picnic park featured bowling alleys, a rifle range and a dance pavilion. Joseph Wagner was the proprietor. The park lost its entire front lawn when Castle Hill Avenue was widened.

WAKEFIELD. This village, named after the birthplace of George Washington, was surveyed in 1853. Its bounds were from present-day East 215th Street to East 233rd

Street, from the Bronx River to approximately Laconia Avenue. Part of Westchester County until June, 1895, it was annexed to New York City. Its northern boundary was then extended to absorb Jacksonville (East 233rd Street to East 238th Street) and then to absorb Washingtonville (East 238th Street to East 243rd Street).

WAKEFIELD CASINO. This was a well-known tavern at East 239th Street and White Plains Road that was owned by Joe Furst for many years before World War II. An adjoining glade was used for picnics, beer parties and clambakes, and this pleasant annex survived until 1950. It was then converted into a parking lot.

WAKEFIELD FARM, The. Mapped in 1863, this tract was approximately 25 acres in extent, and was owned by Alexander and Mary Thompson. The western boundary of the farm ran parallel to White Plains Road along present-day Furman Avenue, and the eastern boundary was Edenwald Avenue from East 233rd Street to Baychester Avenue, then along Baychester Avenue to Nereid Avenue, its northern bound. The farmhouse was located near what is today the junction of Bussing Avenue and East 233rd Street and its situation was a lofty one, for the land fell off to the East, giving the Thompsons a clear view of the Hutchinson River valley and Eastchester Bay. The Thompsons had twelve children, and the best known was William J., who specialized in hauling fine-cut stones for the first mausoleums and monuments in newly-opened Woodlawn Cemetery.

WALDO PLACE. This was the former name of Plymouth Avenue from Middletown Road to Roberts Avenue, according to a 1919 map. See: Bradford Avenue.

WALDO SQUARE. This square was laid out on part of the original Waldo Hutchins estate in Riverdale. In 1940, the name was changed to Brust Square.

WALDORF PLACE. The surrounding land was owned by John Jacob Astor whose family originated in Waldorf, Germany. The name was changed to Ferncliff Place in 1899, but in 1955 the Place itself was wiped out by the Bronx County Jail at East 151st Street and Exterior Avenue, facing the Harlem River. See: Ferncliff Place, Vineyard Bathing Beach.

WALKER AVENUE. This was the former name of East Tremont Avenue from West Farms Square to Castle Hill Avenue. Tremont Avenue, incidentally, is the only avenue that crosses the entire Bronx from the Harlem River to the East River. As it passed through each locality, in the 19th century, it had a different name. See: Bear Swamp Road, Fort Road, Locust Avenue, Main Street, Morris Street, Throggs Neck Road, Walker Street, Waverly Street.

WALKER STREET. Mapped in 1869, this was the former name of Bryant Avenue from Boston Road to Bronx Park. Thomas Walker, landowner and son of a wealthy Quaker, lived on the northeastern corner of Chestnut and Centre Streets (Vyse Avenue and East 179th Street). In the Quaker records, his mother was described: "Elizabeth Hoyland was born in England on the 26th of Eighth-month, 1761. In the autumn of 1798 she came to the United States, and in the First-month of 1802, married Thomas Walker and soon after, removed with her husband to West Farms." The family is buried in the Quaker Plot of St. Peter's churchyard, off Westchester Square. See: Walker Avenue.

WALKER'S SWAMP. This was a boggy stretch that lay in the vicinity of Southern Boulevard and East 176th Street that was fed by two converging streams. The Walkers were a wealthy Quaker family who became landowners there in 1803. See: Hassock Meadow, Walker Street.

WALKLEY PLACE. It once ran from White Plains Road to Garden Place (East 235th Street) but is now eliminated.

WALL STREET. According to the Morrisania Directory of 1868, Thomas S. Wall lived on Prospect Avenue, and this street was the former name of East 165th Street from Boston Road to Prospect Avenue. See: Guttenberg Street, Shingle Plain, Third Street (2).

"WALNUT RIDGE." This was the name of an estate owned by W. L. Hunt, mapped in 1868, in the vicinity of Unionport Road and Morris Park Avenue, which later became the settlement of Van Nest. The name was evidently descriptive of the topography.

WALNUT STREET (1). From Weeks Avenue to Topping Avenue, this was the former name of Mt. Eden Avenue. Walnut trees were plentiful in the area. See: Belmont Street (2), Jane Street (1).

WALNUT STREET (2). An earlier name of Mt. Eden Avenue from the Grand Concourse down to Webster Avenue. See: East Belmont Street, Jane Street (2), Wolf Street (1).

WALNUT STREET (3). On maps, this was the 19th-century name of Gunther Avenue, south of Boston Road for one block. See: Fox Avenue.

WALNUT STREET (4). Once this was a short street that ran from Reid's Mill Lane to Oak Street, a section now covered by the New England Thruway. Daisy Webb, an old-timer, remembers the walnut trees.

WALSH FARM, The. This tract was purchased by the Walsh family in 1831 from Elbert Anderson. It extended along the Throggs Neck Road (now East Tremont Avenue) from Puritan Avenue to Baisley Avenue, opposite St. Raymond's cemetery. It was once part of a larger farm called the Willow Lane farm. See: Hawthorn Farm, Willow Lane Farm.

WALSH STREET. This was the former name of East 188th Street in Tremont from the Grand Concourse to Third Avenue. See: Bayard Street (1).

WALTERS STREET. Noted in 1895, this was an earlier name of Watson Avenue from White Plains Road to Pugsley Avenue, a distance of two blocks. A former landowner was a Walter Larkin. See: Larkin Avenue, Ninth Street (3).

WARD'S HOTEL. It was described in the 1850s as being on the banks of the Harlem River "within earshot of the inhabitants of Harlem," and most likely below East 138th Street in Mott Haven. There, the 40-man Protection Fire Company No. 5 of Melrose was officially and formally organized. Matthias Haffen was one of the charter members. See: Haffen's Brewery.

WARDSVILLE. It was platted in 1888, and embraced the area east of St. John's College (now Fordham University) in the town of West Farms. It was first surveyed by Andrew Findlay in November, 1851. See: Adamsville (1).

WARE CREEK. Mentioned in a deed of 1787, this is a variant of Weir Creek that is now under the Throggs Neck Expressway. See: Weir Creek.

WARNER'S. An early 1800 name of a small settlement which later was given the name of Mosholu. It was on the west side of Broadway near West 242nd Street. The Warner family were prominent in the area before the Civil War. See: Mosholu, Warnerville.

WARNERVILLE. This settlement at West 242nd Street and Broadway was noted on an 1851 map. The Warner mansion stood on the site of Croke Park (formerly Gaelic Park) and mail addressed to the storekeepers and residents of the crossroads community was written: "Warnerville, Westchester County." See: Mosholu, Warner's.

WARREN AVENUE. This was the former name of Kappock Street from Henry Hudson Park to Arlington Avenue. See: Washington Avenue (4).

WARREN STREET. This was an earlier name of East 173rd Street from the Concourse to Webster Avenue. See: Eleventh Street (1).

WASHINGTON AVENUE (1). In Melrose, the original name of Elton Avenue from East 153rd Street to East 161st Street when it was surveyed in 1848 by Andrew Findlay.

WASHINGTON AVENUE (2). Herschell Street near Zerega Avenue (1898 map).

WASHINGTON AVENUE (3). The former name of Overing Street from Westchester Avenue to East Tremont Avenue when it was the southern boundary of the Adee estate.

WASHINGTON AVENUE (4). An earlier name of Kappock Street. See: Warren Avenue.

WASHINGTON HALL. This prominent social center of Morrisania was built in 1855 by John Eisele, between Washington and Third Avenues on East 166th Street. It was the headquarters of the Bronx Old Timers, and for other social clubs of Morrisania, for almost 75 years. Famous lecturers, musicians, "Jubilee groups" (Negro singers), and Tom Thumb and Minnie Warren, accompanied by P. T. Barnum, appeared there.

WASHINGTON PALACE. Listed in 1882, this was a social hall situated on the west side of Washington Avenue between East 169th Street and East 170th Street. Directly behind it, Morris Place ran westward to Park Avenue.

WASHINGTON PARK. This former amusement park and picnic grounds was bounded by Starling Avenue, Odell Street and Purdy Street in the early 1900s. A Mr. Sehring was the proprietor. See: Purdy Street.

WASHINGTON PLACE (1). This was the earlier name of Ritter Place.

WASHINGTON PLACE (2). Before 1900, this was the name of Mt. Hope Place. See: Popham Street.

WASHINGTON PLACE (3). In 1888, this was the name of Reynolds Avenue as plotted (or platted) across the estate called "Pennyfield."

WASHINGTON STREET (1). This was the former name of Purdy Street from Westchester Avenue to Parkchester, a distance of five blocks. See: Purdy Street.

WASHINGTON STREET (2). In Van Nest, this was the original name of White Plains Road for approximately six blocks. See: Cobb Avenue, Third Street (8).

WASHINGTON STREET (3). On early maps, this was the name of Horton Street on City Island. It is not named for the Father of our Country, but for George Washington Horton who owned the southern end of the island until the 1850s.

WASHINGTONVILLE. This was an early name for the small settlement between the Bronx River and White Plains Road, from East 238th Street up to the city line. It got its name from the fact it faced the pre-Revolutionary Hyatt farmhouse on the opposite side of the Bronx River. It was known as "Washington's Gun House"as local folklore had it that the General stored guns in the homestead at one time. Upon annexation to New York City in June, 1895, the community was absorbed by Wakefield - the expanding settlement immediately to the south. A reason given for this move was that there was postal confusion with another Washingtonville in Orange County. See: Wakefield.

WATER STREET. See: Corlear Avenue.

WATERBURY AVENUE. From Layton Avenue to Stadium Avenue, this was the 19th-century name of Potter Place. The Waterburys owned a nearby estate.

WATERBURY LANE. On old maps, the former name of Middletown Road from Bruckner Boulevard to Eastchester Bay on which the Waterbury estate was located in the 1880s. An earlier name for this tree-shaded carriage road was Furman's Lane. See: "Plaisance."

WATERBURY'S HORSE CEMETERY. Some dozen tombstones can be seen in Pelham Bay Park on what had once been Squire Waterbury's estate. Horses and some dogs that were favorites were buried in a double row near the present-day Police Department stables. See: "Plaisance."

WATERS AVENUE. Since the early 1900s a street was planned behind Westchester Square and Blondell Avenue and was to be given the name of an early settler. However, the New York City Transit Authority's train yard now occupies the site. See: Waters Place (Current).

WATSON STREET. This was the former name of Hammersley Avenue from Eastchester Road to East 222nd Street (that overlays Black Dog Brook). The area was the Givan estate, and the Watsons were neighbors and lifelong friends. When laying out the lands in streets and avenues preparatory to a real estate sale, Mrs. LeRoy (a granddaughter of Robert Givan) requested that one street should be named Watson to commemorate a century of friendship between the families.

WATSON'S LANE (1). This was a former carriage road that ran from William Watson's mansion, near present-day James Monroe High School eastward to Westchester Avenue, crossing it at approximately Elder Avenue and terminating as a hard-packed unpaved path near Bruckner Boulevard of today. In the early 1920s and 1930s, Italians used this isolated stretch as a bowling alley, using discs of cheese weighing up to 40 lbs. The contestant would stretch a tape, about 2 inches wide, to his wrist and then wind it around the cheese. He would then toss it as a bowling ball, and the tape unwound with a loud snap. The man whose cheese rolled farthest was the winner. The name of the game was, variously, *Il Terso della Forma*, *Tiro del Formaggio*, or *Giocco della Forma*. See: Fox's Land, "Wilmount."

WATSON'S LANE (2). This was once the western carriage lane leading from William Watson's mansion, near today's James Monroe High School, to Morris Park Avenue at East Tremont Avenue. See: Liberty Street, Saratoga Avenue, "Wilmount."

WATSON'S WOODS. Once the wooded section of the William Watson estate, called "Wilmount," that lay on both sides of Westchester Avenue between the Bronx River and Morrison Avenue. Dr. Virden, writing the history of St. Peter's P.E. church, mentioned a treasured Bible that was stolen, later to be found in Watson's Woods.

WAVERLEY STREET (1). In 1848 when Melrose was laid out, this was the first name given to what is today East 159th Street. Sir Walter Scott wrote the *Waverley Novels*, and also tales of the Abbey of Melrose.

WAVERLEY STREET (2). On early maps, this was the former name of East Tremont Avenue from the Concourse to Third Avenue. See: Walker Avenue, Waverley Place.

WAVERLY PLACE. This was the former name of Fairmount Place on an 1874 map of West Farms. All the previous landowners had been English, and Sir Walter Scott's Waverley Novels enjoyed wide popularity in that period. Streets in adjoining Tremont and Melrose also bore the name of Waverley.

WEAR CREEK. In 1795, Abijah Hammond bought land of Glorianna, Edward and Benjamin Stephenson for 5,500 Pounds (dollars not yet being in currency) and the 360 acres extending from the East River to Wear Creek, a variant spelling of Weir Creek that is now filled in, and is under the Throgs Neck Expressway. See: Weir Creek.

WEBB'S SHIPBUILDING ACADEMY & HOME. This was an imposing red stone building on the northwestern side of Fordham Road and Sedgwick Avenue where shipbuilding was taught and old seamen were housed. William H. Webb was a naval architect and shipbuilder of Civil War days, and his institution was a landmark for over 60 years, but was dismantled in the 1930s to make way for apartment buildings. See: Dutch Reformed Church of Fordham.

WEBER'S LANE. This was the former name of West 232nd Street at Kingsbridge Avenue running to Broadway.

WEBSTER AVENUE. The former name of East 189th Street in the village of Belmont. See: Welch Street.

WEBSTER HOTEL, The. Advertisements in the New York *Eagle* of the 1890s noted this hotel being managed by J. W. Katzenberger at East 175th Street and Webster Avenue.

WEBSTER THEATRE, The. This silent movie house was built in 1910 and had a seating capacity of 1,189. It was listed in 1922 at Webster Avenue and East 167th Street.

WEIR CREEK. This once-important waterway extended from Long Island Sound (Eastchester Bay) north of the Park of Edgewater, inland and up to Middletown Road with its northern branch - and to Lawton Avenue with its southern branch. The early settlers called it Weir Creek as the Indian had showed them how to construct weirs of plaited reeds for fish traps. These nets were strung across the mouth of the creek, and the outgoing tide carried the fish into the trap. Over the centuries it was called Ware

Creek, Wear Creek and Wire Creek. In the 19th century, where the stream passed Robert Turnbull's land (Dewey Avenue to Lafayette Avenue) it was mapped as Turnbull's Creek. It was a saltwater run up to Layton Avenue, where it was fed by a freshwater brook that had originated in Middletown, where it was called Middle Brook. In the 1880s, Weir Creek was cut off at Layton Avenue (then called Town Dock Road). In the 1920s, it was again shortened at Lafayette Avenue, and in the 1950s, the creek was eliminated entirely by the Throgs Neck Expressway. The southern branch was likewise eliminated by the Cross-Bronx Expressway at the same time. The mouth of Weir Creek was blotted out by landfill to form a park. In 1961, it acquired the name of Weir Creek park through this author's effort to preserve the Indian past, but in May, 1976, it was renamed Bicentennial Veterans Memorial Park.

WEIR CREEK PARK. This park was laid out, since 1960, over Weir Creek, a mixed salt- and freshwater course that originated in the settlement of Middletown. It flowed southward and curved into Eastchester Bay at Schley Avenue where, in centuries past, the Indians had once stretched woven "weirs" across the creek to catch fish. When the creek was filled in, during construction of the Throgs Neck Expressway, a park was laid out and this author suggested it carry the name of Weir Creek Park. A local law, number 258, in 1961, was passed to that effect. This is the former name of Bicentennial Veterans Memorial Park. The name change occurred in 1976.

WELCH STREET. This was the former name of East 189th Street from Webster Avenue to Park Avenue, a distance of one block. Mr. Stephen Welch lived there around 1900. The name was changed in 1911. See: Webster Avenue.

WELLBROOK'S HOTEL. An 1899 notice called attention to a new hotel and cafe on Third Avenue and East 149th Street owned and operated by Martin Wellbrook.

WELLESLEY STREET. This was the former name of East 196th Street from Jerome Avenue to the Concourse. See: Donnybrook Street, Ridge Street, Sherwood Street.

WELLINGTON AVENUE. The 19th-century name for Munn Avenue from Seabury Avenue to Commerce Street. A record of a Quaker family by that name was dated 1806. In 1856, a family named Munn occupied a farm on the Westchester Turnpike, and its name was changed in 1916 when some of the Munns were still in the area.

WENDOVER PLACE. See: Claremont Parkway.

WENMAN STREET. James F. Wenman was president of the Board of Commissioners of the Department of Public Parks in 1881-1886. His name was given to a street on Hunts Point, but it was renamed Barry Street.

WEST CREEK. An 1823 deed lists this as the western bound of Castle Hill Point. It was also shown on an 1809 map. See: Clason's Creek, Cromwell's Creek, Wilkins's Creek.

WEST FARMS CREEK. Surveyor D. B. Taylor used this name for the Bronx River in 1834.

WEST FARMS TURNPIKE. See: West Farms Road.

WEST FARMS VILLAGE. This was a separate township until it was annexed to New York City in 1874. In the 1840s the town limits took in all Morrisania and Fordham. See: DeLancey's Mills, Hive Town, Twelve Farms, Twenty-Fourth Ward (1).

WEST FIFTH STREET. Nereid Avenue's former name from Barnes Avenue to the city line.

WEST MORRISANIA. An 1868 Atlas shows this as a portion of the Morris Manorlands, west of Park Avenue from approximately East 161st Street to East 167th Street. It was laid out in 1851 from the northern part of an estate called "Cedar Grove."

"WEST NECK" (1). This was the name of the Pell family's estate prior to the American Revolution, which was located in what is today Pelham Bay Park. Specific mention is made of the house and land when it belonged to Joseph Pell II and his wife Mary (Honeywell) Pell. The 17th-century manor house was ransacked by "patriots" when Pell was arrested as a Loyalist. The vast tract overlooked Hunter Island, and was traversed by the Shore Road.

"WEST NECK" (2). This was the estate of Philip Schuyler in the years 1800-1840, which was a part of the Pell family's "West Neck" of earlier times. Schuyler's smaller estate took in Split Rock golf course and the land north of the New England Thruway.

WEST ROAD. In the 1860s, this was the former name of Kearney Avenue in the Country Club area. It was the western boundary of the various estates there.

WEST SIXTH STREET. On maps of the Bathgate Estate of 1900, this was the earlier name of Pitman Avenue from Barnes Avenue to the city line.

WEST STREET. The former name of Hornaday Place. The name was changed in 1911.

WESTCHESTER AVENUE (1). On early maps of Wakefield, this was the original name of East 240th Street.

WESTCHESTER AVENUE (2). Old name of Riverdale Avenue because it led to Westchester County.

WESTCHESTER BAY. The mouth of Westchester Creek between Castle Hill Point and Ferry Point Park was charted as Westchester Bay on the 1708 map of the Royal Patent of Westchester.

WESTCHESTER DISTRICT, The. The official name of that area between the Harlem River and the Bronx River that was annexed to New York City in 1874, comprising the towns of Kingsbridge, Morrisania and West Farms. It was so called, as it had been part of Westchester County. See: Annexed District, North Riding, North Side.

WESTCHESTER EXEMPT FIREMEN'S HOSE COMPANY. The firehouse, before 1900, was situated off Dock Street in the village of Westchester. In earlier times, this was called the Empire Engine Company, No. 1. An 1899 listing showed Samuel S. Miller, Frank Gass, Owen F. Dolen and Joseph Polchinski on the rolls.

WESTCHESTER GOLF CLUB. The links were laid out over the former estate of William Watson, extending from the Bronx River to Morrison Avenue, as shown on a 24th Ward map of 1905. See: "Wilmount."

WESTCHESTER HEIGHTS. This was the name of a real estate development, planned in 1892, by two men named Pinchot and Morrell of the Westchester & Van Nest Land Company. The tract was bounded by Hering Avenue, Morris Park Avenue, Pelham Parkway and part of the hospital grounds of Van Etten and Jacobi hospitals. Some of the sections opened in 1902, 1910 and even as late as 1917. The avenues were named after fashionable resorts such as Brighton, Tuxedo, Saratoga, Newport, Lakewood and Narragansett. Some of the streets were never cut through. See: Morrell Park.

WESTCHESTER HOMESTEAD, The. A popular tavern on the Westchester Turnpike at Green Lane (now Zerega Avenue), built in the 1880s. In 1890 it was owned by a Christian Moltzen who could fish off his backyard into Seabury Creek (now Munn Avenue) and smoked his own eels and fish for the customers. The daughters often rowed downstream to Westchester Creek. The back lot was used as a Summer garden, although the land was periodically flooded by Seabury Creek. In 1913, the Sicilia family bought the Westchester Homestead and the third generation still operates the restaurant.

WESTCHESTER HOTEL, The. An 1898 map of Westchester village shows this hotel situated on the southwestern corner of the Turnpike and Union Avenue (Westchester Avenue and St. Peter's Avenue).

WESTCHESTER PARK. See: Sulzer's Westchester Park.

WESTCHESTER PATH. This was an earlier name of a portion of Eastchester Road that ran from present-day Boston Road to Williamsbridge Road. The Westchester Path is specifically mentioned in the patent of Colonel Nicolls in 1667 to the Ten Farms. It was an Indian path in those days, running along the west bank of the upper Westchester Creek - hence it was so called by the colonists. IL, was when the settlement of Eastchester came into existence - 1788 -- that the Indian path took on the name of Eastchester Road. See: Corsa Lane.

WESTCHESTER R.R. STREET. This was the former name of East 149th Street from Third Avenue to Southern Boulevard. It referred to the horse car railroad that ran its length. See: Benson Street.

WESTCHESTER THEATRE, The. This small, silent movie house was opened in 1904, and is believed to be the earliest in The Bronx. It was operated by Adam and George Hoffmann near Parker Street on Westchester Avenue on the west side of the avenue. See: Lyons Nicolodeon.

WESTCHESTER TURNPIKE. This was the Colonial name of Westchester Avenue. It was laid out in 1727 as the main road from Morrisania to the village of Westchester.

WESTCHESTER. The earliest settlement in what is today The Bronx, having become a "Burrow Town" in 1683. Still earlier, it had been a Dutch outpost called Oostdorp. Westchester remained a separate village until it was annexed to New York City in 1895. See: Oostdorp, Eastdorp, North Riding, Twenty-Fourth Ward (2).

WESTERN RESERVE. Prior to the Civil War, the Mount Hope sector was known as the Western Reserve of Upper Morrisania. It meant the land was reserved for future development. The name of Upper Morrisania was subsequently changed to Tremont.

WESTERVELT AVENUE. Mickle Avenue's original name above Boston Road. See: Birch Street.

WESTRAY PLACE. This was the former name of Eagle Avenue from East 156th Street to East 163rd Street. The name does not occur in any real estate transactions, nor in directories dating back to the 1870s. A small island in the Scottish Orkneys bears the name of Westray and may have some connection with Andrew Findlay, of Scottish descent, who surveyed the area, or with Robert Elton, also Scottish, whose property was nearby. See: "Eltona."

"WETHERBEE FARM." The 1860 name of the Sackett estate, later sold to the Pierce family. Its bounds were Bear Swamp Road (now Bronxdale Avenue), the railroad, Williamsbridge Road and approximately Morris Park Avenue. See: Sackett Estate.

WETMORE AVENUE. William C. Wetmore had real estate dealings with Gouverneur Morris II from 1854 to 1857, and had his name affixed to this street on Hunts Point. Sometime around 1890, it was changed to Garrison Avenue.

WETMORE STREET. The former name of East 185th Street from Park Avenue to Third Avenue. William C. Wetmore was a 19th-century financier dealing in real estate.

WHITE CROSS DIVISION. A Volunteer Station of the Life Saving Service of New York City that was erected in 1922 at the White Cross Fishing Club on Eastchester Bay. Captain Paul Blum supervised the construction, having transferred the unit from Jack's Rock Station No. 9. Ten surfmen were listed. See: Jack's Rock Division.

WHITE GHOST, The. The fire engine that belonged to the Lady Washington Engine Company No. 1 bore this colorful name, and was the darling of the Morrisania "fire laddies." The firehouse was on Fulton Avenue and East 168th Street. The engine was built in 1851 by James Smith, and was a hand-drawn pumper. It was painted a glossy white (hence its nickname) with wheels of magenta brown, and shiny brass hubs. Trimmings were of brass, surmounted by a gamecock. This wagon is on display in the museum of the New York Historical Society. See: Lady Washington.

WHITE PLAINS ROAD. The former name of Provost Avenue from Boston Road to the city line, originally a drovers' road to White Plains, according to Bromley's Map of 1905.

WHITE'S LANE. The White estate was mapped in 1867 when it was known as "Longwood Park,."the 19th-century name of what had been Leggett's Lane of the Revolutionary times, and now overlaid by Leggett Avenue. See: Dennison's Lane, Leggett's Lane.

"WHITEHEAD PLACE." Former name of William Whitehead's estate, mapped in 1872 and 1874, on the East River. Silver Beach Gardens occupy the land today.

WHITLOCK POINT. The former name of Barretto Point, when Benjamin Whitlock, a wealthy merchant of the 1850s, owned an estate there which he called "Rose Bank."

WHITMAN STREET. Once part of "Longwood Park," it was given this name in 1867. Later, it was changed to Hewitt Place.

WI-KI-SON. An island in the East River noted in Beauchamp's *Aboriginal Place Names of New York* and the name is believed to refer to "reeds." There is one supposition as to which island was meant: South Brother Island still has a reedy shoreline.

WIEGAND PLACE. This Place was laid out across the original Archer manorlands, which passed to the Schwab and Mali families around 1850. Called Mount Fordham, the district was subdivided in 1898, and Wiegand Place was laid out from West 180th Street to Bronx Community College. Albert C. Wiegand was an Assemblyman of that district. The Place is now a service road on the grounds of the college.

WIGWAM BROOK. In 18th-century records this creek wound in from the East River, through the Debatable Lands. Their ownership was a source of conflict between the Twelve Farms (Hunts Point) and the Morris family (Port Morris). See: Leggett's Brook.

WILGUS STREET. This was part of the extensive Lorillard property of the 19th century along the Bronx River. The land was surveyed and sold around 1904, and Wilgus Street was laid out from Bronx Park East to White Plains Road. Originally called Bridge Street, it was renamed to honor W. J. Wilgus, an engineer on the Committee of Regional Planning of New York. Wilgus Street disappeared in 1949, when Parkside Houses were erected on it.

WILKINS AVENUE. In the 1880s to 1900 the Wilkens [spelled this way] family had five acres at the junction of fashionable Prospect Avenue and the Westchester Turnpike, and also property directly south of Crotona Park. In 1905 Wilkins Avenue (misspelled) was laid out, but in 1985 its name was changed to Louis Nine Boulevard.

WILKINS PLACE. On 19th-century maps, this was the name of Wilkins Avenue. It became an Avenue on March 7, 1905.

WILKINS' CREEK. An earlier name of Pugsley Creek. Isaac Wilkins owned what is now Castle Hill Point in the 1770s. See: Clason's Creek, Cromwell's Creek, West Creek.

WILLETT'S POINT. William Willett, grandson of Thomas Cornell, received his patent in 1667 to the peninsula now known as Clason Point. See: Cornell's Neck.

WILLIAM STREET (1). This was the former name of East 161st Street in Melrose, as it was the boundary of the William H. Morris estate to the north. It was mapped in 1860. See: Cedar Street (2), Cliff Street.

WILLIAM STREET (2). On old maps of Tremont, this was the original name of Carter Avenue from Claremont Park to Richman Park (Echo Park). William H. Carter was a landowner there, listed in 1846.

WILLIAM STREET (3). This was the earlier name of East 186th Street from Arthur Avenue to Hughes Avenue according to an 1874 map of West Farms. Several residents answered to the name of William, so it is uncertain which one of them was honored.

WILLIAM STREET (4). This area was surveyed in the 1870s by William Rosa, and the name was given to present-day East 197th Street from Bainbridge Avenue to Decatur Avenue. See: Isaac Street, Rosa Street.

WILLIAMS AVENUE. Originally the name of Jarvis Avenue from Middletown Road to Buhre Avenue. Williams was Mrs. Buhre's maiden name. See: Country Club Road.

WILLIAMS' BRIDGE. In the early 1700s, a John Williams had a farm on the east bank of the Bronx River in the vicinity of Gun Hill Road and White Plains Road. He is credited with building the first span over the Bronx River, but this story is disputed by some historians who say he was one of a band of men who were hired by landowners Betts, Tippett and Hadden. They were anxious to facilitate passage from their holdings in what is today the West Bronx to Eastchester. At any rate, Williams' farm was the nearest to the bridge and the name "Williams' Bridge" was already in usage at the time of the Revolution. The cluster of homes around the bridge became the 19th-century settlement of Williamsbridge.

WILLIAMSBRIDGE INDEPENDENT ENGINE COMPANY NO. 1. This hose company was mentioned in an 1889 Firemen's Journal, but its location is not given.

WILLIAMSBRIDGE SCHUETZENPARK. For some 30 years, this picnic park was located at Gun Hill Road near the Bronx River. It was demolished in 1920 when the Bronx River Parkway was built. See: Voelker's Schuetzenpark.

WILLIAMSBRIDGE. This district was incorporated on November 23rd, 1888. It was a separate unit until it was annexed to New York City in 1895. See: Twenty-Fourth Ward (2), Williams' Bridge.

WILLIARD STREET. The former name of East 235th Street from Van Cortlandt Park to Webster Avenue. E. K. Williard helped lay out the village of Woodlawn Heights in 1873.

WILLIS PLACE. This was the former name of Bruner Avenue north from Bussing Avenue. See: Oakes Avenue.

WILLIS, The. It opened in 1923 as a silent movie house at East 138th Street on the east side of Willis Avenue. It had a seating capacity of 2,166.

WILLOW LANE. Originally an Indian trail that ran along present-day Bruckner Boulevard and Hutchinson River Parkway from East Tremont Avenue to Ferry Point Park. Later on, Willow Lane was extended from the Unionport bridge to Pelham Bay Park, and renamed Eastern Boulevard. In 1942, the road was renamed Bruckner Boulevard.

"WILLOW LANE FARM." This was a 30-acre farm that was mapped in 1829 when Elbert Anderson sold it to D. Austen and William Cooney. It was the forerunner of the settlement of Schuylerville at East Tremont Avenue and Bruckner Boulevard. The farm extended back to LaSalle Avenue from Bruckner Boulevard, then known as Willow Lane. Elbert Anderson also owned adjoining farms, "The Homestead" of 32 acres, "Centu Farm" of 26 acres and "The Doric Farm" of 47 acres.

WILLOW STREET. This was the former name of East 152nd Street from Westchester Avenue to Prospect Avenue, when willow trees were plentiful. See: Elton Street, Kelly Street, Rose Street (1).

WILMA POINT. See: Outlook Avenue.

"WILMOUNT." This was the name of the William Watson estate that stretched from a point opposite West Farms Square over to Westchester Avenue up to Morrison Avenue. In the 1890s, it became the Westchester Golf Course. See: Watson's Lane (1).

WILSON PLACE. On 19th-century maps, the former name of Lester Street from Barker Avenue to White Plains Rd.

WILTON HILL. This would be the southernmost rise of St. Mary's Park, and in the 1868 Directory of Morrisania, it was listed as the residence of Gouverneur Morris II.

WILTON. This was a small village east of present-day St. Mary's Park that was surveyed in 1854-1857 by 1. C. Buckhout, and parcelled from the Morris Manorlands. It was described as being comprised of small estates and picturesque homes, surrounded by shady dells and flower pots. Tennis courts (a rarity in those days) were at East 135th and East 136th Streets east of St. Ann's Avenue. Writers, artists and musicians liked it for its countrified air and its proximity to New York City. It was locally nicknamed "Actorville" as it attracted members of that profession. See: "Actorville."

WINDMILL STREET. This was a name, noted on an 1853 map of City Island, that is now Belden Street. Evidently a windmill once stood there.

WINGHART HILL. This was a prominent social hall in Bedford Park in the 1890s that was later subdivided into stores. The Bedford Park Presbyterian church held its first services in 1900 in one such shop, the pulpit being a former oyster counter.

WINIK PLACE. In the West Bronx, this was the former name of Phelan Place and also an interim name of Osborne Place. Mr. Winik was a public-minded citizen whose friends and neighbors tried in vain to honor him. In April, 1923, they managed to have Osborne Place renamed Winik Place, but their victory was short-lived. In less than a year, the short Place was once again Osborne Place. Undaunted, the citizens saw to it that in 1925 another short street, nearby, became Winik Place. Before long, it was renamed Phelan Place after Thomas A. Phelan who had helped develop the area in the early 1900s.

WINNEMAK STREET. This street was laid out prior to 1898, running from the East River to Bruckner Boulevard, parallel to East 149th Street. It was never cut through, but is under the Oak Point railroad yards. The origin of "Winnemak" is unknown.

WINONA HOTEL. The New York *Eagle* of 1891 and 1892 carried advertisements for this hotel on East 156th Street and Courtlandt Avenue.

WINTER STREET. This is a very old name in the annals of Westchester County, of which City Island was a part. John Winter's signature is on the Patent for the Town of West Chester in 1686, and Gabriel Winter was a Commissioner in 1820. George Winter was a City Island resident in 1907.

WIRE CREEK. A variant of Weir Creek, as noted in an 1852 deed. See: Weir Creek.

WISSMANN COACH HOUSE. See Wissman Street.

WOLF STREET (1). This was the former name of Mt. Eden Avenue from Jerome Avenue to Walton Avenue. See: East Belmont Street, Jane Street (2), Walnut Street (2).

WOLF STREET (2). In Highbridgeville, the early name of West 167th Street from Sedgwick Avenue to University Avenue, on an 1876 map. It follows the course of the boundary brook between Daniel Turneur's land and John Archer's Fordham Patent. It was named after the Wolf family whose ancestor was Anton Woolf, a Hessian soldier, who acquired title to a farm there after the Revolution. See: Beach Street, Woolf Street.

WOLFE'S HILL. On Throggs Neck, this was the mid-19th-century name for Schley Avenue from East Tremont Avenue down to Weir Creek (now covered by the Throgs Neck Expressway). John David Wolfe bought land around 1830, and his daughter Catherine Lorillard Wolfe later inherited the estate. See: Brown's Hill, Doric Farm.

WONDERLAND, The. This was a silent movie house on Third Avenue and East 153rd Street on the east side of the avenue. It was in operation from around 1917 to 1930.

WOOD STREET. This was the former name of Stephen Avenue on Clason Point.

WOOD'S SHIPYARD. Around 1877, Benjamin Franklin Wood established his shipbuilding concern at the western end of Pilot Street facing Eastchester Bay on City Island. He then moved to the eastern end of Marine Street, facing the Sound, and the business remained in the family for three generations -- now as the City Island Yacht Basin.

WOODBINE HOTEL. Newspapers advertised this inn that stood just south of High Bridge at Depot Place in the 1890s. The Major Deegan Expressway runs over the spot today.

"WOODLAWN." An 1868 map shows this estate alongside the Clason Point Road owned by E. H. Ludlow. It was to the east of "Black Rock Farm." See: "Black Rock Farm."

WOODLAWN AVENUE. This was an early name of Cauldwell Avenue from East 163rd Street to Boston Road, and thought to be the name of a small estate of the 1850s.

WOODLAWN HEIGHTS. This was a village that was laid out in 1873 by George Opdyke on a part of the former Gilbert Valentine farm. E. K. Williard extended the settlement northward to the Yonkers line. Rudolph Rosa was the surveyor both times - in 1873 and 1881. In 1889, the Hyatt farm was surveyed by H. H. Speidler, and the subdivision was added to Woodlawn Heights. See: Opdyke Street, Hyatt Farm, Williard Street.

WOODLAWN INN, The. Old-time wayside tavern on Bronx River Road and East 233rd Street catering to funeral parties. Horses could be watered there and rested, while the mourners refreshed themselves with clam chowder, a specialty of Mr. and Mrs. Oetjens. The inn flourished from around 1895 to 1935 when it was razed to make way for the Woodlawn railroad station.

WOODLAWN ROAD (1). In Bedford Park, this was the former name of East 204th Street from the Concourse to the Botanical Garden. The name was changed in 1911. See: Potter Place, Scott Place.

WOODLAWN ROAD (2). This was the former name of Bainbridge Avenue as it was mapped in 1893 from East 205th Street north to Jerome Avenue at Woodlawn Cemetery.

WOODMANSTEN INN, The. A landmark of Morris Park, this was a famous roadhouse in the first quarter of the 20th century off Williamsbridge Road, below Morris Park Avenue. Well-known dance orchestras and entertainers were featured there in the 1920s and 1930s. It was razed by fire sometime prior to World War II. Today, a row of brick homes occupies the site on Tomlinson Avenue.

"WOODMANSTEN." A large estate that was listed in the Atlas of 1868, belonging to Denton Pearsall. Its bounds were Van Nest Avenue (on the south), Williamsbridge Road (on the west), Allerton Avenue (on the north) and Eastchester Road (on the east).

WOODROW WILSON BOULEVARD. In 1927, the Claremont Heights Civic Committee petitioned the City Council, unsuccessfully, to rename the Grand Boulevard and Concourse in honor of the deceased President. See: Memorial Parkway, Woodrow Wilson Parkway.

WOODROW WILSON PARKWAY. In 1918, the Civic Associations of Mott Haven, West Morrisania and Fordham initiated a drive to rename the Grand Concourse to honor the President whose efforts had ended World War I. See: Memorial Parkway, Woodrow Wilson Boulevard.

WOODRUFF STREET. This was the former name of East 176th Street from Crotona Park to the Bronx River, where August Woodruff was the landowner in 1852. It was the 1880 boundary line of the Thomas Minford estate. See: Mapes Lane, Mole Street, Orchard Street, Prospect Street (2).

"WOODSIDE." E. G. Faile's former Hunts Point estate. In an 1868 Atlas it was bounded by Southern Boulevard, Hunts Point Road, Randall Avenue and Intervale Avenue.

WOODSTOCK. Shingle Plain was an 18th-century name for that portion of the Morris Manorlands in the present-day area of East 165th Street and Forest Avenue. In the 1850s, a sale of land mentioned "The land of Gouverneur Morris, Esq. being now part of the Village of Woodstock, situated about one and one-half miles from the railroad depot in the Manor of Morrisania." Forest Houses cover part of the Shingle Plain. Woodstock was a 19th-century village that was centered around Jackson Avenue and East 163rd Street. The limits were Trinity Avenue to Union Avenue, as mapped in 1849. Sir Walter Scott, poet and novelist, had wide popularity at that time and many of his titles and scenes were given to localities in America, as well as in Canada and Australia. One of his novels, written in 1826, was *Woodstock*, which was a royal park near Oxford. Since Peter P. Decker owned property at East 161st Street and Forest Avenue in 1872, Woodstock was also known as Deckerville in the 1880s and 1890s. This cognomen was frequently used by the newspaper, New York *Eagle*. See: Forest Avenue (1).

WOODSTOCK HOUSE. A two-storied hotel, listed in the 1853 Westchester County Journal, and situated on Union Avenue and East 164th Street in the village of Woodstock.

WOODSTOCK PARK. This was a picnic grounds once located at the junction of Union Avenue and Westchester Avenue in the 1890s. George Dettner was the owner who advertised he stocked only beers brewed in The Bronx. Six breweries were within walking distance of this picnic grove.

"WOODY CREST." This 60-acre estate was mapped in 1875 and belonged to Edgar and Angelica Ketchum in the village of Highbridgeville. She had inherited the land from her grandfather, James Anderson, who had farmed the tract back in the 1790s. The name was apt because of its high ground and wooded character. See: Anderson Farm.

WOOLET'S POINT. As found on a Revolutionary map drafted by Major John Andre, depicting present-day Clason Point. As it was known as Willett's Point in the 1770s, it is evident the Major misunderstood the pronunciation of "Willett." See: Cornell's Neck.

WOOLF STREET. It was named after the Wolf family who lived in that section of Highbridgeville. They were descendants of a Hessian soldier, Anton Woolf, who had remained in America after the Revolutionary War. In 1876, the Wolf family asked the town fathers to change the spelling to Wolf Street, which was done. In 1893, the name was changed to West 167th Street. See: Beach Street, Union Street, Wolf Street.

WORDEN STREET. See: Craven Street.

WORTH STREET. This was the former name of Webster Avenue where it passed Claremont Park, a distance of four blocks. It was mapped in 1868 and is believed to have been named in honor of a General in the Mexican War. See: Berrian Avenue, Bronx River Road, Thomas Street.

WRAY'S HALL. This was a meeting place and the center of social activities of West Farms in the 1850s and 1860s. It stood on the corner of West Farms Square and Boston Road. In 1889, the Catholics held their first meeting there to organize the St. Thomas Aquinas parish. Stephen Wray, Jr., (1890-1934) wrote a history of the area, called "The Village of West Farms."

WRIGHT AVENUE. This was the former name of Duryea Avenue in Edenwald.

WRIGHT STREET. In Van Nest, this was the earlier name of Matthews Avenue from Van Nest Avenue to Neill Avenue. See: Adee Avenue, Rose Street.

WRIGHT'S CORNERS. This was the former name of the junction of Pennyfield Avenue and Harding Avenue. From there, a causeway led to Wright's Island (Locust Point) joining the property of Captain George Wright, on the island, to the estate of John T. Wright on the mainland. See: Wright's Island.

WRIGHT'S DOCK. In the 1880s, this wharf was mapped on the eastern bank of Cromwell's Creek where it entered the Harlem River. It is now covered by the Yankee Stadium.

WRIGHT'S ISLAND. Three generations of the Weight family lived on this island from approximately 1830 to 1920. Captain George Wright's home was situated off East 177th Street and Glennon Place on what is now Locust Point. Upon demolition of Blackman's general store to make way for the Throgs Neck Bridge, this writer dug into what had been a miniature golf course and found an unmarked marble headstone. An early map had denoted the Wright burial ground at that spot. See: Horse Neck, Horse Point, Locust Island.

YATES STREET. In 1868, this was the former name of Canal Place in Mott Haven.

YIDDISH ART. See: Art. (movie theatre).

YONKERS AVENUE (1). An alternate 19th-century name was Palisade Avenue, not to be confused with the present day thoroughfare of that name.

YONKERS AVENUE (2). Independence Avenue's 19th-century name as it led to Yonkers.

YONKERS RIVER. This was another name for Mosholu Creek or Tibbett's Brook as noted in an 1815 deed to Robert Macomb. See: Mosholu, Uncas.

YORK, The. In Van Nest, an old silent movie house that was located on Hunt Avenue and Morris Park Avenue. It was built in 1928, and had a seating capacity of 1,360.

ZBOROWSKI ESTATE. An estate mapped in May, 1888, extending from Morris Avenue to Brook Avenue, from East 170th Street to East 174th Street. It was called "Claremont," owned by Elliott and Anna Zborowski de Montsaulain. Their mansion, overlooking Webster Avenue, was built in 1859 and was famous for its raised sculpture in white marble. Beautifully landscaped lawns descended in terraces to the Mill Brook (before Webster Avenue was cut through). Because of a "curse" no male member ever died peacefully in bed: Martin Zborowski died in a wheelchair; Elliott Zborowski was killed by a train; Francis Zborowski drowned in the Bronx River; Max Zborowski was thrown from his horse and died of his injuries; and Elliott Zborowski, Jr., fell from his motorcar and was killed. The Zborowski mansion was used by the Parks Department once the estate became Claremont Park, and it was razed in 1938. See: "Claremont"(2).

ZEABREY CREEK. A variant of Seabury Creek, on an 1868 map. See: Bain's Creek.

ZELTNER'S BREWERY. This establishment was on the northeastern corner of East 170th Street and Third Avenue, in continuous operation by the Zeltner family for 40 years. Heinrich (Henry) Zeltner arrived from Germany in 1854, a third-generation brewer In 1860 he purchased the small brewery in Morrisania from Wilhelm (William) Jaeger. He gradually enlarged it to include a picnic park and dance pavilion. In 1897, the business passed to his son, William Zeltner. See: Jaeger's Brewery.

ZELTNER'S PARK. This was a tree-shaded beer garden and picnic grounds that once flourished along Third Avenue at East 170th Street eastward to Fulton Avenue. See: The Lyceum, Morrisania Park, Niblo's Garden.

ZELTNER'S POND. The brewing family leased a small pond in Crotona Park from the Parks Department and used it to cut ice. It was named Indian Lake by the local boys. See: Indian Rock (3), Zeltner's Brewery.

ZEREGA POINT. An alternate name of Lorillard Point when both wealthy families shared the peninsula now called Ferry Point Park, *circa* 1850s-1900s. See: Ferris Neck.

ZEREGA'S DOCK. This was a private dock that belonged to Augustus Zerega diZerega on Westchester Creek. He lived on a spacious estate that is now Ferry Point Park, and operated a fleet of clipper ships that flew the Red Z flag. In Civil War times, these ships came in to dock at the Zerega wharf, and some of the sea captains lived in Westchester village. Landfill has covered this dock. See: "Island Hall."

ZEREGORS POINT. This was a misspelling in the 1860 *Historical & Statistical Gazetteer of New York*, that obviously refers to Zerega Point.

ZION AVENUE. This was a carriage path through the Bensonia Cemetery which formerly was located between Rae Street and Carr Street before St. Ann's Avenue was extended in 1875. Zion Avenue was bisected by six footpaths that were named after six saints. See: Bensonia Cemetery, Morrisania Cemetery.

BIBLIOGRAPHY

Books

American Council of Learned Societies. *Dictionary of American Biography*. New York, 1928-1937.

Bolton, Reginald Pelham. Indian *Paths in the Great Metropolis*, New York, 1922.

Bolton, Robert. *The History of the Several Towns, Manors, and Patents of the County of Westchester*.... New York, 1881.

Comfort, Randall, with Charles D. Steurer and Charles A. D. Meyerhoff. *History of Bronx Borough*. New York, 1906.

Cook, Harry Tecumseh. *The Borough of the Bronx, 1639-1913* .... New York, 1913.

Costello, Augustine. *Our Police Protectors: History of the New York Police*. Montclair, 1972; reprint of 1885 edition.

Jenkins, Stephen. *The Greatest Street in the World: The Story of Broadway Old and New*.... New York, 1911.

- - - - - -- *The Story of the Bronx* .... New York, 1912.

King, Moses. *Notable New Yorkers of 1896-1899*. Boston, 1899.

Manhattan Company of New York. *Manna-hatin*. New York, 1929.

O'Callaghan, Edmund B. *The Documentary History of the State of New York* .... Upper Saddle River, 1967; reprint of 1849-1851 edition.

- - - - - - and B. Fernow, eds. *Documents Relative to the Colonial History of the State of New York*. Albany, 1853- 1887.

Pelletreau, William S. *Early Wills of Westchester County, New York from 1664 to 1784*. New York, 1898.

Postal, Bernard and Lionel Coppman. *Jewish Landmarks in New York*. New York, 1976.

Ross, Colon. *Unser Amerika*. Leipsig, 1936.

Scharf, John Thomas, ed. *History of Westchester County, N. Y*.... Philadelphia, 1886.

Scoville, Joseph A. *Old Merchants of New York City*. Greenwood, 1968; reprint of 1870 edition.

Seymann, Jerrold. Colonial *Charters, Patents and Grants to the Communities Comprising the City of New York*. New York, 1939.

Shattuck, Lemuel. A *History of the Town of Concord* .... Boston, 1835.

Smith, James Ruell. *Springs and Wells of Manhattan and the Bronx at the End of the Nineteenth Century*. New York, 1938.

Tieck, William A. *God's House and the Old Kingsbridge Road* .... n.p., 1948.

Valentine's Manual of the Corporation of the City of New York. New York, 1841-1870.

Valentine's Manual of Old New York. New York, 1916/17-1928.

Van Pelt, Daniel. *Leslie's History of the Greater New York*, New York, 1898.

Van Rensselaer, Mrs. John King. *The Goede Vrouw of Mana-ha-ta*. New York, 1898.

Vosburgh, Rayden Woodward, ed. *Records of the Reformed Dutch Church of Fordham in the Borough of the Bronx* .... New York, 1921.

Wells, James Lee, Louis F. Haffen and Josiah A. Briggs, ed. *The Bronx and its People: A History*. New York, 1927.

Who's Who. London, 1849 - -.

Annual Report on Exempt Firemen and Ere Houses, 1860.
The Bronx County Historical Society Research library.
Bronx *Homes News* files, 1907-1948.
Board of Alderman and City Council, City Ordinances, 1806 -
Lists of Freemen and Physicians, 1803.
list of Original Lot Owners of Morrisania, 1854.
Marquee: Journal of the Theatre Historical Society files.
Memorials Concerning Deceased Friends (Quakers) 1895.
Municipal Reference library files.
New York *Eagle* files, 1887-1892.
Records of Westchester, 1684-1900, in White Plains.
Records of Westchester, 1684-1970, in The Bronx.
Records of the Holland Society.
Riverdale Press, articles by George Younkheere.
Roster of the Union Club of New York.
Roster of Civil War dead in the West Farms Cemetery.
Westchester County Agricultural Society membership, 1831.
Westchester Journals, 1858.
West Farms records, Hallock files, 1837-1920.

Maps

Visscher Map of New Netherlands, 1651
Dutch map of New Netherlands, 1656.
A small Draught of Fordham and ye Meadows, 1669.
Nicoll's Patent for West Farms, 1669 .
Map of the Boundaries of Long Reach, 1706.
A Draught of New York, 1749.
A Survey of Frog's Neck, 1776.
Division and Survey of Frog's Neck, 1792.
Map of the Town of Pelham, 1789.
Map of Lands belonging to Abijah Hammond on Throgs Point, 1823.
U.S. Army Engineers' Map of Throgs Point, 1833.
Map of Westchester, 1851.
Map of the Lower Part of Westchester County, 1853.
Survey of Morrisania, 1854.
Map of Land purchased from Charlton Ferris, 1863.
Beers, F. W. *Atlas of New York and Vicinity.* New York, 1867.
Robinson, E. and R. H. Pidgeon, *Robinson's Atlas of the City of New York ....* New York, 1885.
Bromley's Atlas of the City of New York. 23rd and 24th Wards. Philadelphia, 1897.
Certified Copies of the Field Maps of the Annexed Districts, Filed in the Register's Office at White Plains. n.d.

Atlas of the Borough of the Bronx 23rd Ward. n.p., 1900,1906.

Certified Copies of Important Maps Appertaining to the 23rd and 24th Wards. New York, 1888-90.

U.S. Department of Commerce. *Coast and Geodetic Survey. 1920.*

Cemeteries Visited

Joseph Rodman Drake Cemetery, Hunts Point.

Ferris family plot, Commerce Street near Westchester Square.

Huguenot cemetery, New Rochelle.

Jesuit cemetery, Fordham University.

Sisters of Charity Cemetery, Mount St. Vincent.

Pell family plot, Pelham Bay Park.

Pelham Cemetery, City Island.

Potters Field, Hart Island.

St. Ann's P.E. churchyard, St. Anns Avenue and East 140th Street.

St. Paul's P.E. churchyard, Mount Vernon.

St. Peter's P.E. churchyard, Westchester Avenue near Westchester Square.

St. Raymond's R.C. churchyard, East Tremont and Castle Hill Avenues.

St. Raymond's R.C. cemetery, East Tremont Avenue, Throggs Neck.

Sleepy Hollow Cemetery, North Tarrytown.

Waterbury's horse and dog cemetery, Pelham Bay Park.

Westchester Methodist churchyard, East Tremont Avenue.*

West Farms Soldiers' Cemetery, East 180th Street and Bryant Avenue.

Woodlawn Cemetery.

Wright family plot, Locust Point.*

---

* No longer in existence.

JOHN McNAMARA, a life-long resident of The Bronx, has spent more than forty years in research to uncover the origin of Bronx street and place names. He is also the author of *McNamara's Old Bronx*. A founding member of The Bronx County Historical Society, he is the only individual to be twice awarded The Society's most prestigious award, the William C. Beller Award For Excellence and Achievement. He also served as Corresponding Secretary of The Society from 1956 to 1971, and, for over three decades has written the weekly "Bronx in History" column for the *Bronx Press-Review*. He has lectured on all aspects of Bronx history, has written articles on the subject, and for many years served as an editor of *The Bronx County Historical Society Journal*.

THE BRONX COUNTY
HISTORICAL SOCIETY

The Bronx County Historical Society was founded in 1955 for the purpose of promoting knowledge, interest and research in The Bronx. The Society administers The Museum of Bronx History, Edgar Allan Poe Cottage, a Research Library, and The Bronx County Archives; publishes a varied series of books, journals and newsletters; conducts historical tours, lectures, courses, school programs, archaeological digs and commemorations; designs exhibitions; sponsors various expeditions; and produces the "Out of the Past" radio show and cable television programs. The Society is active in furthering the arts, preserving the natural resources of The Bronx, and in creating the sense of pride in the Bronx Community.

For additional information, please contact:

THE BRONX COUNTY HISTORICAL SOCIETY
3309 Bainbridge Avenue, The Bronx, New York
10467
Telephone: (718) 881-8900

Edgar Allan Poe Cottage
c. 1812

Poe Cottage is administered by
The Bronx County Historical Society
in agreement with the Department of Parks and Recreation
of the City of New York.

•Guided Tours •

Poe Park, Grand Concourse and East Kingsbridge Road
The Bronx, New York
Telephone: (718) 881-8900

VALENTINE-VARIAN HOUSE
C. *1758*

MUSEUM OF BRONX HISTORY

The Valentine-Varian House is owned and administered by
The Bronx County Historical Society.

•Guided T ours •

3266 Bainbridge Avenue at East 208th Street
The Bronx, New York 10467
Telephone: (718) 881-8900

*The Bronx County Historical Society is supported in part with public funds
and semices provided through The Department of Cultural Affairs and
The Department of Parks and Recreation of The City of New York,
The Office of The Bronx Borough President,
The Bronx City Council Delegation,
The New York State Council on the Arts,
The New York State Office of Parks, Recreation and Historic Preservation,
The New York State Library,
and The Institute of Museum Services.*